T0211187

Lecture Notes in Bioinformatics 11465

Subseries of Lecture Notes in Computer Science

More information about this series at http://www.springer.com/series/5381

Ignacio Rojas · Olga Valenzuela ·
Fernando Rojas · Francisco Ortuño (Eds.)

Bioinformatics and Biomedical Engineering

7th International Work-Conference, IWBBIO 2019
Granada, Spain, May 8–10, 2019
Proceedings, Part I

 Springer

Editors
Ignacio Rojas
Department of Computer Architecture
and Computer Technology Higher Technical
School of Information Technology
and Telecommunications Engineering
CITIC-UGR
Granada, Spain

Fernando Rojas
CITIC-UGR
University of Granada
Granada, Spain

Olga Valenzuela
ETSIIT
University of Granada
Granada, Spain

Francisco Ortuño
Fundacion Progreso y Salud
Granada, Spain

University of Chicago
Chicago, IL, USA

ISSN 0302-9743   ISSN 1611-3349   (electronic)
Lecture Notes in Bioinformatics
ISBN 978-3-030-17937-3   ISBN 978-3-030-17938-0   (eBook)
https://doi.org/10.1007/978-3-030-17938-0

LNCS Sublibrary: SL8 – Bioinformatics

This Springer imprint is published by the registered company Springer Nature Switzerland AG
The registered company address is: Gewerbestrasse 11, 6330 Cham, Switzerland

Preface

We are proud to present the set of final accepted full papers for the 7th edition of the IWBBIO conference—International Work-Conference on Bioinformatics and Biomedical Engineering—held in Granada (Spain) during May 8–10, 2019.

IWBBIO 2019 sought to provide a discussion forum for scientists, engineers, educators, and students about the latest ideas and realizations in the foundations, theory, models, and applications for interdisciplinary and multidisciplinary research encompassing disciplines of computer science, mathematics, statistics, biology, bioinformatics, and biomedicine.

The aims of IWBBIO are to create a friendly environment that could lead to the establishment or strengthening of scientific collaborations and exchanges among attendees, and therefore IWBBIO 2019 solicited high-quality original research papers (including significant work-in-progress) on any aspect of bioinformatics, biomedicine, and biomedical engineering.

New computational techniques and methods in machine learning; data mining; text analysis; pattern recognition; data integration; genomics and evolution; next-generation sequencing data; protein and RNA structure; protein function and proteomics; medical informatics and translational bioinformatics; computational systems biology; modeling and simulation and their application in the life science domain, biomedicine, and biomedical engineering were especially encouraged. The list of topics in the successive Call for Papers has also evolved, resulting in the following list for the present edition:

1. **Computational proteomics**. Analysis of protein–protein interactions; protein structure modeling; analysis of protein functionality; quantitative proteomics and PTMs; clinical proteomics; protein annotation; data mining in proteomics.
2. **Next-generation sequencing and sequence analysis**. De novo sequencing, re-sequencing and assembly; expression estimation; alternative splicing discovery; pathway analysis; Chip-seq and RNA-Seq analysis; metagenomics; SNPs prediction.
3. **High performance in bioinformatics**. Parallelization for biomedical analysis; biomedical and biological databases; data mining and biological text processing; large-scale biomedical data integration; biological and medical ontologies; novel architecture and technologies (GPU, P2P, Grid etc.) for bioinformatics.
4. **Biomedicine**. Biomedical computing; personalized medicine; nanomedicine; medical education; collaborative medicine; biomedical signal analysis; biomedicine in industry and society; electrotherapy and radiotherapy.
5. **Biomedical engineering**. Computer-assisted surgery; therapeutic engineering; interactive 3D modeling; clinical engineering; telemedicine; biosensors and data acquisition; intelligent instrumentation; patient monitoring; biomedical robotics; bio-nanotechnology; genetic engineering.
6. **Computational systems for modeling biological processes**. Inference of biological networks; machine learning in bioinformatics; classification for

biomedical data; microarray data analysis; simulation and visualization of biological systems; molecular evolution and phylogenetic modeling.

7. **Health care and diseases**. Computational support for clinical decisions; image visualization and signal analysis; disease control and diagnosis; genome–phenome analysis; biomarker identification; drug design; computational immunology.

8. **E-health**. E-health technology and devices; e-Health information processing; telemedicine/e-health application and services; medical image processing; video techniques for medical images; integration of classical medicine and e-health.

After a careful peer review and evaluation process (each submission was reviewed by at least two, and on average 3.2, Program Committee members or additional reviewer), 97 papers were accepted for oral, poster, or virtual presentation, according to the recommendations of reviewers and the authors' preferences, and to be included in the LNBI proceedings.

During IWBBIO 2019 several special sessions were held. Special sessions are a very useful tool to complement the regular program with new and emerging topics of particular interest for the participating community. Special sessions that emphasize multi-disciplinary and transversal aspects, as well as cutting-edge topics, are especially encouraged and welcome, and in this edition of IWBBIO they were the following:

– **SS1. High-Throughput Genomics: Bioinformatic Tools and Medical Applications**
   Genomics is concerned with the sequencing and analysis of an organism's genome. It is involved in the understanding of how every single gene can affect the entire genome. This goal is mainly afforded using the current, cost-effective, high-throughput sequencing technologies. These technologies produce a huge amount of data that usually require high-performance computing solutions and opens new ways for the study of genomics, but also transcriptomics, gene expression, and systems biology, among others. The continuous improvements and broader applications on sequencing technologies is generating a continuous new demand of improved high-throughput bioinformatics tools. Genomics is concerned with the sequencing and analysis of an organism genome taking advantage of the current, cost-effective, high-throughput sequencing technologies. Continuous improvement of genomics is in turn leading to a continuous new demand of enhanced high-throughput bioinformatics tools. In this context, the generation, integration, and interpretation of genetic and genomic data are driving a new era of health-care and patient management. Medical genomics (or genomic medicine) is this emerging discipline that involves the use of genomic information about a patient as part of the clinical care with diagnostic or therapeutic purposes to improve the health outcomes. Moreover, it can be considered a subset of precision medicine that is having an impact in the fields of oncology, pharmacology, rare and undiagnosed diseases, and infectious diseases. The aim of this special session is to bring together researchers in medicine, genomics, and bioinformatics to translate medical genomics research into new diagnostic, therapeutic, and preventive medical approaches. Therefore, we invite authors to submit original research, new tools or

pipelines, or their update, and review articles on relevant topics, such as (but not limited to):

- Tools for data pre-processing (quality control and filtering)
- Tools for sequence mapping
- Tools for the comparison of two read libraries without an external reference
- Tools for genomic variants (such as variant calling or variant annotation)
- Tools for functional annotation: identification of domains, orthologs, genetic markers, controlled vocabulary (GO, KEGG, InterPro, etc.)
- Tools for gene expression studies
- Tools for Chip-Seq data
- Integrative workflows and pipelines

Organizers: **Prof. M. Gonzalo Claros**, *Department of Molecular Biology and Biochemistry, University of Málaga, Spain*

Dr. Javier Pérez Florido, *Bioinformatics Research Area, Fundación Progreso y Salud, Seville, Spain*

Dr. Francisco M. Ortuño, *Bioinformatics Research Area, Fundación Progreso y Salud, Seville, Spain*

- **SS2. Omics Data Acquisition, Processing, and Analysis**
  Automation and intelligent measurement devices produce multiparametric and structured huge datasets. The incorporation of the multivariate data analysis, artificial intelligence, neural networks, and agent-based modeling exceeds the experiences of classic straightforward evaluation and reveals emergent attributes, dependences, or relations. For the wide spectrum of techniques, genomics, transcriptomics, metabolomics, proteomics, lipidomics, aquaphotomics, etc., the superposition of expert knowledge from bioinformatics, biophysics, and biocybernetics is required. The series of systematic experiments have to also deal with the data pipelines, databases, sharing, and proper description. The integrated concepts offer robust evaluation, verification, and comparison.
  In this special section a discussion on novel approaches in measurement, algorithms, methods, software, and data management focused on the omic sciences is provided. The topic covers practical examples, strong results, and future visions.

Organizer: **Dipl-Ing. Jan Urban**, *PhD, Head of laboratory of signal and image processing. University of South Bohemia in Ceské Budejovice, Faculty of Fisheries and Protection of Waters, South Bohemian Research Center of Aquaculture and Biodiversity of Hydrocenoses, Institute of Complex Systems, Czech Republic.*

Websites:
www.frov.jcu.cz/en/institute-complex-systems/lab-signal-image-processing

- **SS3. Remote Access, Internet of Things, and Cloud Solutions for Bioinformatics and Biomonitoring**
  The current process of the 4th industrial revolution also affects bioinformatic data acquisition, evaluation, and availability. The novel cyberphysical measuring

devices are smart, autonomous, and controlled online. Cloud computing covers data storage and processing, using artificial intelligence methods, thanks to massive computational power. Laboratory and medical practice should be on the apex of developing, implementing, and testing the novel bioinformatic approaches, techniques, and methods, so as to produce excellent research results and increase our knowledge in the field.

In this special section, results, concepts, and ongoing research with novel approaches to bioinformatics, using the Internet of Things (IoT) devices is presented.

Organizer: **Antonin Barta**, *Antonin Barta, Faculty of Fishery and Waters Protection, Czech Republic*

– **SS4: Bioinformatics Approaches for Analyzing Cancer Sequencing Data**
In recent years, next-generation sequencing has enabled us to interrogate entire genomes, exomes, and transcriptomes of tumor samples and to obtain high-resolution landscapes of genetic changes at the single-nucleotide level. More and more novel methods are proposed for efficient and effective analyses of cancer sequencing data. One of the most important questions in cancer genomics is to differentiate the patterns of the somatic mutational events. Somatic mutations, especially the somatic driver events, are considered to govern the dynamics of clone birth, evolution, and proliferation. Recent studies based on cancer sequencing data, across a diversity of solid and hematological disorders, have reported that tumor samples are usually both spatially and temporally heterogeneous and frequently comprise one or multiple founding clone(s) and a couple of sub-clones. However, there are still several open problems in cancer clonality research, which include (1) the identification of clonality-related genetic alterations, (2) discerning clonal architecture, (3) understanding their phylogenetic relationships, and (4) modeling the mathematical and physical mechanisms. Strictly speaking, none of these issues is completely solved, and these issues remain in the active areas of research, where powerful and efficient bioinformatics tools are urgently demanded for better analysis of rapidly accumulating data. This special issue aims to publish the novel mathematical and computational approaches and data processing pipelines for cancer sequencing data, with a focus on those for tumor micro-environment and clonal architecture.

Organizers: **Jiayin Wang, PhD, Professor,** *Jiayin Wang, PhD, Professor, Department of Computer Science and Technology, Xian Jiaotong University, China*

Xuanping Zhang, PhD, Associate Professor, *Xuanping Zhang, PhD, Associate Professor, Department of Computer Science and Technology, Xian Jiaotong University, China.*

Zhongmeng Zhao, PhD, Professor, *Zhongmeng Zhao, PhD, Professor, Department of Computer Science and Technology, Xian Jiaotong University, China*

- **SS5. Telemedicine for Smart Homes and Remote Monitoring**

  Telemedicine in smart homes and remote monitoring is implementing a core research to link up devices and technologies from medicine and informatics. A person's vital data can be collected in a smart home environment and transferred to medical databases and the professionals. Most often different from clinical approaches, key instruments are specifically tailored devices, multidevices, or even wearable devices respecting always individual preferences and non-intrusive paradigms. The proposed session focused on leading research approaches, proto-types, and implemented hardware/software co-designed systems with a clear networking applicability in smart homes with unsupervised scenarios.

  Organizers: Prof. Dr. Juan Antonio Ortega. Director of the Centre of Computer Scientific in Andalusia, Spain, Head of Research Group IDINFOR (TIC223), University of Seville, ETS Ingeniería Informática, Spain

  Prof. Dr. Natividad Martínez Madrid. Head of the Internet of Things Laboratory and Director of the AAL-Living Lab at Reutlingen University, Department of Computer Science, Reutlingen, Germany

  Prof. Dr. Ralf Seepold. Head of the Ubiquitous Computing Lab at HTWG Konstanz, Department of Computer Science, Konstanz, Germany

- **SS6. Clustering and Analysis of Biological Sequences with Optimization Algorithms**

  The analysis of DNA sequences is a crucial application area in computational biology. Finding similarity between genes and DNA subsequences provides very important knowledge of their structures and their functions. Clustering as a widely used data mining approach has been carried out to discover similarity between biological sequences. For example, by clustering genes, their functions can be predicted according to the known functions of other genes in the similar clusters. The problem of clustering sequential data can be solved by several standard pattern recognition techniques such as k-means, k-nearest neighbors, and the neural networks. However, these algorithms become very complex when observations are sequences with variable lengths, like genes. New optimization algorithms have shown that they can be successfully utilized for biological sequence clustering.

  Organizers: Prof. Dr. Mohammad Soruri Faculty of Electrical and Computer Engineering, University of Birjand, Birjand, Iran. Ferdows Faculty of Engineering, University of Birjand, Birjand, Iran

- **SS7. Computational Approaches for Drug Repurposing and Personalized Medicine**

  With continuous advancements of biomedical instruments and the associated ability to collect diverse types of valuable biological data, numerous recent research studies have been focusing on how to best extract useful information from the 'big biomedical data' currently available. While drug design has been one of the most essential areas of biomedical research, the drug design process for the most part has not fully benefited from the recent explosion in the growth of biological data and bioinformatics algorithms. With the incredible overhead associated with the traditional drug design process in terms of time and cost, new alternative methods, possibly based on computational approaches, are very much needed to propose innovative ways for effective drugs and new treatment options. As a result, drug repositioning or repurposing has gained significant attention from biomedical researchers and pharmaceutical companies as an exciting new alternative for drug discovery that benefits from the computational approaches. This new development also promises to transform health care to focus more on individualized treatments, precision medicine, and lower risks of harmful side effects. Other alternative drug design approaches that are based on analytical tools include the use of medicinal natural plants and herbs as well as using genetic data for developing multi-target drugs.

  *Organizer: **Prof. Dr. Hesham H. Ali**, UNO Bioinformatics Core Facility College of Information Science and Technology University of Nebraska at Omaha, USA*

It is important to note, that for the sake of consistency and readability of the book, the presented papers are classified under 14 chapters. The organization of the papers is in two volumes arranged basically following the topics list included in the call for papers. The first volume (LNBI 11465), entitled "Bioinformatics and Biomedical Engineering. Part I," is divided into eight main parts and includes contributions on:

1. High-throughput genomics: bioinformatic tools and medical applications
2. Omics data acquisition, processing, and analysis
3. Bioinformatics approaches for analyzing cancer sequencing data
4. Next-generation sequencing and sequence analysis
5. Structural bioinformatics and function
6. Telemedicine for smart homes and remote monitoring
7. Clustering and analysis of biological sequences with optimization algorithms
8. Computational approaches for drug repurposing and personalized medicine

The second volume (LNBI 11466), entitled "Bioinformatics and Biomedical Engineering. Part II," is divided into six main parts and includes contributions on:

1. Bioinformatics for health care and diseases
2. Computational genomics/proteomics
3. Computational systems for modeling biological processes
4. Biomedical engineering
5. Biomedical image analysis
6. Biomedicine and e-health

This seventh edition of IWBBIO was organized by the Universidad de Granada. We wish to thank to our main sponsor and the institutions, the Faculty of Science, Department of Computer Architecture and Computer Technology, and CITIC-UGR from the University of Granada for their support and grants. We wish also to thank to the editors of different international journals for their interest in editing special issues from the best papers of IWBBIO.

We would also like to express our gratitude to the members of the various committees for their support, collaboration, and good work. We especially thank the Organizing Committee, Program Committee, the reviewers and special session organizers. We also want to express our gratitude to the EasyChair platform. Finally, we want to thank Springer, and especially Alfred Hofmann and Anna Kramer for their continuous support and cooperation.

May 2019   Ignacio Rojas
                                                       Olga Valenzuela
                                                       Fernando Rojas
                                                       Francisco Ortuño

Organization

Steering Committee

Miguel A. Andrade	University of Mainz, Germany
Hesham H. Ali	University of Nebraska, EEUU
Oresti Baños	University of Twente, The Netherlands
Alfredo Benso	Politecnico di Torino, Italy
Giorgio Buttazzo	Superior School Sant'Anna, Italy
Gabriel Caffarena	University San Pablo CEU, Spain
Mario Cannataro	Magna Graecia University of Catanzaro, Italy
Jose María Carazo	Spanish National Center for Biotechnology (CNB), Spain
Jose M. Cecilia	Universidad Católica San Antonio de Murcia (UCAM), Spain
M. Gonzalo Claros	University of Malaga, Spain
Joaquin Dopazo	Research Center Principe Felipe (CIPF), Spain
Werner Dubitzky	University of Ulster, UK
Afshin Fassihi	Universidad Católica San Antonio de Murcia (UCAM), Spain
Jean-Fred Fontaine	University of Mainz, Germany
Humberto Gonzalez	University of the Basque Country (UPV/EHU), Spain
Concettina Guerra	College of Computing, Georgia Tech, USA
Roderic Guigo	Center for Genomic Regulation, Pompeu Fabra University, Spain
Andy Jenkinson	Karolinska Institute, Sweden
Craig E. Kapfer	Reutlingen University, Germany
Narsis Aftab Kiani	European Bioinformatics Institute (EBI), UK
Natividad Martinez	Reutlingen University, Germany
Marco Masseroli	Polytechnic University of Milan, Italy
Federico Moran	Complutense University of Madrid, Spain
Cristian R. Munteanu	University of Coruña, Spain
Jorge A. Naranjo	New York University (NYU), Abu Dhabi
Michael Ng	Hong Kong Baptist University, SAR China
Jose L. Oliver	University of Granada, Spain
Juan Antonio Ortega	University of Seville, Spain
Julio Ortega	University of Granada, Spain
Alejandro Pazos	University of Coruña, Spain
Javier Perez Florido	Genomics and Bioinformatics Platform of Andalusia, Spain
Violeta I. Pérez Nueno	Inria Nancy Grand Est (LORIA), France

Horacio Pérez-Sánchez	Universidad Católica San Antonio de Murcia (UCAM), Spain
Alberto Policriti	University of Udine, Italy
Omer F. Rana	Cardiff University, UK
M. Francesca Romano	Superior School Sant'Anna, Italy
Yvan Saeys	VIB, Ghent University, Belgium
Vicky Schneider	The Genome Analysis Centre (TGAC), UK
Ralf Seepold	HTWG Konstanz, Germany
Mohammad Soruri	University of Birjand, Iran
Yoshiyuki Suzuki	Tokyo Metropolitan Institute of Medical Science, Japan
Oswaldo Trelles	University of Malaga, Spain
Shusaku Tsumoto	Shimane University, Japan
Renato Umeton	CytoSolve Inc., USA
Jan Urban	University of South Bohemia, Czech Republic
Alfredo Vellido	Polytechnic University of Catalonia, Spain
Wolfgang Wurst	GSF National Research Center of Environment and Health, Germany

Program Committee and Additional Reviewers

Hisham Al-Mubaid	University of Houston, USA
Hesham Ali	University of Nebraska Omaha, USA
Rui Alves	Universitat de Lleida, Spain
Georgios Anagnostopoulos	Florida Institute of Technology, USA
Miguel Andrade	Johannes Gutenberg University of Mainz, Germany
Saul Ares	Centro Nacional de Biotecnología (CNB-CSIC), Spain
Hazem Bahig	Ain Sham, Egypt
Oresti Banos	University of Twente, The Netherlands
Ugo Bastolla	Centro de Biologia Molecular Severo Ochoa, Spain
Alfredo Benso	Politecnico di Torino, Italy
Paola Bonizzoni	Università di Milano-Bicocca, Italy
Larbi Boubchir	University of Paris 8, France
David Breen	Drexel University, USA
Jeremy Buhler	Washington University in Saint Louis, USA
Gabriel Caffarena	CEU San Pablo University, Spain
Mario Cannataro	Magna Graecia University of Catanzaro, Italy
Rita Casadio	University of Bologna, Italy
Francisco Cavas-Martínez	Technical University of Cartagena, Spain
José M. Cecilia	Catholic University of Murcia, Spain
Keith C. C. Chan	The Hong Kong Polytechnic University, SAR China
Ting-Fung Chan	The Chinese University of Hong Kong, SAR China
Nagasuma Chandra	Indian Institute of Science, India
Bolin Chen	University of Saskatchewan, Canada
Chuming Chen	University of Delaware, USA
Jeonghyeon Choi	Georgia Regents University, USA
M. Gonzalo Claros	Universidad de Málaga, Spain

Darrell Conklin	University of the Basque Country, Spain
Bhaskar Dasgupta	University of Illinois at Chicago, USA
Alexandre G. De Brevern	INSERM UMR-S 665, Université Paris Diderot—Paris 7, France
Fei Deng	University of California, Davis, USA
Marie-Dominique Devignes	LORIA-CNRS, France
Joaquin Dopazo	Fundacion Progreso y Salud, Spain
Beatrice Duval	LERIA, France
Christian Esposito	University of Naples Federico II, Italy
Jose Jesus Fernandez	Consejo Superior de Investigaciones Cientificas (CSIC), Spain
Gionata Fragomeni	Magna Graecia University of Catanzaro, Italy
Pugalenthi Ganesan	Bharathidasan University, India
Razvan Ghinea	University of Granada, Spain
Oguzhan Gunduz	Marmara University, Turkey
Eduardo Gusmao	University of Cologne, Germany
Christophe Guyeux	University of Franche-Comté, France
Juan M. Gálvez	University of Granada, Spain
Michael Hackenberg	University of Granada, Spain
Nurit Haspel	University of Massachusetts Boston, USA
Morihiro Hayashida	National Institute of Technology, Matsue College, Japan
Luis Herrera	University of Granada, Spain
Ralf Hofestaedt	Bielefeld University, Germany
Vasant Honavar	The Pennsylvania State University, USA
Narsis Kiani	Karolinska Institute, Sweden
Dongchul Kim	The University of Texas Rio Grande Valley, USA
Tomas Koutny	University of West Bohemia, Czech Republic
Istvan Ladunga	University of Nebraska-Lincoln, USA
Dominique Lavenier	CNRS/IRISA, France
José L. Lavín	CIC bioGUNE, Spain
Kwong-Sak Leung	The Chinese University of Hong Kong, SAR China
Chen Li	Monash University, Australia
Shuai Cheng Li	City University of Hong Kong, SAR China
Li Liao	University of Delaware, USA
Hongfei Lin	Dalian University of Technology, China
Zhi-Ping Liu	Shandong University, China
Feng Luo	Clemson University, USA
Qin Ma	Ohio State University, USA
Malika Mahoui	Eli Lilly, USA
Natividad Martinez Madrid	Reutlingen University, Germany
Marco Masseroli	Politecnico di Milano, Italy
Tatiana Maximova	George Mason University, USA
Roderick Melnik	Wilfrid Laurier University, Canada
Enrique Muro	Johannes Gutenberg University/Institute of Molecular Biology, Germany

Kenta Nakai	Institute of Medical Science, University of Tokyo, Japan
Isabel Nepomuceno	University of Seville, Spain
Dang Ngoc Hoang Thanh	Hue College of Industry, Vietnam
Anja Nohe	University of Delaware, USA
José Luis Oliveira	University of Aveiro, Portugal
Juan Antonio Ortega	University of Seville, Spain
Francisco Ortuño	Clinical Bioinformatics Aea, Fundación Progreso y Salud, Spain
Motonori Ota	Nagoya University, Japan
Joel P. Arrais	University of Coimbra, Portugal
Paolo Paradisi	ISTI-CNR, Italy
Javier Perez Florido	Genomics and Bioinformatics Platform of Andalusia (GBPA), Spain
Antonio Pinti	I3MTO Orléans, Italy
Hector Pomares	University of Granada, Spain
María M. Pérez	University of Granada, Spain
Jairo Rocha	University of the Balearic Islands, Spain
Fernando Rojas	University of Granada, Spain
Ignacio Rojas	University of Granada, Spain
Jianhua Ruan	Utsa, USA
Gregorio Rubio	Universitat Politècnica de València, Spain
Irena Rusu	LINA, UMR CNRS 6241, University of Nantes, France
Michael Sadovsky	Institute of Computational Modelling of SB RAS, Russia
Jean-Marc Schwartz	The University of Manchester, UK
Russell Schwartz	Carnegie Mellon University, USA
Ralf Seepold	HTWG Konstanz, Germany
Xuequn Shang	Northwestern Polytechnical University, China
Wing-Kin Sung	National University of Singapore, Singapore
Prashanth Suravajhala	Birla Institute of Scientific Research, India
Yoshiyuki Suzuki	Tokyo Metropolitan Institute of Medical Science, Japan
Martin Swain	Aberystwyth University, UK
Sing-Hoi Sze	Texas A&M University, USA
Stephen Tsui	The Chinese University of Hong Kong, SAR China
Renato Umeton	Massachusetts Institute of Technology, USA
Jan Urban	Institute of Complex Systems, FFPW, USB, Czech Republic
Lucia Vaira	Set-Lab, Engineering of Innovation, University of Salento, Italy
Olga Valenzuela	University of Granada, Spain
Alfredo Vellido	Universitat Politècnica de Catalunya, Spain
Konstantinos Votis	Information Technologies Institute, Centre for Research and Technology Hellas, Greece
Jianxin Wang	Central South University, China

Contents – Part I

Next Generation Sequencing and Sequence Analysis

Structural Bioinformatics and Function

Telemedicine for Smart Homes and Remote Monitoring

Clustering and Analysis of Biological Sequences with Optimization Algorithms

Computational Approaches for Drug Repurposing and Personalized Medicine

Contents – Part II

Biomedical Engineering

Biomedical Image Analysis

High-throughput Genomics: Bioinformatic Tools and Medical Applications

A Coarse-Grained Representation for Discretizable Distance Geometry with Interval Data

Antonio Mucherino[1], Jung-Hsin Lin[2(✉)], and Douglas S. Gonçalves[3]

[1] IRISA, University of Rennes 1, Rennes, France
antonio.mucherino@irisa.fr
[2] Research Center for Applied Sciences, Academia Sinica, Taipei, Taiwan
jhlin@gate.sinica.edu.tw
[3] DM-CFM, Federal University of Santa Catarina, Florianópolis, Brazil
douglas@mtm.ufsc.br

Abstract. We propose a coarse-grained representation for the solutions of discretizable instances of the Distance Geometry Problem (DGP). In several real-life applications, the distance information is not provided with high precision, but an approximation is rather given. We focus our attention on protein instances where inter-atomic distances can be either obtained from the chemical structure of the molecule (which are exact), or through experiments of Nuclear Magnetic Resonance (which are generally represented by real-valued intervals). The coarse-grained representation allows us to extend a previously proposed algorithm for the Discretizable DGP (DDGP), the branch-and-prune (BP) algorithm. In the standard BP, atomic positions are fixed to unique positions at every node of the search tree: we rather represent atomic positions by a pair consisting of a feasible region, together with a most-likely position for the atom in this region. While the feasible region is a constant during the search, the associated position can be refined by considering the new distance constraints that appear at further layers of the search tree. To perform the refinement task, we integrate the BP algorithm with a spectral projected gradient algorithm. Some preliminary computational experiments on artificially generated instances show that this new approach is quite promising to tackle real-life DGPs.

1 Introduction

Let $G = (V, E, d)$ be a simple weighted undirected graph, where vertices v represent certain objects (such as the atoms of a molecule), and the existence of an edge $\{u, v\}$ between two vertices u and v indicates that the distance between the two corresponding objects is known [8]. The weight $d(u, v)$ is a real interval delimiting the lower and the upper bound on the distance values; *exact* distances are represented by degenerated intervals. In this paper, we will use the compact notation d_{uv} for the distances in the graph.

© Springer Nature Switzerland AG 2019
I. Rojas et al. (Eds.): IWBBIO 2019, LNBI 11465, pp. 3–13, 2019.
https://doi.org/10.1007/978-3-030-17938-0_1

Definition 1.1. *Given a simple weighted undirected graph $G = (V, E, d)$ and a positive integer K, the Distance Geometry Problem (DGP) asks whether a function*

$$x : v \in V \longrightarrow x_v \in \mathbb{R}^K$$

exists such that

$$\forall \{u, v\} \in E, \quad ||x_u - x_v|| \in d_{uv}. \tag{1}$$

The function x is called a *realization* of the graph G. We say that a realization x that satisfies all constraints in Eq. (1) is a *valid realization*.

The DGP is NP-hard [27], and has several different applications. In this paper, we focus on the problem of determining the three-dimensional structure of a molecule from available inter-atomic distances. This is a common problem arising when molecules are subject to experiments of Nuclear Magnetic Resonance (NMR), because they are able to provide estimations of distances between atom pairs [1]. The reader can find in [8], for example, more details about the biological application.

Other interesting applications of the DGP include multi-dimensional scaling [15] and the manipulation of distance-guided animations [26].

The constraints in Definition 1.1 can naturally be specialized to particular cases, on the basis of the nature of the distances. When the distance d_{uv} is exact, the symbol "$\in$" can be replaced by an equality. When the distance d_{uv} is instead represented by an interval $[\underline{d}_{uv}, \bar{d}_{uv}]$, then the generic constraint can be replaced by two inequalities:

$$\underline{d}_{uv} \leq ||x_u - x_v|| \leq \bar{d}_{uv}.$$

Let E' be the subset of E containing exact distances; as a consequence, the subset $E \setminus E'$ contains all approximated distances, the ones represented by a real-valued interval.

Many approaches to the DGP reformulate it as a global unconstrained optimization problem. A penalty function is employed, which is able to measure the distance violations in a given realization x. One commonly used penalty function for the DGP with only exact distances is the Mean Distance Error (MDE):

$$MDE(x) = \frac{1}{|E|} \sum_{\{u,v\} \in E} \frac{|\,||x_u - x_v|| - d_{uv}\,|}{d_{uv}}, \tag{2}$$

whose value is 0 for all valid realizations. This penalty function can be generalized to DGPs with interval distances.

In recent times, a multidisciplinary group of mathematicians, physics, computer scientists and biologists has been working on a discretization process that allows to represent the set of possible solutions for the DGP as a discrete domain having the structure of a tree [5,7,16,21,25]. Particular emphasis is given to the application in structural biology. We give the following definition of discretizable subclass of DGP instances [19].

Definition 1.2. *A simple weighted undirected graph G represents an instance of the Discretizable DGP (DDGP) if and only if there exists a vertex ordering on V such that the following two assumptions are satisfied:*

(a) $G[\{1, 2, \ldots, K\}]$ *is a clique whose edges are in E';*
(b) $\forall v \in \{K + 1, \ldots, |V|\}$, *there exist K vertices $u_1, u_2, \ldots, u_K \in V$ such that*
    (b.1) $u_1 < v, u_2 < v, \ldots, u_K < v$;
    (b.2) $\{\{u_1, v\}, \{u_2, v\}, \ldots, \{u_{K-1}, v\}\} \subset E'$ *and* $\{u_K, v\} \in E$;
    (b.3) $\mathcal{V}_S(u_1, u_2, \ldots, u_K) > 0$ *(if $K > 1$),*

where $G[\cdot]$ is the subgraph induced by a subset of vertices of V, and $\mathcal{V}_S(\cdot)$ is the volume of the simplex generated by a valid realization of the vertices $u_1, u_2, \ldots, u_K$.

In the following, we will refer to assumptions **(a)** and **(b)** as the *discretization assumptions*. Such assumptions can be verified only if a vertex ordering is associated to V. The problem of finding a suitable vertex order for V, which allows to satisfy the discretization assumptions, was previously discussed, for example, in [23] and [12].

Assumption **(a)** ensures the existence of an initial clique in G: the first K vertices in the ordering belong to this clique, where all edges are related to distances in E'. A unique realization for this clique can be computed (modulo total translations, rotations and reflections). Assumption **(b)** is the one that in fact allows us to reduce the search space for the DGP instance to a discrete domain having the structure of a tree, where the positions of vertices are organized layer by layer [21].

The branch-and-prune (BP) algorithm is based on the idea to recursively construct the search space (*branching phase*), and to immediately verify the feasibility of newly generated branches (*pruning phase*) by using ad-hoc pruning devices [25]. This algorithm has been proved to work very efficiently on instances consisting of only exact distances [20]. Even though the BP algorithm can deal with instances containing interval data [17], the presence of such interval distances turns the algorithm into a heuristic, and some issues related to the solution of this class of DDGP instances were recently pointed out in [14].

The main aim of this short paper is to propose a coarse-grained representation for the DDGP class, to be exploited in conjunction with the BP algorithm's main framework. The basic idea is to combine the global search performed by the BP algorithm over the search domain with a local search capable of correcting the unavoidable approximation errors introduced by the presence of interval distances. The coarse-grained representation allows us to bound the (continuous) search space of the local search, so that every DDGP solution, belonging to a particular branch of the tree, cannot lead to any other solution belonging to a different branch.

This paper is organized as follows. In Sect. 2, we will give a quick overview on the discretizable class of DGP instances, and we will briefly describe the BP algorithm. Our coarse-grained representation of DDGP solutions will be introduced in Sect. 3, where we will discuss the potential benefits in using such a representation. Although in this short paper we focus on a particular problem in dimension $K = 3$ and we will present the theory for $K = 3$ to simplify the notations, the entire discussion can be generalized to any dimensions $K \geq 1$. In Sect. 4, we will give the details of an implementation of a Spectral Projected Gradient (SPG) descent method, which we will couple with a BP algorithm implementing the coarse-grained representation. Some preliminary computational experiments on protein instances will be presented in Sect. 5, while Sect. 6 will conclude the paper.

2   The Branch-and-Prune Algorithm

The branch-and-prune (BP) algorithm was formally introduced in [18]. It performs a systematic exploration of the search tree obtained with the discretization process (which is possible when the discretization assumptions (a) and (b) are satisfied, see Introduction). When instances containing interval distances are considered, the original search tree is replaced with an approximated tree where arc nodes are substituted with a predefined number D of nodes representing sample positions (which are extracted from the arcs, see [17] for more details).

The tree of possible solutions can be explored starting from its top, where the first vertex belonging to the initial clique is placed. Subsequently, all other vertices in the initial clique can be placed in their unique positions, and the search can actually start with the vertex having rank $K = 3$ in the associated discretization order. At each step, the candidate positions for the current vertex v are computed, and the search is branched. Depending on the available distance information, the set of candidate positions may contain either two singletons, or two disjoint arcs. In the latter case, every arc can be approximated with a subset of D sample positions.

Pruning devices can be employed for discovering infeasible positions that are computed for the current vertex. The main pruning device exploits the so-called "pruning distances" of DDGP instances, which correspond to all distances that are not involved in the discretization. As soon as a vertex position is found to be infeasible, then the corresponding branch is pruned and the search is backtracked. Thanks to this pruning mechanism, the size of the tree, in terms of nodes, remains relatively small, so that an exhaustive search becomes feasible for certain instances. When this is the case, the BP algorithm is actually able to enumerate all possible solutions to the problem, and not only to provide one of the possible solutions.

As extensively discussed in [14], however, the approximation introduced in the tree when it is necessary to deal with interval distances turns the BP algorithm into a heuristic. Moreover, sampling points from the obtained arcs is not the most efficient strategy to select vertex positions, because the probability to sample two points for two different vertices which are supposed to satisfy the same distance constraint is very low.

For lack of space, we omit the sketch of the BP algorithm. The interested reader can make reference to one of the given citations, and in particular to [18] and [25].

3   A Coarse-Grained Representation

In the BP algorithm, a subset of distances is used for generating new vertex positions (discretization distances), while the complementary subset is used for verifying the feasibility of such newly generated positions (pruning distances). The introduction of approximation errors, due to the presence of interval distances, can have the important consequence of invalidating the feasibility tests.

The first feasibility test that might fail occurs at the same layer of the current vertex v, when the pruning distances are verified. In order to avoid situations where vertex positions are generated and immediately discarded by the pruning device, some adaptive branching schemes for BP have already been proposed in [2,13]. However, even if all distances at the current layer v are satisfied, we cannot guarantee all such selected positions for v will also satisfy the constraints involving vertices associated to further layers of the tree.

This is the main motivation for a coarse-grained representation of DDGP solutions. Instead of fixing, on every branch of the tree, all vertices in unique positions, the idea is to rather associate a small *region* of the search space to every vertex, together with a most-likely position. The shape of the region can be chosen on the basis of the methods that are implemented for their manipulation. In a similar representation [28], spheres are used to define a region around given vertex positions; we will rather use box-shaped regions, because they are more convenient for the local optimization solver that we consider in the experiments proposed in this paper. We recall that our local solver is briefly described in Sect. 4.

Before attempting the identification of a valid realization x for a given DDGP instance, we propose to preliminary look for a feasible realization in the following coarse-grained representation:

$$z : v \in V \longrightarrow (x_v, B_v) \in \mathbb{R}^3 \times \mathbb{R}^6,$$

where B_v is a box defined in the Cartesian system given by the initial clique (see Sect. 2). We point out that B_v has 6 dimensions (in dimension $K = 3$, the position of one vertex of the box, plus the corresponding depth, length and height values are necessary for its unique definition). When a new vertex position x_v is generated for the current vertex v, the function z does not only allow to assign a position x_v to v, but also to keep track of the feasible region where it belongs to. On further layers, in fact, the position x_v may not be feasible w.r.t. some other distances, and it could therefore be *slightly* modified in order to ensure global feasibility. This can be done by employing solvers for local optimization (see next section). The position x_v is naturally constrained to stay in the original box B_v for two main reasons. Firstly, the (continuous) search space of the local solver is in this way reduced; secondly, the situation where the local solvers can move to solutions belonging to other tree branches is avoided.

We motivate the choice of employing a local solver with the fact that, at every layer of the tree where an infeasibility is discovered, there are, in general, only a few distances that are not satisfied. This makes the corresponding subproblem to consider easier to tackle. Naturally, an important point concerning the use of a local solver is its fast converge: in fact, when attempting the solution of harder instances, we expect the local solver to be invoked at almost all recursive BP calls.

When the BP algorithm reaches a leaf node, a valid realization x can be extracted from z by simply taking the set of positions x_v, for every $v \in V$.

4   A Spectral Projected Gradient

During a recursive call of BP at layer $\bar{v}$, the new pair $z_{\bar{v}}$ consisting of a position $x_{\bar{v}}$ and a box $B_{\bar{v}}$ may be feasible or not with the pruning distances. If not, this infeasibility may be due either to the current position $x_{\bar{v}}$, or to the position of some previous vertices $v \in \{1, \dots, \bar{v} - 3\}$. We employ therefore a Spectral Projected Gradient (SPG) to *attempt* the *refinement* all computed positions x_v belonging to the current branch of the search tree. If this attempt fails, then we have the confirmation that the newly generated $z_{\bar{v}}$ is infeasible. If it does not fail, then SPG outputs a new conformation where all distance constraints are satisfied; the exploration can therefore continue on further layers.

The optimization problem that SPG solves is a global optimization reformulation of a DGP related to the subgraph $G[\{1, 2, \dots, \bar{v}\}] \equiv G(V_{\bar{v}}, E_{\bar{v}})$. Moreover, the constraint that all x_v need to be contained in their corresponding box regions B_v is added. Notice, however, that the boxes B_v related to vertices v such that no pruning distance $\{w, \bar{v}\}$ exists with $w \leq v < \bar{v} - 3$ are shrunk so that only the current position of x_v is admitted (these vertices are not involved in the feasibility check at current level). Finally, we remark that we use, for this global optimization problem, a differentiable objective function σ that is different from the MDE function in Eq. (2) (the MDE function is not differentiable on its entire domain). The resulting optimization problem, which is strongly inspired by the works in [10], is the following:

$$
\begin{aligned}
\min_{X,y} \quad & \frac{1}{2} \sum_{\{u,v\} \in E_{\bar{v}}} \pi_{uv} \left(\|x_u - x_v\| - y_{uv} \right)^2 := \sigma(X, y), \\
\text{s.t.} \quad & \forall \{u, v\} \in E_{\bar{v}}, \quad \underline{d}_{uv} \leq y_{uv} \leq \bar{d}_{uv}, \\
& \forall v \in \{1, 2, \dots, \bar{v}\}, \quad x_v \in B_v,
\end{aligned}
\tag{3}
$$

where $X \in \mathbb{R}^{\bar{v} \times 3}$ is a matrix whose rows correspond to the vertex positions $x_v \in \mathbb{R}^3$, $y \in \mathbb{R}^{|E_{\bar{v}}|}$ and π_{uv} is a non-negative weight of the distance constraint related to the edge $\{u, v\}$.

As shown in [9] and discussed in [10,11], the function $\sigma(X, y)$ is differentiable at (X, y) if and only if $\|x_u - x_v\| > 0$ for all $\{u, v\} \in E_{\bar{v}}$ such that $\pi_{uv} y_{uv} > 0$.

In such case, the gradient, with respect to X, can be written as

$$\nabla_X \sigma(X, y) = 2(AX - B(X, y)X), \tag{4}$$

where the matrix A is defined by

$$a_{uv} = \begin{cases} -\pi_{uv}, & \text{if } u \neq v, \\ \sum_{w \neq u} \pi_{uw}, & \text{otherwise.} \end{cases}$$

In expression (4), the matrix $B(X, y) = [b_{uv}(X, y)]$ is a function of (X, y) defined by

$$b_{uv}(X, y) = \begin{cases} -\dfrac{\pi_{uv} y_{uv}}{\|x_u - x_v\|}, & \text{if } u \neq v \text{ and } \|x_u - x_v\| > 0, \\ 0, & \text{if } u \neq v \text{ and } \|x_u - x_v\| = 0, \\ -\sum_{w \neq u} b_{uw}(X, y), & \text{otherwise.} \end{cases}$$

The constraints defining the feasible set

$$\left\{ (X, y) \in \mathbb{R}^{n \times 3} \times \mathbb{R}^{|E_{\bar{v}}|} \ : \ \forall \{u, v\} \in E, \underline{d}_{uv} \leq y_{uv} \leq \bar{d}_{uv}; \forall v \in \{1, \ldots, \bar{v}\}, x_v \in B_v \right\}$$

of the optimization problem are box constraints on X and y. The projection of the pair (X, y) in the feasible set is therefore trivial to perform.

We solve the optimization problem (3) with an implementation of the non-monotone SPG method proposed in [6], where the current BP solution, up to layer $\bar{v}$, is given as a starting point. In our implementation, a spectral parameter is employed to scale the negative gradient direction before the projection onto the feasible set [3]. Then, a non-monotone line-search is performed in order to ensure a sufficient decrease of the objective function after some iterations [30]. More information about the implementations of SPG in this context can be found in [24].

5   Preliminary Computational Experiments

We present in this section some preliminary computational experiments on a set of artificially generated instances. All codes were written in C programming language and all experiments were carried out on an Intel Core i7 2.30 GHz with 8 GB RAM, running Linux. The codes have been compiled by the GNU C compiler v.4.9.2 with the -O3 flag.

We selected the protein conformations that were considered in the experiments presented in [7]. However, in these preliminary experiments, we do not use real NMR data, but we rather generate our protein instances from known models of the selected proteins. The three considered proteins, having codes 2jmy, 2kxa and 2ksl in the Protein Data Bank (PDB) [4], have been experimentally determined by NMR experiments, and, as it is usually the case, more than

Table 1. Some computational experiments on artificially generated instances.

Protein	$|V|$	$|E|$	$|E'|$	MDE	Time
2jmy	77	428	219	2.39161e−05	32 s
2kxa	121	700	367	1.30272e−05	1 m 4 s
2ksl	254	1388	684	2.51421e−05	5 m 15 s
2jmy*	264	2449	787	3.13574e−06	6 m 54 s

one model for each protein was deposited. In our instance generation, we have simply considered the first model that appears in the corresponding PDB file.

Our instances are generated in a way to resemble NMR data. From the initial conformation model, we compute all inter-atomic distances, and we include in our instance the following distances:

– distances between bonded atoms (distances considered as exact);
– distances between atoms bonded to a common atom (distances considered as exact);
– distances between the first and the last atom forming a torsion angle (distances represented by an interval);
– distances between hydrogen atoms that are shorter than 5 Å (distances represented by an interval).

In order to define the interval distances, we create an interval of range 0.1 Å for the distances related to torsion angles, and an interval of range 0.5 Å for distances related to hydrogens, and we place the true distance randomly inside such an interval. The atoms are sorted accordingly to the order proposed in [22], which ensures the discretizability of the instance. The instance 2jmy* also contains the atoms belonging to the side chains of the amino acids forming the molecule.

We run our extended version of the BP algorithm which makes use of the coarse-grained representation and of the SPG algorithm on the set of generated instances. SPG is invoked at each recursive call of BP where at least one pruning distance is not satisfied. It can terminate because of different criteria: either when the objective function value becomes smaller than 10^{-6}, or when the norm of the search direction becomes smaller than 10^{-6}, or when it reaches the maximum number of allowed interactions, which is set to 20000 in our experiments.

Table 1 gives some details about the performed experiments. The MDE function (see Eq. (2)) indicates, for all experiments, that the overall set of distance constraints is satisfied. It is important to remark that the standard BP implementation (see for example the experiments in [25]) can provide results where the MDE function can decrease to 10^{-10} or more. However, this standard implementation is only able to deal with exact distances, and it would not be able to provide any feasible conformation for the instances considered in this work. Considering that the maximal error over a distance is of order 10^{-1} in our instances (when interval distances are generated), the fact that the final MDE value is of order 10^{-5} is a quite promising result for our preliminary experiments.

6   Conclusions

We have presented a new extension of the BP algorithm for a discretizable class of DGP instances, which is based on a new coarse-grained representation of the solutions. At every layer of the search tree, new candidate positions for the current vertex are computed. However, they are not fixed for all subsequent layers, but rather let free to "move" inside a feasible box. This movement possibility allows us to adjust the actual position of the vertex inside the box in order to make it compatible with the new distances that are considered at further tree layers. For adjusting the positions, we employ SPG as a local solver, which appears to be able to perform well the refinement task in our proposed experiments.

This is a very initial step for the extension of the BP algorithm to more general problems arising in the field of the DGP. In order to make it work with real NMR data, we plan in the near future to tailor the SPG algorithm to this particular class of problems (the optimization problem the SPG needs to solve is a DGP problem where, in most of the cases, only one distance is not satisfied at the starting point) and to control in a more efficient way the exploration of the different tree branches. Moreover, our research does not have only real NMR data as a final target: we intent to extend the entire discretization methodology to harder DGP problems, such as, for example, the ones which are based on genomics data [29].

Acknowledgments. AM and JHL wish to thank the CNRS and MoST for financial support (PRC project "Rapid NMR Protein Structure Determination and Conformational Transition Sampling by a Novel Geometrical Approach"). AM and DSG wish to thank CAPES PRINT for financial support. DSG also thanks CNPq for financial support (Grant n. 421386/2016-9).

References

1. Almeida, F.C.L., Moraes, A.H., Gomes-Neto, F.: An overview on protein structure determination by NMR, historical and future perspectives of the use of distance geometry methods. In: Mucherino, A., Lavor, C., Liberti, L., Maculan, N. (eds.) Distance Geometry: Theory, Methods and Applications, pp. 377–412. Springer, New York (2013). https://doi.org/10.1007/978-1-4614-5128-0_18
2. Alves, R., Lavor, C.: Geometric algebra to model uncertainties in the discretizable molecular distance geometry problem. Adv. Appl. Clifford Algebra **27**, 439–452 (2017)
3. Barzilai, J., Borwein, J.: Two-point step size gradient methods. IMA J. Numer. Anal. **8**, 141–148 (1988)
4. Berman, H., et al.: The protein data bank. Nucleic Acids Res. **28**, 235–242 (2000)
5. Billinge, S.J.L., Duxbury, Ph.M., Gonçalves, D.S., Lavor, C., Mucherino, A.: Recent results on assigned and unassigned distance geometry with applications to protein molecules and nanostructures. Ann. Oper. Res. (2018, to appear)
6. Birgin, E.G., Martínez, J.M., Raydan, M.: Nonmonotone spectral projected gradient methods on convex sets. SIAM J. Optim. **10**, 1196–1211 (2000)
7. Cassioli, A., et al.: An algorithm to enumerate all possible protein conformations verifying a set of distance restraints. BMC Bioinform. **16**, 23 (2015). p. 15

8. Crippen, G.M., Havel, T.F.: Distance Geometry and Molecular Conformation. Wiley, Hoboken (1988)
9. de Leeuw, J.: Convergence of the majorization method for multidimensional scaling. J. Classif. **5**, 163–180 (1988)
10. Glunt, W., Hayden, T.L., Raydan, M.: Molecular conformations from distance matrices. J. Comput. Chem. **14**(1), 114–120 (1993)
11. Glunt, W., Hayden, T.L., Raydan, M.: Preconditioners for distance matrix algorithms. J. Comput. Chem. **15**, 227–232 (1994)
12. Gonçalves, D.S., Mucherino, A.: Optimal partial discretization orders for discretizable distance geometry. Int. Trans. Oper. Res. **23**(5), 947–967 (2016)
13. Gonçalves, D.S., Mucherino, A., Lavor, C.: An adaptive branching scheme for the Branch & Prune algorithm applied to distance geometry. In: IEEE Conference Proceedings, Federated Conference on Computer Science and Information Systems (FedCSIS 2014), Workshop on Computational Optimization (WCO 2014), Warsaw, Poland, pp. 463–469 (2014)
14. Gonçalves, D.S., Mucherino, A., Lavor, C., Liberti, L.: Recent advances on the interval distance geometry problem. J. Global Optim. **69**(3), 525–545 (2017)
15. Gramacho, W., Mucherino, A., Lin, J.-H., Lavor, C.: A new approach to the discretization of multidimensional scaling. In: IEEE Conference Proceedings, Federated Conference on Computer Science and Information Systems (FedCSIS 2016), Workshop on Computational Optimization (WCO 2016), Gdansk, Poland, pp. 601–609 (2016)
16. Lavor, C., Liberti, L., Maculan, N., Mucherino, A.: Recent advances on the discretizable molecular distance geometry problem. Eur. J. Oper. Res. **219**, 698–706 (2012)
17. Lavor, C., Liberti, L., Mucherino, A.: The interval Branch-and-Prune algorithm for the discretizable molecular distance geometry problem with inexact distances. J. Global Optim. **56**(3), 855–871 (2013)
18. Liberti, L., Lavor, C., Maculan, N.: A Branch-and-Prune algorithm for the molecular distance geometry problem. Int. Trans. Oper. Res. **15**, 1–17 (2008)
19. Liberti, L., Lavor, C., Maculan, N., Mucherino, A.: Euclidean distance geometry and applications. SIAM Rev. **56**(1), 3–69 (2014)
20. Liberti, L., Lavor, C., Mucherino, A.: The discretizable molecular distance geometry problem seems easier on proteins. In: Mucherino, A., Lavor, C., Liberti, L., Maculan, N. (eds.) Distance Geometry: Theory, Methods and Applications, pp. 47–60. Springer, New York (2013). https://doi.org/10.1007/978-1-4614-5128-0_3
21. Liberti, L., Lavor, C., Mucherino, A., Maculan, N.: Molecular distance geometry methods: from continuous to discrete. Int. Trans. Oper. Res. **18**(1), 33–51 (2011)
22. Mucherino, A.: On the identification of discretization orders for distance geometry with intervals. In: Nielsen, F., Barbaresco, F. (eds.) GSI 2013. LNCS, vol. 8085, pp. 231–238. Springer, Heidelberg (2013). https://doi.org/10.1007/978-3-642-40020-9_24
23. Mucherino, A.: A pseudo de Bruijn graph representation for discretization orders for distance geometry. In: Ortuño, F., Rojas, I. (eds.) IWBBIO 2015. LNCS, vol. 9043, pp. 514–523. Springer, Cham (2015). https://doi.org/10.1007/978-3-319-16483-0_50
24. Mucherino, A., Gonçalves, D.S.: An approach to dynamical distance geometry. In: Nielsen, F., Barbaresco, F. (eds.) GSI 2017. LNCS, vol. 10589, pp. 821–829. Springer, Cham (2017). https://doi.org/10.1007/978-3-319-68445-1_94
25. Mucherino, A., Lavor, C., Liberti, L.: The discretizable distance geometry problem. Optim. Lett. **6**(8), 1671–1686 (2012)

26. Mucherino, A., Omer, J., Hoyet, L., Giordano, P.R., Multon, F.: An application-based characterization of dynamical distance geometry problems. Optim. Lett. (2018, to appear)
27. Saxe, J.: Embeddability of weighted graphs in k-space is strongly NP-hard. In: Proceedings of 17th Allerton Conference in Communications, Control and Computing, pp. 480–489 (1979)
28. Sit, A., Wu, Z.: Solving a generalized distance geometry problem for protein structure determination. Bull. Math. Biol. **73**, 2809–2836 (2011)
29. Sulkowska, J.I., Morcos, F., Weigt, M., Hwa, T., Onuchic, J.N.: Genomics-aided structure prediction. Proc. Natl. Acad. Sci. (PNAS) U.S.A. **109**(26), 10340–10345 (2012)
30. Zhang, H., Hager, W.W.: A nonmonotone line search technique and its applications to unconstrained optimization. SIAM J. Optim. **14**(4), 1043–1056 (2004)

Fragment-Based Drug Design to Discover Novel Inhibitor of Dipeptidyl Peptidase-4 (DPP-4) as a Potential Drug for Type 2 Diabetes Therapy

Eka Gunarti Ningsih, Muhammad Fauzi Hidayat, and Usman Sumo Friend Tambunan$^{(\boxtimes)}$ (ID)

Department of Chemistry, Faculty of Mathematics and Natural Sciences, Universitas Indonesia, Kampus UI Depok, Depok, West Java 16424, Indonesia
usman@ui.ac.id

Abstract. Diabetes mellitus is among the highest cause of death in the world. Medicinal treatment of diabetes mellitus can be achieved by inhibiting Dipeptidyl Peptidase-4 (DPP-4). This enzyme rapidly inactivates incretin, which acts as a glucoregulatory hormone in the human body. Fragment-based drug design through computational studies was conducted to discover novel DPP-4 inhibitors. About 7,470 fragments out of 343,798 natural product compounds were acquired from applying Astex Rule of Three. The molecular docking simulation was performed on the filtered fragments against the binding site of DPP-4. Fragment-based drug design was carried out by growing new structures from the potential fragments by employing DataWarrior software. The generated ligand libraries were evaluated based on the toxicity properties before underwent virtual screening, rigid, and induced-fit molecular docking simulation. Selected ligands were subjected to the pharmacological and toxicological property analysis by applying DataWarrior, Toxtree, and SWISSADME software. According to the ligand affinity, which based on the ΔG binding value and molecular interaction along with the pharmacological properties of the ligand, two best ligands, namely FGR-2 and FGR-3, were chosen as the novel inhibitor of DPP-4. Further in vitro, in vivo, and clinical trial analysis must be executed in order to validate the selected ligands therapeutic activity as drug candidates for type 2 diabetes.

Keywords: Type 2 diabetes · Dipeptidyl Peptidase-4 · Natural compounds · Fragment-based drug design · In silico

1 Introduction

Diabetes mellitus is a chronic hyperglycemic disease caused by a disturbance in carbohydrate, fat, and protein metabolism due to lack or insensitivity of insulin [1]. The emergence of diabetes in adult have been increased significantly and become one of the ten highest causes of death in the world [2]. According to the International Diabetes Federation (IDF), in 2017, about 425 million people (age 20–79 years) living with

© Springer Nature Switzerland AG 2019
I. Rojas et al. (Eds.): IWBBIO 2019, LNBI 11465, pp. 14–24, 2019.
https://doi.org/10.1007/978-3-030-17938-0_2

diabetes worldwide and will probably reach 629 million by 2045 [3]. The majority of diabetic patients are type 2 (non-insulin dependent) which frequently arises because of unhealthy lifestyles such as smoking habit, lack of physical activity, consuming alcohol and high carbohydrate foods, and also supported by the genetic factor [4, 5]. Prolonged hyperglycemia can produce complications such as amputation, kidney failure, cardiovascular, and stroke [1]. Therefore, diabetes emerges as a challenge for international public health that is expected to prevent and control.

Medicinal treatment of diabetes can be carried out by inhibiting Dipeptidyl Peptidase-4 (DPP-4). This enzyme has capable of degrading incretin hormones which act as a glucoregulatory in the human body [6]. Some commercial DPP-4 inhibitors such as sitagliptin, saxagliptin, linagliptin, and alogliptin may cause joint pain and lead to paralysis [7]. Consequently, the discovery of the novel DPP-4 inhibitors which have low toxicity is necessary.

Fragment-Based Drug Discovery (FBDD) was developed to start the lead generation process in drug discovery. FBDD involves the modification of small molecules (<300 Da) using three main approaches such as fragment merging, linking, and growing [8]. Natural products and their derivatives still an important source of new drugs for many years [9]. In drug development, natural products were explored into various lead compounds, which can be used as templates for the discovery of new drugs by the pharmaceutical industry [10]. In this study, FBDD through fragment growing approach was conducted to discover novel DPP-4 inhibitors from natural product compounds.

2   Research Methodology

2.1   Preparation of DPP-4 Protein Structure and Standard Ligands

The preparation process was performed based on the drug design pipeline that was developed by Tambunan et al. [11]. The three-dimensional structure of DPP-4 protein was acquired from Protein Data Bank in the Research Collaboratory for Structural Bioinformatics (RCSB PDB) with PDB ID: 4A5S [12]. The protein structure was prepared using Molecular Operating Environment (MOE) 2014.09 software to remove the unwanted molecules except for the unique ligand. The optimization method was performed using Amber10: EHT forcefield. Then, the prepared protein was saved in moe format. Standard ligands were collected from ZINC15 database and also prepared with the same software using MMFF94x forcefield. The ligand database was saved in MDB format.

2.2   Pharmacophore Generation

By using MOE 2014.09 software, pharmacophore construction was generated through docking simulation of DPP-4 protein to standard ligands with 'Triangle Matcher' as the docking placement methodology. Protein-Ligand Interaction Finger-prints (PLIF) from the simulation result was analyzed to acquire the pharmacophore features. Afterward, the feature was saved in the ph4 file format for the following pharmacophore-based molecular docking simulation.

2.3   Selecting and Growing Potential Fragment

Prior to selecting fragment, natural product compounds from ZINC15 database [13] were screened by applying Astex Rule of Three to evoke the inadequate fragments [8]. Furthermore, the fragments which possessed drug-likeness score below than zero and toxic properties were removed from the database. Then, the fragment database was prepared using the same parameters as the previous ligands. The rigid receptor docking simulation was performed on the filtered fragments against the binding site of DPP-4. From the docking results, potential fragments were selected subsequently for fragment growing process. DataWarrior software was used to create the evolutionary library based on FragFp descriptor structural similarity and Lipinski Rule of Five [14, 15]. The generated ligands were filtered based on the toxicity properties, and drug-likeness score before underwent the virtual screening simulation.

2.4   Virtual Screening and Molecular Docking

Virtual screening and molecular docking simulation were performed by utilizing MOE software with AMBER10: EHT as the forcefield in R-Field solvation. Before docking simulation, the ligand libraries underwent the virtual screening process. The molecular docking simulation of the ligands and standard drug were carried out based pharmacophore with some protocols, such as rigid receptor retain 30-1, 100-1 and induced-fit retain 100-1.

2.5   ADME and Toxicity Prediction

The ten best ligands and standard molecules from induced-fit molecular docking simulation were subjected to the ADME and toxicological properties prediction by applying DataWarrior [14] and Toxtree [16] software. Furthermore, the medicinal chemistry properties of the ligands were predicted using SWISSADME [17].

3   Results and Discussions

3.1   Visualization of Protein and Pharmacophore Features

DPP-4 is a serine protease (UniProtKB-P27487) consist of 766 residues which are classified as a glycoprotein. The crystal structure of the protein was obtained from RCSB PDB with PDB ID: 4A5S, which have resolution 1.62 Å and a novel heterocyclic DPP-4 inhibitor (N7F) inside the binding site [12]. In Fig. 1a, the binding site of the protein was shown in green and purple colors, which indicate polar and nonpolar regions, respectively. The binding site consist of Arg125, Trp201, Glu205, Glu206, Ser209, Phe357, Asp545, Tyr547, Gly549, Lys554, Trp627, Gly628, Trp629, Ser630, Tyr631, Gly632, Val656, Trp659, Tyr662, Asp663, Tyr666, Arg669, Asn710, Val711, His740, and Tyr752 amino acid residues. Two amino acid residues also act as active sites such as Ser630 and His740, which plays an important catalytic function in the degradation of the peptide hormone such as Glucagon-Like Peptide-1 (GLP-1) [18–20].

The protein structure was prepared using MOE software with potential setup Amber10: EHT as forcefield in R-Field solvation. AMBER is an appropriate parameter for the common nucleosides and amino acid [21]. Besides removing unwanted molecules, chain B was also removed because it has a fewer number of hydrogen bond interactions with inhibitors than chain A [20]. Method of optimization was initiated by 'Protonate 3D' feature with selected parameters of 300K in temperature, pH equal to 7 and salt concentration of 0.1 mol/L. Then, it was continued with checked "adjust hydrogens and lone pairs as required" option in 'Partial Charge' and unchecked "allow ASN/GLN/HIS 'flips' in Protonate3D" option. In 'LigX,' the receptor strength of 100,000 and Root Mean Square (RMS) gradient of 0.05 kcal/mol Å were applied. On the other hand, standard ligands were prepared by using default parameters in 'wash' and 'energy minimization' features with RMS gradient of 0.001 kcal/mol Å. MMFF94x forcefield in Gas Phase solvation was suitably parameterized for small organic molecules as the ligands in this research.

In pharmacophore generation, molecular docking of DPP-4 protein to standard ligands was performed using two different protocols, namely rigid receptor and induced-fit, with 'Triangle Matcher' as the docking placement methodology. The docking result was analyzed to acquire pharmacophore features. PLIF method compares ligand-protein interaction that exposed the similarity between the interaction of docking pose and that of a reference protein-ligand complex, through fingerprints [22]. Visualization of the protein revealed that there are three pharmacophore features in the binding site such as Cat&Don (Purple), HydA (Green), and Acc (Blue) (Fig. 1b).

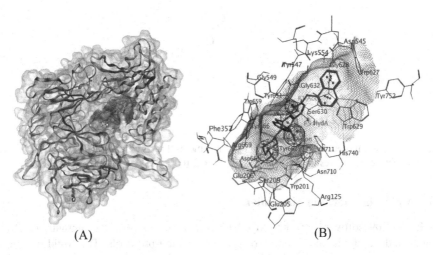

(A)   (B)

Fig. 1. Binding site visualization of DPP-4 with pharmacophore features (Color figure online)

3.2 Fragment Growing Process

In this research, as much as 343,798 natural product compounds were screened by applying Astex Rule of Three, which using parameters such as molecular weight <300 Da, c Log P <3, the number of hydrogen donor <3, the number of hydrogen

acceptor <3, rotatable bond <3 and polar surface area (TPSA) <60 Å^2. These parameters were chosen because of its relevance to fragment-based drug design [8]. Besides,
the compounds which exhibited drug-likeness score lower than zero, and showed the
toxicity properties such as mutagenic, tumorigenic, reproductive effect and irritant were
also omitted.

From the initial screening, 7,470 selected fragments were then subjected to rigid
docking with 'Pharmacophore' as the docking placement. At first rigid docking retain
30-1, there were 493 fragments that bound to the pharmacophore feature in the binding
pocket and having the RMSD value <2.0 Å. Then, from the second docking process
retain 100-1, only 445 fragments that fulfilled the requirement. About 18 potential
fragments that bound to the binding site were chosen based on the observed number of
hydrogen bonds formed in the molecular interaction.

Fragment growing was conducted by utilizing DataWarrior software to create the
evolutionary library which restricted to the FragFp descriptor, Lipinski's Rule of Five,
Veber rules, drug-likeness above than zero and toxicity properties. Potential fragments
were grown based on the Lipinski's Rule of Five (RO5) and Veber rules such as
molecular weight <500 Da, rotatable bonds $\leq$ 10, TPSA <140 Å^2, and logP between
−0.5 and 5.6, number of hydrogen donor <5, and number of hydrogen acceptor <10
[23]. At the end fragment growing process, around 62,392 generated ligands were
stored in MDB format for the molecular docking simulation.

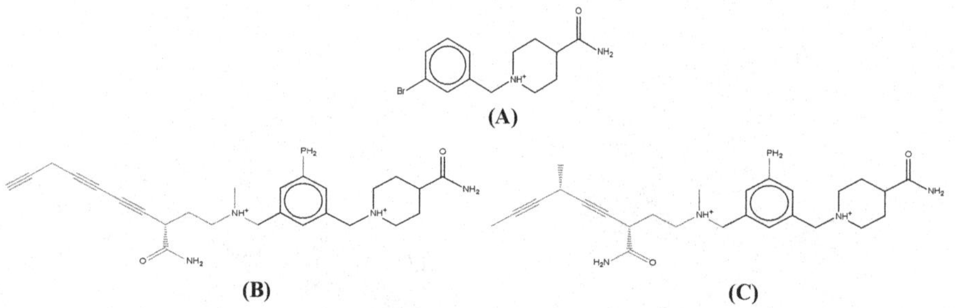

Fig. 2. Structure molecules of 1-[(3-bromophenyl) methyl] piperidine-4-carboxamide (A) along
with fragment growing result, FGR-2 (B) and FGR-3 (C).

3.3 Molecular Docking Simulation

MOE 2014.09 software was employed for all molecular docking simulations. Both
ligands and standards underwent two types of docking protocols. The 'rigid receptor'
protocol was used for the first docking simulation with retain 30 and the repetition with
retain 100. Whereas in the second docking, 'induced-fit' protocol with retain 100 was
performed. In the rigid docking, only 572 ligands attached to the binding site that have
RMSD value <2.0 Å and ΔG binding value lower than standard. Meanwhile, only 527
ligands that result from induced-fit docking which comply with the requirement. As the
standard ligands, we used linagliptin (ZINC03820029), saxagliptin (ZINC13648755),
and alogliptin (ZINC14961096), because they are commercial FDA-approved drug as

the DPP-4 inhibitor [7, 24]. The ten best ligands with the lowest ΔG binding value and the standard are shown in Table 1. These ligands are then analyzed for toxicity and pharmacological properties to select the best drug candidates as shown as in the research flowchart (Fig. 3).

Table 1. The ΔG binding and RMSD value of the best ligand and standard, from induced-fit docking simulation

No.	Molecule names	H-Bond	ΔG (Kcal/mol)	RMSD
1.	(3-(methyl((1Z,12E)-nonadeca-1,3,4,6,7,9,10,12,14,15,17,18-dodecaen-1-yl) amino)-5-(1H-pyrazol-1-yl) phenyl) methanaminium	1	−13.4851	1.2968
2.	4-carbamoyl-1-(3-((((S)-3-carbamoyldeca-4,6,9-triyn-1-yl) (methyl)ammonio) methyl)-5-phosphaneylbenzyl) piperidin-1-ium	5	−12.5597	1.0192
3.	4-carbamoyl-1-(3-((((3S,6S)-3-carbamoyl-6-methylnona-4,7-diyn-1-yl) (methyl)ammonio) methyl)-5-phosphaneylbenzyl) piperidin-1-ium	6	−12.2777	0.8318
4.	1-(3-bromo-5-((((3S,6S)-3-ethyl-6-methylnona-4,7-diyn-1-yl) (methyl)ammonio) methyl) benzyl)-4-carbamoylpiperidin-1-ium	5	−12.2051	1.1632
5.	4-carbamoyl-1-(2-((3S)-3-carbamoyl-6-(isopentyl (methyl) ammonio) hexa-1,4-diyn-1-yl)-6-chlorobenzyl) piperidin-1-ium	2	−12.1617	1.1996
6.	1-(4-((4S,7S)-11-ammonio-4-(ammoniomethyl)-7-(fluorocarbonyl) undeca-1,5,8-triyn-1-yl) benzyl)-4-carbamoylpiperidin-1-ium	5	−12.1471	1.0965
7.	(3-(((1Z,9E)-heptadeca-1,3,4,6,7,9,11,12,14,15-decaen-1-yl) (methyl)amino)-5-(1H-pyrazol-1-yl) phenyl) methanaminium	1	−11.9514	0.7797
8.	1-(2-((3R,6S)-10-ammonio-6-carbamoyl-3-methyldeca-1,4,7-triyn-1-yl)-6-chlorobenzyl)-4-carbamoylpiperidin-1-ium	3	−11.9257	0.7796
9.	1-(2-((3R)-6-((4-amino-4-oxobut-2-yn-1-yl) (ethyl)ammonio)-3-ethylhexa-1,4-diyn-1-yl)-6-chlorobenzyl)-4-carbamoylpiperidin-1-ium	5	−11.8711	1.5126
10.	4-carbamoyl-1-(3-((((S)-3-carbamoylundeca-4,6,9-triyn-1-yl) (methyl)ammonio) methyl)-5-phosphaneylbenzyl) piperidin-1-ium	5	−11.8654	1.2684
Std[1]	Linagliptin	0	−9.7260	1.6075
Std[2]	Saxagliptin	1	−7.8669	0.6046
Std[3]	Alogliptin	5	−7.3038	1.6650

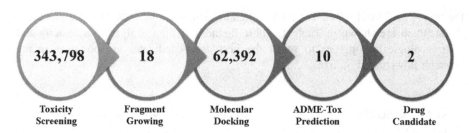

Fig. 3. Research flowchart. Inside the circle marks are the number of ligands that have been used in every step, respectively.

3.4   ADME and Toxicity Prediction

In this research, ADME (Adsorption, Distribution, Metabolism, and Excretion) and toxicity were determined by computational prediction. Using DataWarrior and Toxtree which are presented in Table 2. DataWarrior was used to predict mutagenic, tumorigenic, reproductive effect and irritant [14], while Toxtree was used to analyzed Genotoxic Carcinogenicity (GC), Non-Genotoxic Carcinogenicity (NGC), Potential Salmonella typhimurium TA100 Mutagen (PSM) based on QSAR, and Potential Carcinogen (PC) based on QSAR [16]. From the table, five ligands have carcinogenic properties such as FGR-4, FGR-5, FGR-6, FGR-8, and FGR-9.

Table 2. Toxicity prediction using DataWarrior and Toxtree

No.	Ligands	Mut	Tum	RE	Irr	GC	NGC	PSM	PC
1.	FGR-1	No	No	No	No	No	No	No	No
2.	FGR-2	No	No	No	No	No	No	No	No
3.	FGR-3	No	No	No	No	No	No	No	No
4.	FGR-4	No	No	No	No	No	Yes	No	No
5.	FGR-5	No	No	No	No	No	Yes	No	No
6.	FGR-6	No	No	No	No	Yes	No	No	No
7.	FGR-7	No	No	No	No	No	No	No	No
8.	FGR-8	No	No	No	No	No	Yes	No	No
9.	FGR-9	No	No	No	No	No	Yes	No	No
10.	FGR-10	No	No	No	No	No	No	No	No
Std[1]	Linagliptin	No	No	No	No	No	No	No	No
Std[2]	Saxagliptin	No	No	No	No	No	No	No	No
Std[3]	Alogliptin	No	No	No	No	No	No	No	No

Medicinal chemistry of the ligands like PAINS (Pan-assay interference compounds) and synthetic accessibility was predicted using SWISSADME together with bioavailability score, gastrointestinal absorption, and CYP450 inhibitor [18], which the data are shown in Table 3.

Table 3. Pharmacokinetics and medicinal chemistry prediction using SWISSADME

No.	Pharmacokinetics				MedChem
	GI Absorption	CYP inhibitor	PAINS	Bioavailability	Synthetic Accessibility
1.	High	CYP2C19, CYP2C9, CYP2D6	0	0,55	3,94
2.	Low	No	0	0,55	4,75
3.	High	No	0	0,55	5,05
4.	High	No	0	0,55	4,74
5.	High	No	0	0,55	5,04
6.	High	No	0	0.55	4.94
7.	High	CYP2C19, CYP2C9, CYP3A4	0	0,55	3,86
8.	High	No	0	0,55	4,69
9.	High	No	0	0,55	4,98
10.	High	No	0	0,55	4,86
Std[1]	High	CYP2C9, CYP3A4	0	0,55	4,44
Std[2]	High	No	0	0,55	5,05
Std[3]	High	No	0	0,55	3,54

Bioavailability score and PAINS were identical for all of the ligands which indicated that the ligand was similarly absorbed in our body and not likely to produce false-positives in the high-throughput screen test. From Table 3, some ligands were Cytochrome P450 (CYP) inhibitor including the standard (linagliptin). Irreversible inhibition of CYP can trigger an autoimmune response. Therefore, it is essential to study the inhibition of CYP during the drug discovery process [25].

Although FGR-2 showed good results on almost all ADME-Tox prediction, gastrointestinal absorption was still relatively low compared to other ligands, which indicated that FGR-2 was not suitable for oral administration. On the other hand, the prediction results of FGR-3 indicated that the ligand was approved as an oral drug candidate that can be continued for the next stage testing.

According to all of the data analysis, two drug candidates, such as FGR-2 and FGR-3 have lower ΔG binding value, more hydrogen bond interaction, and pharmacological properties are preferred than the standards. Both drug candidates are the fragment growing product from a natural product compound that has a popular name as 1-[(3-bromophenyl) methyl] piperidine-4-carboxamide. The modification results from this compound can be seen in Fig. 2.

As shown in Fig. 4, FGR-2 interacts with 21 amino acid residues, while FGR-3 interacts with 19 residues. On the other hand, both drug candidates have five important residues when attaching to a protein that forms hydrogen bond interactions at Glu205, Glu206, Tyr631, His740 residues, and π–π stacking interaction at Tyr666 residue. Molecular properties of each drug candidates and standard were presented in Table 4.

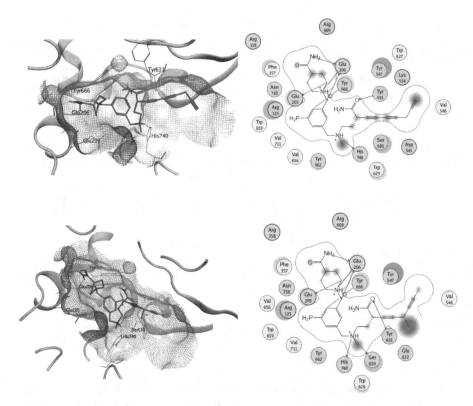

Fig. 4. The 3D (left) and 2D (right) visualization of ligand interaction FGR-2 (top) and FGR-3 (bottom)

Table 4. Molecular properties of candidates and standard drugs

No.	Ligands	Physico-chemical Properties				
		Weight	TPSA	H-Don	H-Acc	log P
1.	FGR-2	466.564	95.06	4	6	−1.8233
2.	FGR-3	470.596	95.06	4	6	−1.2259
Std[1]	Linagliptin	473.559	115.10	1	10	−1.1404
Std[2]	Saxagliptin	316.423	91.97	2	5	−2.8211
Std[3]	Alogliptin	340.406	95.29	1	7	−2.1591

4   Conclusion

The drug candidates were chosen as the novel inhibitor of DPP-4 are FGR-2 and FGR-3 based on their low Gibbs free energy (ΔG) and favorable interactions. However, based on the pharmacological properties of the ligands that was predicted by using DataWarrior, Toxtree v2.6.13, and SWISSADME software, we only recommended FGR-3, as an oral drug candidate. This ligand exhibited non-toxic, non-carcinogenic, non-mutagenic, high gastrointestinal absorption, and no bad effect for metabolism in the human body. Further in vitro, in vivo, and clinical trial analysis must be executed in order to validate the therapeutic activity of drug candidates for type 2 diabetes.

Acknowledgment. The research was funded by Hibah Publikasi Internasional Terindeks 9 (PIT 9) Universitas Indonesia Tahun Anggaran 2019 No: NKB-0036/UN2.R3.1/HKP.05.00/2019. The authors would like to sincerely thank Ahmad Husein Alkaff, Mutiara Saragih, and Satya Anindita for proofreading the article. All authors were responsible for the writing of the manuscript. Herewith, the authors declare that there is no conflict of interest regarding the manuscript.

References

1. Alberti, K.G., Zimmet, P.Z.: Definition, diagnosis and classification of diabetes mellitus and its complications. Part 1: diagnosis and classification of diabetes mellitus. Provisional report of a WHO Consultation. Diab. Med. **15**, 539–553 (1998)
2. World Health Organization: The top 10 causes of death (2018). http://www.who.int/news-room/fact-sheets/detail/the-top-10-causes-of-death. Accessed 18 July 2018
3. Cho, N.H., et al.: IDF diabetes atlas: global estimates of diabetes prevalence for 2017 and projections for 2045. Diab. Res. Clin. Pract. **138**, 271–281 (2018). https://doi.org/10.1016/j.diabres.2018.02.023
4. Hu, F.B.: Globalization of diabetes: the role of diet, lifestyle, and genes. Diab. Care **34**, 1249–1257 (2011). https://doi.org/10.2337/dc11-0442
5. Unnikrishnan, R., Pradeepa, R., Joshi, S.R., Mohan, V.: Type 2 diabetes: demystifying the global epidemic. Diabetes **66**, 1432–1442 (2017). https://doi.org/10.2337/db16-0766
6. Sebokova, E., Christ, A.D., Boehringer, M., Mizrahi, J.: Dipeptidyl peptidase IV inhibitors: the next generation of new promising therapies for the management of type 2 diabetes. Curr. Top. Med. Chem. **7**, 547–555 (2007). https://doi.org/10.2174/156802607780091019
7. Food & Drug Administration: FDA Drug Safety Communication: FDA warns that DPP-4 inhibitors for type 2 diabetes may cause severe joint pain (2015). https://www.fda.gov/Drugs/DrugSafety/ucm459579. Accessed 26 July 2018
8. Abell, C., Scott, D.E., Coyne, A.G., Hudson, S.A.: Fragment-based approaches in drug discovery and chemical biology. Biochemistry **51**, 4990–5003 (2012). https://doi.org/10.4137/DTI.S31566
9. Newman, D.J., Cragg, G.M.: Natural products as sources of new drugs from 1981 to 2014. J. Nat. Prod. **79**, 629–661 (2016). https://doi.org/10.1021/acs.jnatprod.5b01055
10. Khazir, J., Mir, B.A., Mir, S.A., Cowan, D.: Natural products as lead compounds in drug discovery. J. Asian Nat. Prod. Res. **15**, 764–788 (2013). https://doi.org/10.1080/10286020.2013.798314
11. Tambunan, U.S.F., Siregar, S., Toepak, E.P.: Ebola viral protein 24 (VP24) inhibitor discovery by In silico fragment-based design. Int. J. GEOMATE **15**, 59–64 (2018). https://doi.org/10.21660/2018.49.3534
12. Sutton, J.M., et al.: Novel heterocyclic DPP-4 inhibitors for the treatment of type 2 diabetes. Bioorg. Med. Chem. Lett. **22**, 1464–1468 (2012). https://doi.org/10.1016/j.bmcl.2011.11.054
13. Sterling, T., Irwin, J.J.: ZINC 15 - ligand discovery for everyone. J. Chem. Inf. Model. **55**, 2324–2337 (2015). https://doi.org/10.1021/acs.jcim.5b00559
14. Sander, T., Freyss, J., von Korff, M., Rufener, C.: DataWarrior: an open-source program for chemistry aware data visualization and analysis. J. Chem. Inf. Model. **55**, 460–473 (2015). https://doi.org/10.1021/ci500588j

15. Benet, L.Z., Hosey, C.M., Ursu, O., Oprea, T.I.: BDDCS, the rule of 5 and drugability. Adv. Drug Deliv. Rev. **101**, 89–98 (2016). https://doi.org/10.1016/j.addr.2016.05.007
16. Patlewicz, G., Jeliazkova, N., Safford, R.J., Worth, A.P., Aleksiev, B.: An evaluation of the implementation of the Cramer classification scheme in the Toxtree software. SAR QSAR Environ. Res. **19**, 495–524 (2008). https://doi.org/10.1016/j.apsusc.2012.12.143
17. Daina, A., Michielin, O., Zoete, V.: SwissADME: a free web tool to evaluate pharmacokinetics, drug-likeness and medicinal chemistry friendliness of small molecules. Sci. Rep. **7** (2017). https://doi.org/10.1038/srep42717
18. Prajapat, R., Bhattacharya, I.: In-silico structure modeling and docking studies using dipeptidyl peptidase 4 (DPP4) inhibitors against diabetes type-2. Adv. Diab. Metab. **4**, 73–84 (2016). https://doi.org/10.13189/adm.2016.040403
19. Zhong, J., Maiseyeu, A., Davis, S.N., Rajagopalan, S.: DPP4 in cardiometabolic disease. Circ. Res. **116**, 1491–1504 (2015). https://doi.org/10.1161/CIRCRESAHA.116.305665
20. Pantaleão, S.Q., et al.: Structural dynamics of DPP-4 and its influence on the projection of bioactive ligands. Molecules **23**, 1–10 (2018). https://doi.org/10.3390/molecules23020490
21. Aduri, R., Psciuk, B.T., Saro, P., Taniga, H., Schlegel, H.B., SantaLucia, J.: AMBER force field parameters for the naturally occurring modified nucleosides in RNA. J. Chem. Theory Comput. **3**, 1464–1475 (2007). https://doi.org/10.1021/ct600329w
22. Da, C., Kireev, D.: Structural protein-ligand interaction fingerprints (SPLIF) for structure-based virtual screening: method and benchmark study. J. Chem. Inf. Model. **54**, 2555–2561 (2014). https://doi.org/10.1021/ci500319f
23. Lipinski, C.A.: Lead- and drug-like compounds: the rule-of-five revolution. Drug Discov. Today Technol. **1**, 337–341 (2004). https://doi.org/10.1016/j.ddtec.2004.11.007
24. Berger, J.P., et al.: A comparative study of the binding properties, dipeptidyl peptidase-4 (DPP-4) inhibitory activity and glucose-lowering efficacy of the DPP-4 inhibitors alogliptin, linagliptin, saxagliptin, sitagliptin and vildagliptin in mice. Endocrinol. Diab. Metab. **1**, e00002 (2018). https://doi.org/10.1002/edm2.2
25. Fontana, E., Dansette, P.M., Poli, S.M.: Cytochrome p450 enzymes mechanism based inhibitors: common sub-structures and reactivity. Curr. Drug Metab. **6**, 413–454 (2005)

Discovery of Novel Alpha-Amylase Inhibitors for Type II Diabetes Mellitus Through the Fragment-Based Drug Design

Yulianti, Agustinus Corona Boraelis Kantale,
and Usman Sumo Friend Tambunan[✉] [iD]

Department of Chemistry, Faculty of Mathematics and Natural Sciences,
Universitas Indonesia, Kampus UI Depok, Depok, West Java 16424, Indonesia
usman@ui.ac.id

Abstract. Diabetes mellitus is a metabolic disorder leading to hyperglycemia and organ damage. In 2017, the International Diabetes Federation (IDF) reported that about 425 million people living with diabetes, most of which suffer from type 2 diabetes mellitus. The drug development for controlling glucose level is crucial to treat people with type 2 diabetes mellitus. Alpha-amylase plays an imperative role in carbohydrate hydrolysis. Hence, the inhibition of alpha-amylase, which halt the glucose absorption, can be a promising pathway for developing type 2 diabetes mellitus drugs. Natural product has been known as the lead drugs for various diseases. In this research, the fragment merging drug design was performed by employing both the existing drug, voglibose, as the template and the natural product compounds to generate newly constructed ligands. The fragments were acquired from ZINC15 natural product database and then were screened according to Astex's Rules of Three, pharmacophore properties, and molecular docking simulation. The 482 selected fragments were evaluated under Lipinski's Rule of Five and toxicity effects using DataWarrior software. The ligands underwent molecular flexible docking simulation followed by the ADME-Tox prediction by using Toxtree, AdmetSAR, and SwissADME software. In the end, two lead compounds showed the best properties as an alpha-amylase inhibitor based on their low $\Delta G_{binding}$, acceptable RMSD score, favorable pharmacological properties, and molecular interaction.

Keywords: Type 2 diabetes mellitus · Alpha-amylase ·
Fragment-based drug design

1 Introduction

Type 2 diabetes mellitus is diabetes mellitus which is non-insulin-dependent. Diabetes mellitus is caused by changes in carbohydrate metabolism which leads to hyperglycemia [1, 2]. The number of people with diabetes mellitus in the world has quadrupled in the past three decades [3]. According to data from the International Diabetes Federation (2017), around 425 million people suffer from diabetes and most of them are residents living in urban areas. This number is expected to continue increasing every year [4].

© Springer Nature Switzerland AG 2019
I. Rojas et al. (Eds.): IWBBIO 2019, LNBI 11465, pp. 25–35, 2019.
https://doi.org/10.1007/978-3-030-17938-0_3

Diabetes mellitus treatment is focused on securing the quality of life through controlling and decreasing blood glucose levels to normal levels [5]. Inhibition of alpha-amylase enzymes, enzymes that play an imperative role in the digestion of starch and glycogen, can be targeted for the treatment of type 2 diabetes mellitus [6]. The inhibition of alpha-amylase can significantly reduce the increase in post-prandial blood glucose by blocking the hydrolysis of carbohydrates in the human body and reducing glucose absorption [7].

Natural compounds have become a potential source of drug development. The use of natural compounds in drug discovery is related to their presence in various species in nature, their complex chemical structures, and the existence of supporting methods and technologies. Also, the discovery of natural product-based drugs has enormous potential to utilize the chemical diversity of natural products [8]. Numerous methods have been developed to create and optimize lead compounds which have the potential as new therapeutic agents, one on which is fragment-based drug design (FBDD). Fragment-based drug design allows the identification of active fragments that can reach the subpocket within the active site. In addition, a higher hit rate and more efficient optimization capacity are interesting features offered by this method. The construction of drug-like molecule utilizing fragment-based drug design methods can be achieved by merging, linking, or growing the fragment [9–11]. Fragment merging is the incorporation of structural parts of molecules which overlap with elements of a protein substrate or inhibitor which are known to form a molecular complex [12]. In this research, fragment merging was accomplished between the natural product and the existing drug, voglibose, to acquire lead compounds as alpha-amylase inhibitors through molecular docking simulations and ADME-Tox assay.

2 Material and Method

This research was conducted by utilizing the Molecular Operating Environment (MOE) 2014.09, DataWarrior v04.06.01, Toxtree v2.6.13, ChemBioDraw Ultra 14.0, SwissADME, and AdmetSAR software. The 3D structure of alpha-amylase and natural product fragments were acquired from RCSB Protein Data Bank and ZINC15 database, respectively.

2.1 Preparation of Alpha-Amylase

The 3D structure of alpha-amylase with PDB ID: 1HNY was obtained from RCSB protein databank. The 3D structure preparation was conducted using MOE 2014.09 software by applying the Amber10: EHT parameter as forcefield and R-field as sol-vation mode. Structure optimization was done by removing unwanted water molecules and ions. Then, the LigX protocol was implemented with default settings.

2.2 Preparation of Natural Product Fragments and Standard Molecules

In this research, the fragments were obtained from ZINC15 natural product database. These fragments were prepared through MOE 2014.09 with the default parameters in

Wash and Energy Minimization. MMFF94x forcefield and RMS gradient of 0.001 kcal/mol.Å were used as the optimization parameters. Standard molecules such as acarbose, miglitol, voglibose, and metformin were also prepared in the same steps with the natural product fragments. Thereafter, the prepared natural product fragments were screened according to Astex's Rule of Three (RO3) and toxicity test by utilizing DataWarrior software.

2.3 Molecular Docking Simulation of Standard Ligand and Protein-Ligand Interaction Fingerprint (PLIF)

Molecular docking simulations of alpha-amylase and standard ligands were conducted by performing rigid docking and flexible docking utilizing MOE 2014.09 software. Molecular docking simulation results were used to determine the pharmacophore features by operating the stages of Protein-Ligand Interaction Fingerprints (PLIF). PLIF procedure was done according to the default parameter of MOE 2014.09 software.

2.4 Molecular Docking Simulation of Natural Product Fragments and Fragment Merging

Natural product fragments were docked two times using virtual screening protocol and followed by rigid receptor protocol. Molecular docking was operated into the active site of alpha-amylase. In this simulation, parameters used include Pharmacophore/London dG rescoring as placement and Forcefield/GBVI-WSA dG as refinement.

The fragments with proper Gibbs binding energy ($\Delta G_{binding}$) and RMSD value were chosen to be merged with voglibose as the template. The selected fragment should overlap with standard ligand and comply with the Lipinski's Rule of Five (RO5). This process was conducted by utilizing MOE 2014.09, ChemBioDraw Ultra 14.0, and DataWarrior software.

2.5 Molecular Docking of Ligands

Molecular docking of ligands was conducted by performing flexible docking with potential setup AMBER 10: EHT and R-field as forcefield and solvation mode, respectively. The best ligands were chosen according to the ΔGbinding, RMSD value, and molecular interaction.

2.6 Pharmacological Prediction

The potential ligands were analyzed its pharmacological characteristic. Mutagenicity and carcinogenicity were predicted using Toxtree software. The toxicity effects of the ligands were evaluated utilizing DataWarrior and AdmetSAR, while the health effect of human was determined using SwissADME.

3   Results and Discussion

3.1   Visualization of Alpha-Amylase

In this research, the 3D structure of alpha-amylase was acquired from RCSB PDB. The structure of human pancreatic alpha-amylase was determined using X-ray crystallography with the resolution of 1.8 Å. Human pancreatic amylase consists of three domains. The largest domain is domain A which serves as the location of active site residues Asp197, Glu233, and Asp300 [13]. Energy minimization was also implemented in the process of protein preparation to obtain stable conformations, configurations with minimum energy [14], and gradient energy close to zero [15]. After the 3D alpha-amylase structure was optimized, the ideal binding site of alpha-amylase was predicted using the 'Site Finder' feature. The binding site used in this study based on research of Brayer et al. in 1995. The visualization of the binding site alpha-amylase is shown in Fig. 1.

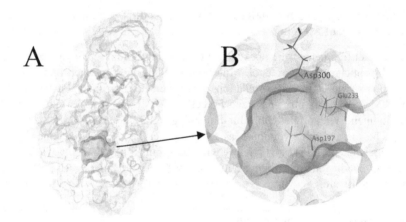

Fig. 1. The 3D structure of human pancreatic alpha-amylase taken from PDB ID: 1HNY (A) and visualization of the alpha-amylase binding site (B)

3.2   Initial Screening of Natural Product Fragments

About 343,798 compounds were collected from ZINC15 database and saved in the .sdf format file. Initial screening of natural product fragments was conducted using Data Warrior software. The compounds were screened according to the Rule of Three (RO3) with the following parameters: molecular weight <300 Da, the cLogP ≤ 3, rotatable bonds ≤ 3, the number of hydrogen bond acceptors ≤ 3, the number of hydrogen donor ≤ 3, and polar surface area ≤ 60 Å^2. Rule of three was used when constructing fragment into lead compound [16]. Also, the natural products with druglikeness score higher than 0 and did not show the mutagenic, tumorigenic, reproductive effect, and irritant characteristic was selected and prepared for the generation of the lead compound. Around 7,470 fragments were retrieved from the initial screening process.

3.3   Protein-Ligand Interaction Fingerprints (PLIF)

The standard ligands used were metformin, miglitol, and acarbose. The standard ligands were subjected to molecular docking simulation with rigid receptor protocol with retain 30 and 100 respectively and induce fit with retain 100 and 300. The results of induce fit retain 300 were superposed using MOE 2014.09 software. This process was also called the Protein-Ligand Interaction Fingerprints (PLIF).

PLIF is the encryption of structural information which shows all the similarities in protein-ligand interactions. The assessments are accomplished quantitatively which informed the interaction between the docking pose and the target protein similar to the known ligand [17]. PLIF concises the interaction between ligands and protein through fingerprint design. The similarity of these interactions can increase the likelihood of finding hits over structure-based screening.

In addition, pharmacophore features were also generated from the PLIF process. Pharmacophore feature was used to determine the interaction point of protein-ligands as binding sites and biological activities [18]. Pharmacophore features have active sites which are used to bind ligands to specific targets. In such a case, only the ligand which has the desired characteristics in the drug molecule will be filtered [19]. Pharmacophore feature constructed in this study is displayed in Fig. 2. Each color of the feature pharmacophore recognizes different characteristic. The first feature is F1: Don&Acc which has characteristic as hydrogen donors and acceptors. The second feature F2: Don acts as hydrogen donors, and the third feature is F3: HydA which means hydrophobic [20]. Pharmacophore was combined with molecular docking simulation to improve success in finding lead compounds.

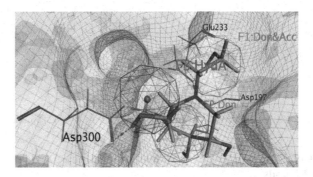

Fig. 2. The pharmacophore feature that was used in this study

3.4   Molecular Docking Simulation of Alpha-Amylase and Fragments

Molecular docking is an essential step in drug discovery and design of the in silico method. Molecular docking simulation predicts the ligand orientation in the target proteins [21]. Docking simulation of fragments was done by using two protocols, virtual screening, and rigid receptor. From the virtual screening protocol simulation, 588 compounds were obtained. Rigid docking was performed two times. The first rigid docking was retained 30. Only 502 compounds bind the pharmacophore feature in the pocket. While the second rigid docking applied retain 100. About 482 compounds which pass the second rigid docking simulation could be used in the fragment merging process.

3.5 Preparation of Natural Product Ligands

The fragments were merging was executed by employing MOE 2014.09 software. The fragments were connected to the element of the voglibose, the existing drug of type 2 diabetes mellitus, by maintaining its crucial structure. The new ligands were constructed under the Lipinski's Rule of Five (RO5) and Veber rule. The RO5 rule is molecular weight lower than 500 Da, the number of hydrogen donor lower than 5, the number of hydrogen acceptor lower than 10, LogP between −0.5 and 5.6 [22]. The TPSA no more than 140 $\mathring{A}^2$ and the rotatable bond lower than 10 are the Veber rule parameters [23]. The ligands must also be screened based on the druglikeness and toxicity properties following mutagenic, tumorigenic, reproductive effect, and irritant. A total of seven ligands were constructed from the process of fragment merging. The best ligands were shown in Fig. 3, with the pink color, is part of voglibose, and the black color is the fragment.

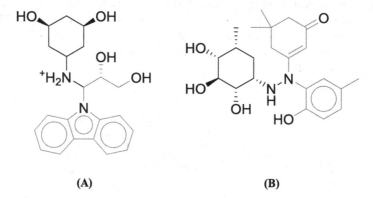

Fig. 3. The selected fragments and the merging position of (A) YS2156 and (B) UT3261

3.6 Analysis of Molecular Docking Simulation of Alpha-Amylase and the Ligands

A total of seven ligands and two standard molecules underwent flexible docking simulation utilizing MOE 2014.09. Flexible docking was employed two times with the retain 100 and 300, respectively. In this step, only 2 ligands have RMSD value lower than 2.0 Å and $\Delta G_{binding}$ lower than standard. The result of best ligands and the standards from flexible molecular docking simulation is presented in Table 1.

Table 1. Pharmacological prediction results of ligands

Ligand	$\Delta G_{binding}$	RMSD	Weight	LogP	Hydrogen acceptor	Hydrogen donor	TPSA
YS2156	−9.0916	1.0376	371.455	0.1029	6	5	102.46
UT3261	−10.0346	1.0880	404.505	1.2611	7	5	113.26
Voglibose*	−7.4587	1.2389	268.285	−5.1960	8	8	158.22
Acarbose*	−9.0856	1.2720	646.631	−8.4018	19	14	325.75

$\Delta G_{binding}$ in Kcal/mol, RMSD in Å, TPSA in Å^2, weight in Da, *standard.

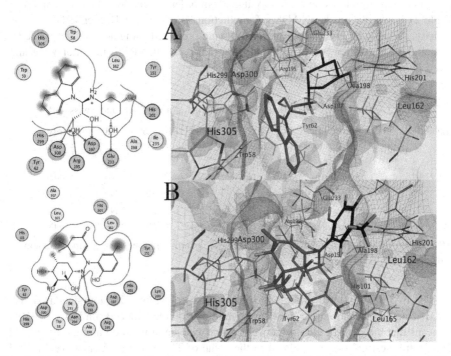

Fig. 4. Molecular interaction visualization of (A) YS2156 and (B) UT3261 in 2D (left) and 3D (right)

As the best ligand, UT3261 exhibited the lowest $\Delta G_{binding}$ value than YS2156 ligand. It also bound on the best position in the binding pocket. Figure 4 shows the molecular interaction between YS2156 and UT3261 with amino acids residues in the binding site of alpha-amylase. YS2156 has 14 interactions with the amino acid residues. Seven hydrogen bonds were binding the ligand in the Asp197, Glu233, Asp300, Arg195, His101, and His299. Trp59, Trp58, His305, Leu162, Tyr151, Ile235, Ala198, and Tyr62 bind through van der Walls interaction. While UT3261 has 18 interaction with the amino acids residue. Three hydrogen bonds were binding the ligands in the Glu233 and Asp300. On the other hand, His101, Ala307, Leu165, His305, Leu162, Tyr151, Lys200, His201, Asp197, Arg195, Ala198, Asn298, Ile235, Trp58, His299, and Tyr62 interacted through van der Walls interaction. YS2156 has interaction with

Arg195, Asp197, Glu233, His201, and His299 on O hydroxyl atom and Asp300 on C carboxyl. While UT3261 has interaction with Asp300 on O hydroxyl atom and Glu233 on O hydroxyl atom and N atom. It shows that Glu233 and Asp300 have an essential role in inhibiting the alpha-amylase enzyme.

3.7 Pharmacological Prediction

The two ligands were analyzed for their pharmacological properties and toxicity by employing admeSAR and SwissADME online software. ADMET test is one of the important steps that must be passed by ligands in order to function as a drug candidate. Table 2 shows the result of ADMET prediction. Blood-Brain Barrier (BBB), Ames toxicity, carcinogen, and acute oral toxicity were predicted using admetSAR, while the subcellular localization, gastrointestinal absorption, and CYP450 inhibitor were analyzed by using SwissADME.

Table 2. ADME-Tox prediction of the ligand using admetSAR and SwissADME

Ligand	BBB	Subcellular localization	GI abs.	CYP450 inhibitor	hERG I inhibitor	Ames Tox.	Carcinogen	Acute oral toxicity
YS2156	+	Nucleus	High	No	No	No	No	III
UT3261	−	Mitochondria	Low	No	No	No	No	III
Voglibose*	−	Lysosome	Low	No	No	No	No	IV
Acarbose*	−	Lysosome	Low	No	No	No	No	IV

BBB: Blood-Brain Barrier, GI abs.: Gastrointestinal absorption, Ames Tox.: Ames Toxicity, *standard

The two ligands have difference subcellular localization with the standards. The UT3162 and standard ligands show negative effects on the Blood-Brain Barrier (BBB). It means that the ligands did not affect transport between the blood and the brain [24]. The gastrointestinal effect is related to the value of TPSA and molecular weight [25]. All ligands showed no inhibition potency of cytochrome P450 (CYP450) enzyme. Major isoforms of the enzyme such as CYP1A2, CYP2C19, CYP2CG, CYP2D6, and CYP3A4 plays a crucial role in the biotransformation of drugs. The most abundant isoenzyme in the liver and involved in drug metabolism is CYP3A4, which is around 30–40% of drugs [26]. In addition, it is also crucial to analyze the ligand properties of the hERG. The hERG I is a potassium ion channel that plays a role in cardiac repolarization. Compounds that inhibit the heart ion channel can cause cardiac arrhythmias [27]. Both ligands and standards did not inhibit hERG I. Toxicity effect of ligands was analyzed by determining the Ames toxicity and carcinogen. All ligands were not showing toxicity effect based on those parameters.

The Ames test was used as an initial screening to analyze the mutagenic probability of new ligands for hazard description by using *Salmonella* bacteria [28]. If the ligands significantly induce the growth of the revertant colonies in at least one of the common five strains, the ligands are categorized as Ames positive. While Ames negative if it does not induce significant revertant colony growth in any strain [29]. All of the ligands have no Ames toxicity properties. It means that all of the ligands did not cause mutagenic in human.

Acute toxicity prediction aims to acquire data on the biologic activity of a chemical and obtain insight into its mechanism of action [30]. Table 2 shows that the ligands in category III of acute oral toxicity, it means that the ligands were slightly toxic, while the standard ligands in IV category [31]. It means the Voglibose and Acarbose were generally considered to be practically nontoxic.

Table 3. The druglikeness properties of ligands using SwissADME

Ligand	Lipinski	Veber	Egan	Bioavailability score	Synthetic accessibility
YS2156	Yes	Yes	Yes	0.55	4.12
UT3261	Yes	Yes	Yes	0.55	5.18
Voglibose*	Yes	No	No	0.55	3.69
Acarbose*	No	No	No	0.17	7.38

The druglikeness of the ligands and standards were checked. As displayed in Table 3, all ligands have good druglikeness based on the Lipinski's, Veber's, and Egan's Rule. The ligands have the highest oral bioavailability score compared to acarbose and same as voglibose. It assumed that the ligands could be absorbed properly in the body. The prediction of synthetic accessibility of the ligands using SwissADME can describe molecule synthetic accessibility as a score between I (easy to make) and 10 (very difficult to make). The synthetic accessibility score of UT3261 was lower than YS2156, which means UT3261 was more easy to synthesize than YS2156, but all the ligands were easier to synthesize compared the standard ligands [32].

4 Conclusion

Fragment-based drug design method has been done on the natural product compounds to find a potential inhibitor of alpha-amylase. YS2156 and UT3261 have lower Gibss binding score than the existing drug, acarbose. UT3261 has greater potential as an inhibitor according to low $\Delta G_{binding}$, RMSD value, and ADME-Tox prediction. Thus, the additional in silico method is needed to analyze the stability of the ligand in alpha-amylase and also is continued to in vitro and in vivo experiment.

Acknowledgment. This research is financially supported by the Direktorat of Research and Community Engagement of Universitas Indonesia (DRPM UI) by Hibah Publikasi Internasional Terindeks 9 (PIT9) Project. Also, I would like to thank you to Mutiara Saragih and Ahmad Husein Alkaff for proofreading this manuscript.

References

1. Hyun, T.K., Eom, S.H., Kim, J.: Molecular docking studies for discovery of plant-derived α-glucosidase inhibitors. Plant Omi. J. **7**, 166–170 (2014)
2. Pontes, J.P.J., Mendes, F.F., Vasconcelos, M.M., Batista, N.R.: Evaluation and perioperative management of patients with diabetes mellitus. A challenge for the anesthesiologist. Rev. Bras. Anestesiol. **68**, 75–86 (2018)
3. Zheng, Y., Ley, S.H., Hu, F.B.: Global aetiology and epidemiology of type 2 diabetes mellitus and its complications. Nature **14**, 88–98 (2018)
4. International Diabetes Federation: IDF Diabetes Atlas Eighth edition. Dipresentasikan pada (2017)
5. Kaku, K.: Pathophysiology of type 2 diabetes and its treatment policy. Jpn. Med. Assoc. J. **53**, 41–46 (2010)
6. Sharifuddin, Y., Chin, Y.-X., Lim, P.-E., Phang, S.-M.: Potential bioactive compounds from seaweed for diabetes management. Mar. Drugs **13**, 5447–5491 (2015)
7. Shankaraiah, P., Reddy, Y.N.: Alpha amylase expression in Indian type - 2 diabetic plants. J. Med. Sci. **11**, 280–284 (2011)
8. Lahlou, M.: The success of natural products in drug discovery. Pharmacol. Pharm. **4**, 17–31 (2013)
9. Kumar, A., Voet, A., Zhang, K.Y.J.: Fragment based drug design: from experimental to computational approaches. Curr. Med. Chem. **19**, 1–19 (2012)
10. Setlur, A.S., Naik, S.Y., Skariyachan, S.: Herbal lead as ideal bioactive compounds against probable drug targets of Ebola virus in comparison with known chemical analogue: a computational drug discovery perspective. Interdiscip. Sci. Comput. Life Sci. **9**, 254–277 (2017)
11. Scoffin, R., Slater, M.: Virtual elaboration of fragment ideas: growing, merging and linking fragments with realistic chemistry. Drug Discov. Dev. Deliv. **7**, 36–40 (2015)
12. Erlanson, D.A., Mcdowell, R.S., Brien, T.O.: Fragment-based drug discovery. J. Med. Chem. **47**, 3463–3482 (2004)
13. Brayer, G.D., Luo, Y., Withers, S.G.: The structure of human pancreatic a-amylase at 1.8 A resolution and comparisons with related enzymes. Protein Sci. **4**, 1730–1742 (1995)
14. Bhasin, M., Raghava, G.P.S.: Computational methods in genome research. Appl. Mycol. Biotechnol. **6**, 179–207 (2006)
15. Rydzewski, J., Jakubowski, R., Nowak, W.: Communication: entropic measure to prevent energy over-minimization in molecular dynamics simulations. J. Chem. Phys. **143** (2015)
16. Congreve, M., Carr, R., Murray, C., Jhoti, H.: Fragment-based lead discovery. Drug Discov. Today **8**, 876–877 (2003)
17. Da, C., Kireev, D.: Structural protein − ligand interaction fingerprints (SPLIF) for structure-based virtual screening: method and benchmark study. J. Chem. Inf. Model. **54**, 2555–2561 (2014)
18. Hu, B., Lill, M.A.: Exploring the potential of protein-based pharmacophore models in ligand pose prediction and rangking. J. Chem. Inf. Model. **53**, 1179–1190 (2014)
19. Zuccotto, F.: Pharmacophore features distributions in different classes of compounds. J. Chem. Inf. Comput. Sci. **43**, 1542–1552 (2003)
20. Qing, X., et al.: Pharmacophore modeling: advances, limitations, and current utility in drug discovery. J. Recept. Ligand Channel Res. **7**, 81–92 (2014)
21. Machado, K.S., Schroeder, E.K., Ruiz, D.D., Cohen, E.M.L., Norberto de Souza, O.: FReDoWS: a method to automate molecular docking simulations with explicit receptor flexibility and snapshots selection. In: BMC Genomics, pp. 1–13 (2011)

22. Lipinski, C.A.: Lead- and drug-like compounds: the rule-of-five revolution. Drug Discov. Today Technol. **1**, 337–341 (2004)
23. Veber, D.F., Johnson, S.R., Cheng, H., Smith, B.R., Ward, K.W., Kopple, K.D.: Molecular properties that influence the oral bioavailability of drug candidates. J. Med. Sci. **45**, 2615–2623 (2002)
24. Daneman, R., Rescigno, M.: Review the gut immune barrier and the blood-brain barrier: are they so different? Immunity **31**, 722–735 (2009)
25. Nasution, M.A.F., Toepak, E.P., Alkaff, A.H., Tambunan, U.S.F.: Flexible docking-based molecular dynamics simulation of natural product compounds and Ebola virus Nucleocapsid (EBOV NP): a computational approach to discover new drug for combating Ebola. BMC Bioinform. **19**, 137–176 (2018)
26. Badyal, D.K., Dadhich, A.P.: Cytochrome P450 and drug interactions. Indian J. Pharmacol. **33**, 248–259 (2001)
27. Danker, T., Möller, C.: Early identification of hERG liability in drug discovery programs by automated patch clamp. Front. Pharmacol. **5**, 1–11 (2014)
28. Hakura, A., Shimada, H., Nakajima, M., Sui, H., Kitamoto, S., Suzuki, S.: Salmonella/ human S9 mutagenicity test: a collaborative study with 58 compounds. Mutagenesis **20**, 217–228 (2005)
29. Hansen, K., et al.: Benchmark data set for in silico prediction of Ames mutagenicity. J. Chem. Inf. Model. **49**, 2077–2081 (2009)
30. Walum, E.: Acute oral toxicity. Environ. Health Perspect. **106**, 497–503 (1998)
31. Li, X., et al.: In silico prediction of chemical acute oral toxicity using multi-classification methods. Chem. Inf. Model. **54**, 1061–1069 (2014)
32. Ertl, P., Schuffenhauer, A.: Estimation of synthetic accessibility score of drug-like molecules based on molecular complexity and fragment contributions. J. Cheminform. **11**, 1–11 (2009)

Compression of Nanopore FASTQ Files

Guillermo Dufort y Álvarez[1], Gadiel Seroussi[1,2], Pablo Smircich[3,4],
José Sotelo[3,4], Idoia Ochoa[5(✉)], and Álvaro Martín[1(✉)]

[1] Facultad de Ingeniería, Universidad de la República, Montevideo, Uruguay
`almartin@fing.edu.uy`
[2] Xperi Corp., San Jose, CA, USA
[3] Facultad de Ciencias, Universidad de la República, Montevideo, Uruguay
[4] Departamento de Genómica,
Instituto de Investigaciones Biológicas Clemente Estable,
Montevideo, Uruguay
[5] Electrical and Computer Engineering,
University of Illinois at Urbana-Champaign, Urbana, IL, USA
`idoia@illinois.edu`

Abstract. The research and development of tools for genomic data compression has focused so far on data generated by second-generation sequencing technologies, while third-generation technologies, such as nanopore technologies, have received little attention in the data compression research community. In this paper, we investigate compression schemes for nanopore FASTQ files. We propose a nanopore quality scores compressor, called *DualCtx*, which yields significant improvements in compression performance with respect to the state-of-the-art. We also extend DualCtx to a full FASTQ compressor, termed *DualFqz*, by substituting DualCtx for the quality score compression module in a variant of Fqzcomp. We tested DualFqz and various existing compressors on a large nanopore data set. The results show that DualFqz achieves the best compression performance. The experiments also show that most current implementations of compressors fail to execute correctly on files with long variable length reads.

DualCtx and DualFqz are freely available for download at: https://github.com/guidufort/DualFqz.

Keywords: Genomic data compression · FASTQ compression · Nanopore sequencing technology

1 Introduction

The rapid evolution of *High-Throughput Sequencing (HTS)* technologies over the past few years has led, among other consequences, to accelerating reductions in cost and sequencing time. In this context, there is a broad consensus that the amount of genomic information that will be generated globally will see explosive

© Springer Nature Switzerland AG 2019
I. Rojas et al. (Eds.): IWBBIO 2019, LNBI 11465, pp. 36–47, 2019.
https://doi.org/10.1007/978-3-030-17938-0_4

growth, leading to increasingly large needs for processing, storage, and transmission resources, which motivates the development of efficient compression tools for these data [24].

The usual theoretical framework to study this topic consists of considering the data to be compressed as emitted by a source of information, which generates symbols according to some probability law. In this setting, the goal in data compression is to minimize the expected length of an encoding of the data generated by the source, where expectation is taken with respect to the probability distribution governing the source. In many cases, from this probability distribution (or an estimation of it), it is possible to efficiently implement a theoretically optimal compressor by making use, for example, of an arithmetic encoder [20]. As a consequence, designing a good data compressor amounts, in essence, to finding good statistical models for the data to be compressed.

For HTS technologies, the result of a sequencing process is a set of readings of genome fragments, called *reads*, which are generally stored in text data files in FASTQ format. For each read, a FASTQ file contains a *base call sequence* (also referred to as a *read*), a *quality score sequence*, and a (possibly duplicated) *identifier string*. The base call sequence is a string of letters from the set $\{A, C, G, T\}$ that represents the nitrogenous bases in the read DNA fragment. In addition, the base call sequence may contain special symbols that represent specific situations that occur during the sequencing process; for example, a letter N in certain position of a base call sequence indicates that the sequencer failed to determine the correct base at that position. The quality score sequence is a string of symbols, of the same length as the base call sequence, where the i-th symbol encodes an estimated probability of the i-th base call being incorrect. The alphabet of symbols used to represent quality scores depends on the sequencing technology; in general, the size of this alphabet is larger than the alphabet of symbols used for base call sequences. Finally, the identifier string is a free text, generally short, which identifies the read.

The set of reads stored in FASTQ files is generally the starting point for a so-called *pipeline* of processing steps. The intermediate results throughout this series of steps are represented by files in specific formats for each case. For example, the result of aligning a set of reads with respect to a reference genome is usually represented by files in SAM (text) or BAM (binary) formats.

The HTS technologies in most common use today are *second-generation sequencing technologies* (also referred to as *Next Generation Sequencing (NGS) technologies*), which produce short reads (a few hundreds base-pair long) generally of fixed length. For these technologies, the quality of the readings is generally high, and quality scores have little correlation to the base call sequence. The alphabet of the quality scores in this case ranges from 4 values to about 40, depending on the specific technology. The *Single Molecule Real-Time (SMRT)* technology, developed by Pacific Biosciences (PacBio), is different in the following sense: it produces long reads with comparatively high error rates. Similarly, the recently developed *nanopore sequencing technology*, mainly driven by Oxford Nanopore Technologies (ONT), also generates very long variable length reads of relatively low quality. In contrast to other technologies, dependencies between the quality score sequence and the base call sequence have been observed for nanopore sequencing [9]. In addition, the alphabet size of the quality scores is 94 (Sanger format using ASCII codes 33 to 126).

Many algorithms have been proposed in the literature and implemented as specific tools for compression of different types of genomic data. Recent surveys are available in [16,17]. For compression of SAM/BAM files, for example, some of the most recent developments are presented in [2,3,7,14,18]. When the data to be compressed are reads in FASTQ format that have not been aligned, but there is a similar reference genome, some compression tools, such as [1,3,8,11,12,25], perform a fast alignment of reads with respect to the reference as a step prior to compression. The base call sequences are then encoded by describing how they vary relative to the reference genome, which is generally more economical than an independent encoding. The majority of compressors for FASTQ files, however, are reference-free, in that they do not use any reference sequence for compression. These compressors, which we focus on in this paper, may be preferred when a reference genome is not available or a self-contained encoding of the data is desirable. Nevertheless, some reference-free compressors, such as Quip [11], Leon [1], and KIC [25], still obtain a reference genome by constructing and encoding a draft assembly from the reads in the FASTQ file. Other tools, such as SCALCE [6], FaStore [22] and Spring [4], reorder the reads in a FASTQ file by base call sequence similarity; this reordering improves the performance of the compression itself, which takes place in a second stage. Finally, compressors like DSRC [21], Fqzcomp and Fastqz [3], and Slimfastq[1] do not apply any pre-processing to the data prior to compression.

As mentioned, all compression algorithms rely, either explicitly or implicitly, on some statistical model, which determines what data is expected to be seen more often than other. A common implementation of such a model consists of capturing statistical characteristics of the data through *context models*. In a context model, the probability distribution for a data symbol x is conditioned on the values of other *previously encoded* symbols, which are referred to as the *context* in which x occurs. On the decompressor side, context symbols have been decoded and are available when decoding x, so the same probability distribution for x can be determined in lockstep with the encoder. For example, in DSRC, the probability distribution for a base call symbol x depends on the nine bases immediately preceding x. Other compressors that make use of context models are Fqzcomp, Fastqz, and Slimfastq. Fqzcomp, in particular, determines a context for each quality score q as a function of the three quality scores immediately preceding q. According to [17], Fqzcomp achieves the best quality score compression performance among an extensive collection of compressors. The experimental data for the evaluation in [17] covers different HTS technologies, but does not include nanopore data. The reason is that most of the HTS data available today are generated by second-generation technologies, and hence, most of the compression algorithms introduced above are optimized for these data. As such, they obtain

---

[1] https://sourceforge.net/projects/slimfastq/.

their best performance when applied to genomic files containing short reads of fixed length, and many fail to work on data containing reads of variable length, or on data produced by other sequencing technologies.

In particular, data compression schemes optimized for nanopore data have, so far, received little attention in the research community. However, data produced by nanopore technologies is becoming increasingly popular, as the long reads have the potential to decrease the ambiguity associated to short reads, and help in the detection of large structural variants, including copy number variants (CNVs), medium- and large-sized insertions and deletions (INDELs), duplications, inversions, and translocations, among others [10,13,23].

With this in mind, in this paper we focus on compression schemes tailored to nanopore data in FASTQ format. Our technical contributions are three-fold. First, we propose a context model lossless compression scheme for nanopore quality scores, called *DualCtx*, which exploits the statistical dependency among neighbour quality scores and also between quality scores and the base call sequence. We show that the proposed scheme results in a significant improvement in compression performance with respect to the state of the art. Second, we extend DualCtx to a full FASTQ compressor, referred to as *DualFqz*, by substituting DualCtx for the quality score compression module in a variant of Fqzcomp. To the best knowledge of the authors, DualFqz is the first FASTQ file compressor optimized for nanopore data. In our experiments, DualFqz shows the best compression performance on our experimental nanopore data set. Third, we provide an evaluation of the performance of existing compression tools on a large data set of nanopore FASTQ files, and show that most implementations fail to execute correctly on files with long variable length reads.

2 Methods

Next, we introduce the proposed compressor for nanopore quality scores DualCtx and its extension to a full FASTQ compressor DualFqz.

DualCtx: Dual Context Quality Score Compression

To compress a sequence of quality scores in a FASTQ file, DualCtx constructs a context model that is comprised of both quality scores and base call symbols. Denote by q_i the quality score at position i within a read, and let $\hat{q}_i$ be a Q-bits quantized version of q_i. The quantized quality score, $\hat{q}_i$, is obtained from q_i as $\hat{q}_i = \lfloor (q_i + 1)/2^{R-Q} \rfloor$, where R is the dynamic range of the quality scores in bits, $R \geq Q$. Denote also by x_i the base call symbol at position i. The context for compressing q_i, denoted ctx_i and depicted in Fig. 1, is determined by the K quantized quality scores to the left of q_i, $\hat{q}_{i-K}, \ldots, \hat{q}_{i-1}$, and the L base call symbols closest to q_i, $x_{i-(L-1)/2}, \ldots, x_i, \ldots, x_{i+(L-1)/2}$, with prescribed conventions for border cases,[2] where K, L are positive integers and L is odd.

---

[2] For $i \leq K$, we arbitrarily let $q_j = 0$ and $x_j = A$ for all negative values of j. Similarly, we let $x_j = A$ for all values of j that surpass the end of a read.

Choices of values for the parameters Q, K, and L are discussed in the sequel. We assume that the base call sequence is described *before* the quality score sequence, so it is available to the decompressor when decoding q_i, so base call symbols at and following position i can be used for the context.

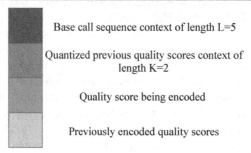

Base call sequence	⋯	T	C	A	T	T	G	C	T	A	⋯
Quantized quality scores	⋯	40	30	38	35	46	36	33	28	34	⋯
Unquantized quality scores	⋯	80	59	75	69	91	72	66	55	68	⋯

Base call sequence context of length L=5

Quantized previous quality scores context of length K=2

Quality score being encoded

Previously encoded quality scores

Fig. 1. Example of a context for R = 7, Q = 6, K = 2 and L = 5.

The compression of quality scores takes place in a single pass through the data. As the FASTQ data is read, we collect, for each context, the number of occurrences of each value of a quality score q_j, $j < i$, in the sequence seen so far in that context. This statistical information is used to compress q_i, for each position i, by feeding an adaptive arithmetic encoder with the statistics collected in context ctx_i through the sequence of quality scores that occur in positions previous to i.

In our implementation of DualCtx the parameters are set to $Q = 6$, $K = 2$ and $L = 5$. This choice has been made as a good trade-off between compression performance and complexity, and it is based on experimental results. Section 3 shows the effect of varying parameters K and L.

DualFqz

For an evaluation of the impact of DualCtx in an overall compression process of FASTQ files, we construct DualFqz, a full FASTQ compressor adapted from Fqzcomp.

When tested on nanopore FASTQ files, we found that the current implementation of Fqzcomp failed to execute correctly on several files. One of the reasons is that the alphabet size of the quality scores and the length of the reads in nanopore data is much larger than that of second-generation technologies. Hence, before constructing DualFqz, we fixed bugs, enlarged internal buffers, and widened the range of quality scores values in Fqzcomp. We still refer to this fixed version of the compressor as Fqzcomp. In particular, the experimental results presented in Sect. 3 refer to this version.

In addition, Fqzcomp relies on the assumption that quality score values that are equal to the lowest possible value correspond to unknown base calls (letter N in the base call sequence) and vice-versa. Although there is a well grounded rationale behind this assumption, we observe that for various nanopore FASTQ files this is not the case. In these cases, Fqzcomp alters the quality scores, resulting in an encoding scheme that is not perfectly lossless and, therefore, not directly comparable to other lossless compressors (i.e., compressors for which the original file is exactly the same as the file obtained after decompression). Hence, for a direct comparison, we generated a variant of Fqzcomp, which we call FQZm, which exactly preserves the quality scores in the original FASTQ file. We obtained DualFqz from FQZm by substituting DualCtx for its quality score compression module. The source code of DualFqz is available online[3].

3   Experimental Results

In this section we report on a set of experiments performed on a large data set of nanopore FASTQ files. The data set is described in Sect. 3.1. In Sect. 3.2 we evaluate the performance of DualCtx for quality score compression on nanopore FASTQ files; for comparison purposes, we also report on the compression of files from other HTS technologies. In Sect. 3.3 we analyze the impact of the parameters K and L on the compression performance of DualCtx. In Sect. 3.4 we compare the compression performance of various compression tools, including DualFqz; we evaluate both the robustness and the compression ratio achieved by each compressor. All experiments were conducted in a desktop PC with 32 GB of RAM, an Intel I7 (3.4 GHz) processor, and Ubuntu OS (14.04.5 LTS).

3.1   Nanopore Data Set

To assess the performance of different compression algorithms on nanopore data we created a data set consisting of nanopore FASTQ files, which we denote *NP DS*. For this purpose, we downloaded a large set of publicly available sequencing files, all generated by Oxford Nanopore technology, from the National Center for Biotechnology Information[4] (NCBI) database. The data set is comprised of 336 different files, with sizes ranging from 7.2 KB to 3.5 GB, including reads that are up to hundreds of thousands base-pair long. The total size of the data set amounts to 114.2 GB and the dynamic range of quality scores is 7 bits. The sequenced samples correspond to viruses, bacteria, fungi, humans, animals, and

---

[3] https://github.com/guidufort/DualFqz.
[4] https://www.ncbi.nlm.nih.gov/.

metagenomic material. The list of SRA Ids of the files that compose the data set is available online[5].

3.2   Evaluation of DualCtx

As mentioned, comparative studies of current FASTQ data compression algorithms report that Fqzcomp achieves the best quality scores compression ratios on data sets from various HTS technologies [16,17]. Hence, we compare the performance of DualCtx with that of Fqzcomp on both our nanopore data set and data sets tested in [16,17]. To this end, we compress each data file separately and calculate, for each data set, the *quality score compression ratio*, CR_{qs}, defined as $CR_{qs} = C_{qs}/T_{qs}$, where T_{qs} is the total size in bytes of the quality score sequences of all files in the data set, and C_{qs} is the total size in bytes of the compressed streams for these sequences. Notice that smaller values of CR_{qs} correspond to better compression performance.

Table 1 shows the quality scores compression ratios obtained with Fqzcomp, the variant FQZm, and DualCtx for the tested data sets. The table also shows the *percentage relative difference*, $\frac{CR_1 - CR_2}{CR_1} \times 100$, between the compression ratios CR_1 and CR_2 obtained by Fqzcomp and DualCtx, respectively, with respect to CR_1. Negative values (highlighted in green) correspond to better performance of DualCtx compared to Fqzcomp. Notice that the proposed dual context scheme does not improve the compression performance for Illumina data sets. This is expected, as there is little correlation between base calls and quality scores obtained with Illumina. On the other hand, for the nanopore data set (and, to less extent, for PacBio), DualCtx yields significantly better results.

Table 1. Comparison of quality scores compression ratio obtained by Fqzcomp, FQZm, and DualCtx for various HTS technologies. In bold we mark the best CR for each data set, and in green we highlight the cases in which DualCtx outperforms Fqzcomp.

Sample	MH0001.081026	SRR554369	ERR174310	SRR327342	SRR1284073	NP DS
Tech.	Illum. GA	Illum. GAIIx	Illum. HiSeq	Illum. GAII	PacBio	Nanopore
Size(GB)	1.9	0.8	102.3	6.0	1.3	114.2
Fqzcomp	**.404**	**.292**	**.290**	.369	.384	.436
FQZm	**.404**	**.292**	**.290**	.369	.391	.436
DualCtx	.407	.313	.313	**.368**	**.369**	**.410**
Rel. diff.	0.6 %	7.4%	7.9%	-0.5%	-3.8%	-6.0%

3.3   Impact of the Context Size on the CR of DualCtx

The choice of values for the parameters Q, K and L in the definition of DualCtx determine the number of distinct context patterns in the statistical model under consideration. Larger models can potentially capture more complex statistical

---
[5] https://github.com/guidufort/DualFqz.

dependencies than simpler models, which may result in better compression performance. However, since the model parameters are adjusted from the same data that is compressed (simultaneously), for relatively small data files, large models may suffer from a large *model cost* [19], which may render a poor compression performance.

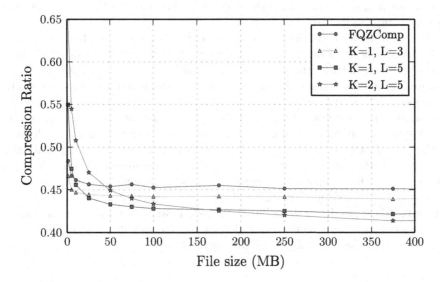

Fig. 2. Dependence of CR_{qs} on the size of the file being compressed, for different combinations of values of K and L.

The impact of the model cost in DualCtx is illustrated in Fig. 2, where we show the value of CR_{qs} for different combinations of values of K and L and for various file sizes.[6] We also show the value of CR_{qs} for Fqzcomp as a reference.

Notice from Fig. 2 that our choice of $K = 2$ and $L = 5$ becomes competitive for relatively large files (about 200 MB). Small files of the data set are better compressed with simpler models.

3.4 Robustness and Performance of FASTQ Compression Tools on Nanopore Data Files

We evaluate several FASTQ compression tools, which are commonly considered in the specific bibliography, on our nanopore data set. Namely, DSRC [21], Fqzcomp [3], Fastqz [3], Slimfastq, FQC [5], LFQC [15], SCALCE [6], Quip [11], Leon [1], and KIC [25]. Most of these tools are not specifically designed to compress long variable length reads and, as a consequence, many of them fail to

---

[6] We obtained files of specific sizes from the original data set by splitting, when necessary, large files into smaller pieces.

execute successfully on some of the files in our nanopore data set. In particular, LFQC and the prototype version of SCALCE tailored to long variable length reads fail on every file, and Fastqz specifically requires fixed length reads. Therefore, we do not report specific results for LFQC, SCALCE, and Fastqz. For comparison to other lossless compressors, we use the version FQZm of Fqzcomp so that every tested compressor is perfectly lossless. In the comparison we also include our proposed compressor DualFqz and, as baseline reference, the general purpose compressors Gzip and Lzma.

To facilitate a comparison among the evaluated tools, we built a nested sequence of subsets of our nanopore data set, starting with the full set of files, and successively deleting the set of files that make a compressor fail, taking these compressors in decreasing order of the number of files that they can successfully compress. This process results in a profile of data sets in which smaller sets are comprised, in general, of smaller files. For each data subset, we compressed each file separately and calculated the overall *compression ratio (CR)*, defined as $CR = C/T$, where T is the sum of the sizes in bytes of all the original data files, and C is the sum of the sizes in bytes of all the compressed files in the data subset. Notice that smaller CR correspond to better compression performance.

The CR obtained by the evaluated tools on each of the successively larger data subset is shown in Table 2. The table also shows the number of files in each data subset, and the percentage fraction of the accumulated size of these files with respect to the total size of the full data set.

Table 2. CR for the evaluated compressors on subsets of the nanopore data set (smaller is better). The leftmost column shows the number of files of each data subset and the percentage fraction, between brackets, of the accumulated file sizes with respect to the full data set. In bold we highlight the best CR for each data subset.

Data subset size	Compression ratio									
	FQC	Leon	Quip	KIC	Slim	Gzip	Lzma	DSRC	FQZm	DualFqz
119 (4%)	**.282**	.332	.289	.309	.283	.381	.319	.286	.290	.286
128 (7%)	-	.323	.279	.299	**.272**	.370	.310	.276	.279	.275
150 (11%)	-	-	.290	.310	.284	.381	.322	.284	.288	**.278**
289 (76%)	-	-	-	.368	.353	.442	.369	.347	.344	**.325**
330 (95%)	-	-	-	-	.353	.445	.372	.349	.345	**.327**
336 (100%)	-	-	-	-	-	.444	.371	.347	.344	**.326**

In addition to DualFqz, the only FASTQ specific compressors that successfully compress the full data set are DSRC and FQZm. Comparing FQZm to DSRC, we notice that DSRC outperforms FQZm in the three smallest data subsets. The substitution of DualCtx for the quality score compression algorithm in FQZm reverts these cases, and the CR obtained by DSRC is never better than that obtained by DualFqz. More generally, DualFqz yields the best CR compared to any other compressor for the four largest data subsets. Indeed, we observe that the advantage in CR of DualFqz compared to other algorithms becomes more significant as the size of the data set increases, consistently with our analysis in Sect. 3.3.

Table 3. Comparison of the CR obtained by DualFqz to that obtained by other compressors on the nanopore data set. Each compressor is tested on the data subset consisting of all the files that it compresses successfully. In bold we mark the best CR for each case.

Compressor	FQC	Leon	Quip	KIC	Slim	Gzip	Lzma	DSRC	FQZm	Fqzcomp
Comp. CR	**.321**	.368	.294	.366	.353	.444	.371	.347	.344	.336
DualFqz CR	.326	**.315**	**.280**	**.324**	**.327**	**.326**	**.326**	**.326**	**.326**	**.326**
Rel. diff.	1.4%	−16.8%	−5.1%	−13.2%	−7.9%	−36.2%	−13.9%	−6.5%	−5.6%	−3.3%
Num. of files	194	185	157	295	330	336	336	336	336	336
% of full DS	10%	13%	15%	81%	95%	100%	100%	100%	100%	100%

An alternate view for a comparative analysis of the performance of DualFqz is shown in Table 3, which includes a direct comparison between the CR obtained by DualFqz and that obtained by other compressors, individually. In this case, for each compressor we build a data subset that contains all the files that it successfully compresses, and we calculate the CR obtained on this data subset by both the competing compressor (first row) and DualFqz (second row). The relative difference between both CRs is shown in the third row. The last two rows show the number of files in each data subset, and the fraction of the size of the full data set represented by these files. Except for the comparison with FQC, which involves just 10% of the full data set, DualFqz achieves the best CR in all cases. In particular, compared to FQZm, the use of DualCtx for quality score compression is responsible for a 5.6% improvement in overall CR. Moreover, DualCtx achieves a better CR than Fqzcomp, even though the latter is not perfectly lossless.

4   Conclusions and Future Work

Currently available implementations of several FASTQ compression algorithms are unable to perform robustly on files with very long variable length reads, such as those generated by nanopore technologies. More importantly, we have presented a new lossless compressor for nanopore FASTQ files, termed DualFqz, and shown that it significantly improves the overall compression for these files by exploiting statistical dependencies between quality scores and base call sequences, in addition to inherent dependencies within the sequences. As an example of the performance, we are able to reduce the size of our nanopore dataset from 114.2 GB to 37.2 GB.

We have identified several improvements to the compressors we have proposed that we plan to investigate in the near future; for now, the matter of fact is that a simple model, such as the one in DualCtx, suffices to booster the performance of FASTQ compression algorithms when applied to nanopore data.

One line for such future work consists of applying model aggregation techniques, for model cost reduction, and context mixing techniques for combining different models. The former would allow for constructing richer models with no increment in model cost, and the latter for achieving better compression performance independently of the data file size. Moreover, mixing models suitable for different sequencing technologies could yield good compression performance with no a priori knowledge of the source that generated the data.

Acknowledgments. This work was partially funded by Comisión Sectorial de Investigación Científica, Udelar, and grant number 2018-182799 from the Chan Zuckerberg Initiative DAF, an advised fund SVCF, and an SRI grant from UIUC.

References

1. Benoit, G., et al.: Reference-free compression of high throughput sequencing data with a probabilistic de Bruijn graph. BMC Bioinform. **16**, 288:1–288:14 (2015)
2. Bonfield, J.K.: The scramble conversion tool. Bioinformatics **30**(19), 2818 (2014)
3. Bonfield, J.K., Mahoney, M.V.: Compression of FASTQ and SAM format sequencing data. PLOS One **8**(3), 1–10 (2013). https://doi.org/10.1371/journal.pone.0059190
4. Chandak, S., Tatwawadi, K., Ochoa, I., Hernaez, M., Weissman, T.: Spring: a next-generation compressor for FASTQ data. Bioinformatics, bty1015 (2018)
5. Dutta, A., Haque, M.M., Bose, T., Reddy, C.V.S.K., Mande, S.S.: FQC: a novel approach for efficient compression, archival, and dissemination of fastq datasets. J. Bioinform. Comput. Biol. **13**(3), 1541003 (2015)
6. Hach, F., Numanagić, I., Alkan, C., Sahinalp, S.C.: SCALCE: boosting sequence compression algorithms using locally consistent encoding. Bioinformatics (Oxford, England) **28**(23), 30513057 (2012). https://doi.org/10.1093/bioinformatics/bts593
7. Hach, F., Numanagić, I., Sahinalp, S.C.: DeeZ: reference-based compression by local assembly. Nat. Methods **11**, 1082–1084 (2014)
8. Huang, Z.A., Wen, Z., Deng, Q., Chu, Y., Sun, Y., Zhu, Z.: LW-FQZip 2: a parallelized reference-based compression of FASTQ files. BMC Bioinform. **18**(1) (2017). https://doi.org/10.1186/s12859-017-1588-x
9. Ip, C., et al.: MinION analysis and reference consortium: phase 1 data release and analysis [version 1; referees: 2 approved]. F1000Research **4**(1075) (2015)
10. Jain, M., et al.: Nanopore sequencing and assembly of a human genome with ultra-long reads. Nat. Biotechnol. **36**(4), 338 (2018)
11. Jones, D.C., Ruzzo, W.L., Peng, X., Katze, M.G.: Compression of next-generation sequencing reads aided by highly efficient de novo assembly. Nucleic Acids Res. **40**(22), e171 (2012). https://doi.org/10.1093/nar/gks754
12. Kingsford, C., Patro, R.: Reference-based compression of short-read sequences using path encoding. Bioinformatics **31**(12), 1920–1928 (2015). https://doi.org/10.1093/bioinformatics/btv071
13. Laver, T., et al.: Assessing the performance of the Oxford nanopore technologies MinION. Biomol. Detect. Quantification **3**, 1–8 (2015)

14. Long, R., Hernaez, M., Ochoa, I., Weissman, T.: Genecomp, a new reference-based compressor for SAM files. In: 2017 Data Compression Conference (DCC), pp. 330–339. IEEE (2017)
15. Nicolae, M., Pathak, S., Rajasekaran, S.: LFQC: a lossless compression algorithm for FASTQ files. Bioinformatics **31**(20), 3276–3281 (2015). https://doi.org/10.1093/bioinformatics/btv384
16. Numanagić, I.: Efficient high throughput sequencing data compression and genotyping methods for clinical environments. Ph.D. thesis, Simon Fraser University (2016)
17. Numanagić, I., et al.: Comparison of high-throughput sequencing data compression tools. Nat. Methods **13**(12), 1005–1008 (2016)
18. Ochoa, I., Hernaez, M., Weissman, T.: Aligned genomic data compression via improved modeling. J. Bioinform. Comput. Biol. **12**(06), 1442002 (2014)
19. Rissanen, J.: Universal coding, information, prediction, and estimation. IEEE Trans. Inf. Theory **30**(4), 629–636 (1984). https://doi.org/10.1109/TIT.1984.1056936
20. Rissanen, J.: Generalized Kraft inequality and arithmetic coding. IBM J. Res. Dev. **20**(3), 198–203 (1976)
21. Roguski, L., Deorowicz, S.: DSRC 2-Industry-oriented compression of FASTQ files. Bioinformatics **30**(15), 2213–2215 (2014). https://doi.org/10.1093/bioinformatics/btu208
22. Roguski, Ł., Ochoa, I., Hernaez, M., Deorowicz, S.: FaStore-a space-saving solution for raw sequencing data. Bioinformatics **1**, 9 (2018)
23. Sović, I., Šikić, M., Wilm, A., Fenlon, S.N., Chen, S., Nagarajan, N.: Fast and sensitive mapping of nanopore sequencing reads with graphmap. Nat. Commun. **7**, 11307 (2016)
24. Stephens, Z.D., et al.: Big data: astronomical or genomical? PLoS Biol. **13**(7), e1002195 (2015)
25. Zhang, Y., Patel, K., Endrawis, T., Bowers, A., Sun, Y.: A FASTQ compressor based on integer-mapped k-mer indexing for biologist. Gene **579**(1), 75–81 (2016). https://doi.org/10.1016/j.gene.2015.12.053

De novo Transcriptome Assembly of *Solea senegalensis* v5.0 Using TransFlow

José Córdoba-Caballero[1], Pedro Seoane-Zonjic[1], Manuel Manchado[2], and M. Gonzalo Claros[1(✉)]

[1] Department of Molecular Biology and Biochemistry,
Universidad de Málaga, Malaga, Spain
claros@uma.es
[2] IFAPA Centro El Toruño, Consejería de Agricultura y Pesca,
11500 El Puerto de Santa María, Cadiz, Spain

Abstract. Senegalese sole is an economically important flatfish species in aquaculture. Development of new bioinformatics resources allows the optimization of its breeding in fisheries. Sequencing data from larvae in different development stages obtained from different sequencing platforms (more than 270 M of Illumina paired-end reads and more than 3 M of Roche/454 reads) were used. Due to the high complexity of the samples, an optimized version of *TransFlow*, an automated, reproducible and flexible framework for *de novo* transcriptome assembly, was used to get the most complete *de novo* transcriptome assembly. Best transcriptome selection was based on the principal component analysis provided by *TransFlow*. Two transcriptomes, one all-Illumina and other reconciling Illumina and Roche/454, were selected and annotated using *Full-LengtherNext*, and the tentative transcripts were filtered by alignment to partial genomic sequences to avoid artifacts. The reconciled non-redundant assembly composed by Illumina and Roche/454 reads seems to be the best strategy. It consists of 55 440 transcripts of which 22 683 code for 17 570 different proteins described in databases. The obtained v5.0 reduces the number of tentative transcripts by 79,33% compared v4.0, what will increase the precision of future transcriptomic studies.

Keywords: Flatfish · Larvae · Transcriptome assembly · Illumina · Roche/454

1 Introduction

Solea senegalensis has become in an important cultivated flatfish species last years producing more than 8 700 M\$ in Spain [1]. Attached to improvements of the cultivation system, bioinformatics studies have been crucial to achieve this. There were several attempts to sequence and assembly *S. senegalensis* transcriptome using Roche/454 and Illumina data, but they have not been entirely

I. Rojas et al. (Eds.): IWBBIO 2019, LNBI 11465, pp. 48–59, 2019.
https://doi.org/10.1007/978-3-030-17938-0_5

satisfactory because they produced fragmented and redundant assemblies [2]. The 4th version of *S. senegalensis* transcriptome was *de novo* assembled by a complex strategy using Roche/454 and Illumina sequencing data [2]. This transcriptome comprises >690 000 transcripts of which 9.91% has similarity with protein sequences of UniProtKB but the remaining transcripts lack of any protein correspondence. The great number of sequences of this transcriptome produces unreliable results in differential expression analysis because of low sensitivity (M. Manchado, in preparation) and is useless for genome structural characterization due to the presence of assembly artifacts and chimerical sequences.

The aim of this study is to build an improved *S. senegalensis* transcriptome in order to use as reference for gene expression analysis and gene identification in genomic sources. Furthermore, in our laboratory, we have developed a versatile *de novo* assembly workflow called *TransFlow* [12], that can produce up to 180 different assemblies using customised strategies. This tool, recently reviewed by RNA-seq Blog, is able to use Roche/454 and Illumina read data and select the best assembling strategy by principal components analysis (PCA) using a targeted reference transcriptome of another related species. With this framework and the read data of the 4th version [2] we have build the 5th version of *S. senegalensis* transcriptome with the following methodology: (1) read library selection and raw read pre-processing, (2) optimal assembling using *TransFlow*, and (3) assembly filtering to get the most reliable set of tentative transcripts. The reconciled assembly including Illumina and Roche/454 data was finally selected as *Solea senegalensis* transcriptome v5.0. It contains 55 440 tentative transcripts with a mean length of 1 661 bp and N50 of 2 769 bp, coding for at least 17 570 different proteins.

2 Materials and Methods

2.1 Sequencing Data

An ongoing *S. senegalensis* draft genome was used to filter artefacts [11]. Regarding the transcriptome data, 156 Illumina paired-read libraries with a mean read count of 4 574 098 were inspected (Library description is listed on [2]). The 5 663 225 Roche/454 reads of the same study were used in the new assembly process. Since the 156 Illumina libraries include at least three replicates from a series of treatments [2], only one replicate was selected per experiment set. When the experiment consisted of a time series, only first and last time points were used. Using this criteria, we use only 30 Illumina libraries from the original pool of 156 (a total of 454 422 926 paired reads). Roche/454 long single-reads were pre-processed using *SeqTrimNext* [5] and Illumina paired-end reads were pre-processed using *SeqTrimBB*, program developed in our laboratory based on *BBmap* suite [3]. In each case, the specific default profile was used.

Illumina reads shorter than 60 bp were discarded, whereas the threshold was set to 90 for Roche/454 reads. The final amount of useful reads were 360 529 552 Illumina paired reads and 3 104 734 Roche/454 long single reads after preprocessing.

Finally, the selected Illumina paired read libraries were mapped onto the genome reference with *Bowtie2* [7], discarding unpaired alignment with option `--no-mixed`, in order to identify the libraries with sequence artifacts manifested by high mapping ratios in spite of use a mapper that not split reads across the exon junctions. Libraries with mapping percentages higher than 70% were discarded.

2.2   Transcriptome Assembly After TransFlow Optimization

The new Senegalese sole transcriptome was assembled from useful reads using the automated and modular framework *TransFlow* [12] with the following parameters: *k*mer length of 45 and 55 (due to the great amount of Illumina reads), minimal coverage of 10 reads for Illumina assemblies, and *k*mer length 29 for Roche/454 assemblies. The *Danio rerio* GRCz11 build was used as reference transcriptome and the reads downloaded from SRA accession ERR216329 was used for its evaluation. The 180 original assembling strategies [12] were simplified discarding Illumina primary assemblies (they were always beat by their scaffolded counterparts), resulting in a total of 80 assemblies. A sequence redundancy removal using *CD-HIT-EST* [10] step at 95% identity was added before the combination of all generated assemblies into a general unique assembly.

2.3   Transcriptome Annotation and Removal of Unexpected Sequences

The annotation was performed with *Full-LengtherNext* (P. Seoane and M.G. Claros, in preparation). It was configured to use the UniProtKB *Actinopterigii* taxon sequences as user database and the Uniprot Vertebrate division as main database. Unexpected sequences were removed based on tentative transcript alignment against the *S. senegalensis* draft genome reference. Transcripts were aligned onto genome using `splice` mode of *Minimap2* [8] configured for finding canonical splicing sites CT-AG in transcript strand (option -uf) and penalising non-canonicals in 5 points (option -C5). Unmapped transcripts and supplementary alignments were removed by *SAMtools* `view` (operation bit encoding code 2052) [9]. Filtered SAM was converted to PAF using `sam2paf` option of `paftools.js` (a script provided by *Minimap2* authors in https://github.com/lh3/minimap2/tree/master/misc) and SAM type CIGAR string was merged to PAF. Coverage percentage, identity and exon numbers were calculated for each transcript using PAF fields (https://github.com/lh3/miniasm/blob/master/PAF.md). Coverage percentage was calculated as the relationship between number of residues matches and transcript length, identity was calculated as the relationship between number of residues matches and full alignment length and exons was counted after CIGAR string splitting by N operation. Coverage, identity and number of exons were used to remove transcripts with one exon and more than 90% of identity, since they are suspected to be fragments or chimeras that contains intronic sequences. Then transcripts with identity and coverage <70% were removed since they are considered low quality transcripts.

The described parameter calculation and filtering were programmed in-house using Ruby scripting language (v2.4.1).

2.4   Gene Expresion Analysis

Last transcriptome version and new transcriptomes, before and after filtering unexpected sequences, were used as reference for gene expression analysis, using six Illumina libraries from [2] (three control and three treatment), with the aim of evaluate which transcriptome produces more consistent results. Useful reads were mapped on each transcriptome with *Bowtie2* [7]. Reads that failed to align and unpaired alignments records were suppressed from SAM file with options `--no-unal` and `--no-mixed` respectively. SAM file was sorted and converted to BAM by *SAMtools* `sort` [9]. Reads mapped per transcript were counted by *sam2counts* (https://github.com/vsbuffalo/sam2counts) and the count matrix was processed with *DEGenesHunter* [6].

3   Results and Discussion

3.1   Illumina Libraries Selection for Assembly

From the initial 156 Illumina libraries, 30 were selected using the selection criteria described in methods. Selected libraries shown a broad mapping ratio distribution (Fig. 1) from 55.56 to 75.43%. Of these libraries, 3 present a mapping ratio higher that 70% and are part of the same experiment/batch (PRJNA255461). The four selected libraries of this experiment were discarded due to the highly probability of artifact presence. Finally, 26 Illumina paired read libraries (275 501 704 paired reads) were used for assembly.

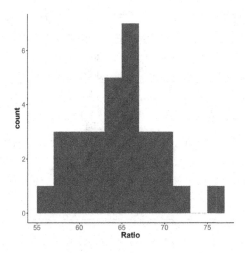

Fig. 1. Distribution of Illumina read mapping ratio for the libraries tested when aligned to the draft genome

3.2   Assembly Selection

From a total of 80 different assembling strategies, *TransFlow* ranked as the best transcriptome (PCA distance to *D. rerio* of 5.3e−3) the Illumina assembly using *Oases* with *k*mer 45 (Oases_*k*45). To take into account Roche/454 long reads (they belong to different tissues than Illumina reads), the best reconciled transcriptome (PCA distance to *D. rerio* of 0.1126) is also selected since it is close to the zebrafish reference too. This reconciliation was performed by merging with *Minimus2* the non-redundant transcripts obtained from Illumina with *Oases* *k*mers 45 and 55 and Roche/454 transcripts produced by *Mira3* and *Euler-SR* using *Cap3* (Min2_Oases_Cap3). This reconciled assembly contains almost 6 000 different orthologues IDs more than the one only with Illumina reads. Therefore, we select the Illumina only and the Illumina-Roche/454 assemblies as putative transcriptomes in this study for further analysis.

Table 1. Assembly features after annotation of old v4.0 and new candidates to v5.0 using *Full-LengtherNext*

Feature	v4.0	Min2_Oases_Cap3	Oases_*k*45
Transcripts	697 125	153 847	87 362
Full transcriptome length	366 337 327	147 562 177	142 218 402
Indeterminations (%)	1.31	0.67	0.45
Indeterminations mean length	22.60	5.22	184.96
Transcripts >200	376 899	134 854	81 375
Transcripts >500	153 963	69 705	64 305
Longest transcript	40 163	41 091	23 792
N50	1 292	2 050	2 659
N90	180	360	826
Tentative transcripts(%)	81.98	99.45	93.14
Artifacts	125 614	842	105
Misassembled	305	124	20
Unmapped	125 309	718	85
With orthologue	69 108	30 973	27 712
Different orthologues IDs	37 438	22 868	16 412
Complete transcripts	27 254	14 657	15 687
Different complete transcripts	15 390	11 181	9 056
ncRNA	9 855	3 765	3 172
Without orthologue	492 547	118 267	56 373
Coding	57 854	25 091	20 540
Unknown	434 693	93 176	35 833

3.3   Transcriptome Annotations and Removal of Unexpected and Low Quality Transcripts

Full-LengtherNext based annotations show that although Oases_*k*45 was the best assembly, Min2_Oases_Cap3 has more different orthologues. Moreover, both new transcriptomes contain a high percentage of sequences without protein annotation and without predictable coding region (labeled as 'Unknown'; Table 1). This hampered the selection of the reference transcriptome in spite of (i) the important reduction of total transcripts with respect to *S. senegalensis* transcriptome v4.0 (included in Table 1 for convenience), (ii) the lower ratio of indeterminations (from 1.31 in v4 to 0.67-0.4 in current putative transcriptomes), and (iii) important increase in N50 (nearly 2-fold) and N90 (from 2- to 4-fold).

Table 2. Transcripts classification depending on filtering criteria. Remaining transcripts appertanins to low quality transcripts that did not aligned onto genome or they did it irregularly.

Conditions	Unexpected (%) (I > 90)	High quality transcripts (%) (I > 70 & C > 70)					
Exons Transciptomes	1	1	2	3	4	5	>5
Min2_Oases_Cap3	55,18	1,01	8,44	5,02	3,78	2,92	14,87
Oases_*k*45	43,29	0,78	8,08	4,87	4,15	3,65	28,22
v4.0	54,98	1,10	6,53	2,88	1,91	1,42	6,84

I: identity, C: coverage. A high quality transcript is considered when its sequence is splitted onto several exons on the draft genome and has coverage and identity parameters of al least 70%

When the transcriptomes were aligned with the genome draft, a high percentage of highly identical transcripts and absence of introns was obtained (Table 2, 'Unexpected' column, only one exon). Furthermore, the low quality transcript filtering is applied (Table 2, 'High quality transcripts' column) the v4.0 transcripts presents, in general, lower percentages for each number of exons than the candidate transcriptomes. In fact, when the percentage of removal is calculated, the *S. senegalensis* v4.0 lose a 24.34% whereas the putative transcriptomes lose only a 8.78% for Min2_Oases_Cap3 and a 6.69% for Oases_*k*45 assemblies. Other remarkable aspect of the results is that the sequences of the putative transcriptomes presents more putative exons than the v4.0 transcriptome (Table 2, High quality transcripts, >5 column).

We consider the removal of the unexpected sequences since may not correspond to any known biological sequence, in contrast to high quality transcripts that can map over a variable number of exons. In fact, only a small fraction of high quality transcripts has no introns (exon = 1).

Supporting this transcript removal is that most unexpected sequences where those qualified as 'Unknown' in Table 1, as can be clearly seen in Fig. 2). More

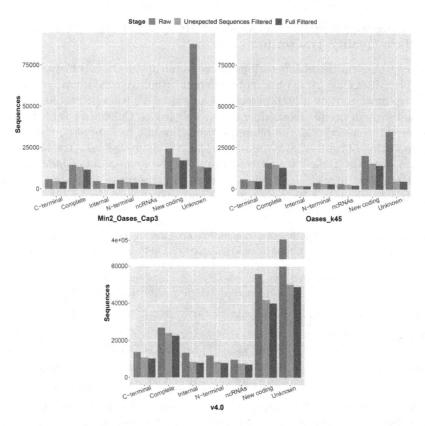

Fig. 2. Annotation statistics of v4.0 transcriptome and selected transcriptomes during different filtering steps.

importantly, annotated transcripts were nearly invariable before and after this filtering (Fig. 2).

The total number of transcripts after filtering was very similar in both transcriptomes comparing with the raw (55 440 transcripts of Min2_Oases_Cap3 and 43 453 of Oases_k45), but Oases_k45 has more transcripts containing a complete coding sequence and less unknown sequences (Fig. 2). Min2_Oases_Cap3 has however 17 570 different orthologue IDs, 3 928 more than Oases_k45, and has less sequence redundancy due to less Redundant Transcripts in all filtering stages (Fig. 3). *S. senegalensis* v4.0 transcriptome also was filtered for comparing purposes. This version have more than 400 000 unexpected sequences (Fig. 2) and it have a greater number of Redundant Transcripts than putative transcriptomes along the filtering stages (Fig. 3). This suggests that the v4 transcriptome is a non optimal assembly and that the strategy proposed here, improves the resulting transcriptome.

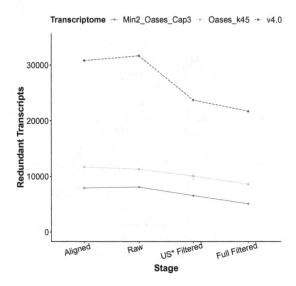

Fig. 3. Evolution of the amount of Redundant Transcripts computed as the difference between *Full-LengtherNext* measures "Total orthologues" and "Different orthologues IDs" in both transcriptomes during filtering process. US*: Unexpected Sequences

3.4 Impact of Transcriptome Quality in Gene Expression Analysis

A differential expression analysis using six Illumina libraries [2], 3 controls and 3 treatments, was performed using each candidate transcriptome, their filtered counterparts and the version 4.0. For convenience, an alignment ratio was defined as the percentage of reads aligned per transcript. This alignment ratio increases during filtering (Fig. 4) indicating that there were more reads per transcript, a desirable feature in RNA-seq experiments to improve the statistical power. This suggests that transcripts with low alignment rates were discarded with the previous filtering processes. It is interestingly to note that, even filtered, the v4 transcriptome presents a mapping ratio lower that any of the putative transcriptomes of this study. In terms of differential expressed genes (DEGs), the Table 3 shows that the filtered transcriptomes presents lower number of DEGs than the raw versions. Also, the density of discovered DEGs is greater in filtered transcriptomes, with the exception of the Oases_k45 transcriptome. When the P-value distribution for each DEG analysis are inspected in the Fig. 5 (first column) it can be observed that the Min2_Oases_Cap3 transcriptome presents the lower P-values for the raw transcriptomes and the v4 presents higher P-values. When the transcriptomes are filtered, Fig. 5 (second column), the Min2_Oases_Cap3 transcriptome has the same distribution but the v4 transcriptome improves in its distribution. NOISeq P-values distribution is useful for transcriptome evaluation because the major changes of the distribution it is shown in the range of 0.75–1 (Fig. 5d). This P-value range indicates non DEGs and when the filtering

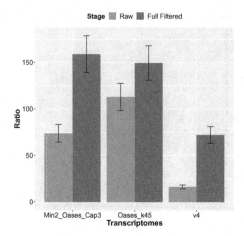

Fig. 4. Alignment ratios for candidate transcriptomes before and after filtering unexpected sequences

Table 3. Number of DEGs discovered using different transcriptomes.

	Stage	Discovered DEGs	% of trasncripts that are DEGs
Oases_k45	Raw	207	0.23
	Filtered	65	0.15
Min2_Oases_Cap3	Raw	511	0.33
	Filtered	290	0.52
v4	Raw	145	0.02
	Filtered	126	0.09

process in applied, the v4 transcriptome filtered presents a high peak whereas the putative transcriptomes have a lower amount of transcripts in this range.

3.5 Selection of the Best Transcriptome

Summarizing the results for the two filtered transcriptomes, (i) filtered Min2_-Oases_Cap3 has less redundant sequences (Fig. 3); (ii) filtered Min2_Oases_-Cap3 has the highest alignment ratio despite of filtered Oases_k45 has less sequences (Fig. 4); (iii) filtered Min2_Oases_Cap3 also produces more DEGs than their Oases_k45 counterpart and v4.0 transcriptome (Fig. 3) with lower P-values (Fig. 5); (iv) high percentage of Oases_k45 annotated sequences were removed and Oases_k45 DEGs percentage decreased after filtering (Fig. 3).

Since, filtering steps reduced DEGs discovery in both transcriptomes and reduces DEGs percentage in Min2_Oases_Cap3 (Table 3), but this DEGs reduction is supported by more frequent low P-values in three different gene expression analysis algorithms after filtering (Fig. 5), results presented here support

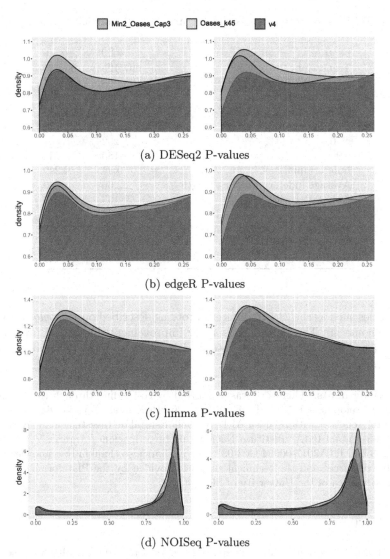

(a) DESeq2 P-values

(b) edgeR P-values

(c) limma P-values

(d) NOISeq P-values

Fig. 5. P-value distribution for each DEG analisys algorithm and each transcriptome. From top to bottom, results for each DEG algorithm and from left to right, results for raw and filtered transcriptomes.

that filtering process as a good strategy to remove unexpected sequences and obtain high quality transcriptomes (Table 4).

Taking together, those results prompt to the choice of filtered Min2_Oases_-Cap3 as the fifth version of *S. senegalensis* transcriptome, containing 55 440 transcripts (Fig. 4) including 22 683 sequences coding for 17 570 different protein orthologues, a reasonable number compared with phylogenetically well-studied

Table 4. *Solea senegalensis* transcriptome v4.0 and v5.0 features.

Feature	v4.0	v5.0
Transcripts	697 125	55 440
Trancripts > 500 bp	153 963	41 645
Mean lengths	525	1 661
Longest transcript length	40 163	23 792
Sum lengths (bp)	366 337 327	92 121 772
N50	1 292	2769
N90	180	749
Indeterminations (%)	0.46	0.299
Average indeterminations length	22.60	3.00
Annotated with protein	69 108	22 683
Different orthologues IDs	37 438	17 570
ncRNAs	9 855	2 567
New coding	57 854	17 277
Unknown	434 693	12 914

near species such as the 21 516 different proteins described on *Cynoglossus semi-laevis* genome [4]. The Illumina assembly improvement due to the presence of Roche/454 reads was previously described in other species [12], and was confirmed in this study for sole.

Acknowledgments. This work was supported by co-funding through the European Regional Development Fund (ERDF) 2014–2020 "Programa operativo de crecimiento inteligente" together with Spanish AEI "Agencia Estatal de Investigación" (AGL2017-83370-C3-3-R) and INIA "Instituto Nacional de Investigaciones Agrarias" (RTA2013-00068-C03 and RTA2017-00054-C03-03). The authors also thankfully acknowledge the computer resources and the technical support provided by the Plataforma Andaluza de Bioinformatica of the University of Malaga.

References

1. APROMAR: La Acuicultura en España 2018. Ministerio de agricultura y pesca, alimentación y medioambiente (2018). https://doi.org/10.1017/CBO9781107415324.004

2. Benzekri, H., et al.: De novo assembly, characterization and functional annotation of Senegalese sole (Solea senegalensis) and common sole (Solea solea) transcriptomes: integration in a database and design of a microarray. BMC Genom. (2014). https://doi.org/10.1186/1471-2164-15-952

3. Bushnell, B.: BBmap Suite (2014). https://sourceforge.net/projects/bbmap/

4. Chen, S., et al.: Whole-genome sequence of a flatfish provides insights into ZW sex chromosome evolution and adaptation to a benthic lifestyle. Nat. Genet. (2014). https://doi.org/10.1038/ng.2890

5. Falgueras, J., Lara, A.J., Fernández-Pozo, N., Cantón, F.R., Pérez-Trabado, G., Claros, M.G.: SeqTrim: a high-throughput pipeline for pre-processing any type of sequence read. BMC Bioinform. (2010). https://doi.org/10.1186/1471-2105-11-38

6. González Gayte, I., Bautista Moreno, R., Seoane Zonjic, P., Claros, M.G.: DEgenes hunter - a flexible R pipeline for automated RNA-seq studies in organisms without reference genome. Genom. Comput. Biol. (2017). https://doi.org/10.18547/gcb. 2017.vol3.iss3.e31

7. Langmead, B., Salzberg, S.L.: Fast gapped-read alignment with Bowtie 2. Nat. Methods (2012). https://doi.org/10.1038/nmeth.1923

8. Li, H.: Minimap2: fast pairwise alignment for long DNA sequences. arXiv (2017). https://doi.org/10.1101/169557

9. Li, H., et al.: The sequence alignment/map format and SAMtools. Bioinformatics (2009). https://doi.org/10.1093/bioinformatics/btp352

10. Li, W., Godzik, A.: CD-HIT: a fast program for clustering and comparing large sets of protein or nucleotide sequences. Bioinformatics (2006). https://doi.org/10. 1093/bioinformatics/btl158

11. Manchado, M., Planas, J.V., Cousin, X., Rebordinos, L., Claros, M.G.: Genetic and genomic characterization of soles. In: Muñoz-Cueto, J.A., Mañanós-Sánchez, E.L., Sánchez-Vázquez, F.J. (eds.) The Biology of Sole, pp. 361–379. No. B6.1. CRC Press (2019)

12. Seoane, P., et al.: TransFlow: a modular framework for assembling and assessing accurate de novo transcriptomes in non-model organisms. BMC Bioinform. (2018). https://doi.org/10.1186/s12859-018-2384-y

Deciphering the Role of PKC in Calpain-CAST System Through Formal Modeling Approach

Javaria Ashraf[1], Jamil Ahmad[2(✉)], and Zaheer Ul-Haq[3]

[1] Research Center for Modeling and Simulation,
National University of Sciences and Technology, Islamabad, Pakistan
[2] Department of Computer Science and Information Technology,
University of Malakand, Chakdara, Pakistan
`jamil.ahmad@uom.edu.pk`
[3] Dr. Panjwani Center for Molecular Medicine and Drug Research,
International Center for Chemical Sciences, University of Karachi, Karachi, Pakistan

Abstract. Calcium-activated calpain has critical role in a variety of calcium regulated processes. Calcium activates two other proteins, Calpastatin (CAST) and Protein Kinase C (PKC) to make a regulatory network which is pivotal in cell physiology. CAST binds with calpain to form complex for hampering its hyperactivation. PKC phosphorylates CAST while calpain proteolyzes active PKC and increases calcium influx. Based on biological knowledge, a qualitative (discrete) model is constructed that provides new insights into the dynamics of calpain-CAST and PKC relationship. The model predicts that PKC maintains calpain-CAST complex by interacting with both active calpain and CAST. It is also observed that in physiological condition, there is a homeostatic behavior between calcium, CAST and PKC. Some significant discrete cycles are also identified by analyzing betweenness centralities of the discrete states. There is one stable state in the model in which calpain and calcium are hyperactivated while CAST and PKC are inactivated. The model is validated through the stochastic Petri Net model that further reveals its quantitative dynamical behaviors. Physiology is perturbed by hyperactivation of calpain which results in the deregulation of homeostasis. Both models suggest that inhibition of calpain by CAST is a better therapeutic strategy which requires healthy assistance from PKC. In conclusion, homeostasis of calcium, CAST and PKC is pivotal for a healthy state.

Keywords: Calpain · Calpastatin (CAST) · Calcium homeostasis · Protein Kinase C (PKC) · Stochastic Petri net · Qualitative modeling

1 Introduction

Calpains (**CAL**cium ion-dependent pa**PAIN**-like cysteine) proteases are notorious enzymes. There are multiple members in this family; in which, some are ubiquitous and some are tissue specific [1]. The ubiquitous calpains mu(μ)-Calpain

© Springer Nature Switzerland AG 2019
I. Rojas et al. (Eds.): IWBBIO 2019, LNBI 11465, pp. 60–71, 2019.
https://doi.org/10.1007/978-3-030-17938-0_6

(Calpain1) and m-Calpain (Calpain2) require micro-molar concentration of calcium (Ca^{2+}) (10–50 μM) and mili-molar concentration of Ca^{2+} (250–350 μM) [2], respectively. Their main function is to cleave membrane proteins or membrane linked proteins. Hyper-activity of these enzymes can lead to acute inflammatory processes [3], neuro-degeneration [4], muscular dystrophy [5] and cardiovascular disorders [6]. To maintain cell physiology, calpains are tightly regulated by calpastatin (CAST). This ubiquitous enzyme is described as an endogenous and sole suicide substrate for calpains. CAST interacts calpain at two sites; first it hinders pro-calpain at membrane (at low Ca^{2+} influx) and then it controls concentration of active calpains at cytosol by forming reversible complex with it [7]. In cytosol, active calpain modulates CAST slowly by proteolysing it into smaller inactive fragments which results in hyperactivation of calpain in the cell which contributes in patho-physiology (Fig. 1). CAST is also phosphorylated by Protein Kinase C (PKC) to lower its inhibitory efficiency towards calpain, [8]. Inactive PKC is converted to Ca^{2+}-bound activated form in the presence of diacylglycerol (DAG). Meanwhile, active calpain regulates PKC by converting it into constitutive active enzyme [9]. Ca^{2+} ions also play important physiological role in a cell; the magnitude of Ca^{2+} ions inside a cell is very low (between 50–100/50–300 nM) and it can rise to several micromoles after activation [10]. Higher level of efflux and lower influx are maintained through multiple homeostatic apparatuses (Fig. 1). The influx channels include voltage-gated channels (VGCs) and receptor gated channels (RGCs). Different ATP-dependent membrane pumps such as plasma membrane calcium ATPase channel (PMCA) and sodium-calcium exchanger (NCX) which are dependent on sodium-potassium ATPase (NKA) are used for Ca^{2+} efflux. Both types of apparatuses are working in harmony to maintain homeostasis of Ca^{2+} ions. When there is homeostasis, Ca^{2+} can perform its function properly such as mediation of hormones, neurotransmitters and other stimuli. To understand the above stated mechanism (Fig. 1), a biological regulatory network (BRN) is constructed using qualitative modeling (Fig. 2). Qualitative modeling of Calp-CAST BRN results in a state graph, there are state representing respective entities and they are evolving into homeostatic states and stable state (disease state). The qualitative model is then converted into stochastic Petri Net (SPN) for further validation. This extensive study of Calp-CAST system provides useful insights; Over-activation of calpain leads to epigenetics by disrupting Ca^{2+} homeostasis in the cytosol. The system remains healthy when calpain is under the direct influence of CAST and CAST is regulated by PKC. They all together maintain Ca^{2+} homeostasis that is pivotal for many cellular and neuronal functions.

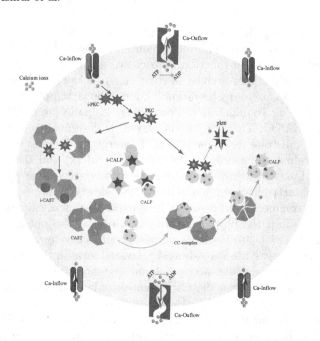

Fig. 1. In the cell, calcium channels are establishing homeostasis and regulating three important proteins Calpain, CAST and PKC which are interlinked. Calcium homeostasis is controlled by Ca-Inflow and Ca-Outflow channels. Inactive CAST, PKC and calpain (i-CAST, i-PKC, i-CALP) are converted into active CAST, PKC and CALP by the addition of calcium. CAST is also regulated by PKC through phosphorylation (i-CAST). CAST restricts activity of CALP by forming a complex (CC-complex) with it in cytosol and the complex is degraded gradually by CALP. CALP also converts PKC into protein kinase m (pkm).

2   Methods

The qualitative modeling of BRN is performed using GINsim [11] and Genotech [12] tools which are based on the kinetic logic formalism of Thomas [13]. GINsim is a tool for qualitative modeling and analysis; it has a graphical user interface (GUI), a simulation core and a graph analysis tool [11]. Genotech has a simple GUI and it takes a directed graph with logical parameters as an input. It generates a state graph as output, which can predict stable states and cyclic trajectories. The behavior of entities of BRN depicted in state graph depends on a particular set of logical parameters represented as $K_{ent}\{resources\}$ where ent is a protein or gene which has set of regulators $\{resources\}$ associated with it. This parameter plays a major role in observing the dynamics of model and most often they are not experimentally measurable. This can be solved using the modeling tool SMBioNet [14]. It needs two inputs, BRN with unspecified parameters and computational tree logic (CTL) formulas to express the biological observations. The output generates set of parameters verified by CTL formulas. The tool help in accurate estimation of the parameters and saves time. The BRN with K-parameters form a logical regulatory graph which can be transformed into a Petri Net (PN) through GINsim [11] using the method established by [15]. Molecular processes can be best simulated by stochastic PN (SPN) as they are random in nature. A detailed SPN model of neuronal network in Alzheimer's was studied by [16], which concluded that calpain-CAST system plays important role in neuronal degradation. Here, role of calpain-CAST system was further explored in neuronal and other biological processes using qualitative modeling and SPN models. Formal definitions and examples of PNs are provided in [16] which can be referred for understanding.

2.1   Construction of Calp-CAST System

From the existing literature on calpain CAST regulatory system, an abstracted qualitative BRN was constructed. It consists of three proteins (calpain, CAST and PKC) and one ion Channel (Ca^{2+}) as shown in Fig. 2. The BRN is an abstraction of three important pathways namely, calpain-CAST system, Calcium influx efflux channel and PKC signaling pathway [16] that represents a specific functionality of the system in human brain. The key protein of the BRN is calpain; it is regulated by CAST that is the sole inhibitor of both calpains (Calpain1 and Calpain2). Activation of both the substrate and inhibitor (Calpain and CAST, respectively) takes place in the presence of Ca^{2+} ions in the cytosol (Fig. 2). Ca^{2+} influx and efflux is the main event in regulating this system. In addition, CAST is inhibited by PKC whose activation is also dependent on Ca^{2+} ions. Normal working of the calpain-CAST system ensures healthy functioning of human organs through homeostasis. Deregulation of this system may lead to disease.

2.2   Logical Parameter Estimation

Estimation of logical parameters was performed by SMBioNet tool [17] on the basis of biological observations extracted from literature [5,8,10,18]. These observations were converted into CTL formulas, described in the form of equations.

$$\phi_1 = (((PKC = 0 \wedge Calp = 0 \wedge CAST = 0 \wedge Ca = 0) \implies EX(EF(PKC = 0$$
$$\wedge Calp = 0 \wedge CAST = 0 \wedge Ca = 0))))$$
$$(1)$$

$$\phi_2 = ((PKC = 0 \wedge CAST = 0 \wedge Calp = 0 \wedge Ca = 0) \implies EF(AG(PKC = 0 \wedge$$
$$Calp = 1 \wedge CAST = 0 \wedge Ca = 2)))$$
$$(2)$$

ϕ_1 (Eq. 1) states that in a particular homeostatic behavior, all entities with zero (0) expression levels after reaching next qualitative state (represented by CTL X) will eventually reach the same expression levels (represented by CTL F). ϕ_2 (Eq. 2) describes the epigenetic condition that all entities with zero (0)

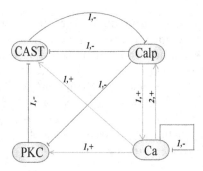

Fig. 2. Biological Regulatory graph of *Calpain-CAST* regulatory system consists of four entities namely, Protein Kinase C as **PKC**, Calpastatin as **CAST**, Calpain as **Calp** and Calcium as **Ca**. The red arc represents inhibition while the green arc represents activation. There is a negative self loop on **Ca** which regulates the calcium inflow and outflow in the BRN. (Color figure online)

expression levels depicting homeostatic state will reach to pathological condition in future from all paths and from all states of a path (represented by CTL A and G, respectively). In pathological state, $Ca = 2$ and $Calp = 1$ are over-expressed while $PKC = 0$ and $CAST = 0$ are degraded by *Calp*.

3 Results

3.1 Selection of Qualitative Model

Equations 1 & 2 were verified in five qualitative models of Calp-CAST system. Only one qualitative model (Fig. 2) was selected which shows more relevance to biological inferences. The set of logical parameters for this model was given in Table 1. The biological observations were verified in the state graph (Fig. 3) in the form of an initial state (*is*):(Calp, CAST, Ca, PKC) →(1, 1, 0, 1), cycles and a stable state (*ss*). Biological plausible *ss*: (Calp, CAST, Ca, PKC) →(1, 0, 2, 0) depicted hyperactivation of Calpain, depletion of CAST, deregulation of Ca^{2+} homeostasis through high Ca^{2+} ions influx and inactivation of PKC. The collective behavior of proteins involved in Calp-CAST system can only be determined by analyzing pathway transitions in the state transition graph. The state graph (Fig. 3) showed complete behavior of Calp-CAST system in the cytosol by showing important trajectories from *is* to cycles (homeostasis) and from *is* to *ss* which are considered as the most lethal transitions. On the basis of betweenness centrality, important cycles are also computed (Fig. 4) by using cytoscape tool [19] that sort all states of the transition graph based on betweenness centralities [20]. States with larger diameter represent higher betweenness centrality. Cycles with maximum betweenness centrality are: (Calp, CAST, Ca, PKC)

Table 1. Resources and logical parameters of the Calp-CAST BRN

No.	Logical Parameter	Resource	Range	Selected Values	No.	Logical Parameter	Resource	Range	Selected Values
1	K_{CAST}	{}	{0}	0	11	K_{PKC}	{Ca}	{0,1}	0
2	K_{CAST}	{PKC}	{0,1}	0	12	K_{PKC}	{Calp, Ca}	{0,1}	1
3	K_{CAST}	{Calp}	{0,1}	0	13	K_{Ca}	{}	{0}	0
4	K_{CAST}	{Ca}	{0,1}	0	14	K_{Ca}	{Calp}	{0,1,2}	2
5	K_{CAST}	{Ca, PKC}	{0,1}	0	15	K_{Ca}	{Ca}	{0,1,2}	2
6	K_{CAST}	{Calp, PKC}	{0,1}	0	16	K_{Ca}	{Calp, Ca}	{0,1,2}	2
7	K_{CAST}	{Calp, Ca}	{0,1}	0	17	K_{Calp}	{}	{0}	0
8	K_{CAST}	{Calp, Ca, PKC}	{0,1}	1	18	K_{Calp}	{CAST}	{0,1}	1
9	K_{PKC}	{}	{0}	0	19	K_{Calp}	{Ca}	{0,1}	1
10	K_{PKC}	{Calp}	{0,1}	0	20	K_{Calp}	{CAST, Ca}	{0,1}	1

$(0,0,0,0) \rightarrow (0,0,1,0) \rightarrow (0,0,0,0)$
$(0,1,0,0) \rightarrow (0,1,1,0) \rightarrow (0,1,0,0)$
$(0,0,0,0) \rightarrow (0,0,1,0) \rightarrow (0,1,1,0) \rightarrow (0,1,1,1) \rightarrow (0,1,0,1) \rightarrow (0,0,0,1) \rightarrow$
$(0,0,0,0)$ and
$(0,1,0,0) \rightarrow (0,1,1,0) \rightarrow (0,1,1,1) \rightarrow (0,1,0,1) \rightarrow (0,1,0,0)$

Cycles are shown with circles having large diameter and lighter color. These cycles represent the normal physiology of the calpain-CAST system. All the crucial proteins are oscillating such as calcium ions showing equilibrium in influx and efflux, regulation of CAST and PKC. States with smaller diameter and darker shades are rendered as *ss*. There are also some states deviating from or coming into the cycles, they have comparatively larger diameters than *ss* and low betweenness centrality (Fig. 4).

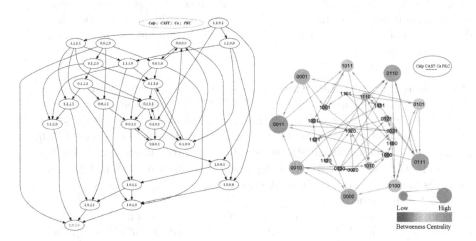

Fig. 3. State graph of *Calpain-CAST* BRN.

Fig. 4. Betweenness centrality extracted from the state graph of *Calpain-CAST* BRN.

3.2 Petri Net Modeling and Analysis

After studying the regulatory graph (combination of BRN and logical parameters), it is converted into discrete PN (Fig. 5) using GINsim which implemented the logic described in [21]. The PN is further converted into SPN using Snoopy Petri net tool [22]. In the SPN (Fig. 5), each entity (*Calp*, *CAST*, *PKC* and *Ca*) has its complementary state (inhibitory state) such as *cCalp*, *cCAST*, *cPKC* and *cCa*, respectively. There are also two types of transitions labeled as *p* and *n* for representing activation and inhibition, respectively (Fig. 5). *Calp* is the second entity in the SPN, which is connected by one *n*-labeled and two *p*-labeled transitions. The protein PKC is represented by place *PKC* and complementary place *cPKC* (Fig. 5). PKC has three transitions associated with it: **p_PKC**, **n1_PKC** and **n2_PKC**. The last protein in the pathway is CAST represented by places *CAST* and *cCAST* (Fig. 5). The associated transitions are **p_CAST**, **n1_CAST**, **n2_CAST** and **n3_CAST**. All the transitions have respective rates associated with them which is necessary to convert standard PN into SPN. The kinetic rates were adjusted manually using available biological observations and experiments (such as rate of *cCa* is higher than *Ca*, rate of *Calp* activation slowly degrades *CAST* and *Calp* increases rate of *Ca*). These rate parameters successfully reciprocated the qualitative model and biological inferences. The value of rates were tabulated for both homeostatic (C1 & C2) and stable state conditions. It is note worthy that in Table 2, rate of all the transitions related to places *Ca* and *cCa* were constant in all cycles and cases. It can be inferred that PKC, calpain and CAST effected the cell physiology. Simulation with these tabulated values are in agreement with already known biological observations. Results show that in Calp-CAST system, CAST and PKC are crucial in regulating calpain in cell which will have positive effect on cell by maintaining calcium homeostasis. Irregular calpain production lead to deregulation of Ca^{2+} ions, PKC and CAST which will eventually cause death of cell through apoptosis.

Homeostasis and Stable State. The state graph of the Calp-CAST BRN (Fig. 2) and simulations of the SPN model (Fig. 5) predicted the same cyclic conditions for homeostasis. Ca^{2+} is the most important ion in cell. It is maintained in homeostatic state before and after the activation of calpain (*Calp*). Both conditions describe the physiology of cell. Before calpain activation, Ca^{2+} inside cell (*Ca*) is in homeostasis as depicted by the cycle no. 1 (C1). Ca^{2+} homeostasis is also maintained with oscillation of CAST and PKC regulation in cycle no. 2 (C2). The homeostasis is disturbed when Ca^{2+} influx level rises (level = 2), system moves to diseased state due to calpain hyperactivation (Fig. 3). The simulation graphs (Fig. 6) show the oscillation of Ca^{2+} influx and efflux represented by places *Ca* and *cCa* respectively. The relative level of *cCa* is higher than *Ca*, as concentration of Ca^{2+} ions in extracellular space is higher than concentration of Ca^{2+} in cytosol. The simulation graph (Fig. 6a) is generated from following

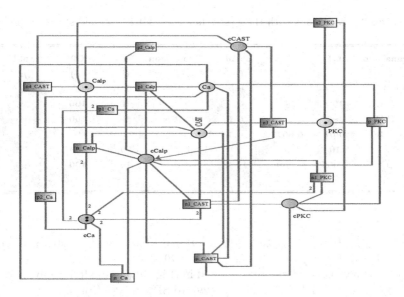

Fig. 5. SPN model of Calp-CAST system derived from the BRN. Each protein is depicted by two places (oval), one normal protein (green) and its counter part (red) which shows the inhibited protein. When the normal protein e.g., *Calp* is inhibited the token moves to counterpart (*cCalp*) and reciprocal happens in activation. The places are connected by transitions (rectangle). *p*-labeled transitions represent activation and *n*-labeled transitions show inhibitions. (Color figure online)

transition rates of *C1* of Table 2. It shows oscillation of *Ca* and *cCa* while *Calp* is absent or in infinitesimal concentration. Second cycle *C2* has four variant, all entities are oscillating except for calpain (*Calp*). This cycle, Fig. 6b, is closer to biological phenomenon. All entities (proteins and ions) are playing their roles in maintaining physiological state but with the passage of time, system eventually moves towards *ss* or diseased state. After studying the state graph of the Calpain-CAST BRN (Fig. 3), it can be observed that sooner or later all the states move to *ss*. It can also be observed that system either directly move to *ss* (in fatal case) or first enters and stay in homeostasis then traverse to *ss*.

When calpain (*Calp*) inhibits PKC (*PKC*), system either moves to *ss* directly (Fig. 3): (Calp, CAST, Ca, PKC)

(1,1,0,1) $\rightarrow$ (1,1,0,0) $\rightarrow$ (1,1,1,0) $\rightarrow$ (1,1,2,0) $\rightarrow$ (*1,0,2,0*).

Or enters homeostasis for a short time and then proceeds to *ss* (Fig. 3): (Calp, CAST, Ca, PKC)

(1,1,0,1) $\rightarrow$ (1,1,0,0) $\rightarrow$ (0,1,0,0) $\longleftrightarrow$ (0,1,1,0) $\rightarrow$ (0,1,1,1) $\rightarrow$ (0,0,1,1) $\rightarrow$ (1,0, 1,1) $\rightarrow$ (1,0,2,1) $\rightarrow$ (*1,0,2,0*).

Table 2. Transitions with their rates in cycles (C1 & C2) and stable state

Transition	C1 (μ)	C2 (μ)	Stable (μ)	Transition	C1 (μ)	C2 (μ)	Stable (μ)
p_CAST	0.01	100	0.5	p_PKC	0.01	10	5
$n1\_CAST$	100	0.5	0.5	$n1\_PKC$	100	100	100
$n2\_CAST$	100	1	0.1	$n2\_PKC$	100	100	100
$n3\_CAST$	0.01	0.007	0.05	$p1\_Calp$	0.001	0	0.1
$p1\_Ca$	10	10	10	$p2\_Calp$	0.001	0.001	0.001
$p2\_CA$	0.009	0.009	0.009	n_Calp	100	100	100
n_Ca	1	1	1				

$(1,1,0,1) \rightarrow (1,1,0,0) \rightarrow (1,1,1,0) \rightarrow (0,1,1,0) \rightarrow (0,1,0,0) \longleftrightarrow (0,1,1,0) \rightarrow (0,1,1,1)$
$\rightarrow (0,0,1,1) \rightarrow (1,0,1,1) \rightarrow (1,0,2,1) \rightarrow (1,0,2,0)$.

Both homeostasis and **ss** are simulated in (Fig. 6b and c) after applying rates of Table 2 of column labeled as *Stable state* to SPN model (Fig. 5).

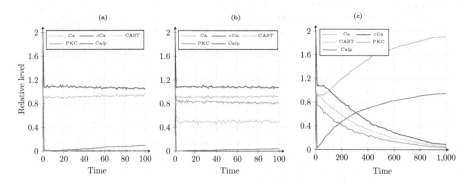

Fig. 6. *a* shows calcium homeostasis, (**Ca** and **cCa** are maintained at threshold, high efflux rate and low influx rate). In **b**, **PKC** and **CAST** both are oscillating with Calcium. In **c**, initially, system moves to homeostasis when **Calp** inhibits **PKC** for short time. As **Calp** concentration level increases, the system moves to stable state by breaking the homeostasis.

4   Discussion

There are some interesting inferences deduced from both the models. Firstly, Calp-CAST BRN and SPN model have shown states and dynamics of underlying mechanism. Formation of Calpain-CAST complex takes the system to physiological state i.e., homeostasis, Ca^{2+} influx and efflux are in equilibrium with CAST and PKC while calpain is in dormant form. The simulations showed that there was a basal concentration of calpain present in (homeostasis) cell. CAST limits calpain activation without interfering with basal calpain activity, which is required for physiology [23]. Homeostasis is primarily maintained through the formation of calpain-CAST complex and it is further strengthened by homeostasis of Ca^{2+} [24]. It is also necessary to keep PKC and CAST in homeostasis to maintain calpain-CAST system for longer time. Phosphorylation of CAST by PKC is the step through which localization of CAST takes place and is also a mechanism to regulate CAST in homeostasis [25]. The homeostatic condition is disturbed as active calpain degrades the substrate CAST to set itself free [26]. Under pathological condition, tight regulation of calpain through CAST is destroyed and Ca^{2+} homeostasis is lost as studied in ischemia/reperfusion, Alzheimer's (neuronal disorder), muscular and cardiovascular disorders [27]. Increase in calpain activity heightens Ca^{2+} sensitivity and sustained rise in Ca^{2+} levels causes cellular damage. Over activation of calpain, deregulation of Ca^{2+} ions, attenuation of CAST and degradation of PKC into pkm is observed in neurodegeneration and coronary vascular diseases [5]. Moreover, it is observed that association of calpain and CAST in the form of complex is also influenced by PKC (simulation graphs Fig. 6). It has hormetic effect in calpain-CAST system, suitable concentration level of PKC is beneficial in the form of homeostasis. Both models also showed that calpain has biphasic effect on PKC; at low calpain level, Ca^{2+} ions favor PKC activation while at high level, calpain causes down regulation of PKC [28]. One trajectory observed in the state graph is $(1,1,0,1) \rightarrow (1,1,1,1) \rightarrow (0,1,1,1) \rightarrow (0,0,1,1) \rightarrow (0,0,0,1) \rightarrow (0,0,1,1) \rightarrow (1,0,1,1) \rightarrow (1,0,1,0) \rightarrow (1,0,2,0)$. (Green states and transitions are representing homeostasis)

This trajectory can possibly be used to explain the interaction of calpain and PKC activation in ischemic tissue at high Ca^{2+} [29]. Kang et al., conducted the study to observe the role of both protein in muscular heart tissues. This trajectory and study of whole state graph can be useful in assessing the role and functionalities of these protein in mayocardium and it would help in answering questions which remained unexplored in that study. A recent study carried out on heart failure using human heart samples and animal models also proved elevation of calpain is the main cause of heart failure [30]. Elevated activity of calpain degrades PKC and CAST [30,31]. The devastating effect of calpain activity can be lowered or controlled if calpain-CAST association remains everlasting/eternal. This complex is long lasting during Ca^{2+} homoestasis and in the presence of PKC. This formal modeling study will also be helpful in designing in vitro experiments on all these crucial proteins.

References

1. Ferreira, A.: Calpain dysregulation in Alzheimer disease. ISRN Biochem. **2012**, 12 (2012)
2. Ryu, M., Nakazawa, T.: Calcium and calpain activation. In: Nakazawa, T., Kitaoka, Y., Harada, T. (eds.) Neuroprotection and Neuroregeneration for Retinal Diseases, pp. 13–24. Springer, Tokyo (2014). https://doi.org/10.1007/978-4-431-54965-9_2
3. Lokuta, M.A., Nuzzi, P.A., Huttenlocher, A.: Calpain regulates neutrophil chemotaxis. Proc. Natl. Acad. Sci. **100**(7), 4006–4011 (2003)
4. Battaglia, F., Trinchese, F., Liu, S., Walter, S., Nixon, R.A., Arancio, O.: Calpain inhibitors, a treatment for Alzheimer disease. J. Mol. Neurosci. **20**(3), 357–362 (2003)
5. Higuchi, M., et al.: Mechanistic involvement of the calpain-calpastatin system in Alzheimer neuropathology. FASEB J. **26**(3), 1204–1217 (2012)
6. Letavernier, E., et al.: Targeting the calpain/calpastatin system as a new strategy to prevent cardiovascular remodeling in angiotensin ii-induced hypertension. Circ. Res. **102**(6), 720–728 (2008)
7. Hanna, R.A., Campbell, R.L., Davies, P.L.: Calcium-bound structure of calpain and its mechanism of inhibition by calpastatin. Nature **456**(7220), 409–412 (2008)
8. Averna, M., De Tullio, R., Passalacqua, M., Salamino, F., Pontremoli, S., Melloni, E.: Changes in intracellular calpastatin localization are mediated by reversible phosphorylation. Biochem. J. **354**(1), 25–30 (2001)
9. Goll, D.E., Thompson, V.F., Li, H., Wei, W.E.I., Cong, J.: The calpain system. Physiol. Rev. **83**(3), 731–801 (2003)
10. Berridge, M.J., Bootman, M.D., Roderick, H.L.: Calcium: calcium signalling: dynamics, homeostasis and remodelling. Nat. Rev. Mol. Cell Biol. **4**(7), 517 (2003)
11. Naldi, A., Berenguier, D., Faure, A., Lopez, F., Thieffry, D., Chaouiya, C.: Logical modelling of regulatory networks with GINsim 2.3. Biosystems **97**(2), 134–139 (2009)
12. Ahmad, J.: Modélisation hybride et analyse des dynamiques des réseaux de régulations biologiques en tenant compte des délais. Ph.D. thesis, Nantes (2009)
13. Thomas, R., Thieffry, D., Kaufman, M.: Dynamical behaviour of biological regulatory networks-i. biological role of feedback loops and practical use of the concept of the loop-characteristic state. Bull. Math. Biol. **57**(2), 247–276 (1995)
14. Comet, J.-P., Richard, A.: SMBioNet: a tool for modeling biological regulatory networks driven by temporal behavior. In: ECCB 2003, 27–30 September 2003, Paris, France (2003)
15. Chaouiya, C., Naldi, A., Thieffry, D.: Logical modelling of gene regulatory networks with GINsim. In: van Helden, J., Toussaint, A., Thieffry, D. (eds.) Bacterial Molecular Networks, vol. 804, pp. 463–479. Springer, Heidelberg (2012). https://doi.org/10.1007/978-1-61779-361-5_23
16. Ashraf, J., Ahmad, J., Ali, A., Ul-Haq, Z.: Analyzing the behavior of neuronal pathways in Alzheimer's using petri net modeling approach. Front. Neuroinformatics **12**, 26 (2018)
17. Khalis, Z., Comet, J.-P., Richard, A., Bernot, G.: The SMBioNet method for discovering models of gene regulatory networks. Genes Genomes Genomics **3**(1), 15–22 (2009)
18. Thibault, O., Gant, J.C., Landfield, P.W.: Expansion of the calcium hypothesis of brain aging and Alzheimer's disease: minding the store. Aging Cell **6**(3), 307–317 (2007)

19. Shannon, P., et al.: Cytoscape: a software environment for integrated models of biomolecular interaction networks. Genome Res. **13**(11), 2498–2504 (2003)
20. Tareen, S.H.K., Ahmad, J., Roux, O.: Parametric linear hybrid automata for complex environmental systems modeling. Front. Environ. Sci. **3**, 47 (2015)
21. Chaouiya, C., Remy, E., Thieffry, D.: Qualitative petri net modelling of genetic networks. In: Priami, C., Plotkin, G. (eds.) Transactions on Computational Systems Biology VI. LNCS, vol. 4220, pp. 95–112. Springer, Heidelberg (2006). https://doi.org/10.1007/11880646_5
22. Heiner, M., Herajy, M., Liu, F., Rohr, C., Schwarick, M.: Snoopy – a unifying petri net tool. In: Haddad, S., Pomello, L. (eds.) PETRI NETS 2012. LNCS, vol. 7347, pp. 398–407. Springer, Heidelberg (2012). https://doi.org/10.1007/978-3-642-31131-4_22
23. Ye, T., et al.: Over-expression of calpastatin inhibits calpain activation and attenuates post-infarction myocardial remodeling. PLoS One **10**(3), e0120178 (2015)
24. De Tullio, R., et al.: Differential regulation of the calpain-calpastatin complex by the l-domain of calpastatin. Biochimica et Biophysica Acta (BBA)-Mol. Cell Res. **1843**(11), 2583–2591 (2014)
25. Blomgren, K., et al.: Calpastatin is up-regulated in response to hypoxia and is a suicide substrate to calpain after neonatal cerebral hypoxia-ischemia. J. Biol. Chem. **274**(20), 14046–14052 (1999)
26. Rao, M.V., et al.: Marked calpastatin (cast) depletion in Alzheimer's disease accelerates cytoskeleton disruption and neurodegeneration: neuroprotection by cast overexpression. J. Neurosci. **28**(47), 12241–12254 (2008)
27. Piper, H.M., Abdallah, Y., Schäfer, C.: The first minutes of reperfusion: a window of opportunity for cardioprotection. Cardiovasc. Res. **61**(3), 365–371 (2004)
28. Cressman, C.M., Mohan, P.S., Nixon, R.A., Shea, T.B.: Proteolysis of protein kinase C: mM and μM calcium-requiring calpains have different abilities to generate, and degrade the free catalytic subunit, protein kinase M. FEBS Lett. **367**(3), 223–227 (1995)
29. Kang, M.-Y., Zhang, Y., Matkovich, S.J., Diwan, A., Chishti, A.H., Dorn, G.W.: Receptor-independent cardiac protein kinase Cα activation by calpain-mediated truncation of regulatory domains. Circ. Res. **107**(7), 903–912 (2010)
30. Wang, Y., et al.: Targeting calpain for heart failure therapy: implications from multiple murine models. JACC: Basic Transl. Sci. **3**(4), 503–517 (2018)
31. Noma, H., Kato, T., Fujita, H., Kitagawa, M., Yamano, T., Kitagawa, S.: Calpain inhibition induces activation of the distinct signalling pathways and cell migration in human monocytes. Immunology **141**(2), 286–286 (2014)

The Application of Machine Learning Algorithms to Diagnose CKD Stages and Identify Critical Metabolites Features

Bing Feng[1], Ying-Yong Zhao[2(✉)], Jiexi Wang[1], Hui Yu[3],
Shiva Potu[4], Jiandong Wang[1], Jijun Tang[1(✉)], and Yan Guo[3(✉)]

[1] Department of Computer Science and Engineering,
University of South Carolina, Columbia, SC, USA
{bingf,wang372,jiandong}@email.sc.edu,
jtang@cse.sc.edu
[2] Key Laboratory of Resource Biology and Biotechnology in Western China,
School of Life Sciences, Northwest University, Xi'an, China
zyy@nwu.edu.cn
[3] Department of Internal Medicine, University of New Mexico,
Albuquerque, NM, USA
{huiyul,YaGuo}@salud.unm.edu
[4] Supply Chain Technology, Walmart Labs, Sunnyvale, CA, USA
spotu@walmartlabs.com

Abstract. Background: Chronic kidney disease (CKD) is a progressive and heterogeneous disorder that affects kidney structures and functions. Now it becomes one of the major challenges of public health. Early-stage detection, specialized stage treatments can significantly defer or prevent the progress of CKDs. Currently, clinical CKD stage diagnoses are mainly based on the level of glomerular filtration rate (GFR). However, there are many different equations and approaches to estimate GFR, which can cause inaccurate and contradictory results.

Methods: In this study, we provided a novel method and used machine learning techniques to construct high-performance CKD stage diagnosis models to diagnose CKDs stages without estimating GFR.

Results: We analyzed a dataset of positive metabolite levels in blood samples, which were measured by mass spectrometry. We also developed a feature selection algorithm to identify the most critical and correlated metabolite features related to CKD developments. Then, we used selected metabolite features to construct improved and simplified CKD stage diagnosis models, which significantly reduced the diagnosis cost and time when compared with previous prediction models. Our improved model could achieve over 98% accuracy in CKD prediction. Furthermore, we applied unsupervised learning algorithms to further validate our models and results. Finally, we studied the correlations between the selected metabolite features and CKD developments. The selected metabolite features provided insights into CKD early stage diagnosis, pathophysiological mechanisms, CKD treatments, and drug development.

Keywords: Chronic Kidney Diseases · Machine learning · Feature selection · CKDs diagnosis · Metabolites

© Springer Nature Switzerland AG 2019
I. Rojas et al. (Eds.): IWBBIO 2019, LNBI 11465, pp. 72–83, 2019.
https://doi.org/10.1007/978-3-030-17938-0_7

1 Introduction

Chronic Kidney Diseases (CKDs) are progressive losses and abnormalities of kidney functions and structures [1]. Now they become the major challenges of public health and affect approximately 10% of the world population [2]. CKDs often result in Acute Kidney Injury (AKI), which appears abruptly and further results in rapid deterioration of kidney function [3, 4]. The final stage (stage 5) of CKD and AKI would cause renal failure and require special treatments such as dialysis and renal transplant [5]. CKDs are also associated with other severe complications, including cardiovascular diseases, hypertension, diabetes [6], cognitive decline, anemia, mineral and bone disorders, and fractures [7]. Screening and stage detection are critical to CKDs prognosis and interventions. Renal failure can be reversed if CKDs are detected and treated properly in the early stage [6, 8]. Prompt treatments and management of specific CKD stages can significantly prevent and delay the developments, as well as the progress of CKD [6–8]. Meanwhile, different CKD stages require different management, treatments, and medicines [9–11]. Proper management strategies can significantly reduce the incidence of final-stage CKDs [7]. However, the early CKD stage (stage 1) is hard to recognize [12] since no symptom could be found initially. Usually, treatments and management for CKDs will not be determined until severe symptoms or accidental findings from tests for other diseases [13].

Recent international guidelines classify CKDs into five stages based on the levels of glomerular filtration rate (GFR) [1, 12]. GFR is computed by estimating equations and other associated exogenous bio-factors [13]. In fact, there are many issues in the current GFR estimation approaches. The equations may lead to inaccurate estimation due to the variation of personal situations, which include age, race, gender, and serum creatinine level [14]. Moreover, various GFR estimating approaches and equations will also result in disagreements and errors for the same type of patients [15]. Recent Bland-Altman analysis showed that CKD-EPI creatinine-cystatin C, BIS2, CKD-EPI cystatin C and Simple cystatin C GFR equations were not accurate in estimating GFR for elderly people [15]. Researchers also discovered that using the Schwartz formula to estimate GFR for children would result in overestimation [16]. Muna et al. found that the BIS equation was not a proper approach to predict the risk of death for older women when compared with the CKD-EPI equations [17]. These inaccurate and contradictory issues might result in misjudging of CKD stages, which might result in further under-diagnoses and under-treatments [6].

Using machine learning algorithms to diagnose diseases has a few advantages. Rather than solely relying on the doctor's experiences and stereotyped formulas, researchers can use learning algorithms to analyze sophisticated, high-dimensional and multimodal biomedical data, as well as construct classification models to make decisions even when some information was incomplete, unknown, or contradictory. Current machine learning studies in CKD diagnostic have already shown high accuracies and reliabilities in CKD diagnosis [18, 19]. Neves et al. used Artificial Neural Networks to build a classification model to classify 558 CKD/non-CKD patients based on 24 features, achieving a 92.3% accuracy [18]. Celik et al. used Decision Tree and Support Vector Machine (SVM) algorithms to classify a CKD dataset with 400 patients and 24 features, which both achieved accuracies of over 96% in classifying the CKD/non-CKD patients [20]. Polat et al. used SVM to classify Celik's dataset and reached an accuracy of 98.5% [19]. Chen et al. applied three different learning algorithms to construct classification models for the datasets with CKD/non-CKD patients, all achieving over 93% accuracies [21]. However, most present CKD diagnoses machine learning studies only focused on the simple yes/no problems and cannot classify among the multiple CKD stages [18–20]. Most of them were only performed on small datasets with limited numbers of physiological features.

In this study, we provided a novel and independent way to diagnose six CKD stages without measuring the GFR, which could overcome the inaccurate and contradictory issues caused by various GFR estimating approaches. We used three supervised learning algorithms to build CKD stage diagnosis models to diagnose the six CKD stages (five CKD stages plus the non-CKD). We analyzed a large volume of metabolites dataset that was obtained from positive ion mass spectrometry of clinical blood samples. We first built three CKD stage diagnosis models based on all metabolite features, achieving very high accuracies and low false positive rates. Then we developed a feature selection algorithm to identify the most critical and informative metabolite features related to CKD development. We then used the selected metabolite features to construct improved, simplified diagnosis models, which only required the selected features to determine the CKD stages in practice. Furthermore, we applied unsupervised learning algorithms to validate that the selected metabolites were the most critical factors correlated with CKDs symptoms and progressive perturbations. Finally, we studied the correlations among selected metabolite features and developments of CKD stages. The selected metabolite features provided insights into CKD stage diagnosis, pathophysiological mechanisms, CKD treatments, and drug development.

2   Result

2.1   CKD Stage Diagnosis Models Built on All Metabolite Features

In this section, we applied Random Forest, SVM, and Decision Tree three algorithms to build CKD stage diagnosis models. To provide systematic evaluations of these diagnosis models, we ran ten parallel experiments with randomized datasets for each model to compute their average performance metrics. Meanwhile, we also performed 10-fold cross-validation on the entire dataset to test their robustness and reliabilities of the constructed models.

As Table 1 showed, all three constructed models achieved very good performances in the CKD stage classification and 10-fold cross-validations, which all reached over 95% accuracies. The model built from Random Forest achieved the best performance, which had an accuracy of 95.6% in classification and an accuracy of 96% in cross-validation. The SVM model had the least good performance with an accuracy of 95.4% and 96%. The performance of the Decision Tree model lied in the middle of the other two models. Table 1 also showed that all CKD stage diagnosis models had very high average precisions, recalls and F-scores. Figure 1(A) showed the confusion matrices of the constructed CKD stage classification/prediction models in predicting the CKD stages of 100 patients. All three models achieved high accuracies with only four mispredictions each. No patient in non-CKD or early stage of CKD was mispredicted into wrong CKD stages. Hence, we provided a novel and independent way to diagnose the CKD stages based on the analyses of blood levels, which didn't require the estimation of GFR and overcame the contradictory issues caused by different GFR estimation approaches.

Table 1. Performances of CKD stage diagnosis models built from all 16382 metabolite features.

Models	Evaluation metrics of different models				
	Accuracy	Precision	Recall	F-score	Cross-validation
Random forest	95.6%	95.1%	95.3%	95.0%	96($\pm$0.5)%
SVM	95.4%	95.0%	95.1%	94.9%	96($\pm$0.4)%
Decision tree	95.5%	95.0%	95.0%	94.9%	95($\pm$0.5)%

2.2 CKD Stage Diagnosis Models Built on the Selected Metabolite Features

In the last section, we constructed three CKD stage diagnosis models based on all 16382 metabolite features, which achieved very good performances. However, its impractical to have all 16382 metabolites features measured properly for each clinical test, due to the cost of data collections, efficiency and timeliness, and errors of data observation and measurements. Therefore, brief, fast, and robust models were required to perform reliable and fast CKD stage diagnoses for patients with potential risks. To identify the critical biomarkers and metabolic components that were related to symptoms and developments of different CKD stages, we developed a feature selection algorithm to identify the most critical and informative metabolite features from the original 16382 features. As a result, we obtained 69 metabolite features, which were utilized to construct simplified CKD stage diagnosis models.

As Table 2 showed, although the new models were significantly simplified, they achieved improvements in all evaluation metrics when compared with the models built with all 16382 features. Random Forest maintained the best performance over all three models. Its accuracy was increased from 95.6% to 98.6%. The SVM model still had the least good performance but displayed some improvement (accuracy increased from 95.4% to 97.2%). The accuracy of the Decision Tree model was increased from 95.5% to 97.3%. Figure 1(B) showed the confusion matrices of the improved CKD stage classification/prediction models built from the same training and testing dataset that used in the last section. All of them obtained higher accuracies in diagnosing the CKD stages of 100 patients. Only three patients were mispredicted by the models built by Random Forest and Decision Tree, and one by the SVM model. Also, no patient in the non-CKD or early stages (stage 1 and stage 2) of CKD was mispredicted into wrong CKD stages in any of the three models.

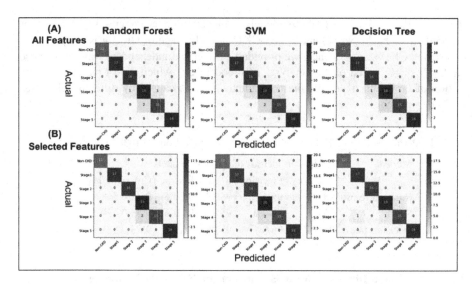

Fig. 1. Confusion matrices of the CKD stage classification/prediction models built from the same training and testing data set.

Table 2. Performances of CKD stage diagnosis models built from the selected 69 metabolite features.

Models	Evaluation metrics of different models				
	Accuracy	Precision	Recall	F-score	Cross-validation
Random forest	98.6%	98.6%	98.7%	98.6%	98(±0.3)%
SVM	97.2%	97.1%	97.0%	97.0%	97(±0.5)%
Decision tree	97.3%	97.1%	97.3%	97.2%	98(±0.4)%

Currently, a few studies had demonstrated that the levels of a few metabolites were associated with the CKD pathophysiological processes. They also affected the CKD stage development [23–25]. However, previous studies cannot determine the CKD stages only by the analyses of a few metabolites. In this section, we identified the most critical and correlated metabolite makers and components from 16382 features, which could distinguish the CKD stages effectively. CKD stage diagnosis models built from the selected metabolite features achieved even better performances than the previous models, indicating the significances and effectiveness of the selected features. Improved CKD stage diagnosis models had obvious practicabilities and feasibilities since the chances of errors, consumption of time, and testing costs were significantly reduced. We could use improved CKD stage diagnosis models to diagnose the CKD stages through a regular blood test, especially for the early stages.

2.3   Validation of Selected Metabolites Features Subset

Previous studies already revealed that metabolite levels were associated with the pathophysiologic processes and development of CKD stages [23, 24]. In this section, we applied unsupervised learning algorithms to further validate the results of selected metabolites features. We demonstrated that the selected metabolite features were the most correlated and informative bio-markers and components, which was associated with the CKD pathogenesis and progressions. First of all, we removed the stage labels and performed the Principal component analysis (PCA) for all 703 patients based on both 16382 metabolite features and selected features. Then we projected all patient's data into three dimensions. After that, we marked the true CKD stages with six different colors and showed stages distributions of all patients in 3D spaces. As Fig. 2(A) and (B) showed, PCA results based on the selected metabolite features set could better discriminate the CKD stages compared with the PCA results from all metabolite feature set.

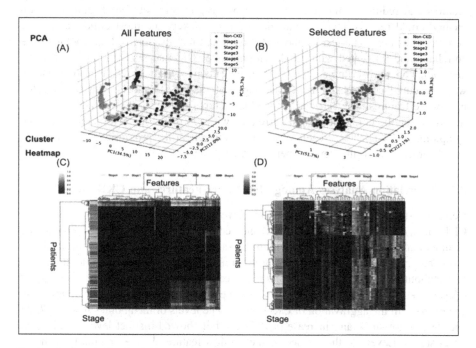

Fig. 2. Analyses of selected metabolites subset based on unsupervised learning algorithms. Figure A and B are the PCA dimension reductions for all 16382 metabolite features dataset and selected 69 metabolite features dataset. Figure C and D are the hierarchical/agglomerative clustering heatmaps based on all 16382 metabolite features and selected 69 metabolite features. The red bar (six color level) represents the original CKD stage of each patient. The green color represents the levels for each feature. The hierarchical tree on the left represents patients sample hierarchical/agglomerative clustering. The hierarchical tree on the top represents the features hierarchical/agglomerative clustering. (Color figure online)

In addition, we applied hierarchical/agglomerative clustering to cluster all patients based on all 16382 metabolite features and the selected features. First, we removed the class labels for all patients. Afterward, we used the Ward variance minimization algorithm to compute the distances among all patients and then constructed hierarchical trees. We further plotted the cluster heat maps to visualize the levels, patients' hierarchical trees, features hierarchical trees, and patients true CKD stage labels for two feature sets. As Fig. 2(C) and (D) shown, the selected metabolites features could better cluster the CKD patients into six stages when compared with the results from all metabolites features. In Fig. 2(D), patients within the same stages could be better clustered into the same hierarchical sub-trees when compared with Fig. 2(C). Patients in Fig. 2(D) could also better match their true CKD stage labels (shown six red color level) when compared with Fig. 2(C). In addition, in Fig. 2(C), there were lots of features fluctuated in similar patterns according to the CKD stage changes. These features were clustered into hierarchical sub-trees and showed as dense clusters in green color. However, in Fig. 2(D), the levels of the metabolite features were not clustered into dense clusters but were fluctuated with different patterns according to the CKD stages changes. These results demonstrated that there were many redundant features and redundant information in the original feature set. Our feature selection method could remove these redundant features and kept the most critical and stage correlated features. The results above demonstrated that the selected metabolites were correlated and fluctuated with the developments of CKD stages, which were also the best indicators to discriminate the CKD stages effectively. These results also explained why we could construct better CKD stages of diagnosis models with the selected metabolite features. The selected metabolite features were the most critical and informative metabolic components that correlated with the metabolic level pathognomonic symptoms of CKD stages. Their variations could reflect the progressive abnormalities and disorders due to the developments of CKD stages.

2.4   Correlation Analyses Among Metabolites Features and CKD Stages

Here, we studied the correlations among levels of selected metabolite features and CKD stage developments. As Fig. 3(A) showed, the levels of these selected metabolic features maintained similar levels within the same CKD stage. On the other hand, the levels had changed significantly among different CKD stages. For example, the levels of metabolites feature 6, 7 and 20 had significantly increased with the aggravation of CKD stages. However, the levels of metabolites feature 58 of 65,66 had significantly decreased with the aggravation of CKD stages. The levels of metabolites feature 17, 22 and 25 did not show any increase or decrease but showed distinct levels in different CKD stages. Levels of these selected metabolites feature changed regularly with the development of CKD stages. Therefore, levels of these selected features can affect, or be affected by the progressive CKD developments and related abnormalities. The variances of these selected metabolite features were essential information for the CKD stage pathological studies and related medicine developments. In addition, Fig. 3(B) showed the correlation coefficients heatmap among the selected metabolite features.

The light color indicated weak correlations, which could be as low as 0. The dark color indicated strong correlations, which was up to 1. Most features showed low correlation levels, which had correlation coefficients between −0.4 and 0.4. The max correlation between any two features is 1, which was the correlation with itself (showed in the diagonal line). Most selected features were irrelevant to each other, indicating that there was no redundant feature or information in the selected metabolite feature set. This result was also consistent with the result in the heatmap of Fig. 2(D) in the section above.

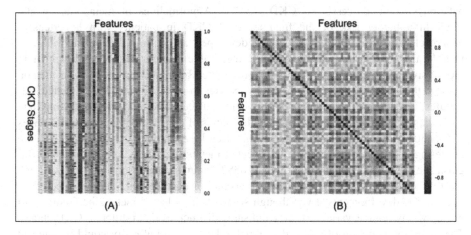

Fig. 3. Correlation analyses among metabolites features and CKD stages. Figure A shows the correlations between CKD stages and the metabolic level of selected metabolite features. Figure B shows the correlation levels among the selected metabolite features.

3 Discussion

Early CKD stage and Non-CKD detection are critical to the prognosis and intervention of CKDs. Specific treatments and management for different CKD stages can significantly prevent and delay the disease promotions and progressions [6–8]. Renal failure can also be reversed if the CKDs were detected and treated in the early stage [6, 8]. However, diagnosis and management of CKD are usually made when symptoms become severe or after accidentally findings from clinical testing for other diseases [13]. Currently, clinical diagnoses are not reliable enough to determine the CKD stage or make accurate decisions only based on the level of GRF and few additional biofactors. Because various GFR equations and approaches can cause inaccurate and contradictory issues [6, 15, 17]. In this study, first, we used supervised learning algorithms to construct high-performance CKD stage diagnosis models based on the blood metabolic level of 16382 features, which overcame the contradictory issues caused by different GFR estimation approaches. In practice, it's impractical to have all 16382 features measured for each clinical test, due to the cost of data collections, efficiency, and timeliness. Therefore, we further constructed improved and simplified diagnosis models by selecting 69 metabolite features from blood samples, which significantly reduced the cost of CKD screening and stage diagnoses. On the other hand, previous machine learning studies about CKD diagnosis focused on distinguishing the CKD and Non-CKD. However, our CKD stages diagnosis models could diagnose CKDs into six stages (five CKD stages and non-CKD) with very high accuracies and low false-positive rate. Also, they had no error in determining the non-CKD or early CKD stages (stage 1 and stage 2), which could be further used in CKD screening. We provided a novel, brief, and feasible machine learning based approach to screening and diagnosing CKDs without measuring the GFR.

Previous studies revealed that metabolite levels were highly associated with CKD pathophysiologic processes of CKD [23–25]. Abnormalities of metabolites would result in a higher risk of the final stage of CKD in patients with hypertensive nephrosclerosis [26]. In this study, we developed a feature selection algorithm to identify the most critical and correlated metabolite features that were involved in progressive perturbations of CKD stage and related biological processes. Our method selected the most critical and correlated 69 features from the original 16382 metabolites features. These selected features not only helped us construct improved and simplified CKD diagnosis models but also revealed the effects of the developments of CKD stages and pathophysiologic mechanisms. Unsupervised PCA and hierarchical/agglomerative clustering analysis demonstrated that selected metabolite features could better distinguish the CKD stages than the original 16382 features. Our results also showed that most selected features had correlated incremental or decremental variations according to the CKD developments. Even though some of the selected metabolite features did not show any regular variations, they did show distinct levels in different CKD stages. Therefore, the variation of selected metabolite levels could affect or could be affected by the progressive developments of CKD. Furthermore, there was very little correlation among selected metabolite features, indicating that there was no redundant information in the selected features. Therefore, these selected metabolites were the critical factors and components that were correlated with CKD stages. These results also explained why these selected features could lead to better CKD stages of diagnosis models. They also provided researchers with the opportunities to gain new insights into metabolic profiling and pathophysiological mechanisms of the developments of CKD stages. Thus, our studies could be further used for CKD early stage diagnosis, CKD treatments, and drug development.

4   Method

4.1   Data

The dataset used in this study contained positive metabolites levels measured by mass spectrometry of patients' blood samples. There were 703 patients' samples in total, including 587 CKD patients of five different CKD stages and 116 healthy subjects (non-CKD). 120 patients were in CKD stage 1; 104 patients were in CKD stage 2; 110 patients were in CKD stage 3; 119 patients were in CKD stage 4, and 134 patients were in CKD stage 5 (the final stage). Each blood sample was analyzed by positive ion mass spectrometry first and then identified total 16382 positive metabolites, which were also referred to as 16382 features. We further used MinMaxScaler algorithm to normalize the metabolite levels into the [0,1] interval.

4.2   Feature Selection

Our feature selection algorithm was based on the ideas of univariate feature selection algorithms, recursively elimination algorithms and embedded supervised learning classifiers. We first applied univariate feature selection filter algorithm Select-K-Best to select n most important features with the highest scores, which was based on the Chi-squared statistics. Next, given the 10-fold cross-validation C-Support Vector Classification as the external estimator, we recursively assigned weights to each available metabolite feature and eliminated the feature with the least weight from the current feature set. This procedure was recursively repeated on the same feature set until the feature number was eventually reached the threshold. Then we recursively pruned features from the current n SelectKBest features and obtained a subset A. Meanwhile, we used the embedded Random Forest classifier as a black box to find the b most important feature subset B from the current n features that selected from SelectKBest. At last, we obtained selected features that were in the intersections of subsets A and B. We parallelly ran the above steps t times and got the unions of the intersections of subsets A and B obtained in the last step. As a result, we selected 69 metabolite features from the original 16382 features. This algorithm was implemented by Python and Scikit Learn package [22].

4.3   Supervised Learning Algorithms

Our Random Forest algorithm used Gini impurity as the supported criteria. The maximum number of features was set to the total number of features in the dataset. The maximum depth of the tree was set to until all leaves are pure or all leaves contain less than two examples. The Support Vector Machine algorithm was implemented by the multi-class classification C-Support Vector Classification with the linear kernel and one-against-one scheme. The penalty parameter C of the error term was set to 1.0. The degree of the polynomial kernel function was set to 3. The coef0 was set to 0. The Decision Tree algorithm used in this study was the Classification and Regression Tree, which was similar to the C4.5 Decision Tree algorithm. This algorithm also used Gini impurity to measure the quality of a split feature. The maximum number of features was set to the total number of features in the dataset. There was no limit on the depth of this Decision Tree. The threshold for early stopping tree growth was set to 1e-7. Before constructing the CKD stage diagnosis models, we normalized all patients' levels for each metabolite features into the interval [0, 1]. Then we randomly selected the first 603 instances as the training set and used the rest 100 instances as testing sets. We also ran 10-folds cross-validation for each classifier to provide a systematic evaluation of the constructed CKD stage diagnosis models. All three supervised learning algorithms used throughout this study were implemented by Python and Scikit Learn package [22].

4.4   Unsupervised Learning Algorithms

Our Principal Component Analysis (PCA), implemented by Python and Scikit Learn package [22], utilized a linear dimensionality reduction by singular value decomposition and projected all patient's data into three dimensions. The hierarchical/agglomerative clustering for all patients was implemented by SciPy package [22, 28], which utilized the Ward variance minimization algorithm to compute the distances among all patients. The correlation study among the selected features was computed by the Kendall Tau correlation coefficient. All calculations in this section were performed by Python, pandas, and Seaborn packages [22, 27, 28].

Authors' Contributions. JT, YG, and BF conceived and designed the project. BF, HY, JW, SP, YG, and JT designed and performed the experiments. All authors analyzed the experiments results of this project. BF wrote the manuscript. All authors reviewed the manuscript. All authors read and approved the manuscript.

Funding. Cancer Center Supporting Grant from the National Cancer Institute (P30CA118100). Availability of data and materials The datasets used and/or analyzed during the current study are available from the corresponding author on reasonable request.

Ethics Approval and Consent to Participate. This study was approved by the Ethical Committee of Northwest University, Xi'an, China. All patients provided informed consent prior to entering the study.

References

1. Levey, A.S., et al.: Definition and classification of chronic kidney disease: a position statement from Kidney Disease: Improving Global Outcomes (KDIGO). Kidney Int. **67**(6), 2089–2100 (2005)
2. Subasi, A., Alickovic, E., Kevric, J.: Diagnosis of chronic kidney disease by using random forest. In: Badnjevic, A. (ed.) CMBEBIH 2017, vol. 62, pp. 589–594. Springer, Heidelberg (2017). https://doi.org/10.1007/978-981-10-4166-2_89
3. Chawla, L.S., Eggers, P.W., Star, R.A., Kimmel, P.L.: Acute kidney injury and chronic kidney disease as interconnected syndromes. New Engl. J. Med. **371**, 5866 (2014)
4. Bellomo, R., Kellum, J.A., Ronco, C.: Acute kidney injury. Lancet **380**, 756766 (2012)
5. Bagshaw, S.M., Berthiaume, L.R., Delaney, A., Bellomo, R.: Continuous versus intermittent renal replacement therapy for critically ill patients with acute kidney injury: a meta-analysis. Critical Care Med. **36**, 610617 (2008)
6. Levey, A.S., et al.: National Kidney Foundation practice guidelines for chronic kidney disease: evaluation, classification, and stratification. Ann. Intern. Med. **139**(2), 137–147 (2003)
7. Jha, V., et al.: Chronic kidney disease: global dimension and perspectives. Lancet **382** (9888), 260–272 (2013)
8. Morton, R., Tong, A., Howard, K., Snelling, P., Webster, A.: The views of patients and carers in treatment decision making for chronic kidney disease: systematic review and thematic synthesis of qualitative studies. BMJ **340**, c112 (2010)
9. Coresh, J., et al.: Prevalence of chronic kidney disease in the United States. Jama **298**(17), 2038–2047 (2007)

10. Lameire, N., Van Biesen, W.: The initiation of renal-replacement therapy: Just-in-time delivery. New Engl. J. Med. **363**, 678680 (2010)
11. Levey, A.S., Coresh, J.: Chronic kidney disease. Lancet **379**, 165180 (2012)
12. Levin, A., et al.: Kidney disease: improving global outcomes (KDIGO) CKD work group. KDIGO 2012 clinical practice guideline for the evaluation and management of chronic kidney disease. Kidney Int. Suppl. **3**, 150 (2013)
13. Webster, A.C., Nagler, E.V., Morton, R.L., Masson, P.: Chronic kidney disease. Lancet **389**, 12381252 (2017)
14. National Kidney Foundation: K/DOQI clinical practice guidelines for chronic kidney disease: evaluation, classification, and stratification. Am. J. Kidney Dis.: Official J. Natl. Kidney Found. **39**, S1 (2002)
15. Bevc, S., et al.: Estimation of glomerular filtration rate in elderly chronic kidney disease patients: comparison of three novel sophisticated equations and simple cystatin C equation. Ther. Apher. Dial. **21**, 126–132 (2017)
16. Levey, A.S., et al.: A new equation to estimate glomerular filtration rate. Ann. Intern. Med. **150**(9), 604–612 (2009)
17. Canales, M.T., et al.: Renal function and death in older women: which eGFR formula should we use? Int. J. Nephrol. **2017**, 10 (2017)
18. Neves, J., et al.: A soft computing approach to kidney diseases evaluation. J. Med. Syst. **39**(10), 131 (2015)
19. Polat, H., Mehr, H.D., Cetin, A.: Diagnosis of chronic kidney disease based on support vector machine by feature selection methods. J. Med. Syst. **41**, 55 (2017)
20. Celik, E., Atalay, M., Kondiloglu, A.: The diagnosis and estimate of chronic kidney disease using the machine learning methods. Int. J. Intell. Syst. Appl. Eng., 27–31 (2016)
21. Chen, Z., Zhang, X., Zhang, Z.: Clinical risk assessment of patients with chronic kidney disease by using clinical data and multivariate models. Int. Urol. Nephrol. **48**, 20692075 (2016)
22. Pedregosa, F., et al.: Scikit-learn: machine learning in Python. J. Mach. Learn. Res. **12**(Oct), 2825–2830 (2011)
23. Weiss, R.H., Kim, K.: Metabolomics in the study of kidney diseases. Nat. Rev. Nephrol. **8**, 2233 (2012)
24. Shah, V.O., Townsend, R.R., Feldman, H.I., Pappan, K.L., Kensicki, E., Vander Jagt, D.L.: Plasma metabolomic profiles in different stages of CKD. Clin. J. Am. Soc. Nephrol. **8**(3), 363–370 (2013)
25. McMahon, G.M., et al.: Urinary metabolites along with common and rare genetic variations are associated with incident chronic kidney disease. Kidney Int. **91**(6), 1426–1435 (2017)
26. Scialla, J.J., et al.: Mineral metabolites and CKD progression in African Americans. J. Am. Soc. Nephrol. **24**, 125135 (2013)
27. Waskom, M., et al.: seaborn: v0.7.1, June 2016. Zenodo. https://doi.org/10.5281/zenodo.54844. Chicago
28. McKinney, W.: Python for Data Analysis: Data Wrangling with Pandas, NumPy, and IPython. OReilly Media Inc, Sebastopol (2012)

Expression Change Correlations Between Transposons and Their Adjacent Genes in Lung Cancers Reveal a Genomic Location Dependence and Highlights Cancer-Significant Genes

Macarena Arroyo[1] , Rafael Larrosa[2,3(✉)] , M. Gonzalo Claros[4] ,
and Rocío Bautista[2]

[1] Unidad de Gestión Clínica de Enfermedades Respiratorias,
Hospital Regional Universitario de Málaga, Avda Carlos Haya s/n, Malaga, Spain
macarroyo@uma.es
[2] Plataforma Andaluza de Bioinformática,
Universidad de Málaga, 29590 Malaga, Spain
{rlarrosa,rociobm}@uma.es
[3] Departamento de Arquitectura de Computadores,
Universidad de Málaga, 29071 Malaga, Spain
[4] Departamento de Biología Molecular y Bioquímica,
Universidad de Málaga, 29071 Malaga, Spain
claros@uma.es

Abstract. Recent studies using high-throughput sequencing technologies have demonstrated that transposable elements seem to be involved not only in some cancer onset but also in cancer development. However, their activity is not easy to assess due to the large number of copies present throughout the genome. In this study *NearTrans* bioinformatic workflow has been used with RNA-seq data from 16 local patients with lung cancer, 8 with adenocarcinoma and 8 with small cell lung cancer. We have found 16 TE-gene pairs significantly expressed in the first disease, and 32 TE-gene pairs the second. Interestingly, some of the genes have been previously described as oncogenes, indicating that normal lung cell compromised on an oncogenic change displays some transposon expression reprogramming that seems to be genome-location dependent. Supporting this is the finding that most differentially expressed transposons change their expression in the same direction than their adjacent genes, and with a similar level of change. The analysis of adjacent genes may reveal or confirm important lung cancer biomarkers as well as new insights in its molecular basis.

Keywords: Transposon · Transposable element · Cancer · RNA-seq · Tool · Workflow

© Springer Nature Switzerland AG 2019
I. Rojas et al. (Eds.): IWBBIO 2019, LNBI 11465, pp. 84–92, 2019.
https://doi.org/10.1007/978-3-030-17938-0_8

1 Introduction

Cancer is one of the leading causes of morbidity and mortality, with lung cancer being the most common malignancy and the most common cause of cancer deaths in the past few decades [6]. Recent studies have demonstrated that, besides the specific somatic or germinal mutations that drive tumor growth, mobile elements, also known as transposable elements (TEs), are involved in the development of cancer. For example, in epithelial cancer, activation of TEs correlates with their mobilisation and genomic drift [8]. TE expression is known to contribute to genomic instability and can cause many genetic disorders. Since nearly 50% of the human genome is composed of TEs, many cells try to avoid these deleterious consequences inducing the inactivation of most TEs. It has been recently shown that some human endogenous viral elements (HEVEs) are still active in somatic tissues and play a crucial role in, for example, placental development in various mammalian species [10].

The study of TEs using high-throughput technologies is very difficult due to the complexity of its measurement and processing, since there is a large number of copies of TEs present throughout the genome. However, our laboratory has recently developed *NearTrans* [9], a bioinformatic workflow that takes advantage of well known software as well as the definition of the genomic positions of individual transposable elements. Therefore, *NearTrans* makes possible the differential expression analyses of TEs and their location in genome, besides integrating the calculation of differential expression of their nearby genes. It was presented last year [9] illustrating prostate cancer studies. In the present study, *NearTrans* is tested in two different lung cancers: lung adenocarcinoma (LAC) and small celular lung cancer (SCLC). A significant number of TEs and nearby genes were found differentially co-expressed (both up-regulated and down-regulated), revealing them as putative lung cancer biomarkers and a source of new insights in the cancer-specific genes.

2 Materials and Methods

2.1 Sample Data

Two lung cancers, LAC and SCLC, have been analysed using RNA-seq data from tumoral and healthy lung tissues from the same patient. Sequencing reads for SCLC can be downloaded from Bioproject EGAS00001000334 [12], containing 2×75 bp paired-reads for total RNA from 17 patients sequenced with Illumina HiSeq2000. Sequencing reads for LAC were obtained from 8 samples from the Regional Hospital of Malaga (Spain) and sequenced in our laboratory as previously described [1].

2.2 Bioinformatic Analyses

NearTrans can perform automatically the differential expression of TEs and genes from the raw data a provide TE-closer gene pairs that are co-expressed in the same of opposite direction. To do so, raw sequencing reads were pre-processed with *SeqTrimNext* [5] with the specific NGS Illumina configuration parameters to remove low quality, ambiguous and low complexity stretches, adaptors, organelle DNA, polyA/polyT tails, and contaminated sequences while keeping the longest (at least >20 bp) informative part of the read. Useful reads were mapped to human genome hg38 using *STAR* v2.5 [3]. Then, GFF of hg38 and a home-made GFF based on gEVE [10] were used with *CuffLinks/Cuffquant/Cuffdiff* tools (v.2.2.1 [13]) to assess expression levels of genes and TEs between healthy and cancer lung tissues, as described in [7]. Differential expression was considered significant when a gene or a TE has $FDR < 0.05$ and $|log_2FC| > 1$. *NearTrans* also includes the co-localisation of genes close to differentially expressed TEs in the genome with the assistance of *BedTools* (v.2.26.0 [11]). Distances between each TE and the closer gene is also given. Results can be filtered by distances between gene and TE, the level of expression, and the log_2FC sign.

3   Results

3.1   Sample and Sequencing Qualities

After preprocessing raw RNA-seq datasets data from the 17 SCLC cancer patients, the mean percentage of useful reads is 98.00%, and in the case of LAC the mean percentage of useful reads is 97.76%. The high mapping rate (>95%) confirms that results will not be affected by inadequate sequencing. However, BCV analysis revealed that one sample in SCLC (S585275) and other in LAC (69160211009) could have an abnormal behavior, so they were filtered out [1,2], resulting in 16 SCLC patients for 32 samples (16 healthy lung, 16 tumoral lung) and 7 LAC patiens for 14 samples (7 from their healthy lung and 7 for the corresponding tumoral lung).

3.2   TE-gene Pairs in LAC

A total of 45 TEs were identified as differentially expressed in LAC, where 22 TEs and their adjacent genes (what we call the TE-gene pair) were found both up-regulated. Between these co-upregulated TE-gene pairs, 14 are LINEs, 7 are LTR and 1 is SINE. Likewise, 18 TE-gene pairs were down-regulated (10 LINEs, 7 LTR and 1 SINE). We have found only 5 TE-gene pairs (where TEs are 3 LINEs and 2 LTR) which provide opposite expression change between gene and TE.

When those TE-gene pairs were filtered for statistical significance in both TE and gene (both having $FDR < 0.05$ and $|log_2FC| > 1$), only 16 TE-gene pairs can be obtained (Table 1). A total of 4 TE-gene pairs were up-regulated (corresponding to 2 LINEs and 2 LTRs), and 9 TE-gene pairs were down-regulated (including 7 LINEs, 1 LTR and 1 SINE). Finally, 3 TE-pairs (including 1 LINE and 2 LTRs) present opposite differential expression. It is worth mentioning that distance between co-upregulated TE-gene pairs are 10× longer (ranging from 21 to 82 kb) than co-downregulated TE-gene pairs (most ranging from 1.5 to 8.7 kb

Table 1. The 16 differentially expressed TE-gene pairs in LAC including the distance between gene and TE. Repeated instances of the same TE are referring to different chromosome locations of the same TE.

TE[a]	log_2FC_{TE}	P_{TE}	Gene	log_2FC_{gene}	P_{gene}	Distance (nt)[b]
Both having positive Log_2FC (co-upregulation)						
HERVIP10F-int	Inf[c]	5e−05	SERINC2	2.57469	5e−05	21863
HERVIP10F-int	Inf[c]	5e−05	SERINC2	2.57469	5e−05	−31352
L1PA4	5.14037	5e−05	LGSN	5.38051	5e−05	38833
L1PB1	2.89586	2e−04	GPR39	2.11901	0.00025	81934
Both having negative Log_2FC (co-downregulation)						
AluY	−1.76328	0.00035	LONRF3	−0.734273	0.04125	−5586
HERVL-int	−3.49372	5e−05	CRISPLD2	−1.52242	0.00015	1770
L1PA13	−3.06684	0.00025	PTPRC	−1.51911	0.00025	2605
L1PA4	−2.38546	5e−05	IL1RL1	−3.41531	5e−05	8627
L1PA4	−2.69614	5e−04	PCAT19	−2.62722	5e−05	4203
L1PA6	−2.81211	2e−04	MGC27382	−2.72387	5e−05	27991
L1PA7	−2.05145	5e−05	SLC39A8	−2.25943	5e−05	−8690
L1PA7	−2.24649	5e−05	SPARCL1	−2.78843	5e−05	1541
L2	−3.07575	1e−04	ANO2	−2.69836	5e−05	1764
Independent expression of gene and TE						
HERV9NC-int	−5.98136	0.00055	CENPU	2.06716	0.017	−6385
HERV9NC-int	−6.96764	0.00045	CENPU	2.06716	0.017	−4092
L1P1	3.27802	2e−04	AGR3	−0.678936	0.0483	170600

[a]Transposable element.

[b]Distance from the TE to its closest, adjacent gene; negative means that gene is upstream the TE, while positive means that gene is downstream TE.

[c]Inf means that TE is not expressed in normal lung, but it is on LAC cells.

and one at 28 kb). However, when TE-gene pairs follow independent expression, the distances are more variant. This suggests that distances may play a role in TE-gene pair expressions.

3.3 TE-gene Pairs in SCLC

Regarding SCLC, the same approach that with LAC was performed. Hence, 83 TEs were identified as differentially expressed, most of them (67) were down-regulated and the remaining 16 were up-regulated. Up-regulated TEs can produce 14 (2 LINEs, 10 LTRs and 2 unknown) co-upregulated TE-gene pairs, while 47 TEs (30 LTRs, 15 LINEs and 2 SINEs) can form co-downregulated Te-gene pairs. A non-negligible group of 22 TEs (15 LTRs, 7 LINEs) presented TE-gene pairs with opposite expression changes.

When the three groups of TE-gene pairs were again filtered for statistical significance in both gene and TE expression change, the number of differentially expressed TE-gene pairs decreases to 36 (Table 2). A total of 9 TE-gene pairs (including 5 LTRs and 2 LINEs) were co-upregulated and 22 (including 13 LTRs,

Table 2. The 36 differentially expressed TE-gene pairs in SCLC including the distance between gene and TE. Repeated instances of the same TE are referring to different chromosome locations of the same TE.

TE[a]	logFC$_{TE}$	P_{TE}	Gene	logFC$_{gene}$	P_{gene}	Distance (nt)[b]
Both having negative Log_2FC (co-upregulation)						
(CAA)n	3.26754	5e−05	PEG10	3.44881	5e−05	0
(CCG)n	3.25802	1e−04	PEG10	3.44881	5e−05	0
HERVH-int	3.36646	0.00265	RIMS2	1.94245	0.00015	−35548
HERVIP10F-int	1.66963	0.01035	LOC101928595	1.77794	0.02145	0
HERVS71-int	2.21337	0.00375	LINC00665	3.08854	5e−05	−2902
HERVS71-int	2.28517	0.0056	LINC00665	3.08854	5e−05	−3514
HERVS71-int	2.30486	0.00585	LINC00665	3.08854	5e−05	3337
L1P1	2.11549	0.00425	TMEM182	2.41653	0.0024	0
L1PA5	3.57967	0.00155	NCAM1	2.91979	5e−05	2678
Both having negative Log_2FC (co-downregulation)						
AluY	−3.51472	9e−04	LONRF3	−1.87593	1e−04	−5586
HERVE_a-int	−3.38255	0.00085	KLF8	−1.48978	0.00205	−207223
HERVH-int	−Inf[c]	5e−05	NTM	−1.20571	0.02165	−113215
HERVH-int	−6.3376	5e−05	LINC01108	−4.353	5e−05	13889
HERVH-int	−8.58592	5e−05	AKAP7	−1.05849	0.0294	−42761
HERVH-int	−9.0489	5e−05	NTM	−1.20571	0.02165	−112757
HERVI-int	−4.80447	0.0123	ATF7IP2	−1.29798	0.01475	4512
HERVI-int	−5.27054	0.0131	ATF7IP2	−1.29798	0.01475	4051
HERVI-int	−5.64963	5e−05	ATF7IP2	−1.29798	0.01475	1987
HERVK3-int	−4.44042	5e−05	ALDH2	−3.47048	5e−05	6828
HERVK9-int	−1.98071	0.01095	IL32	−1.50026	5e−05	8605
HERVL-int	−2.28606	0.00445	ARHGAP29	−1.39685	0.02475	−5513
HERVL-int	−3.3239	0.01425	ARHGAP29	−1.39685	0.02475	−3899
L1MEd	−Inf[c]	5e−05	IL8	−3.15971	5e−05	2401
L1P1	−7.60885	5e−05	MGC27382	−5.83346	0.00265	−2128
L1PA10	−4.88943	0.0026	ATF7IP2	−1.29798	0.01475	−8169
L1PA3	−2.17051	0.0092	AC007743.1	−1.56602	0.0024	0
L1PA3	−5.76049	5e−05	ATF7IP2	−1.29798	0.01475	3023
L1PA4	−2.75346	0.0143	CYP3A5	−1.87053	0.00515	2633
L1PB2	−6.82712	5e−05	RNF145	−2.12989	5e−05	−5943
L1PB4	−3.21371	0.00415	CD33	−2.70422	5e−05	−4994
LTR10A	−6.36865	5e−05	ATF7IP2	−1.29798	0.01475	−10538
Independent expression of gene and TE						
HERVE-int	−3.57984	0.00055	LOC101928803	1.37193	0.04085	40603
HERVH-int	−3.04437	3e−04	MYO16	2.0961	0.0108	−60669
HERVH-int	−6.27208	0.00775	PAWR	0.916712	0.00765	41707
L1PB1	−Inf[c]	5e−05	FAM133A	2.6097	0.00075	569255
L1PB1	−7.93841	5e−05	FAM133A	2.6097	0.00075	586400

[a]Transposable element.
[b]Distance from the TE to its closest, adjacent gene; 0 means overlap; negative means that gene is upstream the TE, while positive means that gene is downstream TE.
[c]Inf means that TE is not expressed in normal lung, but it is on SCLC cells

8 LINEs and 1 SINE) TE-gene pairs were down-regulated, which is clearly biased to down-regulation, as we have previously found using Biobase [2]. Between these TE-gene pairs, some of them are coincident in chromosome location, which completely justifies their co-expression. It also merits mention that most of these co-regulated TE-gene pairs are within a range of 10 kb and only a few (HERVH-in it co-upregulated and the 5 HERVE and HERVH in co-downregulated), indicanting, once again than the distance between gene and TE may play a role in this co-expression. Finally, 5 TE-gene pairs (including 3 LTRs, 2 LINEs) present opposite, differential expression, but the huge distance (ranging from 41 kb to 586 kb) between the TE and the closest gene makes us think that the opposite expression change might to be considered as independent expression.

3.4 TE-gene Pairs in Both SCLC and LAC

Since LAC and SCLC are both lung cancers, we would like to expect, in addition to cancer-specific TE-gene pairs, other TE-gene pairs that are common to both lung cancers that may provide information about some common molecular basis of the disease. Three TE-gene pairs (AluY, L1PA3 and LTR10A) are in common Table 3 and, interestingly, the three were co-downregulated. That means that the adjacent genes to these TEs (LONRF3, AC007743.1 and ATF7IP2) may play a role in the common functional alterations produced by SCLC and LAC.

Table 3. Common TE-gene pairs with the same co-regulation in both SCLC and LAC.

Gene	TE	Distance (nt)	Status
LONRF3	AluY	−5586	Co-downregulated
AC007743.1	L1PA3	0	Co-downregulated
ATF7IP2	LTR10A	−10538	Co-downregulated

4 Discussion

It is widely assumed that specific TEs could play a key role in lung cancer development by controlling the expression of their adjacent genes. This is true in some cases where TEs have LTRs that are known to act as transcriptional enhancers. However, we suspect that most TEs change their expression as a consequence of the expression of their neighbouring genes, and not the reverse, as we have previously found in prostate cancer [9]. To further test this hypothesis, RNA-seq data from total RNA from tumoral and healthy lung tissues of SCLC and LAC has been analysed using our *NearTrans* pipeline. As a result, 45 and 83 genomic locations of TEs where found differentially expressed in LAC and SCLC, respectively. When the differential expression was significant for both the TE and the closest, adjacent gene, the list was reduced to 16 and 36 differentially expressed TE-gene pairs in LAC (Table 1) and SCLC (Table 2), respectively.

Several findings were striking with the two datasets. First, most TE-gene pains were formed by TEs of HERV and LTR classes, which is consistent with what we have previously observed using the transposon repeats in Biobase [2]. Second, the expression changes are not balanced for TEs, the majority of the differentially expressed TEs being down-regulated. This finding made us think that the need of tumour-suppressor gene repression implicates more genome loci than the tumoural gene expression. TE down-regulation is not so widely described in tumours as TE up-regulation, but we must remark that most studies about TE and cancer use to study one or a few specific TEs, while here it is described the overall behaviour of TEs. And third, most co-regulated TE-gene pairs were located within 5 kb, while TE and gene presenting independent expression use to be more far apart. This made us think co-regulated TE-gene pairs are in the same regulatory region of the genome, among not yet identified insulators.

The importance of the TE-gene pairs of Tables 1 and 2 is supported by the nature of the genes. For example, *SERINC2* (Table 1) is flanked by two copies of the HERVIP10F-int, and the three are co-upregulated. *SERINC2* is a member of a transmembrane protein family that incorporate serine into membrane lipids during synthesis. Recent studies have demonstrated that expression levels of *SERINC2* are significantly up-regulated in tumours with respect to healthy tissues in patients with lung adenocarcinoma, providing them proliferation, migration and invasion capabilities [16]. *GPR39* (Table 1) is another example, since it is frequently over-expressed in primary esophageal squamous cell carcinoma, which has been significantly associated with the lymph node metastasis. Functional studies showed that *GPR39* has a strong tumorigenic ability [14]. There are also cases of functional consistency in the co-downregulated TE-gene pais, with special attention to several oncogenes, including *PTPRC* [4]. But even more important, nothing is described for *LGSN* (Table 1) and cancer, which may reveal new or neglected molecular basis for LAC.

Regarding SCLC, *PEG10*, an oncogene implicated in the proliferation, apoptosis and metastasis of tumors [15], is co-upregulated with repetitive elements in its locus. Also, *PEG10* has been found to be positively expressed in a variety of cancers with seemingly complex expression regulation mechanisms.

Regarding the TE-gene pairs in Table 3, TEs AluY, L1PA3 and LTR10A could be suggested as positive marker of lung cancer. A few thing are known about the genes in these TE-gene pairs as revealed by GeneCards (https:// www.genecards.org): *LONRF3* can be involved in protein-protein and protein-DNA interactions but with unknown function), AC007743.1 (uncharacterised protein) or ATF7IP2 (a transcription factor involved in chromatine formation and methylation; related with macular degeneration). The three are expressed, between other tissues, in normal lung, so it would merit the effort to know why they are repressed in LAC and SCLC, since nothing is known about their relation with cancer. Moreover, this supports the utility of a bioinformatic pipeline such as *NearTrans* in the discovery of new biological information.

All those results confirm that *NearTrans* seem to be a suitable and useful tool for correlation studies of TEs and their adjacent genes. It was essential to reveal that a normal lung cell compromised on an oncogenic change displays some TE expression reprogramming, highly biased to down-regulation. And even more important, that the expression change of the adjacent gene had the same sign in most cases. This reprogramming revealed to be genome-location dependent as most differentially expressed TEs change their expression in the same direction that their adjacent genes, and with a similar level of change. The analysis of TE-gene pairs may reveal or confirm important lung cancer biomarkers as well as new insights in its molecular basis. We propose that the study of TEs in cancer could help in the discovery or corroboration of genes involved in cancer, and could be used as specific biomarkers for the diagnosis, prognosis or treatment of cancer.

Acknowledgements. This work was funded by the NeumoSur grants 12/2015 and 14/2016. The authors also thankfully acknowledge the computer resources and the technical support provided by the Plataforma Andaluza de Bioinformatica of the University of Malaga.

References

1. Arroyo, M., Bautista, R., Larrosa, R., de la Cruz, J.L., Cobo, M.A., Claros, M.G.: Potencial uso biomarcador de los retrotransposones en el adenocarcinoma de pulmón. Revista Española de Patología Torácica **30**(4), 224–230 (2018)
2. Arroyo, M., Larrosa, R., Bautista, R., Claros, M.G.: Specifically reprogrammed transposons AluYg6, LTR18B, HERVK11D-Int and UCON88 as potential biomarkers in lung cancer. Submitted
3. Dobin, A., et al.: STAR: ultrafast universal RNA-seq aligner. Bioinformatics **29**(1), 15–21 (2013)
4. Du, Y., Grandis, J.R.: Receptor-type protein tyrosine phosphatases in cancer. Chin. J. Cancer **34**(2), 61–69 (2015)
5. Falgueras, J., Lara, A.J., Fernandez-Pozo, N., Canton, F.R., Perez-Trabado, G., Claros, M.G.: SeqTrim: a high-throughput pipeline for preprocessing any type of sequence reads. BMC Bioinform. **11**(1), 38 (2010)
6. Ferlay, J., et al.: Cancer incidence and mortality worldwide: sources, methods and major patterns in GLOBOCAN 2012. Int. J. Cancer **136**(5), E359–E386 (2015)
7. Ghosh, S., Chan, C.K.K.: Analysis of RNA-Seq data using TopHat and Cufflinks. Methods Mol. Biol. **1374**, 339–361 (2016)
8. Kassiotis, G.: Endogenous retroviruses and the development of cancer. J. Immunol. **192**(4), 1343–1349 (2014). (Baltimore, Md.: 1950)
9. Larrosa, R., Arroyo, M., Bautista, R., López-Rodríguez, C.M., Claros, M.G.: NearTrans can identify correlated expression changes between retrotransposons and surrounding genes in human cancer. In: Rojas, I., Ortuño, F. (eds.) IWBBIO 2018. LNCS, vol. 10813, pp. 373–382. Springer, Cham (2018). https://doi.org/10.1007/978-3-319-78723-7_32

10. Nakagawa, S., Takahashi, M.U.: gEVE: a genome-based endogenous viral element database provides comprehensive viral protein-coding sequences in mammalian genomes. Database **2016**, baw087 (2016)
11. Quinlan, A.R., Hall, I.M.: BEDTools: a flexible suite of utilities for comparing genomic features. Bioinformatics **26**(6), 841–842 (2010)
12. Rudin, C.M., et al.: Comprehensive genomic analysis identifies SOX2 as a frequently amplified gene in small-cell lung cancer. Nat. Genet. **44**(10), 1111–1116 (2012)
13. Trapnell, C., et al.: Differential gene and transcript expression analysis of RNA-seq experiments with TopHat and Cufflinks. Nat. Protoc. **7**(3), 562–578 (2012)
14. Xie, F., et al.: Overexpression of GPR39 contributes to malignant development of human esophageal squamous cell carcinoma. BMC Cancer **11**(1), 86 (2011)
15. Xie, T., et al.: PEG10 as an oncogene: expression regulatory mechanisms and role in tumor progression. Cancer Cell Int. **18**(1), 112 (2018)
16. Zeng, Y., et al.: SERINC2-knockdown inhibits proliferation, migration and invasion in lung adenocarcinoma. Oncol. Lett. **16**(5), 5916–5922 (2018)

Signal Processing Based CNV Detection in Bacterial Genomes

Robin Jugas[✉], Martin Vitek, Denisa Maderankova, and Helena Skutkova

Brno University of Technology, Technicka 10, 61600 Brno, Czech Republic
jugas@feec.vutbr.cz

Abstract. Copy number variation (CNV) plays important role in drug resistance in bacterial genomes. It is one of the prevalent forms of structural variations which leads to duplications or deletions of regions with varying size across the genome. So far, most studies were concerned with CNV in eukaryotic, mainly human, genomes. The traditional laboratory methods as microarray genome hybridization or genotyping methods are losing its effectiveness with the omnipotent increase of fully sequenced genomes. Methods for CNV detection are predominantly targeted at eukaryotic sequencing data and only a few of tools is available for CNV detection in prokaryotic genomes. In this paper, we propose the CNV detection algorithm derived from state-of-the-art methods for peaks detection in the signal processing domain. The modified method of GC normalization with higher resolution is also presented for the needs of the CNV detection. The performance of the algorithms are discussed and analyzed.

Keywords: CNV · Copy number variant · Bacterial genomes · Signal processing · Sequencing

1 Introduction

Copy number variations, which belongs to structural variations, are parts of a genome and with size bigger than 1000 bases and less than 5 megabases. That is widely accepted size range for eukaryotic genomes [15]. However, much smaller genome parts bigger than 50 bases can fit the CNV definition too [1]. These genome parts are either increased or decreased in occurrence when compared with reference. The process of creation of any structural variation needs several steps involved: double-strand breakage at at least two locations, re-ligation of the broken ends and producing new chromosomal arrangement [11].

Copy number variants are changes of the observation frequency of certain DNA sequence, most easily manifested as deletions or duplications (tandem or interspersed). Inversions, insertions or trans-locations are not considered as CNV [11].

Several types of CNV can be described in eukaryotic genomes, but those can not be taken into account when considering prokaryotic genomes with single

© Springer Nature Switzerland AG 2019
I. Rojas et al. (Eds.): IWBBIO 2019, LNBI 11465, pp. 93–102, 2019.
https://doi.org/10.1007/978-3-030-17938-0_9

circular chromosome. Also, the symmetrical organization of prokaryotic genomes leads to also symmetrical rearrangements. Properties involved are theoretically distance of a gene from origin of replication (oriC), difference in replication in leading and lagging strand and constraint of keeping both replichores at the same size, which leads to symmetrical inversions at oriC and terminus of replication (ter) [11]. Large rearrangements, which disrupt the symmetry, have negative impact on prokaryotic organism fitness [11, 12].

Laboratory methods traditionally used for CNV detection are fluorescence in situ hybridization (FISH) and array comparative genomic hybridization (aCGH), but both suffer from low resolution, thus they are unable to detect short CNVs [5]. Algorithm methods consists of several approaches and their combinations - read-pair approach, read-depth approach, split-read approach, de-novo sequence assembly methods and hybrid approaches [10, 11, 13].

Read-pair approach takes the distance and orientation of paired-end reads and cluster the pairs in whose either the distance or orientation doesn't match the reference genome [1]. Several features can be implied, such as deletion exhibited by long distance, insertions exhibited in the opposite way as too close, whereas inconsistent orientation can be manifest of insertion or tandem duplication. Read-pair method is the most used among the other [10, 11, 13].

Read-depth approach is based on the graph of read depth aligned to the reference genome. It typically assumes either Poisson or another random distribution and investigates the divergence from this distribution. Underlying premise is that duplicated regions will manifest higher read depth, whereas the deletions will manifest in lower read depth [10, 11, 13].

Split-read approach is based on broken alignment to the reference genome. The local gapped alignment is performed and gaps are evaluated. A continuous extension of gaps in the read indicate a deletion, oppositely, extension of gaps in the reference indicates an insertion. If the reads are long (longer than MGE), mobile genomic elements can also be discovered [10, 11, 13].

Sequence assembly approach contrary to the other ones does not perform alignment of the reads to the reference genome, hence it is the challenging one. Instead, the fully sequenced genome is assembled and the output is compared with already known high quality reference [10, 11, 13].

None of the approaches covers the whole spectrum of the task. When different tools and approaches are used, the outputs differs and some are unique to specific approach [1]. Read depth is the only methods effective at detecting absolute copy numbers, its low resolution for breakpoints detection is disadvantage. Read-pair approaches are computationally demanding by the process of resolving ambiguous mapping to the reference genome. Split-reads are accurate only in unique regions of the genome. Sequence assembly requires much higher read depth, which is costly, and it is efficient in pairwise comparisons. However, it is biased against the repeats and duplications created in the process of de-novo assembly [1].

2 Methods

2.1 GC Normalization

PCR-based sequencing methods are inducing GC bias [4]. Several models of underlying mechanism were described - fragmentation model (GC counts are tied with the stability of the sequence), read model (GC counts modify the base sequencing), full-fragment models (GC count of the whole fragment determines the amplification of fragments) and global models (consider GC counts at the genome level). GC bias then can be described as proportion between the GC counts in a region and number of fragments mapped to the region [2].

For further understanding, we defined read depth as number of bases aligned to nucleotide in the reference genome.

The GC bias influences following analysis based on the read depth, i.e. copy number detection. Thus, we developed the GC count normalization algorithm suitable for our CNV detection algorithm. GC normalization is derived from the already in use methods based on computing GC count in bin of variable length. Instead of them, the GC counts are not obtained from static regions of sequence, but in sliding overlapping windows. Also, the GC normalization is not applied to the static regions, but for each nucleotide in the sequence. The only drawback is higher computational demand, but the higher resolution is achieved.

The coverage file, as computed by SAMtools (*depth* command) [9], together with consensual genome sequence computed from aligned reads, serve as the input files of the GC normalization. The GC counts are computed in a sliding overlapping window of variable length (70 bp). The property of overlapping window creates the GC count for every base in the consensual sequence with influence of neighboring regions. Only the bases belonging to regions with read depth higher than 5 are calculated because of their noisiness property. Then, the median of the read depth corresponding with certain level of GC count are obtained. Normalization of the GC count is performed according to the modified Eq. 1 from [3]:

$$RC_i^{corr} = RC_i \times mRC/mRC_{GCi}. \qquad (1)$$

Each value of read depth is corrected by weighting its value with division of the overall median of read depth and the median of read depth with the same GC count. The RC_i denotes i-th value of read coverage (RC), mRC is median of the whole read depth, mRC_{GCi} is median value of read depth values with corresponding GC count. Finally, C_i^{corr} is corrected i-th value of read depth with define GC count. The effect of GC normalization is in Fig. 1.

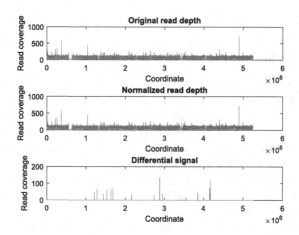

Fig. 1. Effects of GC normalization on read depth signal

2.2   CNV Detection

The coverage data which are GC normalized serves as the input of the CNV detection algorithm. Firstly, the read depth graph taken as signal needs to be smoothed. The average moving windows was implemented for the smoothing operation according to the formula 2, where x are data values, n is size of window:

$$\bar{x}_{MA} = \frac{x_M + x_{M-1} + \cdots + x_{M-(n-1)}}{n} = \frac{1}{n}\sum_{i=0}^{n-1} x_{M-i}. \qquad (2)$$

The smoothing ensures higher robustness of peak detection algorithm implemented for the CNV detection. Default size of moving average window is the same as the GC normalization bin (70 bp). Signal processing based peaks detection is then applied to detect the position and width of the peaks detected. The basic thresholds are applied, based on previous analysis, it was decided that the detected peaks should be higher than double of the average of the read depth along the whole signal and the peaks should be longer than at least 10 bp.

2.3   Dataset

Types of CNVs were artificially simulated in order to test the accuracy and abilities of algorithms. The deletion, tandem duplication and interspersed duplication types of CNVs were simulated. As the source, the already assembled and well-known genome of *Klebsiella pneumoniae* (strain NTUH-K2044, NCBI Ref. NC_012731.1) was used. *Klebsiella pneumoniae* was chosen as an attractive organism from the perspective of drug resistance in many studies [6,14].

The original genome sequence was then modified to simulate the CNV. The gene of suitable length was found in GenBank notation: the gene LysR (family transcriptional regulator; 2450591:2451499, length of 908 bp). This gene was either deleted, or duplicated in a tandem or interspersed way.

Genome sequences in FASTA file generated by this process served as an input for ART next-generation sequencing read simulator [7]. ART package (version 2.5.8) serves as an artificial reads simulator, an have been already used for testing CNV detection [3]. The parameters for ART was specified as follows: paired-end read simulation of HiSeq 2500, length of reads 76, fold of read depth 100, the mean size of DNA/RNA fragments 500, the standard deviation of DNA/RNA fragment size 100. The output FASTQ files were mapped against the original genome sequence of Klebsiella using Burrows-Wheeler aligner BWA [8]. The following files necessary for CNV detection were acquired as mentioned in the methods using SAMtools package. The whole analysis was performed on desktop computer (Core i5-6500, 24 GB RAM) using Ubuntu Linux distribution. The simulated CNV details are in Table 1.

Table 1. The positions of simulated CNVs in artificial genomes

Type of CNV	#	CNV position
Deletion of region	0	2,450,591:2,451,499
Tandem duplication	2	2,450,591:2,452,407
Interspersed duplication	2	2,450,591:2,451,499; 2,454,499:2,455,407
Interspersed duplication	3	2,450,591:2,451,499; 2,454,499:2,455,407; 2,948,499:2,949,407

3 Results and Discussion

The artificial data were created and the implemented algorithm for CNV detection based on signal processing was used to asses the algorithm accuracy and abilities. Because of the foreknowledge about induced CNV, the tool can be assess. The comparison with other state-of-the-art tools was not performed as proposed algorithm is still in the development. The aim of the paper is to demonstrate preliminary abilities and possible drawbacks of the design.

The default read depth signal is in Fig. 2. There are drops in read depth at the beginning and at the end of reference genome coordinates. This is caused by mapping algorithm and can be omitted from further analysis. Except for the drops, there are no visible significant peaks. The average read depth along coordinate is 99.9989 as defined in ART sequencing simulator (depth 100).

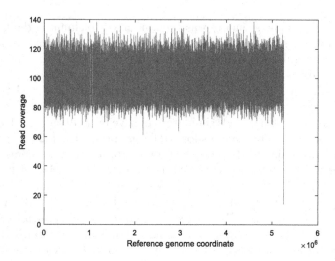

Fig. 2. The read depth signal of reference genome of *Klebsiella pneumoniae* with no CNV introduced

The first demonstrated CNV type is deletion in Fig. 3. The sequence of a gene was deleted from template reference sequence and the process of sequencing was simulated followed by the analysis. There is significant negative peak on coordinates 2,449,683:2,451,573 with a length of 1890. The ending border of the peak is only 73 bp after the real end, however, the start of the peak is detected 908 bp before the CNV region of deletion. It was necessary, to use the negative values of signal, so the peak detection algorithm could detect positive peaks. The negative signal together with detected peak is in Fig. 4.

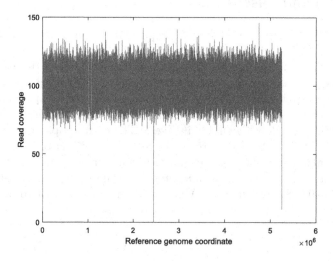

Fig. 3. The read depth signal of deletion CNV

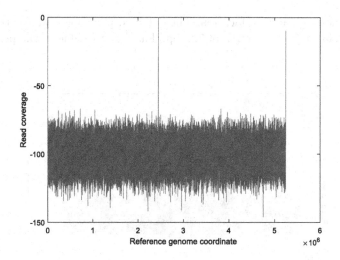

Fig. 4. The read depth signal of deletion CNV - negative signal

CNV type of tandem repetition, the sequence was copied and putted exactly after the original gene position. No gaps between these two were inserted. The presumption is, that peak should be twice as high compared to average depth as twice as reads should be theoretically sequenced from the region. The read depth with detected peak is in Fig. 5. The detected peak is at position 2,450,631:2,451,413 with length of 782 bp. The highest value of the peak is 212.

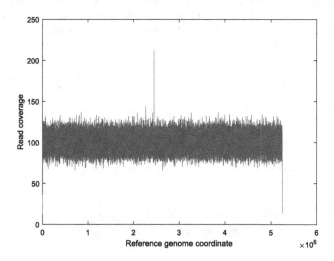

Fig. 5. The read depth signal of tandem CNV

Following type of CNV tested is interspersed duplication. These are randomly place across the genome. The first type is duplicated once and placed closely to the original gene location. The detected peak spans across the region 2,450,741:2,451,469 with length of 728 bp. The highest value of the peak is 216 (Fig. 6).

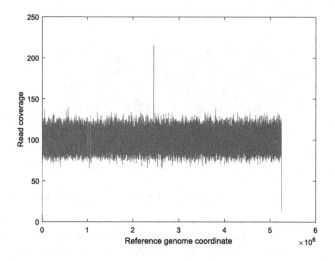

Fig. 6. The read depth signal of interspersed duplication (2x) CNV

The interspersed duplication placed three times across the genome is the last type of induced CNV. Another duplication of gene sequence was added to the previous two ones at the position 0.5 million bp after them. The detected peak spans across the region 2,450,645:2,451,450 with length of 805 bp. The highest value of the peak is 330 (Fig. 7).

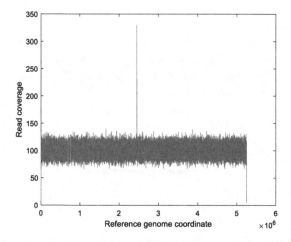

Fig. 7. The read depth signal of interspersed duplication (3x) CNV

Some of the theoretical assumptions were proven during analysis. The copy number count can be denoted from the size of the peaks. The number of times the peak is higher than the average of the read depth, indicates the number of copy number. The general drawback of CNV detection derived from reference sequence is inability to obtain either the position or genetic sequence of CNVs in the sequenced genome. This can be done by employing further laboratory methods, i.e. targeted sequencing or hybridization or using genome assembly method. The further work will be aimed on achieving higher accuracy of the CNV borders. This is challenging task, as the read depth signal has very noisy character and establishing the strict borders of peak, so that it would match the real border of CNV, is algorithmically demanding. The main advantage of the proposed algorithm are low computational demands with accurate indication of copy number variations.

The computational demands are following: the CNV detection algorithm performs the task in average 15 s for the genome of length over 5 million bp. The GC normalization is more computational demanding since of sliding window computing GC counts and HDD operations. The overall computation time of the whole process (including loading the input files) takes in average 350 s. The algorithm was implemented in the Matlab environment.

Acknowledgments. This work was supported by grant project GACR 17-01821S.

References

1. Alkan, C., Coe, B.P., Eichler, E.E.: Genome structural variation discovery and genotyping. Nat. Rev. Genet. **12**(5), 363–376 (2011)
2. Benjamini, Y., Speed, T.P.: Summarizing and correcting the GC content bias in high-throughput sequencing. Nucleic Acids Res. **40**(10), 1–14 (2012)
3. Brynildsrud, O., Snipen, L.G., Bohlin, J.: CNOGpro: detection and quantification of CNVs in prokaryotic whole-genome sequencing data. Bioinformatics **31**(11), 1708–1715 (2015)
4. Dohm, J.C., Lottaz, C., Borodina, T., Himmelbauer, H.: Substantial biases in ultra-short read data sets from high-throughput DNA sequencing. Nucleic Acids Res. **36**(16), e105 (2008)
5. Duan, J., Zhang, J.-G., Deng, H.-W., Wang, Y.-P.: Comparative studies of copy number variation detection methods for next-generation sequencing technologies. PLoS One **8**(3), e59128 (2013)
6. Holt, K.E., et al.: Genomic analysis of diversity, population structure, virulence, and antimicrobial resistance in *Klebsiella pneumoniae* an urgent threat to public health. Proc. Natl. Acad. Sci. **112**(27), E3574–E3581 (2015)
7. Huang, W., Li, L., Myers, J.R., Marth, G.T.: ART: a next-generation sequencing read simulator. Bioinformatics **28**(4), 593–594 (2012)
8. Li, H., Durbin, R.: Fast and accurate long-read alignment with Burrows-Wheeler transform. Bioinformatics **26**(5), 589–595 (2010)
9. Li, H., et al.: The Sequence Alignment/Map format and SAMtools. Bioinformatics **25**(16), 2078–2079 (2009)

10. Medvedev, P., Stanciu, M., Brudno, M.: Computational methods for discovering structural variation with next-generation sequencing. Nat. Methods **6**(11S), S13 (2009)
11. Periwal, V., Scaria, V.: Insights into structural variations and genome rearrangements in prokaryotic genomes. Bioinformatics **31**(1), 1–9 (2015)
12. Rocha, E.P.: The organization of the bacterial genome. Annu. Rev. Genet. **42**(1), 211–233 (2008)
13. Treangen, T.J., Salzberg, S.L.: Repetitive DNA and next-generation sequencing: computational challenges and solutions. Nat. Rev. Genet. **13**(1), 36–46 (2012)
14. Wyres, K.L., Holt, K.E.: Klebsiella pneumoniae as a key trafficker of drug resistance genes from environmental to clinically important bacteria. Curr. Opin. Microbiol. **45**, 131–139 (2018)
15. Zhao, M., Wang, Q., Wang, Q., Jia, P., Zhao, Z.: Computational tools for copy number variation (CNV) detection using next-generation sequencing data: features and perspectives. BMC Bioinform. **14**(Suppl. 1(Suppl. 11)), S1 (2013)

Omics Data Acquisition, Processing, and Analysis

Dependency Model for Visible Aquaphotomics

Vladyslav Bozhynov$^{(\boxtimes)}$, Pavel Soucek, Antonin Barta, Pavla Urbanova, and Dinara Bekkozhayeva

Laboratory of Signal and Image Processing, Faculty of Fisheries
and Protection of Waters, South Bohemia Research Center of Aquaculture
and Biodiversity of Hydrocenoses, Institute of Complex Systems,
University of South Bohemia in České Budějovice,
Zámek 136, 373 33 Nové Hrady, Czech Republic
vbozhynov@frov.jcu.cz
http://www.frov.jcu.cz/en/institute-complex-systems/
lab-signal-image-processing

Abstract. The main idea of this research is the extension of the aquaphotomics method to the visible range of the spectrum. Already known as a fact that each chemical element has a unique pattern in the absorption of electromagnetic radiation. Such a structure is a spectrum bands absorbed by an element and is called its 'fingerprint'. The fingerprint section is presented in a wide range of spectrum, including the visible part. Absorption in the visible spectrum provides unique information about the elements or compounds present in water. This allows to analyze the concentration of microparticles and chemical elements in water due to changes in the molecular water system, presented in the form of a spectral picture of water. The results presented in this paper prove the existence of a correlation between some parameters of water and its spectral characteristics.

Keywords: Spectrum · Aquaphotomics · Spectrophotometry ·
Biomonitoring · Aquaculture · Measurement · Nutrients

1 Introduction

Water is one of the most familiar substances on the Earth. Currently, about 70% of our planet is covered with water. It has been studied using various tools and methods, but its behaviour is still the subject of intensive scientific research. On micro- and nano levels water is not a homogeneous structure, but rather dynamic equilibrium among changing percentages of assemblages of different oligomers and polymers species. The structure and these assemblages or units themselves are dependent on its chemical contents, temperature and pressure [1].

Nowadays there are many different methods for examination the chemical composition, the concentration of micro-particles and the micro-organism, and

© Springer Nature Switzerland AG 2019
I. Rojas et al. (Eds.): IWBBIO 2019, LNBI 11465, pp. 105–115, 2019.
https://doi.org/10.1007/978-3-030-17938-0_10

other characteristics of water. An important place among them is occupied by the methods of spectrophotometry. One of the modern methods of spectrophotometry is Aquaphotomics. Aquaphotomics (aqua - water; photo - light; omics - all about) is a new discipline introduced by Prof. Roumiana Tsenkova from the Kobe University, Japan [2]. Method is based on the knowledge, that water, as a natural biological matrix containing only small molecules with a strong potential for hydrogen bonding, changes its absorbance pattern every time it adapts to physical or chemical change in the environment. In Aquaphotomics near infrared (NIR) light is used to obtain the information about the hydrogen bonding in the water [3,4]. NIR light allows a penetration to a depth of 10 mm in water and even deeper for the short wavelength region (750–1098 nm). Therefore, every absorbance spectrum of water solutions or biological systems contains information at the molecular level with 'one hydrogen bonding' resolution. Aquaphotomics has been successfully applied in various fields from water characterization, food quality control to early diagnosis of disease [5,6].

Studies of the spectral characteristics of chemical elements showed that each element has its own pattern in absorbing of electromagnetic radiation. This pattern is called 'fingerprint' of the element. Each chemical element, or compound, has its own fingerprints in a wide range of the spectrum, including the visible part. Generally, water only reflects in the visible light range. Based on this, it is assumed that the chemical composition of water can be investigated by analyzing the reflected spectrum of sample in visible range.

Water spectral changes permit measurement of small quantities of or structural changes in molecules of the system. Absorption pattern of water sample depend not only on the chemical composition, but also on the physical parameters. In order to obtain more information about the values of the parameters, it is necessary to analyze whole range of the visible spectrum. This article discusses the dependence of the spectral characteristics of water on the values of some of the basic water parameters. The dependence was tested on such parameters: temperature (T), electrical conductivity (EC), pH.

To characterize the water, it is necessary to find the dependence of the spectral characteristics with the parameters of interest. For these purposes, multivariate analysis (MVA) is used. Multivariate analysis is a set of techniques used for analysis of data sets that contain more than one variable, and the techniques are especially valuable when working with correlated variables. The techniques provide an empirical method for information extraction, regression, or classification [7]. Multivariate analysis, due to the size and complexity of the underlying data sets, requires much computational effort. With the continued and dramatic growth of computational power, multivariate methodology plays an increasingly important role in data analysis, and multivariate techniques, once solely in the realm of theory, are now finding value in application [8].

Regression analysis is one of the most common methods used in statistical data analysis. It estimates the relationship between two sets of values: the predicted (independent) and the actual (dependent). Regression analysis helps in understanding how the typical value of the dependent variable changes when any one of the independent variables varies.

2   Correlation Between Spectral Characteristics and Values of Parameters

During the first half of the 19th century, scientists such as John Herschel, For Talbot, and William Swan studied the spectra of different chemical elements in flames. Since then, the idea that each element produces a set of characteristic emission lines has become well-established. Each element has several prominent, and many lesser, emission lines in a characteristic pattern. Today spectrophotometry methods are widely used in the various scientific fields, such as physics, materials science, chemistry, biochemistry, and molecular biology.

Spectral absorption is a measurement of the amount of light absorbed by a water sample of a given path length. There are several different types of matter, such as organics and nitrates that naturally absorb light in different regions of the spectrum. Using absorbance data in combination with laboratory data for a target parameter, it is possible to build a relationship between these two data sets. This type of relationship makes possible to convert the measured absorbance data into value of the parameter of interest [9].

The absorption of light is due to the interaction of light with the electronic and vibrational modes of molecules. Each type of molecule has an individual set of energy levels associated with the makeup of its chemical bonds and nuclei, and thus will absorb light of specific wavelengths, or energies, resulting in unique spectral properties [10]. Based on the fact that water molecules change their absorption pattern each time it adapts to a physical or chemical change, we can conclude that changes in some physical parameters of water can be also estimated using spectral measurements.

In our work, a ColorMunki spectrophotometer (Fig. 1) was used for spectral measurements, which measures the spectrum in the range from 360 to 740 nm.

Fig. 1. Spectrophotometer ColorMunki

The available spectrophotometer from X-Rite company has a precision (min-imal step) in 10 nm. Thus, a spectrum consisting of 36 bars (values) (Fig. 2) is saved for each measurement.

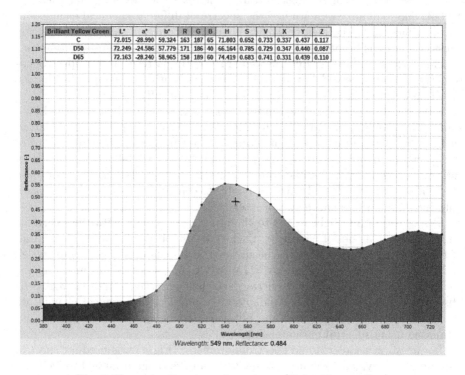

Brilliant Yellow Green	L*	a*	b*	R	G	B	H	S	V	X	Y	Z
C	72.015	-28.990	59.324	163	187	65	71.803	0.652	0.733	0.337	0.437	0.117
D50	72.249	-24.586	57.779	171	186	40	66.164	0.785	0.729	0.347	0.440	0.087
D65	72.163	-28.240	58.965	158	189	60	74.419	0.683	0.741	0.331	0.439	0.110

Wavelength: **549 nm**, Reflectance: **0.484**

Fig. 2. Example of measured spectrum (Color figure online)

Such precision is not sufficient to identify small changes in concentration of a compound or the value of a physical parameter in water. Required patterns differ in much smaller steps. To increase the amount of information obtained from spectral measurements, we measured the spectrum of different color etalons. For this purpose, we used the ColorChecker plate (Fig. 3). It includes 24 colors with known RGB values and serves for the calibration of optical instruments. Only 18 etalons were used in experiments without different shades between black and white colors.

Depending on the value of the parameter, the spectral absorption of water will be different. This means that by changing the value of one of the tested parameters, you will get a radically different spectrum. To find the relationship between the spectral characteristic of a water sample and the parameter value, the differences of 18 spectra were compared with the measured parameter value. This gives information on how the spectrum changes in the entire visible range with changing parameter values.

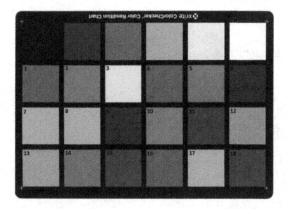

Fig. 3. ColorChecker Classic card (X-Rite)

3   Experiment Setup

The experiment for data collection consisted of two types of measurements: experimental and control. The first was to measure the spectral characteristics of water samples. For these purposes, a spectrophotometer was used. Control measurements were carried out parallel to the experimental for each sample. They included measurements of such parameters: pH, electrical conductivity (EC) and temperature. During the experiment, the parameters varied in such ranges: pH from 2.56 to 10.5; EC from almost 0 to 6.51 mS/cm; temperature from 2 to 50.5 °C.

For the experiments, distilled water (as a zero etalon), pipe water, and water from the aquaponic system were used. Samples were prepared for different types of water, where only one parameter was changed, while the changes of remaining parameters were close to 0. To minimize the error created by the user or device for the same pH, EC, T values, 3 repetitions were performed.

The obtained data were divided into two sets readings of parameters (pH, EC, temperature) and the results of reading spectra (Fig. 4).

For each color etalon, the spectrophotometer provides 36 values for different wavelengths. Having 18 etalons for each water sample, 36 * 18 = 648 values were obtained. And 91 combinations with different values of parameters for each type of water. Total, the data consists of 176 904 values. All data were used for multivariate analysis to confirm the existence of a relationship between the values of the parameters and the spectrum in the visible range.

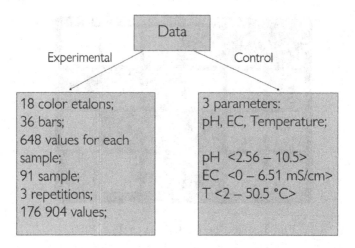

Fig. 4. Experimental data

4   Multivariate Data Analysis

To determine the value of a parameter by the method of spectrophotometry for each parameter, the absorption of one particular wavelength is usually used. Since for our study we use the visible range of the spectrum, to determine the value of the parameter we need to analyze the absorption in the whole range of the visible spectrum. This means that, in our case, univariate analysis is not suitable. For data processing and creating a model that is able to predict the value of a parameter from spectral measurements, we used multivariate analysis.

To understand how the value of the spectral characteristics of water changes when any of the independent variables changes, a regression analysis was used. Regression analysis is a form of predictive modelling technique which investigates the relationship between a dependent (target) and independent (predictor) variables. There are various kinds of regression techniques available to make predictions. These techniques are mostly driven by three metrics: number of independent variables, type of dependent variables and shape of regression line. Within multiple types of regression models, it is important to choose the best suited technique based on type of independent and dependent variables, dimensionality in the data and other essential characteristics of the data [11,12].

In our case, the predictors are the measurements that comprise the spectrum; they number in the hundreds. The targets are the parameter values that we want to predict in future samples. For this reason, we need a method which is the most suitable to predict a set of dependent variables from a large set of independent variables. From all possible regression analysis methods, the partial least square (PLS) regression was choised.

PLS was developed in the 1966 by the Swedish statistician Herman Wold as an econometric technique. Today, PLS regression most widely used in chemometrics and related areas. It is also used in bioinformatics, sensometrics, neuroscience and anthropology [13].

The purpose of the PLS regression is to create a model capable to predict Y (the values of T, EC, pH) from X (the results of spectrum measurements). When Y is a vector and X is full rank, this goal could be accomplished using ordinary multiple regression. When the number of predictors is large compared to the number of observations, X is likely to be singular and the regression approach is no longer feasible (i.e., because of multicollinearity) [14].

The PLS model combines multiple regression and principal component analysis (PCA). However, unlike PCA regression, where the latent factors maximize the variance of the covariates, PLS regression searches for a set of components that performs a simultaneous decomposition of X and Y with the constraint that these components explain as much as possible of the covariance between X and Y. Such set of components is called latent vectors [15].

In this work, a normal PLS regression model was created. As in multiple linear regression, PLS regression is based on the linear model:

$$Y = X\beta + \epsilon \tag{1}$$

where, Y - vector of responses (the values of T, EC, pH);
X - matrix of covariates (the values of spectra);
β - vector of regression coefficients;
ϵ - vector of model errors.

5   Results

During preprocessing, we calculated the medians between three repetitions for each sample (Fig. 5). This was done to avoid operator error.

Before using MVA data were standardized. The PLS regression method, which was used for data analysis, rebuilds our data into a multidimensional space consisting of the principal components.

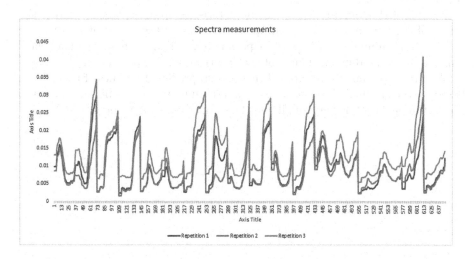

Fig. 5. The difference in spectra measurements for the water sample with the same attributes between three repetitions

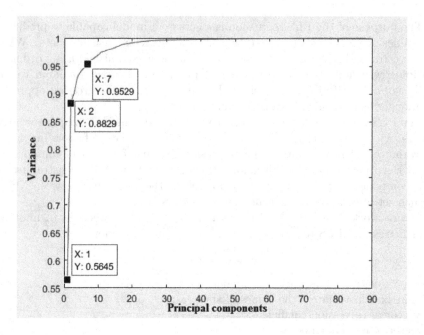

Fig. 6. Graph of data variance versus principal components. The value X indicates the number of principal components, the value Y indicates how many of all possible events describes a set of principal components.

As can be seen from the Fig. 6, the one-dimensional space (consisting of 1 principal component) describes 56% of the data variance, two-dimensional - 88%. To reach the 95%, it is enough to use 7 principal components. This means that to create a dependency model, it will be sufficient to use only 7 principal components out of 88.

Not all of the measured spectral characteristics are necessary to predict the values of the parameters. To speed up the response of the model, it is necessary to leave a minimum set of important parameters. Figure 7 shows the load (how important) of the different bands of the spectrum among all measured.

As a result, a model was created that should predict the values of parameters based on the spectral characteristics of water samples. As can be seen from the Fig. 8, the deviation is very small and does not exceed the value $4*10^{-8}$.

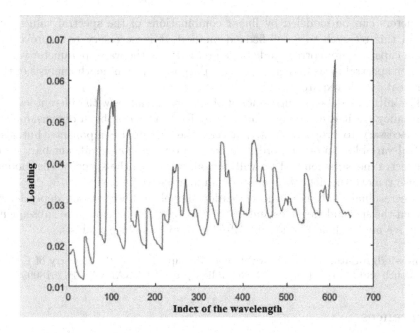

Fig. 7. Loadings of the spectrum bars. The X axis shows the indices of spectral measurements from 1 to 648 (36 bars for each measured spectrum * 18 color etalons). The Y axis represents the importance (load) of these values.

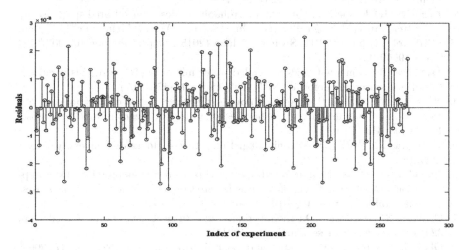

Fig. 8. Residuals of the results predicted by the regression model from the real

6   Conclusion and Discussion

At this stage, was created a regression model with excellent training results. Fitting showed that dependence of the spectral characteristics of water on tested parameters can be modeled by linear combinations of the spectral values. To expand this research to a full-fledged method, it is necessary to improve the system capability for correct and stable prediction of the water parameter values based on spectral measurements. For this purpose, more in-depth analysis of the statistical data is required.

The initial model uses huge amount of spectral variables with different level of significancy (as it is clear from the Fig. 7). To decrease problem dimensionality it is necessary to properly chose not only the principal components, but also spectral variables. In other words, we need to remove non-significant bars from the spectra measurements. This will not only increase the speed of the model, but also remove the distorting effect of uninformative values.

The residual analysis of the regression model promises strong dependency between the spectral variables and water parameters. Therefore, the subsequent predictive model should be feasible after intensive loadings analysis.

Acknowledgments. The study was financially supported by the Ministry of Education, Youth and Sports of the Czech Republic - project CENAKVA (LM2018099).

References

1. Muncan, J.S., Matija, L., Simic-Krstic, J.B., Nijemcevic, S.S., Koruga, D.L.: Discrimination of mineral waters using near infrared spectroscopy and aquaphotomics. Hemijska Industrija **68**(2), 257–264 (2014)
2. Tsenkova, R., Kovacs, Z., Kubota, Y.: Aquaphotomics: near infrared spectroscopy and water states in biological systems. In: Disalvo, E.A. (ed.) Membrane Hydration. SCBI, vol. 71, pp. 189–211. Springer, Cham (2015). https://doi.org/10.1007/978-3-319-19060-0_8
3. Tsenkova, R.: Aquaphotomics: extended water mirror approach reveals peculiarities of prion protein alloforms. NIR News **18**(6), 14–17 (2007)
4. Tsenkova, R.: Introduction: Aquaphotomics: dynamic spectroscopy of aqueous and biological systems describes peculiarities of water. J. Near Infrared Spectrosc. **17**(6), 303–313 (2010)
5. Tsenkova, R.: NIRS for biomonitoring. Ph.D. thesis. Hokkaido University, Japan (2004)
6. Jinendra, B., et al.: Near infrared spectroscopy and aquaphotomics: novel approach for rapid in vivo diagnosis of virus infected soybean. Biochem. Biophys. Res. Commun. **397**(4), 685–690 (2010)
7. Grimnes, S., Martinsen, O.: Bioimpedance and Bioelectricity Basics, 3rd edn, pp. 329–404. Academic Press, Cambridge (2015)
8. Olkin, I., Sampson, A.R.: Multivariate analysis: overview, pp. 10240–10247 (2001)
9. Ninfa, A.J., Ballou, D.P., Benore, M.: Fundamental Laboratory Approaches for Biochemistry and Biotechnology, 2nd edn, p. 65. Wiley, Hoboken (2010)
10. Ninfa, A.J., Ballou, D.P.: Fundamental Laboratory Approaches for Biochemistry and Biotechnology, p. 66. Wiley, Hoboken (2004). ISBN 9781891786006. OCLC 633862582
11. Freedman, D.A.: Statistical Models: Theory and Practice. Cambridge University Press, Cambridge (2009)

12. Bishop, C.M.: Pattern Recognition and Machine Learning. Springer, New York (2006)
13. Tobias, R.D.: An introduction to partial least squares regression. In: Proceedings of the Twentieth Annual SAS Users Group International Conference, pp. 1250–1257. SAS Institute Inc., Cary (1995)
14. Lewis-Beck, M., Bryman, A.E., Liao, T.F.: The SAGE Encyclopedia of Social Science Research Methods. Sage Publications, Thousand Oaks (2003)
15. Huerta, M., Leiva, V., Lillo, C., Rodrguez, M.: A beta partial least squares regression model: diagnostics and application to mining industry data. Appl. Stoch. Models Bus. Ind. **34**, 305–321 (2018)

Image Based Individual Identification of Sumatra Barb (*Puntigrus Tetrazona*)

Dinara Bekkozhayeva$^{(\boxtimes)}$, Mohammademehdi Saberioon, and Petr Cisar

Institute of Complex Systems, Faculty of Fisheries and Protection of Waters, University of South Bohemian in České Budějovice, Zámek 136, 373 33 Nové Hrady, Czech Republic
dbekkozhayeva@frov.jcu.cz

Abstract. The paper deal with the individual fish identification of the same species based on digital image of the fish. The proof of concept of image based individual identification is introduced on the small group fish. The method is completely noninvasive and can overcome the disadvantages of standard invasive identification such as tagging. The experiments proved the hypothesis that the visible patterns on Sumatra Barb (*Puntigrus tetrazona*) body can be used for individual identification. In the first step, the database of 43 fish (was created by the taking of the images of fish in different pose. Images were taken in an aquarium with a water. After data collection, data was processed by the image processing methods to determine the features. The simple nearest neighbor classification was used to test individual identification. The accuracy of classification was 100%. The method proved the hypothesis that the visible pattern on Sumatra Barb can be used for fully automated individual fish identification. It can be substituted current practice of fish identification based on tagging and marking. The long-term stability of the pattern and the classification power for large fish group should be studied in the future.

Keywords: Individual identification · Classification · Image processing · Machine vision · Fish biometric · Fish tagging

1 Introduction

Nowadays the role of aquaculture in the world is significant. The growing population lead to rising of food production. According to FAO (2018), production of all aquatic organisms in the world grow fast and this trend will not decline. Fish behavior monitoring can give an important information in the field of fish nutrition, welfare, health condition and environmental interaction with aquaculture system. All this measurement can be done with using Machine Vision Systems (MVS) (Saberioon et al. 2017). Today MVSs, becoming cheaper, more comfortable for untrained users and less stressful to fish and even more accurate alternative in comparison with traditional methods (Delcourt et al. 2012). It is therefore important that the industry aspires to monitor and control the effects of these challenges to avoid also upscaling potential problems when upscaling production.

© Springer Nature Switzerland AG 2019
I. Rojas et al. (Eds.): IWBBIO 2019, LNBI 11465, pp. 116–119, 2019.
https://doi.org/10.1007/978-3-030-17938-0_11

The fish identification is very important in many fields of aquaculture and is one of the cornerstones in the precise aquaculture. Precision Fish Farming (PFF) (Fore et al. 2017) concept whose aim is to apply control-engineering principles to fish production, thereby improving the farmer's ability to monitor, control and document biological processes in fish farms. The standard procedure of fish individual identification is fish tagging, which is invasive and has several limitations (Cousin et al. 2012)

The main aim of the study is the prove the concept that the visible features on a Sumatra barb (*Puntigrus tetrazona*) body (in our case it is vertical black stripes on a body) can be used for automatized identification of individuals. Fish identification of individual fish of the same species done by human experts from the images of the skin pattern was proved by Hirsch (Hirsch and Eckmann 2015). The study proved the hypothesis that the skin pattern can be used to distinguish the individual fish of the same species and that the pattern can be automatically determined, converted to the features describing the fish and used for classification.

2   Materials and Methods

Sumatra barb small ornamental fish was used in this study. The fish was bought in the local pet shop and kept for one month in the aquarium for acclimatization. The fish was chosen for the study because of unique black vertical stripes on their body. These patterns can be used as visible features for individual identification of commercial important fish with this kind of patterns, for example pike-perch *Sander lucioperca* and European perch *Perca fluviatilis*.

The digital camera (Camera Nikon D90), with controlled lighting, the background and the fish position, was used for data collection. Fixed background and lighting simplify the segmentation task. Data were collected under different angle and position. Images were taken from one side view of all fish. The data were collected from fish inside the aquarium to simulate the real conditions of fish cultivation (43 fish individuals). Four images of each individual were taken.

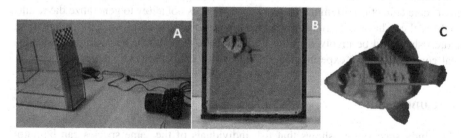

Fig. 1. Data collection design (A) picture of the fish inside aquarium (B), segmented fish with the region used for fish pattern parametrization (C)

Data processing procedure consists of standard image segmentation based on the know model of background, object detection, filtration and parametrization. To be able to perform object classification we have to describe the object by the specific features of it. This process is called parametrization (Šajn and Kononenko (2008)). In our study the parametrization was based on grayscale image of the part of the fish body containing the pattern, see Fig. 1c. First the specific subpart of the fish body was localized based on the fish length and fish height. The selected region was transformed into grayscale and used as parametrization.

The fish database for identification (reference set) was constructed from the two images of each fish. Two images of the fish were taken as a reference set and two images as a test set. The identification was then tested on these two datasets (total number of images from two datasets is 86 images).

The classification was based on the simple nearest neighbor classifier where the measure of the similarity was based on the cross-correlation function of the two images: two template images of the fish and unknown image of the fish. The method of template matching was used to get the best math between the images and the similarity was used to select the best fit to one of the 43 classes.

3   Results and Discussion

The first step of the identification was the image processing. The image processing, consisting of fish object detection, did not negatively influence the result of the identification. The use of standardized background and illumination enables to easily detect the fish in the image and do the parametrization. The only problem of the semi-transparency of the fish tail was solved by removing the tail part from the fish object. Therefore, there was no error in the detection and parametrization step. The identification of individuals was demonstrated as the classification of the unknow fish into one of 43 classes which correspond to the 43 individual images of the fish (2 images per class). The overall accuracy of classification based on the selected part of fish body was 100%. The classification accuracy clearly show that the fish pattern is unique to distinguish between 43 individuals of the same species. The small-scale research studies usually use comparable number of fish; therefore, the method is promising to substitute conventional procedure of fish identification (e.g. tagging or marking). The restriction in the data collection (illumination, background) does not allow to generalize the results to the real conditions. To fully automatize the identification based on the visible pattern more fish should be involved in the study and the data should be collected under the real conditions of the experimental studies or fish cultivation.

4   Conclusion

This study successfully shows that fish individuals of the same species can be automatically identified based on the visible pattern on the body and using computer vision. The overall classification accuracy was 100%. The visible patterns, in our case it is vertical stripes, can be used for identification of Sumatra Barb (*Puntigrus tetrazona*) in the group of 43 individuals. However, this is not just limited to Sumatra Barb. The approach can be applied to any fish species with visible pattern on the body (stripes or dots). The approach could be beneficial mainly for commercial fish species.

Acknowledgments. The study was financially supported by the Ministry of Education, Youth and Sports of the Czech Republic - project "CENAKVA" (LM2018099).

References

Cousin, X., Daouk, T., Pan, S., Lyphout, L., Schwartz, M.-E., Bgout, M.-L.: Methods and Techniques. Electronic individual identification of zebrafish using radio frequency identification. (RFID) microtags. J. Exp. Biol. **215**, 2729–2734 (2012). https://doi.org/10.1242/jeb.071829

Delcourt, J., Denoel, M., Ylieff, M., Poncin, P.: Video multitracking of fish behavior: a synthesis and future perspectives. Fish Fish. **14**(2), 186–204 (2012)

FAO: The State of World Fisheries and Aquaculture 2018 - Meeting the sustainable development goals. Rome. Licence: CC BY-NC-SA 3.0 IGO (2018)

Fore, M., et al.: Precision fish farming: a new framework to improve production in aquaculture. Biosyst. Eng. **173**, 176–193 (2017). https://doi.org/10.1016/j.biosystemseng.2017.10.014

Hirsch, P.E., Eckmann, R.: Individual identification of Eurasian perch Perca fluviatilis by means of their stripe patterns. Limnologica **54**, 1–4 (2015). https://doi.org/10.1016/j.limno.2015.07.003

Saberioon, M., Gholizadeh, A., Cisar, P., Pautsina, A., Urban, J.: Application of machine vision systems in aquaculture with emphasis on fish: state-of-art and key issues. Rev. Aquac. **9**, 369–387 (2017)

Šajn, L., Kononenko, I.: Multiresolution image parametrization for improving texture classification. EURASIP J. Adv. Signal Process. **2008**, 12 p. (2008)

Alignment of Sequences Allowing for Non-overlapping Unbalanced Translocations of Adjacent Factors

Simone Faro[1(✉)] and Arianna Pavone[2]

[1] Dipartimento di Matematica e Informatica, Università di Catania,
Viale Andrea Doria 6, 95125 Catania, Italy
`faro@dmi.unict.it`
[2] Dipartimento di Scienze Cognitive, Università di Messina,
Via Concezione 6, 98122 Messina, Italy
`apavone@unime.it`

Abstract. *Unbalanced translocations* take place when two unequal chromosome sub-sequences swap, resulting in an altered genetic sequence. Such large-scale gene modification are among the most frequent chromosomal alterations, accounted for 30% of all losses of heterozygosity. However, despite of their central role in genomic sequence analysis, little attention has been devoted to the problem of aligning sequences allowing for this kind of modification.

In this paper we investigate the sequence alignment problem when the edit operations are non-overlapping unbalanced translocations of adjacent factors.

Specifically, we present an alignment algorithm for the problem working in $\mathcal{O}(m^3)$-time and $\mathcal{O}(m^3)$-space, where m is the length of the involved sequences. To the best of our knowledge this is the first solution in literature for the alignment problem allowing for unbalanced translocations of factors.

Keywords: Sequence alignment · Unbalanced translocations ·
DNA sequence analysis · Text processing

1 Introduction

Retrieving information and teasing out the meaning of biological sequences are central problems in modern biology. Generally, basic biological information is stored in strings of nucleic acids (DNA, RNA) or amino acids (proteins).

In recent years, much work has been devoted to the development of efficient methods for aligning strings and, despite sequence alignment seems to be a well-understood problem (especially in the edit-distance model), the same cannot be said for the approximate string matching problem on biological sequences.

String alignment and *approximate string matching* are two fundamental problems in text processing. Given two input sequences x, of length m, and y,

© Springer Nature Switzerland AG 2019
I. Rojas et al. (Eds.): IWBBIO 2019, LNBI 11465, pp. 120–131, 2019.
https://doi.org/10.1007/978-3-030-17938-0_12

of length n, the *string alignment* problem consists in finding a set of edit operations able to transform x in y, while the *approximate string matching* problem consists in finding all approximate matches of x in y. The closeness of a match is measured in terms of the sum of the costs of the elementary edit operations necessary to convert the string into an exact match.

Most biological string matching methods are based on the *Levenshtein distance* [11], commonly referred to just as *edit distance*, or on the *Damerau distance* [8], which assume that changes between strings occur locally, i.e., only a small portion of the string is involved in the mutation event. However, evidence shows that in some cases large scale changes are possible [6,7,16] and that such mutations are crucial in DNA since they often cause genetic diseases [10,14]. For example, large pieces of DNA can be moved from one location to another (*translocations*) [6,13,17,18], or replaced by their reversed complements (*inversions*) [1–4].

Translocations can be *balanced* (when equal length pieces are swapped) or *unbalanced* (when pieces with different lengths are moved). Interestingly, unbalanced translocations are a relatively common type of mutation and a major contributor to neurodevelopmental disorders [18]. In addition, cytogenetic studies have also indicated that unbalanced translocations can be found in human genome with a de novo frequency of 1 in 2000 [17] and that it is a frequent chromosome alteration in a variety of human cancers [13]. Hence the need for practical and efficient methods for detecting and locating such kind of large scale mutations in biological sequences.

In the last three decades much work has been made for the alignment and matching problem allowing for chromosomal alteration, especially for non overlapping inversions. Concerning the alignment problem with inversions, a first solution based on dynamic programming, was proposed by Schöniger and Waterman [15], which runs in $\mathcal{O}(n^2m^2)$-time and $\mathcal{O}(n^2m^2)$-space on input sequences of length n and m. Several other papers have been devoted to the alignment problem with inversions. The best solution is due to Vellozo *et al.* [16], who proposed a $\mathcal{O}(nm^2)$-time and $\mathcal{O}(nm)$-space algorithm, within the more general framework of an edit graph.

Regarding the alignment problem with translocations, Cho *et al.* [6] presented a first solution for the case of inversions and translocations of equal length factors (i.e., balanced translocations), working in $\mathcal{O}(m^3)$-time and $\mathcal{O}(m^2)$-space. However their solution generalizes the problem to the case where edit operations can occur on both strings and assume that the input sequences have the same length, namely $|x| = |y| = m$.

In this paper we investigate the alignment problem under a string distance whose edit operations are non-overlapping unbalanced translocations of adjacent factors. To the best of our knowledge, this slightly more general problem has never been addressed in the context of alignment on biological sequences. A related result has been very recently introduced by Cantone et al. [5] who presented a $\mathcal{O}(nm^3)$-time and $\mathcal{O}(m^2)$-space algorithm for the approximate string matching problem with unbalanced translocations based on the dynamic-programming approach, where n is the length of the text and m is the length of the pattern. They also improved their solution by making use of the Directed Acyclic Word Graph of the pattern achieving a $\mathcal{O}(n \log^2 m)$ average time complexity still maintaining the same worst case time complexity.

In this paper we present an alignment algorithm for the same problem working in $\mathcal{O}(m^3)$ worst case time and $\mathcal{O}(m^3)$-space. Given two input equal length sequences x and y, our algorithm is able to establish if x can be transformed in y by way of unbalanced translocations of adjacent factors.

The rest of the paper is organized as follows. In Sect. 2 we introduce some preliminary notions and definitions. Subsequently, in Sect. 3 we present our alignment algorithm running in $\mathcal{O}(m^3)$-time, discussing its worst case time complexity. Finally draw our conclusions in Sect. 4.

2   Basic Notions and Definitions

A string x of length $m \geq 0$, over an alphabet Σ, is represented as a finite array $x[1 .. m]$ of elements of Σ. We write $|x| = m$ to indicate its length. In particular, when $m = 0$ we have the empty string ε. We denote by $x[i]$ the i-th character of x, for $1 \leq i \leq m$. Likewise, the substring of x factor, contained between the i-th and the j-th characters of x is indicated with $x[i .. j]$, for $1 \leq i \leq j \leq m$. We assume that $x[i .. j] = \varepsilon$ when $x > y$.

A string $w \in \Sigma^*$ is a suffix of x (in symbols, $w \sqsupseteq x$) if $w = x[i .. m]$, for some $1 \leq i \leq m$. Similarly, we say that w is a prefix of x (in symbols, $w \sqsubseteq x$) if $w = x[1 .. i]$, for some $1 \leq i \leq m$. Additionally, we use the symbol x_i to denote the prefix of x of length i (i.e., $x_i = x[1 .. i]$), for $1 \leq i \leq m$, and make the convention that x_0 denotes the empty string ε. In addition, we write $x.w$ for the concatenation of the strings x and w.

A string w is a border of x if both $w \sqsubseteq x$ and $w \sqsupseteq x$ hold. The set of the borders of x is denoted by $borders(p)$. For instance, given the string $x = atacgata$, we have that $at \sqsubseteq p$, $gata \sqsubseteq p$, while ata is a border of x. Moreover we have $borders(p) = \{a, ata\}$.

The border set of a string x of length m can be efficiently computed in $\mathcal{O}(m)$-time by using the border-table of x, introduced for the first time in the well known Morris-Pratt algorithm [9,12]. Formally the border table of x is a function $\pi : \{1, \ldots, m\} \rightarrow \{0, 1, \ldots, m - 1\}$ such that $\pi(i)$ is the length of the longest proper prefix of $x[1..i]$ that is also a suffix of $x[1..i]$.

A distance $d : \Sigma^* \times \Sigma^* \rightarrow \mathbb{R}$ is a function which associates to any pair of strings x and y the minimal cost of any finite sequence of edit operations which transforms x into y, if such a sequence exists, ∞ otherwise.

In this paper we consider the unbalanced translocation distance, $utd(x, y)$, whose unique edit operation is the translocation of two adjacent factors of the string, with possibly different lengths. Specifically, in an unbalanced translocation a factor of the form zw is transformed into wz, provided that both $|z|, |w| > 0$ (it is not necessary that $|z| = |w|$). We assign a unit cost to each translocation.

Example 1. Let $x = g\overline{tgac}\overline{cgt}\underline{ccag}$ and $y = gg\overline{atc}\underline{ccag}\overline{cgt}$ be given two strings of length 12. Then $utd(x, y) = 2$ since x can be transformed into y by translocating the substrings $x[3..4] = ga$ and $x[2..2] = t$, and translocating the substrings $x[6..8] = cgt$ and $x[9..12] = ccag$.

When $utd(x, y) < \infty$, we say that x and y have utd-match. If x has utd-match with a suffix of y, we write $x \overset{utd}{\sqsupseteq} y$.

3   A New *utd*-Alignment Algorithm

In this section we present a new solution for the string alignment problem allowing for unbalanced translocations of adjacent factors. In the next sections we start by describing the algorithm, named UNBALANCED-TRANSLOCATIONS-ALIGN (shown in Fig. 2), used for checking whenever an alignment exists between two equal length strings x and y.

The corresponding approximate string matching algorithm allowing for unbalanced translocations of adjacent factors can be trivially obtained by iterating the given procedure for all possible subsequences of the text of length $|x|$.

Our alignment algorithm is composed by a preprocessing and searching phase, which we describe in Sect. 3.1 and in Sect. 3.2, respectively. Then, in Sect. 3.3, we prove the correctness of the algorithm and discuss its worst case time complexity.

3.1   The Preprocessing Phase

During the preprocessing phase of procedure UNBALANCED-TRANSLOCATIONS-ALIGN three functions are computed, in the form of tables, which will be then used during the alignment process.

We first define the *next position function* $\mu_x : \Sigma \times \{1, \ldots, m\} \to \{2, \ldots, m\}$, associated to a given pattern x of length m, as the function which returns the next position (to a given input position i) where a given character $c \in \Sigma$ occurs. Specifically $\mu_x(c, i)$ is defined as the position $j > i$ in the pattern such that $x[j] = c$. If such a position does not exist then we set $\mu(c.i) = m + 1$. More formally

$$\mu_x(c, i) := \min\left(\{j \mid 1 \leq i < j \leq m \text{ and } x[j] = c\} \cup \{m + 1\}\right)$$

The next position function μ_x can be precomputed and maintained in a table of size $m \times \sigma$ in $\mathcal{O}(m\sigma + m^2)$ time by using procedure COMPUTE-NEXT-POSITION depicted in Fig. 1 (on the left).

Example 2. Let $x = gtgtaccgtgt$ be a string of length $m = 11$. We have $\mu_x(g, 1) = 3$, $\mu_x(g, 4) = \mu_x(g, 5) = 8$, $\mu_x(g, 8) = 10$, while $\mu_x(g, 10) = 12$.

COMPUTE-NEXT-POSITION
1. foreach $c \in \Sigma$ do
2.   for $i \leftarrow 1$ to m do
3. $\mu(c, i) \leftarrow m + 1$
4. for $i \leftarrow m$ downto 2 do
5.   for $j \leftarrow i - 1$ downto 1 do
6. $\mu(x[i], j) \leftarrow i$
7. return μ

COMPUTE-BORDER-SET
1. for $i \leftarrow 1$ to m do
2.   for $j \leftarrow i$ to m do
3.   for $k \leftarrow i$ to j do
4. $\Psi[i, j, k] \leftarrow 0$
5. for $i \leftarrow 1$ to m do
6.   for $j \leftarrow i$ to m do
7. $\pi \leftarrow$ COMPUTE-BORDER-TABLE(x, i, j)
8.   for $k \leftarrow 0$ to $j - i + 1$ do
9. $\Psi[i, j, \pi[k]] \leftarrow 1$
10. return Ψ

Fig. 1. Procedure COMPUTE-NEXT-POSITION (on the left) for computing the *next position function* μ_x which returns the next position (to a given input position i) where a given character $c \in \Sigma$ occurs in x; and procedure COMPUTE-BORDER-SET (on the right) for computing the *border set function* ψ_x as the set of the lengths of all borders of a given string x.

We also define the *border set function* ψ_x of a given string x as the set of the lengths of all borders of x. Specifically we define $\psi_x(i, j)$, for each $1 \leq i < j \leq m$, as the set of the lengths of all borders of the string $x[i..j]$, so that $k \in \psi_x(i, j)$ if and only if the string $x[i..j]$ has a border of length k. Formally we have

$$\psi_x(i, j) := \{k \mid 0 < k < j - i \text{ and } x[i..i + k - 1] = x[j - k + 1..j]\}$$

Example 3. Let $x = gtgtaccgtgt$ be a string of length $m = 11$. We have

$\psi_x(1, 11) = \{2, 4, 11\}$,   since the set of borders of $gtgtaccgtgt$ is $\{gt, gtgt, gtgtaccgtgt\}$;
$\psi_x(1, 4) = \{2, 4\}$,   since the set of borders of $gtgt$ is$\{gt, gtgt\}$;
$\psi_x(4, 9) = \{1, 6\}$,   since the set of borders of $taccgt$ is$\{t, taccgt\}$;
$\psi_x(5, 7) = \{3\}$ since the set of borders of acc is$\{acc\}$.

For each i, j with $1 \leq i \leq j \leq m$, we can represent the set $\psi_x(i, j)$ by using a vector of $(j - i + 1)$ boolean values such that its k-th entry is set iff $k \in \psi_x(i, j)$. More formally the function ψ_x can be maintained using a tridimensional bit-table Ψ_x, which we call *border set table of x*, defined as

$$\Psi_x[i, j, k] := \begin{cases} 1 & \text{if } x[i..i + k - 1] = x[j - k + 1..j] \\ 0 & \text{otherwise} \end{cases}$$

for $1 \leq i < j \leq m$ and $k < j - i$.

The border set table Ψ_x can be computed in $\mathcal{O}(m^3)$-time and space by using procedure COMPUTE-BORDER-SET-FUNCTION depicted in Fig. 1 (on the right), where COMPUTE-BORDER-TABLE is the $O(m)$ function used in the Knuth-Morris-Pratt algorithm [9].

Observe that using Ψ_x we can answer in costant-time to queries of the the type "*is k the length of a border of the substring $x[i..j]$?*", which translates to evaluate if $\Psi[i, j, k]$ is set.

In addition we define the *shortest border function* of a string x, as the function $\delta_x : \{1, \ldots, m\} \times \{1, \ldots, m\} \rightarrow \{1, \ldots, m\}$ which associates any nonempty substring of x to the length of its shortest border. Specifically we set $\delta_x(i, j)$ to be the length of the shortest border of the string $x[i \ldots j]$, for $1 \leq i < j \leq m$. More formally we have

$$\delta_x(i, j) := \min\{k \mid 0 \leq k < j - i \text{ and } x[i..i + k - 1] = x[j - k + 1..j]\} = \min(\psi(i, j))$$

It is trivial to observe that, if we already computed the border set function ψ_x, for the pattern x, the shortest border function δ_x of x can be computed in $\mathcal{O}(m^3)$-time using $\mathcal{O}(m^2)$ space.

Example 4. Let $x = gtgtaccgtgt$ be a string of length $m = 11$. According to Example 3, we have

$\delta_x(1, 11) = 2$, since gt is the shortest nonempty border of $gtgtaccgtgt$;
$\delta_x(1, 4) = 2$, since gt is the shortest nonempty border of $gtgt$;
$\delta_x(4, 9) = 1$, since t is the shortest nonempty border of $taccgt$;
$\delta_x(5, 7) = 3$ since acc is the shortest nonempty border of acc.

In what follows we will use the symbols μ, ψ and δ, in place of μ_x, ψ_x and δ_x, respectively, when the reference to x is clear from the context.

3.2   The Searching Phase

The proposed alignment procedure finds a possible *utd*-match between two equal length strings. The pseudocode of our algorithm, UNBALANCED-TRANSLOCATIONS-ALIGN(y,x,m), is presented in Fig. 2 and is tuned to process two strings x and y, of length m, where translocations can take place only in x.

In order to probe the details of the alignment procedure, let x and y be two strings of length m over the same alphabet Σ. The procedure sequentially reads all characters of the string y, proceeding from left to right. While scanning it tries to evaluate all possible unbalanced translocations in x which may be involved in the alignment between the two strings.

We define a *translocation attempt* at position i of y, for $1 \leq i \leq m$, as a quadruple of indexes, (s_1, k_1, s_2, k_2), with all elements in $\{0, 1, 2, \ldots, m\} \cup \{\text{null}\}$ and where, referring to the string x, s_1 and k_1 pinpoints the leftmost position and the length of the first factor (the factor moved on the left), while s_2 and k_2 pinpoints the leftmost position and the length of the second factor (the factor moved on the right). In this context we refer to s_1 and s_2 as the *key positions* of the translocation attempt. In the special case where no translocation takes place in the attempt we assume by convention that $s_1 = i$ and $s_2 = k_1 = k_2 = \text{null}[1]$. During the execution of the algorithm for each translocation attempt, (s_1, k_1, s_2, k_2), at position i, the invariant given by the following lemma[2] holds.

---

[1] We use the value null to indicate the length of an undefined string in order to discriminate it from the length of an empty string whose value is 0 by definition.
[2] In this context we assume that $s + \text{null} = s$, for any s.

Unbalanced-Translocations-Align(y,x,m)

1. $\Gamma^{(0)} \leftarrow \{(0,0,\text{null},\text{null})\}$
2. fo $i \leftarrow 1$ to m do
3.   for each $(s_1, k_1, s_2, k_2) \in \Gamma^{(i-1)}$ do
4.   if ($k_2 = $ null) then Δ Case 1
5.   if ($x[i] = y[i]$) then
6. $\Gamma^{(i)} \leftarrow \Gamma^{(i)} \cup \{(i, \text{null}, \text{null}, \text{null})\}$
7. $r \leftarrow \mu(y[i], i)$
8.   while ($r \leq m$) do
9. $\Gamma^{(i)} \leftarrow \Gamma^{(i)} \cup \{(i-1, 0, r-1, 1)\}$
10. $r \leftarrow \mu(y[i], r)$
11.   else
12.   if ($k_1 = $ null) then Δ Case 2
13.   if ($x[s_2 + k_2 + 1] = y[i]$) then
14. $\Gamma^{(i)} \leftarrow \Gamma^{(i)} \cup \{(s_1, k_1, s_2, k_2 + 1)\}$
15.   else $k_1 = 0$
16.   if ($k_1 \geq 0$) then
17.   while ($s_1 + k_1 < s_2$ and $k_2 > 0$ and $x[s_1 + k_1 + 1] \neq y[i]$) do Δ Case 3.b
18. $b \leftarrow 0$
19.   do $b \leftarrow b + \delta(s_1 + 1, s_2 + k_2 - b)$
20.   while ($s_1 + k_1 + b < s_2$ and $k_2 - b > 0$ and $(k_1 - s_1 + 1) \notin \phi(s_1 + 1, k_1 + b)$)
21. $k_2 \leftarrow k_2 - b$
22. $k_1 \leftarrow k_1 + b$
23.   if ($s_1 + k_1 \geq s_2$ or $k_2 \leq 0$) then break
24.   if ($x[s_1 + k_1 + 1] = y[i]$) then Δ Case 3.a
25.   if ($s_1 + k_1 = s_2$ and $(i, \text{null}, \text{null}, \text{null}) \notin \Gamma^{(i)}$) then
26. $\Gamma^{(i)} \leftarrow \Gamma^{(i)} \cup \{(i, \text{null}, \text{null}, \text{null})\}$
27.   else
28. $\Gamma^{(i)} \leftarrow \Gamma^{(i)} \cup \{(s_1, k_1 + 1, s_2, k_2)\}$
29. if ($\Gamma^{(m)} \neq \emptyset$) then
30.   return true
31. return false

Fig. 2. The pseudocode of the Unbalanced-Translocations-Align(y,x,m) for the sequence alignment problem allowing for unbalanced translocations of adjacent factors.

Lemma 1. *Let y and x be two strings of length m over the same alphabet Σ. Let $\Gamma^{(i)}$ be the set of all translocation attempts computed by procedure* Unbalanced-Translocations-Align *during the i-th iteration. If $(s_1, k_1, s_2, k_2) \in \Gamma^{(i)}$ then it holds that:*

(a) $i = s_1 + k_1 + k_2$;
(b) $x_i \overset{utd}{=} y_i$;
(c) *if $s_2 \neq $ null then $x[s_1 + 1..s_1 + k_1] = y[s_2 + 1..s_2 + k_1]$;*
(d) *if $s_2 \neq $ null then $y[s_2 + 1..s_2 + k_2] = y[s_1 + 1..s_1 + k_2]$;*

$\square$

For each $1 \leq i \leq m$, we define $\Gamma^{(i)}$ as the set of all translocation attempts tried for the prefix $y[1..i]$, and set $\Gamma^{(0)} = \{(0, \text{null}, \text{null}, \text{null})\}$.

However we can prove that it is not necessary to process all possible translocation attempts. Some of them, indeed, leads to detect the same utd-matches so that they can be skipped.

Lemma 2. *Let y and x be two strings of length m over the same alphabet Σ. Let $s \sqsubseteq y$ and $u \sqsubseteq x$ such that $|s| = |x|$ and $s \overset{utd}{=} u$. Moreover assume that*

(i) *$s.w.z \sqsubseteq y$ and $u.z.w \sqsubseteq x$*
(ii) *$s.w'.z \sqsubseteq y$ and $u.z.w' \sqsubseteq x$*

with $|z| > 0$ and $|w'| > |w| > 0$. If we set $i = |s.w.z|$ and $j = |s.w'.z|$ then we have $x[i+1..j] \overset{utd}{=} y[i+1..j]$.   □

The procedure iterates on the values of i, for $1 \le i \le m$, while scanning the characters of y, and during the i-th iteration it computes the set $\Gamma^{(i)}$ from $\Gamma^{(i-1)}$. For each translocation attempt $(s_1, k_1, s_2, k_2) \in \Gamma^{(i-1)}$ we distinguish the following three cases (depicted in Fig. 3):

– Case 1 ($s_2 = k_1 = k_2 = $ null)
  This is the case where no unbalanced translocation is taking place (line 5). Thus we simply know that $x_{i-1} \overset{utd}{=} y_{i-1}$. If $x[i] = y[i]$ the match is extended of one character by adding the attempt $(s_1 + 1, $ null, null, null$)$ to Γ^i (line 7). Alternatively, when possible, new translocation attempts are started (lines 9–12). Specifically for each occurrence of the character $y[i]$ in x, at a position r next to s_1, a new right factor u_r is attempted starting at position r (line 10) by extending $\Gamma^{(i)}$ with the attempt $(s_1, 0, r - 1, 1)$.
– Case 2 ($k_1 = 0$ and $k_2 > 0$)
  This is the case where an unbalanced translocation is taking place and the right factor u_r is currently going to be recognized (line 14). Specifically we know that $x[s_2 + 1..s_2 + k_2] = y[i - k_2..i - 1]$ and that $x[1..s_1] \overset{utd}{=} y[1..i - k_2 - 1]$. If $x[s_2 + k_2 + 1] = y[i]$ the right factor u_r can be extended of one character to the right, thus Γ^i is extended by adding the attempt $(s_1, k_1, s_2, k_2 + 1)$ (line 16). Otherwise, if $x[s_2 + k_2 + 1] \ne y[i]$, the right factor u_r cannot be extended further, thus we start recognizing the left factor u_ℓ. Specifically, in this last case, we update k_1 to 0 (line 17) and move to the following Case 3.
– Case 3 ($k_1 \ge 0$)
  This is the case where an unbalanced translocation is taking place, the right factor u_r has been already recognized and we are attempting to recognize the left factor u_ℓ. Specifically we know that $x[s_1 + 1..s_1 + k_1] = y[i - k_1..i - 1]$, $x[s_2 + 1..s_2 + k_2] = y[i - k_1 - k_2..i - k_1 - 1]$ and that $x[1..s_1] \overset{utd}{=} y[1..i - k_1 - k_2 - 1]$. We distinguish two sub-cases:
  • Case 3.a ($x[s_1 + k_1 + 1] = y[i]$)
    In this case the left factor u_ℓ can be extended of one character to the right (line 24). Thus if the left factor has been completely recognized, i.e. if $s_1 + k_1 = s_2$, Γ^i is extended by adding the attempt $(s_1 + k_1 + k_2, $ null, null, null$)$ (lines 27–28) which indicates that $x_i \overset{utd}{=} y_i$. Otherwise $\Gamma^{(i)}$ is extended by adding the attempt $(s_1, k_1 + 1, s_2, k_2)$ (line 30).

- Case 3.b $(x[s_1 + k_1 + 1] \neq y[i])$

  In this case the right factor u_ℓ cannot be extended. Before quitting the translocation attempt we try to find some new factors rearrangements on the same key positions, s_1 and s_2, but with different lengths, k_1 and k_2. Specifically we try to transfer a suffix w of u_r to the prefix position of u_ℓ, reducing the length k_2 and extending the length k_1 accordingly. This can be done only if we find a suffix w of u_r which is also a prefix of $x[s_1+1..s_2]$ and, in addition, we can move u_ℓ to the right of $|w|$ position along the left factor. More formally, if we assume that $|w| = b$ we must have:
  1. $|w| < |u_r|$, or rather $b < k_2$ (indicating that w is a proper suffix of u_r);
  2. $b \in \phi(s_1+1, s_2+k_2)$ (indicating that w is a border of $x[s_1+1..s_2+k_2]$);
  3. $(k_1 - s_1 + 1) \in \phi(s_1 + 1, k_1 + b)$ (indicating that u_ℓ is a border of $x[s_1 + 1..s_1 + k_1 + |w|]$);
  4. $s_1 + k_1 + |w| < s_2$ (indicating that the updated u_ℓ does not overflow onto u_r);
  5. $x[s_1 + k_1 + |w| + 1] = y[i]$ (indicating that the updated u_ℓ can be extended by $y[i]$).

3.3   Worst-Case Time and Space Analysis

In this section we discuss the worst-case time and space analysis of procedure UNBALANCED-TRANSLOCATIONS-ALIGN presented in the previous section. In particular, we will refer to the implementation reported in Fig. 2.

Let x and y be two nonempty strings of length $m \geq 1$ over the same alphabet Σ and assume to run procedure UNBALANCED-TRANSLOCATIONS-ALIGN(y, x, m). Regarding the space analysis, as stated in Sect. 3.1 we need $O(m\sigma)$ to maintain the next position function, $O(m^3)$ to maintain the border set function and $O(m^2)$ to maintain the shortest border function. Thus the overall space complexity of the algorithm is $O(m^3)$.

Regarding the time analysis, let $\Gamma^{(i)}$ be the set of all translocation attempts computed at iteration i, for $0 \leq i \leq m$.

First of all we observe that each $\Gamma^{(i)}$ contains at most one translocation attempt with $k_2 = $ null (i.e. of the form $(s_1, \text{null}, \text{null}, \text{null})$). We put $\Gamma^{(0)} = \{(0, \text{null}, \text{null}, \text{null})\}$ (line 1), thus the statement holds for $i = 0$. Observe that, if $i > 0$ a translocation attempt of the form $(s_1, \text{null}, \text{null}, \text{null})$ can be added to $\Gamma^{(i)}$ only in line 6 or in line 26. However by condition at line 25, if it is added to $\Gamma^{(i)}$ in line 6, it cannot be added again in line 26.

We now prove that the total number of translocation attempts processed during the execution of the algorithm is bounded by m^2. More formally we have

$$\sum_{i=0}^{m} |\Gamma^{(i)}| \leq m^3. \tag{1}$$

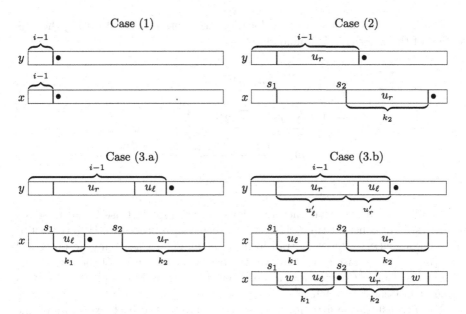

Fig. 3. Three cases of procedure UNBALANCED-TRANSLOCATIONS-ALIGN(y, x, m) while processing the translocation attempt $(s_1, k_1, s_2, k_2) \in \Gamma^{(i-1)}$ in order to extend it by charcater $y[i]$. Character $y[i]$ and its counterpart in x are depicted by a bullet symbol. Case (1): $x_{i-1} \overset{utd}{=} y_{i-1}$ and $x[i] = y[i]$, then the match is extended of one character; Case (2): the right factor u_r is currently going to be recognized and $x[s_2+k_2+1] = y[i]$, then the right factor u_r can be extended of one character; Case (3.a): the left factor u_ℓ can be extended of one character to the right; Case (3.b): the right factor u_ℓ cannot be extended, then we try to transfer a suffix w of u_r to the prefix position of u_ℓ, reducing the length k_2 and extending the length k_1 accordingly (w is a suffix of u_r and also a prefix of $x[s_1+1..s_2]$ and, in addition, we can move u_ℓ to the right of $|w|$ position along the left factor).

To prove that Eq. (1) holds observe that new translocation attempts are added to $\Gamma^{(i)}$ only when we are in Case 1. When we are in Case 2 or in Case 3 a translocation attempt is simply rearranged by extending the right factor (Case 2) or the left factor (Case 3). As observed above only one translocation attempt in $\Gamma^{(i)}$ is in Case 1 and the while cycle of line 8 can add at most $m - i$ new translocation attempts to $\Gamma^{(i+1)}$. In the worst case each translocation attempt added to $\Gamma^{(i+1)}$ will be closed only at iteration m, thus it will be extended along the sets $\Gamma^{(j)}$, for $j > i$. Thus the overall contribute of each translocation attempt added to Γ^{i+1} is $m - i$.

Thus the total number of translocation attempts processed during the execution of the algorithm is bounded by

$$
\sum_{i=0}^{m} |\Gamma^{(i)}| \leq 1 + \sum_{i=1}^{m}(m-i)^2
$$

$$
= 1 + \sum_{i=1}^{m} m^2 - \sum_{i=1}^{m} 2im + \sum_{i=1}^{m} i^2
$$

$$
= 1 + m^3 + \frac{m(m+1)(2m+1)}{6} - \frac{m(m+1)}{2}
$$

$$
= \frac{1}{3}m^3 - \frac{1}{2}m^2 + \frac{1}{3} \leq m^3
$$

Finally we observe that each translocation attempt in Case 2 and Case 3.a is processed in constant time, during the execution of procedure UNBALANCED-TRANSLOCATIONS-ALIGN. A translocation attempt in Case 1 my be processed in $O(m-i)$ worst case time. However the overall contribution of the while cycle at line 9 is at most $O(m^2)$ since, as observer above, there is a single translocation attempt in Case 1 for each $\Gamma^{(i)}$.

For a translocation attempt in Case 3, observe that at each execution of line 19 the value of b is increased of at most 1. Then in line 21 we decrease k_2 by b. Since the value of k_2 is increased only in line 16, this implies that overall number of times the while cycle of line 22 is executed is bounded by k_2, which is at most m. Thus the overall contribution given by the while cycle of line 19 is $O(m^3)$.

We can conclude that the overall time complexity of procedure UNBALANCED-TRANSLOCATIONS-ALIGN is $\mathcal{O}(m^3)$.

4   Conclusions and Future Works

We presented a first solution for the alignment problem allowing for unbalanced translocations of adjacent factors working in $\mathcal{O}(m^3)$ worst case time using $\mathcal{O}(m^3)$-space. As suggested in [5], an alternative solution working in $\mathcal{O}(m^3)$ worst case time can be obtained for the same problem by using a standard dynamic programming approach. However our algorithm uses a constructive approach could be more efficient in practice and which can be easily optimized. It turns out, indeed, by our preliminary experimental results (not included in this paper) that our solution has a sub-quadratic behaviour in practical cases. This suggests us to focus our future works on an accurate analysis of the algorithm's complexity in the average case.

In addition we are also planning to modify the given algorithm in order to compute the minimum number of translocations needed to transform x in y and an additional procedure able to retrieve the correct alignment of the two strings. Finally, we wonder if the problem can be solved in sub-cubical worst-case time complexity by extending the result obtained in Lemma 2 with additional restrictions.

References

1. Cantone, D., Cristofaro, S., Faro, S.: Efficient matching of biological sequences allowing for non-overlapping inversions. In: Giancarlo, R., Manzini, G. (eds.) CPM 2011. LNCS, vol. 6661, pp. 364–375. Springer, Heidelberg (2011). https://doi.org/10.1007/978-3-642-21458-5_31
2. Cantone, D., Cristofaro, S., Faro, S.: Efficient string-matching allowing for non-overlapping inversions. Theor. Comput. Sci. **483**, 85–95 (2013)
3. Cantone, D., Faro, S., Giaquinta, E.: Approximate string matching allowing for inversions and translocations. In: Proceedings of the Prague Stringology Conference, pp. 37–51 (2010)
4. Cantone, D., Faro, S., Giaquinta, E.: Text searching allowing for inversions and translocations of factors. Discrete Appl. Math. **163**, 247–257 (2014)
5. Cantone, D., Faro, S., Pavone, A.: Sequence searching allowing for non-overlapping adjacent unbalanced translocations. Report arXiv:1812.00421. Cornell University Library (2018). https://arxiv.org/abs/1812.00421
6. Cho, D.-J., Han, Y.-S., Kim, H.: Alignment with non-overlapping inversions and translocations on two strings. Theor. Comput. Sci. **575**, 90–101 (2015)
7. Cull, P., Hsu, T.: Recent advances in the walking tree method for biological sequence alignment. In: Moreno-Díaz, R., Pichler, F. (eds.) EUROCAST 2003. LNCS, vol. 2809, pp. 349–359. Springer, Heidelberg (2003). https://doi.org/10.1007/978-3-540-45210-2_32
8. Damerau, F.: A technique for computer detection and correction of spelling errors. Commun. ACM **7**(3), 171–176 (1964)
9. Knuth, D.E., Morris Jr., J.H., Pratt, V.R.: Fast pattern matching in strings. SIAM J. Comput. **6**(1), 323–350 (1977)
10. Lupski, J.R.: Genomic disorders: structural features of the genome can lead to DNA rearrangements and human disease traits. Trends Genet. **14**(10), 417–422 (1998)
11. Levenshtein, V.I.: Binary codes capable of correcting deletions, insertions and reversals. Sov. Phys. Dokl. **10**, 707–710 (1966)
12. Morris, J.H., Pratt, V.R.: A linear pattern-matching algorithm. Technical report 40. University of California, Berkeley (1970)
13. Ogiwara, H., Kohno, T., Nakanishi, H., Nagayama, K., Sato, M., Yokota, J.: Unbalanced translocation, a major chromosome alteration causing loss of heterozygosity in human lung cancer. Oncogene **27**, 4788–4797 (2008)
14. Oliver-Bonet, M., Navarro, J., Carrera, M., Egozcue, J., Benet, J.: Aneuploid and unbalanced sperm in two translocation carriers: evaluation of the genetic risk. Mol. Hum. Reprod. **8**(10), 958–963 (2002)
15. Schöniger, M., Waterman, M.: A local algorithm for DNA sequence alignment with inversions. Bull. Math. Biol. **54**, 521–536 (1992)
16. Vellozo, A.F., Alves, C.E.R., do Lago, A.P.: Alignment with non-overlapping inversions in $O(n^3)$-time. In: Bücher, P., Moret, B.M.E. (eds.) WABI 2006. LNCS, vol. 4175, pp. 186–196. Springer, Heidelberg (2006). https://doi.org/10.1007/11851561_18
17. Warburton, D.: De novo balanced chromosome rearrangements and extra marker chromosomes identified at prenatal diagnosis: clinical significance and distribution of breakpoints. Am. J. Hum. Genet. **49**, 995–1013 (1991)
18. Weckselblatt, B., Hermetz, K.E., Rudd, M.K.: Unbalanced translocations arise from diverse mutational mechanisms including chromothripsis. Genome Res. **25**(7), 937–947 (2015)

Probability in HPLC-MS Metabolomics

Jan Urban[✉]

Laboratory of Signal and Image Processing, Institute of Complex Systems,
South Bohemian Research Center of Aquaculture and Biodiversity of Hydrocenoses,
Faculty of Fisheries and Protection of Waters,
University of South Bohemia in České Budějovice,
Zamek 136, 37333 Nove Hrady, Czech Republic
urbanj@frov.jcu.cz
http://www.frov.jcu.cz

Abstract. This article is pinpointing the importance of the probabilistic methods for the analysis of the HPLC-MS measurement datasets in metabolomics research. The approach presents the ability to deal with the different noise sources and the process of the probability assignment is demonstrated in its general form.

The illustrative examples of the probability functions and propagation into subsequent processing and analysis steps consist of precision correction, noise probability, segmentation, spectra comparison, and biomatrices effects on calibration curve estimation.

The possible advantages of probability propagation in more data handling are also discussed.

Keywords: Liquid chromatography · Mass spectrometry ·
Probability · Noise · Spectra comparison · Entropy

1 Introduction

The interpretation of any measurement often tends to dichotomous decisions, which offer simple understanding, easy overview, and rapid publication. However, the possibility of oversimplification, misunderstanding, or even misinterpretation, could be enormous.

In fact, this is considered as significant crisis in the natural science, affecting both, the trust and the reproducibility [1]. As a major cause behind such abuse is given the continuous attempts to search for deterministic description and consequently, the both underestimation and overestimation of what statistic could describe or what it is really telling.

Probably the most deterrent example is the obsession with the p-value criterion [2,3], which becomes a demanded mantra, often without any justified purpose. Simply, if the p-value of the null hypothesis significance testing (NHST) is not fulfilling the 0.05 criterion, it does not necessarily means that the hypothesis is invalid, it only means that it is not acceptable at the predesigned 5% error probability threshold - which is completely different answer.

© Springer Nature Switzerland AG 2019
I. Rojas et al. (Eds.): IWBBIO 2019, LNBI 11465, pp. 132–141, 2019.
https://doi.org/10.1007/978-3-030-17938-0_13

Therefore, to follow strictly the p-value criterion is decreasing the research decision accuracy. Moreover, it is not possible to consider all possible hypothesis in the NHST. Thus, even the strong results still do not prove that the correct one was chosen and there should always exist more appropriate but omitted models, theories, and explanations [4–6].

Fortunately, the paradigm slowly shifts into the incorporation of the stochastic point of view [7–9]. In other words, we have to accept that there is no exact deterministic dichotomy (binary yes or not), but only weak or strong likelihood (maybe). This becomes crucial for false negative and false positive alarms in drug testing, metabolic pathways investigation, medical treatments, new bioactive compounds discovery, pollution and contamination detections, etc. One measurement, no matter how complex and precise, is telling absolutely nothing about the general event frequency distributions in nature or the specific law instance behind. On the other hand, the knowledge of the probabilistic density function can not predict or fully comprehend the result of a single event.

Especially in high performance liquid chromatography and mass spectrometry (HPLS-MS), it is an unpleasant practice, that one of three repetitions is completely different in some point. Even the perfectioned repetitions are never equal, just congruent.

The stochastic behavior of the nature has to be accepted, studied, and always taken into the account. Since the apriori probabilities are usually unknown or roughly estimated, the confidences in the measurements are acquired by aposteriori (Bayesian) probabilities. As will be shown in this article, the confidence values could be used in many subsequent analysis as an additional but significantly relevant parameter [9–11] to improve the accuracy and precision of the data processing steps, and increase the robustness of the obtained knowledge.

2   Data Acqusition and Precision

It was reported, that precision in value of the measured data is depending on the file format, binary depth, and coding method [12]. Such dependency could cause difficulties with oversegmentation, and therefore mass peak detection and centroid computation. To detect and improve the false precision, it is powerful yet simple the method of relative entropy (Fig. 1):

$$e(d) = \sum_d p_d log_2 p(d), \qquad (1)$$

$$re = e(d) \times k_d \qquad (2)$$

where re is relative entropy, $e(d)$ is entropy of dth binning, k_d is length of the bin in d^{th} binning, $p(d)$ is probability over d^{th} binning, and d is index of binning iteration. The oversegmented mass peaks are improved for the further analysis [12,13].

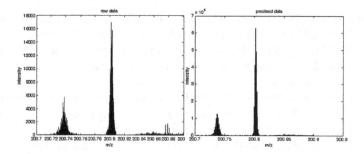

Fig. 1. Example of file format dependency on the peak shape. Left raw data, right relative entropy correction.

3   Probability of the Noise

Generaly, dataset structure of the HPLC-MC measurement consist of three dimensional discrete points defined by axis of retention time (rt), molecular mass (mz) and intensity (counts). Mathematically was the process described as the mapping process from the rt and mz sets into the count set:

$$y : T \times M \to \cup_{t \in T, m \in M} |y(t, m) \in I, I = 0, 1, ..., i_{max}, \tag{3}$$

where I is the indexing set (of natural numbers and zero). The value of mapping $y(t, m) \in I$ means intensity molecular mass $m \in M$ in retention time $t \in T$ [9].

Two basic principals of error occurrence are taken into account. The baseline presence could be considered as a special type of systematic noise $(q(t, m))$, produced by presence of mobile phase and solvents impurities. The random noise $(r(t, m))$ includes all unwanted sources of transient disturbances, mostly during MS ionization process.

The baseline and ionization influences are always presented, even if they are delimited. Both of them affect the signal transparency and could be also formally described as mappings:

$$q : T \times M \to \cup_{t \in T, m \in M} |q(t, m) \in I, I = 0, 1, ..., i_{max}, \tag{4}$$

$$r : T \times M \to \cup_{t \in T, m \in M} |r(t, m) \in I, I = 0, 1, ..., i_{max}. \tag{5}$$

Thus the relation between measured signal $y(t, m)$, real analyte signal $s(t, m)$ and noises is in ideal case as follows

$$y(t, m) = s(t, m) + q(t, m) + r(t, m). \tag{6}$$

The obvious object object of interest is the aposteriori description of $s(t, m)$ to reduce influence of presented noises, which can produce false peak or hide analytes signal under reasonable level. Exact contributions are apriori unknown because of the stochastic characteristics. However, the aposteriori estimation is possible:

$$\tilde{s}(t, m) = y(t, m) - \tilde{q}(t, m) - \tilde{r}(t, m). \tag{7}$$

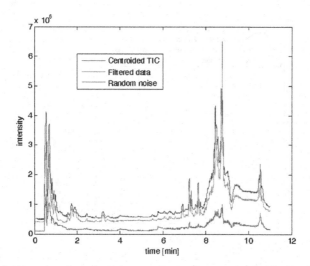

Fig. 2. Example of the dataset decomposition, removal of the random noise, blue curve is for measured Total Ion Current Chromatogram (centroids), green curve for the signal without random noise (so it is the analyte signal + baseline), and red for random noise. Evaluation performed by Expertomica metabolite profiling Hi-Res, www.expertomica. eu. (Color figure online)

Instead of direct value estimation of the analyte intensity $\tilde{s}(t, m)$, it is more appropriate to evaluate probability (confidence, significance) factor $p(t, m)$ that the measurement data output $y(t, m)$ is the analyte signal $\tilde{s}(t, m)$, probability $p_r(t, m)$ that the measurement data output $y(t, m)$ is not produced by random noise $r(t, m)$, the probability $p_r(t, m)$ that the measurement data output $y(t, m)$ is not produced by random noise $r(t, m)$, the probability $p_q(t, m)$ that the measurement data output $y(t, m)$ is not produced by systematic noise $q(t, m)$, and final probaility $p(t, m)$ repectively:

$$p(t, m) = p\left[y(t, m) = s(t, m)|\lambda_q, \lambda_r\right], \tag{8}$$

$$p_r(t, m) = p\left[y(t, m) = s(t, m) + q(t, m)|\lambda_r\right]. \tag{9}$$

$$p_q(t, m) = p\left[y(t, m) = s(t, m) + r(t, m)|\lambda_q\right], \tag{10}$$

$$p(t, m) = p_r(t, m) * p_q(t, m). \tag{11}$$

where λ_x are estimated characteristic of mappings.

Random noise is considered any unwanted influence during the measuring process witch causes inequality of measured data output $y(t, m)$ to analyte intensity $s(t, m)$. There are many sources from small substances eluted from stationary phase in HPLC column, through ionization disturbances, to short term fluctuations in signal intensity on MS detector [14].

Range of detectable molecular mass is wide by 5–6 orders in typical mass spectrum. It is advisable to reduce the intensity range by a compression function, like logarithm [15] in probability computation (Fig. 3):

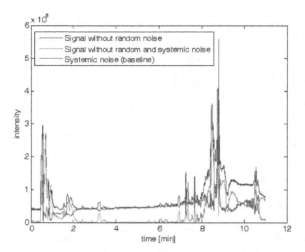

Fig. 3. Example of the dataset decomposition, removal of the systemic noise (mobile phase, baseline), blue curve for signal without random noise (the green one in Fig. 2, green curve for signal without noise contribution (it is blue signal without baseline, therefore signal without both noises, random and systemic), and red curve for baseline. Evaluation performed by Expertomica metabolite profiling Hi-Res, www.expertomica. eu. (Color figure online)

$$ly(m) = ln\,[y(m)]. \tag{12}$$

$$p_r(m) = \frac{p\,[ly(m)|\lambda_{ls+lq}]\,p(s+q)}{p\,[ly(m)|\lambda_{lr}]\,p(r) + p\,[ly(m)|\lambda_{ls+lq}]\,p(s+q)}. \tag{13}$$

$$\mu_r\,[lr(m) + ls(m)] \cong \mu\,[lr(m)]. \tag{14}$$

where mean value is μ_r and standard deviation σ_r. In analogy, the mobile phase contribution:

$$ly(t) = ln\,[y(t)], \tag{15}$$

$$p_q(t) = \frac{p\,[ly(t)|\lambda_{ls+lr}]\,p(s+r)}{p\,[ly(t)|\lambda_{lq}]\,p(q) + p\,[ly(t)|\lambda_{ls+lr}]\,p(s+r)}. \tag{16}$$

Those λ characteristics are a priory unknown and are usually detector type dependent (Gaussian, Lorentzian, Trapezoid, Sync, ... [16]). Proper dataset centroidization is required [16].

4   Generalization of the Signal to Noise Probability

The approach of the probability estimation and subsequent filtration could be generally adopted to many bioinformatic signals, like two-dimensional gel electrophoresis (2DE), where proteins are separated by the isoelectric focusing and molecular weight, using SDS-PAGE for visualization. The representation of the gel information is affected by position uncertainty, inconsistent backgrounds, shifts in both dimensions, and non-linear responses in staining (Fig. 4).

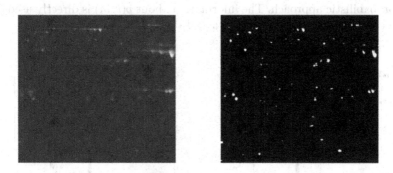

Fig. 4. Example of proteomics 2D gel probalistic analysis. Left: Captured image. Right: Probabilities after filtration process

The 2DE gel could be considered again as 2D mapping generated as intensity (greyscale) image. The partitioning of the data into background/signal contributions could be carried out using previous probabilistic approach. The protein signal is then just a disturbance in the noise probability density function (PDF). The probabilistic approach does not require any supervised parametrization, therefore it is resistant to over-segmentation as well as eliminating blurring effects [13].

5   Spectra Comparison

The estimated probability $p(t, m)$ for all $y(t, m)$ is the only one parameter to characterize quality of the measurement. Illustrative example of the probabilistic $p(t, m)$ information is a simple correlation between independent measured spectra of known known antifungal drug Nystatin ($C_{47}H_{75}NO_{17}$, mol. mass 926.09 [m/z]):

- Pure analyte in concentration 0.5 mg/ml as Reference.
- Pure analyte in concentration 0.5 μg/ml as Pure.
- Mixture of analyte in concentration 0.05 mg/ml and addition of 70% MetOH extract from cyanobacteria Nostoc sp, as Mix.

The correlation criterion was used to show the compared differences:

$$R(Y_1, Y_2) = \frac{C(Y_1, Y_2)}{\sqrt{C(Y_1, Y_1)C(Y_2, Y_2)}}, \tag{17}$$

$$C(Y_1, Y_2) = \frac{1}{M-1} \sum_{m=1}^{M} [y_1(m) - \mu_1][y_2(m) - \mu_2]. \tag{18}$$

where Y_1 is Reference spectrum, Y_2 is probabilistically filtered spectrum, $y_1(m)$ and $y_2(m)$ are molecular mass of spectra Y_1 and Y_2, respectively. μ_1 and μ_2 are average values of spectra intensities.

The example shows correlation criteria for unfiltered data and principle of using probabilistic approach. The information about $p(t,m)$ is directly used during the correlation evaluation (Fig. 5 and Table 1):

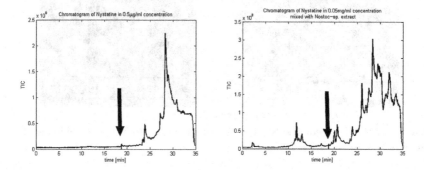

Fig. 5. Tic chromatograms of Nystatin

Table 1. Spectra correlation of Pure and Mix with Reference

$R(Y_1, Y_2)$	Pure	Mix
Unfiltered	98.37%	59.03%
Prob.Corr.	99.67%	95.93%

$$R_p(Y_1, Y_2, P) = \frac{C_p(Y_1, Y_2, P)}{\sqrt{C_p(Y_1, Y_1, P)C_p(Y_2, Y_2, P)}}, \tag{19}$$

$$C_p(Y_1, Y_2, P) = \frac{1}{M-1} \frac{\sum_{m=1}^{M} p(m)[y_1(m) - \mu_1][y_2(m) - \mu_2]}{\sum_{m=1}^{M} p(m)}. \tag{20}$$

where P is $p(t,m)$ in retention time t and $C_p(Y_1, Y_2, P)$ is weighted covariance [17].

6   Estimation of Biomatrix Effect

In order to compare chromatographic methods, it is useful to estimate the response of the given analyte in different biological matrixes. For example in the search for MCYST-LR heptapeptide with a molecular weight of 994.5 Da that is produced by different cyanobacterial taxa e.g., Microcystis, Nostoc, Anabaena, etc. It was proven that MCYST-LR causes hepatosis via the inhibition of protein phosphatases in the liver cells of mammals including humans. Moreover, the stability of microcystin implies that it can accumulate in high concentrations in fish organs [6]. Therefore, it is important to search for the analyte in different extracts and determine/estimate the calibration curves.

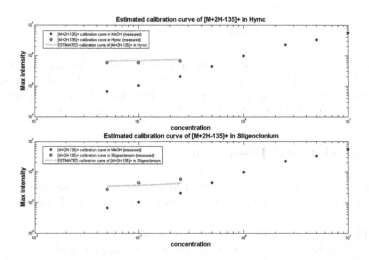

Fig. 6. Calibration curves of the MCYST-LR molecular ion with cleavage of the Adda moiety in MeOH and the estimated calibration curves in food additives matrices (shydrolyzate and Stigeoclonium extract). From top to bottom: the estimated calibration curve in salmon hydrolyzate and the estimated calibration curve in Stigeoclonium extract. The blue stars represent the measured calibration curve of the MCYST-LR molecular ion with cleavage of the Adda moiety in MeOH. The red circles represent the measured calibration curve of the MCYST-LR molecular ion with cleavage of the Adda moiety in a given food additive matrix. The green lines represent the estimated calibration curve of the MCYST-LR molecular ion with cleavage of the Adda moiety in a given food additive matrix. (Color figure online)

The calibration curves for MCYST-LR were constructed from the analysis of 10 MCYST-LR concentration measurements in repetitions, pure MCYST-LR standard (Sigma No. 33893), food additive filamentous green algae Stigeoclonium sp. and salmon meat hydrolyzate. To estimate the effect of the different biological matrices, the pure calibration curve (concentration/signal), and blank matrices were used. The statistical attributes of the target analyte mass are obtained from the blank measurements and hypothetical linear responsivity function:

$$Y1 = s1 * c + o1, \tag{21}$$

$$X = R1 * (s1 * c + o1), \tag{22}$$

$$Y2 = (s1 * c + o1) * \frac{R1}{R2}, \tag{23}$$

$$Y2 = (s1 * c + o1 + CRO2 - CRO1) * \frac{R1}{R2}, \tag{24}$$

$$Y2 = s2 * c + o2, \tag{25}$$

$$s2 = s1 * (R1/R2), \tag{26}$$

$$o2 = (o1 + CRO2 - CRO1) * \frac{R1}{R2}, \tag{27}$$

where s1 is the responsivity slope in matrix B1 and o1 is the responsivity offset in matrix B1, s2 is the responsivity slope in matrix B2, o2 is the responsivity offset in matrix B2. All values, as well as the estimation of calibration curve Y2, have to be computed independently for all target m/z values (Fig. 6).

7   Conclusion and Discussion

As was presented, the simple additional information to the measurement, the probability, is strongly affecting the subsequent analysis of the HPLC-MS dataset.

The illustrative examples showed that it decrease the difficulties of spectra comparison in the presence of heavy baseline background or noise level. This is important especially in the metabolomic screening, where new and apriori unknown bioactive compounds are searched. The probabilistic approach for comparison, once implemented, could significantly increase the search within the databases, even if the purity of the measurements is not ideal.

The filtration and the comparison [9,18,19] are not the only applications of the probability values in metabolomics, these are rather the basic kinds of direct usage. The additional topics cover more advanced tasks, including the time alignment algorithms [20], resolution, accuracy, and precision analysis [12, 13,16], concentration response dependencies on biological matrices [6], or peak decomposition [9,10].

Unfortunately, characterization of the probability distribution functions of different noise sources is not a trivial task, and all performed computations are still the estimations. It is necessary to understand, that no measurement, no processing, and no estimation could be error free.

Therefore, there is still a demand of broad acceptance of stochastic principle of the nature, and finally a time to discard or at least diminish the classical dichotomous, deterministic, or binary point of view.

Acknowledgments. The study was financially supported the Ministry of Education, Youth and Sports of the Czech Republic - projects Biodiversity (CZ.02.1.01.-0.0-0.0-16-025-0007370) and CENAKVA (LM2018099).

References

1. Baker, M.: 1,500 scientists lift the lid on reproducibility. Nat. News **533**(7604), 452 (2016)
2. Chavalarias, D., Wallach, J.D., Li, A.H.T., Ioannidis, J.P.: Evolution of reporting P values in the biomedical literature, 1990–2015. Jama **315**(11), 1141–1148 (2016)
3. Chia, K.S.: Significant-itis-an obsession with the P-value. Scand. J. Work Environ. Health **23**(2), 152–154 (1997)
4. Broadhurst, D.I., Kell, D.B.: Statistical strategies for avoiding false discoveries in metabolomics and related experiments. Metabolomics **2**(4), 171–196 (2006)
5. Price, P.C., Chiang, I.C.A., Jhangiani, R.: Research methods in psychology. BCcampus, BC Open Textbook Project (2015)
6. Urban, J., Hrouzek, P., Štys, D., Martens, H.: Estimation of ion competition via correlated responsivity offset in linear ion trap mass spectrometry analysis: theory and practical use in the analysis of cyanobacterial hepatotoxin microcystin-LR in extracts of food additives. BioMed Res. Int. **2013** (2013)
7. Vivo-Truyols, G.: Bayesian approach for peak detection in two-dimensional chromatography. Anal. chem. **84**(6), 2622–2630 (2012)
8. Ennis, E.J., Foley, J.P.: Stochastic approach for an unbiased estimation of the probability of a successful separation in conventional chromatography and sequential elution liquid chromatography. J. Chromatogr. A **1455**, 113–124 (2016)
9. Urban, J., Vaněk, J., Soukup, J., Štys, D.: Expertomica metabolite profiling: getting more information from LC-MS using the stochastic systems approach. Bioinformatics **25**(20), 2764–2767 (2009)
10. Woldegebriel, M., Vivo-Truyols, G.: Probabilistic model for untargeted peak detection in LC-MS using Bayesian statistics. Anal. Chem. **87**(14), 7345–7355 (2015)
11. Fay, D.S., Gerow, K.: A biologist's guide to statistical thinking and analysis. In: WormBook: The Online Review of C. elegans Biology pp. 1–54 (2013)
12. Urban, J.: False precision of mass domain in HPLC-HRMS data representation. J. Chromatogr. B **1023**, 72–77 (2016)
13. Urban, J.: Resolution, precision, and entropy as binning problem in mass spectrometry. In: Rojas, I., Ortuño, F. (eds.) IWBBIO 2018. LNCS, vol. 10813, pp. 118–128. Springer, Cham (2018). https://doi.org/10.1007/978-3-319-78723-7_10
14. Ardrey, R.E.: Liquid Chromatography-Mass Spectrometry: An Introduction. Wiley, Hoboken (2003)
15. Urban, J., Vaněk, J., Štys, D.: Systems Theory in Liquid Chromatography-Mass Spectrometry. Lap Lambert Academic Publishing (2012)
16. Urban, J., Afseth, N.K., Štys, D.: Fundamental definitions and confusions in mass spectrometry about mass assignment, centroiding and resolution. TrAC Trends Anal. Chem. **53**, 126–136 (2014)
17. Urban, J., Vaněk, J., Štys, D.: Mass spectrometry: system based analysis. IFAC Proc. **43**(6), 269–274 (2010)
18. Gan, F., Ye, R.: New approach on similarity analysis of chromatographic fingerprint of herbal medicine. J. Chromatogr. A **1104**(1–2), 100–105 (2006)
19. Jeong, J., Shi, X., Zhang, X., Kim, S., Shen, C.: An empirical Bayes model using a competition score for metabolite identification in gas chromatography mass spectrometry. BMC Bioinf. **12**(1), 392 (2011)
20. Urban, J.: Blank measurement based time-alignment in lC-MS. arXiv preprint arXiv:1205.1912 (2012)

Pipeline for Electron Microscopy Images Processing

Pavla Urbanova[1,2(✉)], Vladyslav Bozhynov[2], Dinara Bekkozhayeva[2],
Petr Císař[2], and Miloš Železný[1]

[1] Department of Cybernetics, Faculty of Applied Sciences,
University of West Bohemia in Pilsen, Univerzitní 8, 306 14 Plzeň, Czech Republic
urbanovp@kky.zcu.cz
[2] Institute of Complex Systems, South Bohemian Research Center of Aquaculture
and Biodiversity of Hydrocenoses, Faculty of Fisheries and Protection of Waters,
University of South Bohemia in České Budějovice,
Zámek 136, 373 33 Nové Hrady, Czech Republic

Abstract. This article is summarizing the general subtask pipeline during the processing and analysis of electron microscopy images. The overview is going from data acquisition, through noise description, filtration, segmentation, to detection. There are emphasized the difference from the expected conditions in macro-world imaging. The illustrative parameterization and statistical classification are explained on the immunolabeling example.

Keywords: Electron microscopy · Noise · Detection · Segmentation · Classification · Bayes

1 Introduction

Electron microscopy allows us to observe nanostructures within the organisms (cellular organelles, photosystems, membranes). Examination of microsomes of living organisms, ie functional units of biochemical and biophysical processes, requires high quality measurements for correct understanding of function and behavior, and understanding the structure and morphology of protein complex subsystems or specific reactions (e.g. immunological) [1,2]. In the last decade, microscopic methods (both optical and electron) have developed abruptly due to breakthroughs in digital technology (sensitivity & contrast), quantum physics, cryogenic techniques, laser-induced fluorescence and ptychography [3–8].

Electron microscopy has different properties than optical microscopy. The difference is in the replacement of photons by electrons, and thus the limit resolution, which is much larger than the light microscope in the electron microscope, is different, since it is proportional to the wavelength of the radiation used. High voltage (up to 300 keV) accelerated electrons have a significantly shorter wavelength (1.96–3.7 pm) compared to the visible light (380–780 nm). Thanks to

© Springer Nature Switzerland AG 2019
I. Rojas et al. (Eds.): IWBBIO 2019, LNBI 11465, pp. 142–153, 2019.
https://doi.org/10.1007/978-3-030-17938-0_14

higher resolving capability, it can also achieve higher effective magnification. The maximum theoretical resolution is given by the so-called Abbe diffraction limit [9] (for green light it is 210 nm, for 200 keV electrons it is 42 pm). Both these theoretical limits have been virtually overcome (30 nm in optical microscopy at 2014; 39 pm in electron microscopy at 2018) [3,10]. This fundamental contradiction between theory and practice evokes the necessity of fundamentally refining the theory used, in particular, the variance and the diffraction of [11]. The resolution of electron microscopes is theoretically infinite, depending on the applied electron beam voltage. Practically, however, it is heavily limited by two factors:

- high energy deforms the sample (it becomes plastic)
- magnetic lenses suffer from aberrant defects, it is necessary to use the correctors [12] for resolution below 200 pm

For calculating the resolution of electron microscopes, it is also necessary to include relativistic effects.

2   Description of the Task and Relevant Features

The basic output of electron microscopy is typically a two-dimensional signal, correctly referred to as a spatial variation of a sample, but more often only as a photograph of an electron microscope. Thus, the bi-dimensionality of the signal is the same as for a standard photograph given by the spatial coordinates x and y (columns, rows). The signal value on the x, y co-ordinates is generated by electron collisions on a fluorescence or luminescent emulsion emitting visible light photons. Light photons create a contrast of brightness and darkness depending on the intensity of the incident electrons. In 2015, EMPAD [8], an essentially CMOS chip for direct electron detection, was developed at Cornell University. Compared to previous detectors, it achieves a higher dynamic range. It was with the use of this detector that the 2018 record was achieved at 39 pm.

2.1   Probability Distribution

Often, the Central Limit Theory (CLV) does not apply here, at most, the cumulative measurement theorem can be used. This is due to the fact that the distribution functions involved in signal interference are generally of a power character, therefore, they do not have the ultimate central moments. Therefore, their superposition always does not convert to Gaussian normal distribution. Since a variety of signal processing and analysis tools have been developed for macroworld (i.e. with the assumption of CLV validity), their use in microscopy at larger magnifications than $20x$ can produce artifacts. Electron microscopes increase $50\,000x - 10\,000\,000x$, so we are moving in order where this is already very significant. From our point of view, this concerns mainly the distribution of noise. There is no color noise (white, pink, red, $\cdots$) in the microcosm, only shot noise (photon, shot noise) [13]. This noise has a quantum (discrete) nature. Shot noise is dominant in situations where we work with the ultimate number of energy-carrying particles (photons, electrons). This noise is often modeled by the Poisson process. For very large numbers (in the macro world), the observation of shot noise is no longer distinguishable from white noise.

3 Methods

3.1 Distance of the Signal from the Noise

The used bit depths are based on photodetector standards, namely 8, 10 and 12 bits. From the user point of view, the most widely used bit depth of 8 bpc (bits per channel), which we will work with later in the text. While multiple channels can be used in optical microscopy due to the presence of color, only one intensive channel (gray scale) is used in electron microscopy. The photodetector is in principle an analogue-to-digital converter (ADC), so the measured signal is quantized (rounded to the nearest quantization level). The theoretical best possible signal-to-noise ratio (SNR) is therefore:

$$\text{SNR} = \left[\frac{S_0}{N_Q}\right]_{dB} = 10\log_{10} 2^{2b} = 20\log_{10} 2^b \approx 6b, \tag{1}$$

where S_0 is the output signal, N_Q is the quantization noise, b is the bit depth used. For b equal to 8 bpc is therefore:

$$\text{SNR} = 6b = 6 \times 8 = 48\,\text{dB}. \tag{2}$$

Human senses are no longer recognizable by $\text{SNR} \geq 55\,\text{dB}$, which is approximately 10 bpc. Generally, the distance of 46 dB is considered the noise observability threshold. In normal digital photography $\text{SNR} \geq 32\,\text{dB}$ is still considered excellent quality. In electron microscopy, it is virtually impossible to measure the unintentional signal. For the SNR calculation, therefore, an alternative definition is used:

$$\text{SNR} = 3 + 20\log_{10}\frac{S_\mu}{S_\sigma}, \tag{3}$$

where S_μ is the average value of the signal and S_σ is the standard deviation of the signal. The 3 constant represents the thermal offset[1] of the fluctuating dissipation theorem of electric noise on the photodetector [14–16]. The presence of shot noise in electron microscopy therefore fundamentally affects the SNR value, which, regardless of the method used (SEM, TEM), is approximately in the range $\langle 10; 20\rangle$ dB. Often, Rose's 14 dB criteria is not even reached [17]. Telecommunication is considered an unusable distance of 20 and less dB.

3.2 Point Spread Function

The most serious complication in the microscope is the so-called point spread function. In a simplified way, this is the spatial propagation of the wave function of the point source. From a theoretical point of view, Maxwell's description

---

[1] Offset.

of the dispersion of electromagnetic wavefloor for homogeneous spheres in vacuum, using diffraction and interference from splinters of other spheres. However,
physical objects are not homogeneous balls. Rayleigh's variance approximation,
Fraunhofer's diffraction approximation, and Airy surface interpretation (derived
for telescopic star observations) are used to describe the solution in microscopy.
Thus, between the theoretical description and the practical observation, there is
a contradiction again [18]. For example, theory assumes that PSF is always symmetrical in the z axis, which only applies in vacuum. Biological objects always
exist in an optically denser non-homogeneous environment (cytoplasm, solvent)
that causes non-linearity of PSF.

Point spread function is a function in space, but we only see its cut in
the plane of focus. In terms of system theory, it is the impulse characteristic - the response of the imaging system (microscope) to the Dirac impulse.
Since the detector (generally each measuring device) requires some sampling
time to acquire the signal and the plane of field is non-zero, the measured two-
dimensional (x, y) signal is the integral of the point spread function in the time
and space intervals (depth or z-azis), ie the transient characteristic. Point spread
function negatively affects resolving ability, "blurring" the resulting signal. In the
macroworld, it has already been converted to Gaussian normal distribution, and
sharpening can be done easily, for example, by applying the Wiener-Kolmogor
filter. In practice, an optical transmission function (OTF) consisting of two parts
- the modulation (MTF) and phase (PhTF) defined by the ISO standard 9334, is
used to determine the resolution. As the name suggests, OTF is the transmission
characteristic of the microscope and is obtained by Fourier Laplace transformation (or Z-transformation). In general, OTF is a complex function and MTF is
real, is defined as the absolute value (size, magnitude, amplitude) OTF [19]:

$$\text{OTF}(\nu) = \int_0^\infty \text{PSF}(x) e^{-j\ x\nu} dx; \tag{4}$$

$$\text{OTF}(\nu) = A\cos(\nu) + Aj\sin(\nu); \tag{5}$$

$$\text{MTF}(\nu) = \sqrt{A^2\cos^2(\nu) + A^2\sin^2(\nu)} = A; \tag{6}$$

$$\text{PhTF}(\nu) = \arg\left(\text{OTF}(\nu)\right) = \nu; \tag{7}$$

$$\text{OTF}(\nu) = \text{MTF}(\nu)\ e^{j\ \text{PhTF}(\nu)}, \tag{8}$$

where x is a unit of space (nm, pm) and ν is a spatial frequency $(rad \cdot nm^{-1},$
$rad \cdot pm^{-1})$ and variables A, x are real positive numbers. With the exception
of phase contrast microscopy, the phase transition function in microscopy is not
important, and most of the OTF and MTF are confused[2] [20–22]. The simplest
interpretation of MTF is that it expresses the contrast for a given spatial frequency ν with respect to low frequencies. For this reason, MTF is sometimes
reported as a percentage, with 100% taking the lowest cutoff rate [23]. MTF
9% corresponds to Rayleigh's diffraction limit (effective resolution of the microscope).

---

[2] MTF refers to the microscope property, the CTF should be used correctly for each
particular specimen. However, some authors use different definitions.

3.3   Spatial Frequency and Resolution

Due to the current technological impossibility of producing a calibration slide with sufficiently small patterns (under 80 nm), the natural structures of the two-part silicas boxes of single-cell brown diatoms microalgae are used in the electron microscopy to determine the resolution (see Fig. 1). While we define a one-dimensional frequency spectrum for a one-dimensional signal, we need to define a two-dimensional spectrum in a two-dimensional signal. The dimension character matches the character of the signal, or we solve the changes in the x axis and the changes in the axis y:

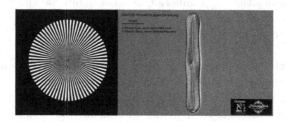

Fig. 1. Left: Star-shaped test scheme for determining effective optical microscope resolution by optical transmission function. Close to the center is an already observable aliasing effect. On the right: Microalgae from subsurface diatoms, individual pores inside the clipboard structure are spaced from 170 nm.

$$\xi(\lambda_x) = \frac{1}{\lambda_x};$$
$$\eta(\lambda_y) = \frac{1}{\lambda_y}; \tag{9}$$
$$\lambda \in (0; \text{FOV}),$$

where $\xi(\lambda_x)$, resp. $\eta(\lambda_y)$ is the frequency spectrum in the axis x resp. y. λ_x, λ_y are variable periods of cycles. Fourier analysis of optical systems is more general than systems analysis in the time domain because objects and images are essentially two-dimensional and therefore the basic set of sinusoids is also two-dimensional. The sinusoid has a space period in the axis x and y, respectively. λ_x, λ_y. If we invert these spatial periods, we get two spatial frequency components that describe this spectrum: $\xi(\lambda_x)$ a $\eta(\lambda_y)$ see Eq. 9. To specify a two-dimensional spatial frequency, therefore, the necessary information is two[3] [19]. For digital images, we mostly use the matrix pixel coordinate system. For the calibrated microscope, the corresponding pixel size is known and, therefore, the FOV in units of length (nm, pm). In this case, the space frequency unit used is the number of cycles (cy) or pairs (lp) per unit of length. Furthermore, with the spatial frequency characteristic, we work analogously to the time frequency characteristic (Fig. 2).

---

[3] In magnetic resonance, the spatial frequency is called k-space and uses the negative frequencies as it depends on the direction of change.

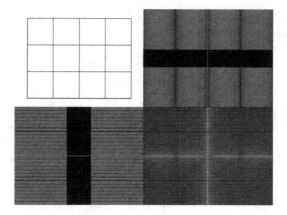

Fig. 2. Example of converting a two-dimensional single-grid image into the frequency domain using a two-dimensional discrete fast Fourier transform (bottom right). For clarity, the transformation in each x, y axis is also shown separately. The resulting frequency map contains all frequencies, their magnitude corresponds to the gray scale intensity in the logarithmic scale.

3.4   Filtration

In electron microscopy images, we have 2 types of unwanted frequencies. Low frequencies produced by inhomogeneity of luminescent emulsion and biological background, and then high frequency caused by shot noise. To remove both types of noise, it is best to use finite impulse response filters (FIR) [24] that differ only in the width of the weight function. FIR filtering can generally be described as a convolution of the signal f and the weight function h:

$$G\left[\xi,\eta\right] = \sum_{k=-V}^{V} \sum_{l=-W}^{W} H_{k,l} F\left[\xi - k, \eta - l\right], \tag{10}$$

$$F\left[\xi,\eta\right] = \mathcal{F}(f\left[x,y\right]),$$

$$H\left[\xi,\eta\right] = \mathcal{F}(h\left[x,y\right]),$$

$$g\left[x,y\right] = \mathcal{F}^{-1}(G\left[\xi,\eta\right]),$$

where g is the resulting image, k, l are window indexes, V, W half the window side sizes rounded down, h filter window, and f is the input image. The functions of G, F, H are the respective Fourier transform of functions g, f, h, $\mathcal{F}$ is Fourier transform, $\mathcal{F}^{-1}$ is inverse Fourier transform. Further, for the dimensions of the u, w filter window, the following applies:

$$u = 2V + 1; \tag{11}$$

$$w = 2W + 1,$$

for symmetric filter windows (matrices of odd number of rows and columns). For the square (regular) windows, then:

$$V = W; \tag{12}$$
$$u = w.$$

Fourier transform implicitly assumes that it is working with an infinite signal. Because we use the final length filter windows, we require a window shape (weight function) that suppresses the amplitude at its edges and focuses the center area. For this reason, Hamming's window of a certain width should be used [25, 26]:

$$h(k) = \frac{25}{46} + \frac{21}{46} \cos \frac{\pi k}{W}, \tag{13}$$

but in our case we have to use a two-dimensional window θ:

$$\theta = h(k) \times h(k)^T, \tag{14}$$
$$\Theta = \mathcal{F}(\theta).$$

Because we can assume that the power distribution is supposed to be a distribution of noise in a microworld, we also have to take into account the proportionality of the filter window size. The background of a two-dimensional signal can be considered a special kind of noise that is reflected on a larger scale. It is therefore a spatial frequency that is low. Assume the proportionality assumption by filter window area m_1:

$$m_1 = \left\lfloor \sqrt{N} + 0.5 \right\rfloor, \tag{15}$$

where N is the number of pixels. This will ensure the same number of non-overlapping image segments as the number of pixels in the segment. Because the window is square, the w_1 filter window side is

$$w_1 = \sqrt{m_1} = \left\lfloor \sqrt[4]{N} + 0.5 \right\rfloor. \tag{16}$$

We use the same logic when choosing a filter window size for noise, which is assumed to be at higher frequencies. In this case, the number of pixels in the filter window m_2 is equal to the number of non-overlapping segments in the one previous segment (Figs. 3 and 4):

$$m_2 = \sqrt{m_1} = \left\lfloor \sqrt[4]{N} + 0.5 \right\rfloor \tag{17}$$

and the square side w_2 is

$$w_2 = \sqrt{m_2} = \sqrt[4]{m_1} = \left\lfloor \sqrt[8]{N} + 0.5 \right\rfloor. \tag{18}$$

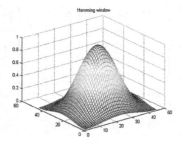

Fig. 3. Example of a large two-dimensional filter window m_1.

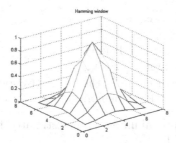

Fig. 4. Example of a small two-dimensional filter window m_2.

4   Bayes Classification

It is based on a conditional probability clause:

$$P(T|O) = \frac{P(O|T)P(T)}{P(O)};$$

$$P(O) = \sum_T P(O|T)P(T),$$

(19)

where O is an object, T is a class of objects, and probabilities:

- $P(T|O)$ is the posterior probability of the class T for a given O object;
- $P(T)$ is the a priori probability of the class T;
- $P(O|T)$ is the conditional probability of objects O (only) for a given T class;
- $P(O)$ is the a priori probability of the O object.

In the defined task we have three types of source data from electron microscopy. Objects can be distinguished from the background and parameterized after the filtering [27,28]. Object can be expressed as the following vector of flags:

$$O = \begin{bmatrix} circumference \\ area \\ roundness \\ eccentricity \\ homogeneity \\ darkness/lightness \\ size \\ concavity/convexity \end{bmatrix} = \begin{bmatrix} o_1 \\ o_2 \\ o_3 \\ o_4 \\ o_5 \\ o_6 \\ o_7 \\ o_8 \end{bmatrix}$$

For classification purposes, it is necessary to define object classes based on the assignment of the task:

- T_1: Photosystem I (PSI) - circular;
- T_2: Photosystem II (PSII) - oval;
- T_3: Gold nanoparticles small 5–10 nm;
- T_4: Gold nanoparticles large 15–20 nm;
- T_5–T_7: 3 various $2D$ projections of the $3D$ restriction enzyme;
- T_8: Residual objects (peptides, membranes, cytochromes).

The relevant conditional probabilities can be estimated from the training set of data, see the following table:

	o_1	o_2	$\cdots$	o_8	$\sum$
T_1	$P(o_1\|T_1)P(T_1)$	$P(o_2\|T1)P(T_1)$	$\cdots$	$P(o_8\|T1)P(T_1)$	$P(T_1)$
T_2	$P(o_1\|T2)P(T_2)$	$P(o_2\|T2)P(T_2)$	$\cdots$	$P(o_8\|T2)P(T_2)$	$P(T_2)$
$\vdots$	$\vdots$	$\vdots$	$\ddots$	$\vdots$	$\vdots$
T_8	$P(o_1\|T8)P(T_8)$	$P(o_2\|T8)P(T_8)$	$\cdots$	$P(o_8\|T8)P(T_8)$	$P(T_8)$
$\sum$	$P(o_1)$	$P(o_2)$	$\cdots$	$P(o_8)$	1

The probability of $P(O)$ then represents the distribution of all O objects independently of the membership of a T class and is therefore the same for all classes. For classification purposes, therefore, it is only a normalization factor and it is not necessary to quantify it [29]. The Bayes classifier then decides to assign an unknown object to the appropriate class based on the multiplication of conditional and posterior probability:

$$P(O|T_\alpha)P(T_\alpha) > P(O|T_\beta)P(T_\beta), \alpha \neq \beta. \tag{20}$$

5   Results and Discussions

Sets of collected images were divided into training (2/3) and test (1/3) sets. The methods described above were performed, compared and tuned on training data. The success of the approach was evaluated on test group data. Two-dimensional signal processing was performed using Matlab development, computing and programming environment. For the purpose of using segmentation image processing methods, it was advisable to work with a negative input signal. The background obtained by the convolution of the negative and the large two-dimensional filter window (low frequencies) can then be subtracted from the negative. The same analogy is used to obtain noise (high frequencies), ie, the convolution of the image after subtracting the background and the small two-dimensional filter window. The resulting filtered signal was further segmented using the intermediate variance [30] to obtain a set of objects for the Bayes classifier. The average detection time was 58 s for the Intel Core2 Duo CPU E8400, 3 GHz, 4 GB.

The method of using FIR for filtering non-homogeneous backgrounds and shot noise appears to be appropriate, computerized and sufficiently effective. The key parameter of the filters used is the size of the filter window. The assumption of proportionality in electron microscopy can be optimized by a series of time-consuming additional measurements and by analyzing shot noise in a particular type of task. Typical sample preparation for the electron microscope, however, takes 1–4 weeks. For immunological labeling, at least 2 weeks are required. The shot noise behavior can be modeled by the Poisson process and its spatial distribution described by Bose-Einstein's statistics [31], but unfortunately this noise is also affected by Fano's detector noise [32] and the detecting noise of the detector. As has already been mentioned, there is a great deal of discrepancy between the applied theory and the practically observed reality in super-resolution microscopy. Analytical models of shot noise distribution only approximate reality with respect to the electron energy, the diffractive limit, and in the frequency domain as well as the Gabor localization limit used [33]. Frequency signal analysis options in this role have not been fully exhausted. The use of wavelet transformation is not the most appropriate because of its high computing abilities and non-invariance to shift in discretely sampled data [34]. Conversely, in particular, cryEM can be expected to be used for modal decomposition or phasor analysis, which is also successfully used, for example, in fluorescence microscopy [35] for the classification of protein complexes.

Acknowledgments. The work has been partially supported by the SGS of the University of West Bohemia, project No. SGS-2019-027; and by the Ministry of Education, Youth and Sports of the Czech Republic - project CENAKVA (LM2018099).

References

1. Bumba, L., Havelková-Doušová, H., Hušák, M., Vácha, F.: Structural characterization of photosystem II complex from red alga porphyridium cruentum retaining extrinsic subunits of the oxygen-evolving complex. Eur. J. Biochem. **271**(14), 2967–2975 (2004)
2. Cyran, N., Klepal, W., von Byern, J.: Ultrastructural characterization of the adhesive organ of idiosepius biserialis and idiosepius pygmaeus (mollusca: Cephalopoda). J. Mar. Biol. Assoc. UK **91**(7), 1499–1510 (2011)
3. Möckl, L., Lamb, D.C., Bräuchle, C.: Super-resolved fluorescence microscopy: nobel prize in chemistry 2014 for Eric Betzig, Stefan Hell, and William E. Moerner. Angewandte Chemie Int. Ed. **53**(51), 13972–13977 (2014)
4. Cressey, D., Callaway, E.: Cryo-electron microscopy wins chemistry nobel. Nat. News **550**(7675), 167 (2017)

5. Dubochet, J., et al.: Cryo-electron microscopy of vitrified specimens. Q. Rev. Biophys. **21**(2), 129–228 (1988)
6. Frank, J.: Single-particle imaging of macromolecules by cryo-electron microscopy. Ann. Rev. Biophys. Biomol. Struct. **31**(1), 303–319 (2002)
7. Henderson, R.: Avoiding the pitfalls of single particle cryo-electron microscopy: Einstein from noise. Proc. Natl. Acad. Sci. **110**(45), 18037–18041 (2013)
8. Tate, M.W., et al.: High dynamic range pixel array detector for scanning transmission electron microscopy. Microsc. Microanal. **22**(1), 237–249 (2016)
9. Abbe, E.: A contribution to the theory of the microscope and the nature of microscopic vision. Proc. Bristol Nat. Soc. **1**, 200–261 (1874)
10. Jiang, Y., et al.: Electron ptychography of 2D materials to deep sub-ångström resolution. Nature **559**(7714), 343 (2018)
11. Rychtáriková, R., et al.: Superresolved 3D imaging of live cells organelles from brightfield photon transmission micrographs. Ultramicroscopy **179**, 1–14 (2017)
12. Zach, J.: Chromatic correction: a revolution in electron microscopy? Philos. Trans. R. Soc. Lond. A: Math. Phys. Eng. Sci. **367**(1903), 3699–3707 (2009)
13. Schottky, W.: Über spontane stromschwankungen in verschiedenen elektrizitätsleitern. Annalen der Physik **362**(23), 541–567 (1918)
14. Constant, M., Ridgeon, P.: The Principles and Practice of CCTV. Miller Freeman, San Francisco (2000)
15. Callen, H.B., Welton, T.A.: Irreversibility and generalized noise. Phys. Rev. **83**(1), 34 (1951)
16. Izpura Torres, J.I., Malo Gomez, J.: A fluctuation-dissipation model for electrical noise. Circ. Syst. **2**(3), 112–120 (2011)
17. Rose, A., Vision, H.: Electronic (1973)
18. Náhlík, T., Štys, D.: Microscope point spread function, focus and calculation of optimal microscope set-up. Int. J. Comput. Math. **91**(2), 221–232 (2014)
19. Boreman, G.D.: Modulation Transfer Function in Optical and Electro-Optical Systems, vol. 21. SPIE Press, Bellingham (2001)
20. Cho, H., Chon, K.: Change of MTF for sampling interval in digital detector. J. Korean Soc. Radiol. **8**(5), 225–230 (2014)
21. Presenza-Pitman, G., Thywissen, J.: Determination of the contrast and modulation transfer functions for high resolution imaging of individual atoms. NSERC summer report, University of Toronto 83 (2009)
22. Coltman, J.W.: The specification of imaging properties by response to a sine wave input. JOSA **44**(6), 468–471 (1954)
23. Sekuler, R., Blake, R.: Perception. alfred a. Kopf NY (1985)
24. Gupta, S.: Performance analysis of fir filter design by using rectangular, hanning and hamming windows methods. Int. J. Adv. Res. Comput. Sci. Softw. Eng. **2**(6), 273–277 (2012)
25. Hamming, R.W.: Digital Filters. 3rd edn (1989)
26. Möbus, G., Necker, G., Rühle, M.: Adaptive fourier-filtering technique for quantitative evaluation of high-resolution electron micrographs of interfaces. Ultramicroscopy **49**(1–4), 46–65 (1993)
27. Urbanová, P., et al.: Unsupervised parametrization of nano-objects in electron microscopy. In: Rojas, I., Ortuño, F. (eds.) IWBBIO 2018. LNCS, vol. 10813, pp. 139–149. Springer, Cham (2018). https://doi.org/10.1007/978-3-319-78723-7_12
28. Urbanová, P., Urban, J., Císař, P., Železný, M.: Parametrization of discrete sphericity in electron microscopy images. In: Technical Computing Prague (2017)
29. Psutka, J.: Automatické rozpoznávání předmětů a jevů

30. Otsu, N.: A threshold selection method from gray-level histograms. IEEE Trans. Syst. Man Cybern. **9**(1), 62–66 (1979)
31. Huang, K.: Introduction to Statistical Physics. Chapman and Hall/CRC, Boca Raton (2009)
32. Fano, U.: Ionization yield of radiations. II. The fluctuations of the number of ions. Phys. Rev. **72**(1), 26 (1947)
33. Gabor, D.: Theory of communication. Part 1: The analysis of information. J. Inst. Electr. Eng.-Part III: Radio Commun. Eng. **93**(26), 429–441 (1946)
34. Bradley, A.P.: Shift-invariance in the discrete wavelet transform. In: Proceedings of VIIth Digital Image Computing: Techniques and Applications, Sydney (2003)
35. Digman, M.A., Caiolfa, V.R., Zamai, M., Gratton, E.: The phasor approach to fluorescence lifetime imaging analysis. Biophys. J. **94**(2), L14–L16 (2008)

A Greedy Algorithm for Detecting Mutually Exclusive Patterns in Cancer Mutation Data

Chunyan Yang[1], Tian Zheng[1], Zhongmeng Zhao[1(✉)], Xinnuo He[1], Xuanping Zhang[1], Xiao Xiao[2], and Jiayin Wang[1(✉)]

[1] Department of Computer Science and Technology,
School of Electronic and Information Engineering,
Xi'an Jiaotong University, Xi'an 710049, China
{zmzhao,wangjiayin}@mail.xjtu.edu.cn
[2] School of Public Policy and Administration,
Xi'an Jiaotong University, Xi'an 710049, China

Abstract. Some somatic mutations are reported to present mutually exclusive patterns. It is a basic computational problem to efficiently extracting mutually exclusive patterns from cancer mutation data. In this article, we focus on the inter-set mutual exclusion problem, which is to group the genes into at least two sets, with the mutations in the different sets mutually exclusive. The proposed algorithm improves the calculation of the score of mutual exclusion. The improved measurement considers the percentage of supporting cases, the approximate exclusivity degree and the pair-wise similarities of two genes. Moreover, the proposed algorithm adopts a greedy strategy to generate the sets of genes. Different from the existing approaches, the greedy strategy considers the scores of mutual exclusion between both the genes and virtual genes, which benefits the selection with the size restrictions. We conducted a series of experiments to verify the performance on simulation datasets and TCGA dataset consisting of 477 real cases with more than 10 million mutations within 28507 genes. According to the results, our algorithm demonstrated good performance under different simulation configurations. In addition, it outperformed CoMEt, a widely-accepted algorithm, in recall rates and accuracies on simulation datasets. Moreover, some of the exclusive patterns detected from TCGA dataset were supported by published literatures.

Keywords: Cancer genomics · Somatic mutation analysis ·
Mutually exclusive pattern · Detection algorithm · Greedy strategy

1 Introduction

Some somatic mutations are reported to present mutually exclusive patterns. Detecting such patterns are considered to facilitate a wide range of researches in cancer genomics, such as exploring the roles of mutational events during tumor occurrence and evolution [1, 2], identifying novel germline variants in cancer predisposition genes [3, 4], etc. Thus, it is a basic computational problem to efficiently extract mutually exclusive patterns from cancer mutation data. There are two types of patterns that the existing approaches mainly focus on: intra-set mutual exclusion and inter-set mutual exclusion.

© Springer Nature Switzerland AG 2019
I. Rojas et al. (Eds.): IWBBIO 2019, LNBI 11465, pp. 154–165, 2019.
https://doi.org/10.1007/978-3-030-17938-0_15

For intra-set mutual exclusion, the existing approaches will generate one or multiple sets of genes, and the genes in the same set are mutually exclusive. On the other hand, for inter-set mutual exclusion, the existing approaches will generate at least two sets of genes, with the genes in the different sets mutually exclusive.

Unfortunately, only a proportion of patients present the mutually exclusive patterns. Normally, when the number of cases increases, it is often impossible to obtain a perfect set of mutual exclusion. To solve this issue, the existing approaches usually adopt a mutual exclusion score to measure the performance. Given a set of cases, the score is often calculated based on the percentage of cases that follow the pattern(s). Then, based on the scores, different approaches use different strategies to generate the sets for the patterns: (1) Some methods, such as PathScan [5] and MEMo [6], incorporate prior knowledge into the searching processes of the mutation combinations. This additional information limits the searching space, which improves the efficiency of the algorithm. However, the prior knowledge is often unavailable or incomplete. (2) Some methods adopt approximate algorithms for speedup. For example, RME [7] designs an improved online learning linear-threshold algorithm. Dendrix [8] uses the greedy strategy and a Markov chain Monte Carlo algorithm to optimize the gene sets. These methods are simple to implement, but may introduce bias when some hotspot genes present high mutation rates. (3) Some other methods follow statistical models. For example, CoMEt [2] adopts a one-sided Fisher test for a pair of mutations. For multiple genes with multi-dimensional contingency tables, it introduces an accurate-tail enumeration algorithm (or an approximate algorithm) to generalize Fisher's exact test. Permutation test, used in MEMo and MEMCover [9] is another idea of calculating p-value to compare the candidate sets to the null model. A recent published method [10] proposed a weighted exclusive test (WExT) to estimate the probabilities of mutations (a priori of mutual exclusion) from the null distribution of a permutation test, which is able to approximate the genes efficiently.

However, most of the existing approaches are designed for intra-set patterns and encounter a great challenge on computational complexity when dealing with inter-set patterns. To overcome this weakness, in this article, we proposed an algorithm for recognizing inter-set patterns. The proposed algorithm incorporated the advantages of the existing scoring models and simplified the searching strategy to reduce the complexity. Based on the Dendrix's measurement, it improved the score calculation by incorporating the measurements of the percentage of supporting cases, the approximate exclusivity degree and the pair-wise similarities of two genes. Both genes and the sets of genes were considered in calculating mutual exclusions. A greedy strategy was then adopted: It traversed each pair of genes and generated two sets by the pair with the highest mutual exclusion score. Then, it iteratively added genes into these sets by updating the scores, until the number of genes in each set reached a pre-set threshold. We conducted a series of experiments to test the performance of the proposed algorithm. The proposed algorithm obtained satisfied results on accuracy, recall and genes aggregation when applied on simulation datasets with different data configurations, and outperformed a popular method, CoMEt, in most of the cases. In addition, the proposed algorithm is applied on a real TCGA mutation dataset that consists of 477 cases with more than 10 million mutations within 28507 genes. Some of the results were in agreement with the literatures.

2 Methods

Suppose that we are given a set of M cases with G genes (or mutational events). The aim is to generate multiple clusters of genes, namely gene sets. Each set consists of a number of genes from G. There is no overlap among all the sets. The genes in the same set have no mutually exclusive patterns, while the genes from different sets are mutually exclusive. In this case, a set of genes can be collapsed to a virtual "gene": For each case, if the genes in the same set carry one or more mutations, the virtual gene carries a mutation. The aim is to generate at least two virtual genes with mutually exclusive patterns. To simplify this problem, in this article, we only consider the condition of two sets.

Based on the existing measurements, here, we introduce an integrated scoring method for measuring mutual exclusion. This measurement considers the percentage of supporting cases, the approximate exclusivity degree and the pair-wise similarities of two genes. The score of mutual exclusion between two sets, G_1 and G_2, is:

$$E = \frac{1}{M} \left(1 - \frac{(a_1 - a_2)^2}{(a_1 + a_2)^2} \right) (a_1 + a_2 - \alpha a_0)$$

Where E represents the score of mutual exclusion, a_1 denotes the number of the cases that carry at least one mutation in G_2 and have no mutation in G_1, a_2 denotes the number of the cases that carry at least one mutation in G_1 and have no mutation in G_2, a_0 denotes the number of samples which carry one or more mutations in both G_1 and G_2, and α is a model parameter to control the tolerance. According to this equation, the score of each pair of genes can be calculated and stored, whose time complexity is $O(n^2)$.

Then, based on the stored scores, we simply adopt a greedy strategy for generating the two sets. The algorithm first sorts the scores and selects the pair with the highest score. For this pair, the two genes are allocated into G_1 and G_2 respectively. Then, the algorithm iteratively selects a gene from the remaining part of the given genes. In one iteration, the algorithm computes the percentage of the mutually exclusive cases between the selected gene and the two sets (virtual genes). If the score for one set, for example G_1, is higher than a preset threshold and the score for another set, G_2, is lower than the threshold, this gene seems to have a mutual exclusive pattern with G_1. In this case, the algorithm adds this gene to G_2, if the number of genes in G_2 does not reach a preset threshold C. Here, we set a threshold C to control the size of the set because in some cases, the size of a set may be limited by prior knowledge, such as pathway information, etc. The algorithm continues until either both sets reach the size of C or no gene satisfies the score threshold.

In the implementation for the case of given C, there is another issue: a mutation carried by only a small number of cases may contribute less than a mutation carried by a large number of cases. Thus, to improve the efficiency of the searching process, we further update the greedy strategy as follows: Suppose that gene i is selected from the remaining part of the given genes. Let $E_{i,j}$ be the score between gene i and gene j, which is selected from a set, E_{set} be the scores between the virtual genes without gene i. Then, gene i is added if one more condition is satisfied: if $E_{i,j}$ reaches a percentage of the highest E_{set}, the algorithm adds it to the set which achieved the highest score. In this way, we have a higher probability of having a limit number of genes to maximize the percentage of the mutual exclusions. Finally, the sets of genes are sorted by the scores, where the top-ranked ones are retained. The pseudo code of the search algorithm is as follows:

Algorithm1

INPUT: a 0-1 matrix

OUTPUT: a pair of gene sets, xset and yset, |ultimate|=U, U is a user-setting threshold

---

1. Begin
2. Initial: |Initial|=1000←{xset}+{yset}:xset={},yset={}
3. for(xset:|xset|=1, yset:|yset|=1←any two genes in the matrix) do
4.   Initial:|Initial|=1000←({xset}+{yset},Initial) with maximum score
5. while({xset}+{yset}←each Initial) do
6.   {already}←{xset}+{yset}
7.   for(Z←each gene in the matrix) do
8.   for(R←each gene in {yset}) do
9.   if(score of {Z,R} < score/3 of {xset,yset}) then goto 13
10.   if(score of {xset+Z,yset} > score of {xset,yset}:|xset+Z| < C and Z maximizes score of {xset+Z,yset}) then
11.   C:gene number in a mutually exclusive gene set
12.   {xset}←{xset}+Z
13.   for(R←each gene in {xset}) do
14.   if(score of {Z,R} < score/3 of {xset,yset}) then goto 17
15.   if(score of {xset,yset+Z} > score of {xset,yset}:|yset+Z| < C and Z maximizes score of {xset,yset+Z}) then
16.   {yset}←{yset}+Z
17.   if(|yset|=C&|xset|=C) then
18.   ultimate:|ultimate|=U({xset}+{yset},ultimate) with maximum score
19.   Initial←{xset}+{yset}:{xset}+{yset}⊆Initial,{xset}+{yset}not in{already}
20. End

---

3 Experiments and Results

We performed experiments on three datasets. The first dataset, named (a) in the following figures, was randomly generated according to the mutation rate 0.15 per gene per case. The second dataset, named (b) in the following figures, was generated by ms software [11]. ms is a widely used simulator for population genomics, whose computational model follows a Wright–Fisher neutral model. The parameters were set as follows: -t 0.02 -r 100 1000 -s 50. Specifically, we planted a pair of mutually exclusive sets with 5 genes in each set. Half of the cases had a mutation rate $q = 0.4$

in one set of genes, while half of the cases had the same q in the other set of genes. The genes with mutation rates of less than 1% were removed. For the simulation datasets, we repeated experiments for 100 times to calculate the average performance. The third dataset, named (c) in the following figures, was a set of real mutation calls from 477 TCGA cases across multiple cancer types, and the total number of the mutation sites was 28507.

We set the size threshold C to 3 and set the percentage threshold for E_{set} to 1/3. For a dataset, we enumerated all of the gene combinations limited by C, and then recorded the top 200 results with the highest scores as the true results. Then, the proposed algorithm was tested on the dataset, and the results were compared to the true results. Based on the comparisons, we calculated the recall rates and accuracies to measure the performance.

3.1 The Performance Tests on Accuracy and Recall Rate

We first verified the performance on the basis of the mutual relationship mining model, and then verified the model advantages in the subsequent experiments. On one hand, to speed up the calculation, the greedy strategy ignored a lot of calculations on mutual exclusions during the selections of genes, which might hurt the recall rates. On the other hand, the greedy strategy calculated the scores of the formed gene sets instead of the final gene sets, and adopted an empirical threshold, which might reduce the accuracies. Thus, we first tested these issues. On the three datasets without exceptions, we obtained the recall rates and accuracies by comparing the results of the proposed algorithm to the true results. Besides, the results were sorted by the scores. Since the acceptances of the results were different between different scores. We computed the recall rates and accuracies of results with score in the top 200 results, and drew the trend line chart shown in Fig. 1(a), (b), (c).

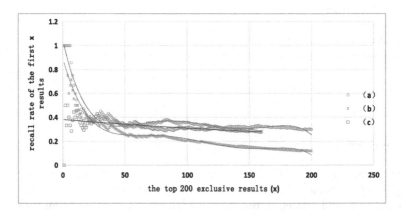

Fig. 1. The recall rates of greedy strategy based on the datasets (a), (b) and (c), respectively.

According to Fig. 1, the score of the xth result was gradually reduced. Along with the decreasing of the scores, the recall rates of the first x results gradually decreased. The trend was particularly evident on datasets (a) and (b). On dataset (c), although the recall rates had relatively larger fluctuations, the overall trend was still declining. We found that as scores of the results declined, the recall rates of the greedy strategy gradually decreased. Moreover, it could even achieve a recall rate close to 100% in some of the superior results. This was reasonable due to the definition of the superior results and the acceptances of results.

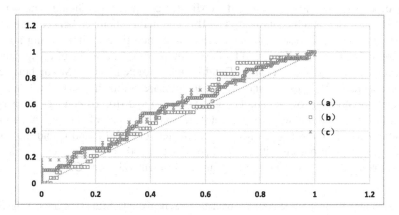

Fig. 2. The ROC curves of the results by greedy strategy on the datasets (a), (b) and (c), respectively.

In addition, we established the receiver operating characteristic (ROC) curve to judge the accuracies of the results. The curves were shown in Fig. 2(a), (b), (c). According to Fig. 2, both the trend line and the standard line passed through points (0, 0) and (1, 1). Based on a brief view of the curves, it was obvious that the curves were almost above the standard line. This indicated that the slopes of the trend lines were declining. Based on the definition, the more convex the curve was, the higher the accuracy the superior results had. The curve indicates the results with the highest score have higher accuracy. On average, the total numbers of the false positive results of the datasets (a) and (b) among the first 200 results were 140 and 155, respectively. However, some of the latter false positives were not completely wrong. Because when we revisited these results, we found that there might be the mutually exclusive patterns among some of those genes, while the patterns were not significant among the other genes. It was also possible that those genes might have mutual exclusions not included in the top 200 true sets.

Note that, we set the size threshold C to 3 before. However, for each dataset, if we increased parameter C repeatedly until the results were stable, then the recall rate for the common problem could be calculated. Here, we used the largest number of exceptions in each result to count the recall rate. On average of 100 datasets, the recall rates of datasets (a) and (b) were closer to 100% and 88.6%, respectively. In many datasets, the parameter C stopped at 5. We also observed that, for dataset (a), when we set the mutation rate $q < 0.2$ or $q > 0.4$, the performance of the algorithm was significantly decreased. For dataset (b), when we set the mutation rate $q < 0.4$, the performance of the algorithm was significantly decreased. We thought that these might be caused by the large interference of the genes outside the exceptions.

3.2 Comparison Experiments with Existing Approaches

CoMEt is a popular method for identifying the intra-set mutually exclusive patterns. To the best of our knowledge, we do not find a suitable algorithm for inter-set mutual exclusion detection. Thus, we adopted CoMEt for comparisons. There is another version, WExT, which is quite similar to CoMEt. The main difference between these two methods is that WExT greatly shortens the computation time by introducing a saddle point approximation. CoMEt and WExT have outperformed a series of computational approaches including Dendrix, Multi-Dendrix, muex, mutex and MEMo on both recall rate and accuracy [2]. Thus, it is meaningful to compare our method to CoMEt. We then conducted the comparison on simulation datasets (a) and (b). To conduct a fair comparison, we set two methods to generate the same amount of the preset mutually exclusive pairs of genes: $k * s$ (a MCMC parameter) in CoMEt, and C and the output number of results in our algorithm. We kept $C * C * 100 = 0.5 * k * (k - 1) * s * 10$. Here, we reserved the top 100 results by using different values of C.

Table 1. Comparisions on true positives on dataset (a) between CoMEt and proposed algorithm

Parameter settings			True positives of pairs of genes		True positives of genes		Aggregated results	
k	C	$k * s$	CoMEt	Our algorithm	CoMEt	Our algorithm	CoMEt	Our algorithm
2	1	NA	NA	**5/5**	NA	**10/10**	NA	(2/2)/(2/2)
	2	2 * 40	2/5	**5/5**	**10/10**	6/10	(2/2)/(2/2)	(2/2)/(2/2)
	3	3 * 30	5/5	5/5	**10/10**	6/10	(2/2)/(2/2)	(2/2)/(2/2)
3	2	2 * 40	3/9	9/9	5/9	**9/9**	(2/3)/(2/3)	(3/3)/(3/3)
	3	3 * 30	6/9	8/9	7/9	**9/9**	(3/3)/(3/3)	(3/3)/(3/3)
	4	4 * 28	**7/9**	6/9	8/9	8/9	(3/3)/(3/3)	(3/3)/(3/3)
4	3	3 * 30	12/18	**18/18**	11/12	**12/12**	(3/4)/(4/4)	(4/4)/(4/4)
	4	4 * 28	11/18	**13/18**	11/12	11/12	(3/4)/(4/4)	(4/4)/(4/4)
	5	5 * 25	12/18	12/18	11/12	11/12	(3/4)/(4/4)	(4/4)/(4/4)

For simulation dataset (a), the dataset was generated following the way of CoMEt: We planted a mutually exclusive set. The set consisted of $k = 2$–4 genes, which was randomly selected. The mutation rates for two genes were (0.15, 0.35), while for three genes and four genes were (0.15, 0.35, 0.5) and (0.15, 0.35, 0.25, 0.25), respectively. Then, for the rest genes, the mutation rates were ranged from 0.15 to 0.67. We compared the two methods according to the averages of the true positive exceptions of pairs of genes pairs, the true positive exception genes and the aggregated exceptions. The results were listed in Table 1. For $k = 2$, the experiments were repeated 5 times, where

5 pairs and 10 genes were planted. For $k = 3$, the experiments were repeated 3 times, where 9 pairs and 9 genes were planted. For $k = 4$, the experiments were repeated 3 times, where 18 pairs and 12 genes were planted. In Table 1, the result of true positives of pairs of genes indicated the number of recalled pairs/the number of preset pairs. The result of true positives of genes indicated the number of recalled genes/the number of preset genes. The result of aggregated result denoted (the best result obtained/k)/(the preset result/k). According to Table 1, our algorithm obtained higher recall rates than CoMEt in most of the cases.

For the ms generated dataset, we updated the command to "./ms 10000 1 -t 0.02 -r 100 1000 -s 1000" because we found that CoMEt performed better under these settings. After ms generated the raw dataset, we first selected 6 genes (sites) randomly, and then removed the cases that did not conform to mutually exclusive patterns between the two gene sets. For the cases carrying the mutually exclusive patterns, if the scores between any pair of the 6 selected genes were positive, the dataset was accepted and sampled with a certain number of cases. Here, we set the percentage of the mutually exclusive cases between any pair of the selected genes not high enough, because we aimed to accommodate and preserve the relationships among other mutations. Otherwise, the genes would be re-selected randomly. The dataset was then roughly similar to a dataset with $k = 6$ or $C = 3$. We kept $C * C * 100 = 0.5 * k * (k - 1) * s * 10$. We collected the results by altering C ranging from 2 to 9, which were shown in Table 2. Similar to Table 1, the result of true positives of pairs of genes indicated the number of recalled pairs/the number of preset pairs. The result of true positives of genes indicated the number of recalled genes/the number of preset genes. The result of aggregated result denoted (the best result obtained/k)/(the preset result/k).

Table 2. Comparisions on true positives on dataset (b) between CoMEt and proposed algorithm

Parameter settings		True positive exceptions of pairs of genes		True positive exceptions of genes		Aggregated exceptions	
C	$k * s$	CoMEt	Our algorithm	CoMEt	Our algorithm	CoMEt	Our algorithm
2	2 * 40	3/15	**9/15**	3/6	**6/6**	(2/6)/(2/6)	(4/6)/(4/6)
3	3 * 30	4/15	**9/15**	4/6	**6/6**	(2/6)/(3/6)	(5/6)/(6/6)
4	4 * 28	4/15	**9/15**	5/6	**6/6**	(2/6)/(4/6)	(5/6)/(6/6)
5	5 * 25	3/15	**9/15**	4/6	**6/6**	(2/6)/(5/6)	(5/6)/(6/6)
6	6 * 24	9/15	9/15	6/6	6/6	(3/6)/(6/6)	(5/6)/(6/6)
7	7 * 23	6/15	**9/15**	5/6	**6/6**	(3/6)/(6/6)	(6/6)/(6/6)
8	8 * 23	11/15	9/15	6/6	6/6	(3/6)/(6/6)	(6/6)/(6/6)
9	9 * 23	9/15	9/15	6/6	6/6	(3/6)/(6/6)	(6/6)/(6/6)

From Table 2, we observed that, when setting the same number of outputting the mutually exclusive pairs, our algorithm achieved higher recall rates than CoMEt, in more than half of the cases. Moreover, the results of our algorithm had more repeated true positives, while CoMEt sometimes had more false positives. In addition, we found that the results were a little better if $\frac{C}{k*s}$ was slightly less than k, thus, empirically, it might be better to choose a smaller $\frac{C}{k*s}$ for unknown datasets.

As CoMEt is an MCMC-based approach, it is important to consider the stability on different parameter configurations. Here we focus on the parameter configuration on $\frac{C}{k*s}$. For the recalled true positives, we tried different settings of $k*s$, the results were shown in Fig. 3. From Fig. 3, we might conclude that CoMEt seemed a little sensitive when the value of $k*s$ was changing. As CoMEt presented fluctuation along with the altering of $k*s$, we used the overall trend of recalls in some results. According to the experimental results, the numbers of genes clustering in one set was increasing along with the increasing of $k*s$.

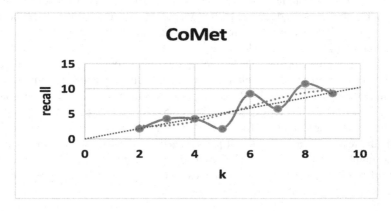

Fig. 3. The recall trend graph of exception maximally aggregated with k*s changing

The results on the datasets (b) were somehow more important than results on the datasets (a) because ms generated the datasets by considering the models in population genetics, which were much closer to the real cases. Moreover, the mutation rates and the percentage of perfect mutually exclusive cases were lower than those in datasets (a), which significantly challenged the computational methods. Therefore, our algorithm was a little better than CoMEt on dealing with the datasets with these features.

3.3   Real Data Application

Finally, we applied our approach on a real data dataset consisting of 477 cases with 28507 mutations. The mutation calls were obtained following the WashU MGI mutation calling pipeline [3, 4]. For each case, the raw whole exome sequencing data were mapped to reference genome hg19 by bwa under default parameter settings. Then, three variant callers, GATK, VarScan2 and Pindel, were adopted to call the candidate mutations. A two-out-of-three filter was applied to all of the candidate mutations, which kept the mutation calls which were supported by at least two of these three callers. The candidate mutations were also filtered by WashU mutation filters [4]. According to this computational pipeline, we obtained a total number of more than 10 million mutational events, including both SNVs and indels shorter than 75 bp, across 477 cases. We downloaded the gene annotations and located the mutations into the genes. Figure 4 gave a brief view on the distribution of the numbers of the mutations in the different numbers of the genes. The horizontal axis referred to the numbers of genes, while the vertical axis referred to the numbers of the mutations harbored. The trend line was similar to a power-rate distribution curve with a long tail.

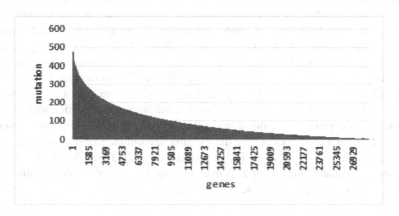

Fig. 4. The distribution of the mutations captured from the TCGA dataset

The ROC curve was shown in Fig. 2(c). From the curve, there were no false positives at the beginning 8 results, while the initial recall rates reached nearly 100%. These results were considered the most important ones and demonstrated significant mutual exclusive patterns across the datasets. The total number of the false positive results was 81. Our algorithm detected multiple sets of genes. Many of them were reported as cancer susceptibility genes. Here, we only highlighted some of the pairs with high scores, such as GPRIN2, KDSR and IDH1, PTEN, which were reported to have mutually exclusive patterns. GPRIN2 is a G protein-regulated inducer of neurite outgrowth 2, which was reported to be mutually exclusive with TP53 and was confirmed in the enrichment of differentially expressed genes in some cancers [12, 13]. KDSR is 3-ketodihydrosphingosine reductase, which was tested to be mutually exclusive with IDH1, PTEN, and was reported to cause recessive progressive symmetric erythrokeratoderma [14]. Some other genes appearing in our results were also have potential mutually exclusive patterns, whose impacts on cancers were already reported in some literatures: IFI44L [15], MTCP1 [16], LIPF [17], ZNF503 [18], NR2F1-AS1 [19] and DSC1 [20]. Moreover, our results show that a large number of pseudogenes, such as RP11-like, might also have the mutual exclusive patterns, whereas researches about their roles in cancers are limited currently [21].

4 Discussion

In this article, we propose an improved algorithm to efficiently detect the mutually exclusive patterns from cancer mutation data. The proposed algorithm aims deal with the inter-set mutual exclusion problem, which herein includes two sets of genes and the genes/mutations in the different sets are mutually exclusive. The proposed algorithm improves the calculation of the score of mutual exclusion. The improved measurement considers the percentage of supporting cases, the approximate exclusivity degree and the pair-wise similarities of two genes. Moreover, the proposed algorithm adopts a greedy strategy to generate the sets of genes. Different from the existing approaches, the greedy strategy considers the scores of mutual exclusion between both the genes and virtual genes, which benefits the selection with the size restrictions. We conducted a series of experiments to verify the performance, using simulation datasets and real data from TCGA cases. According to the results, the proposed algorithm showed good performance under different simulation configurations. The recall rates and accuracies on simulation dataset outperformed the existing algorithm, CoMEt, in most of the cases. Moreover, some of the exclusive patterns detected from TCGA dataset were supported by published literatures. However, the proposed algorithm could only handle two sets of genes, which should be further improved.

Acknowledgement. This work is supported by the National Science Foundation of China (Grant No: 31701150) and the Fundamental Research Funds for the Central Universities (CXTD2017003).

References

1. Dees, N.D., Zhang, Q., Kandoth, C., et al.: MuSiC: identifying mutational significance in cancer genomes. Genome Res. **22**(8), 1589–1598 (2012)
2. Vandin, F., Upfal, E., Raphael, B.J.: De novo discovery of mutated driver pathways in cancer. Genome Res. **22**(2), 375–385 (2012)
3. Huang, K., Mashl, R.J., Wu, Y., et al.: Patheogenic germline variants in 10,389 adult cancers. Cell **173**(2), 355–370 (2018)
4. Lu, C., Xie, M., Wendl, M., et al.: Patterns and functional implications of rare germline variants across 12 cancer types. Nat. Commun. **6**, 10086 (2015)
5. Wendl, M.C., Wallis, J.W., Lin, L., et al.: PathScan: a tool for discerning mutational significance in groups of putative cancer genes. Bioinformatics **27**(12), 1595–1602 (2011)
6. Giovanni, C., Ethan, C., Chris, S., et al.: Mutual exclusivity analysis identifies oncogenic network modules. Genome Res. **22**(2), 398 (2012)
7. Aldape, K.D., Sulman, E.P., Settle, S.H., et al.: Discovering functional modules by identifying recurrent and mutually exclusive mutational patterns in tumors. BMC Med. Genomics **4**(1), 34 (2011)
8. Leiserson, M.D.M., Wu, H.T., Vandin, F., et al.: CoMEt: a statistical approach to identify combinations of mutually exclusive alterations in cancer. Genome Biol. **16**(1), 160 (2015)
9. Kim, Y.A., Cho, D.Y., Dao, P., et al.: MEMCover: integrated analysis of mutual exclusivity and functional network reveals dysregulated pathways across multiple cancer types. Bioinformatics **31**(12), i284 (2015)

10. Leiserson, M.D.M., Reyna, M.A., Raphael, B.J.: A weighted exact test for mutually exclusive mutations in cancer. Bioinformatics **32**(17), i736–i745 (2016)
11. Hudson, R.R.: Generating samples under a Wright-Fisher neutral model of genetic variation. Bioinformatics **18**(2), 337–338 (2002)
12. Wang, C., Li, L., Cheng, W., et al.: A new approach for persistent cloaca: laparoscopically assisted anorectoplasty and modified repair of urogenital sinus. J. Pediatr. Surg. **50**(7), 1236–1240 (2015)
13. Khalilipour, N., Baranova, A., Jebelli, A., et al.: Familial esophageal squamous cell carcinoma with damaging rare/germline mutations in, KCNJ12/KCNJ18, and GPRIN2, genes. Cancer Genet. **221**, 46–52 (2017). S221077621630299X
14. Boyden, L.M., Vincent, N.G., Zhou, J., et al.: Mutations in, KDSR, Cause Recessive Progressive Symmetric Erythrokeratoderma. Am. J. Hum. Genet. **100**(6), 978–984 (2017)
15. Huang, W.-C., Tung, S.-L., Chen, Y.-L., et al.: IFI44L is a novel tumor suppressor in human hepatocellular carcinoma affecting cancer stemness, metastasis, and drug resistance via regulating met/Src signaling pathway. BMC Cancer **18**(1), 609 (2018)
16. Guignard, L., Yang, Y.S., Padilla, A., et al.: The preliminary solution structure of human p13MTCP1, an oncogenic protein encoded by the MTCP1 gene, using 2D homonuclear NMR. J. de Chim. Phys. et de Phys.-Chim. Biol. **95**(2), 454–459 (1998)
17. Kong, Y., Zheng, Y., Jia, Y., et al.: Decreased LIPF expression is correlated with DGKA and predicts poor outcome of gastric cancer. Oncol. Rep. **36**(4), 1852–1860 (2016)
18. Shahi, P., Slorach, E.M., Wang, C.Y., et al.: The transcriptional repressor ZNF503/Zeppo2, promotes mammary epithelial cell proliferation and enhances cell invasion. J. Biol. Chem. **290**, 3803–3813 (2015)
19. Huang, H., Chen, J., Ding, C.M., et al.: LncRNA NR2F1-AS1 regulates hepatocellular carcinoma oxaliplatin resistance by targeting ABCC1 via miR-363. J. Cell Mol. Med. **22**, 3238–3245 (2018)
20. Myklebust, M.P., Fluge, Ø., Immervoll, H., et al.: Expression of DSG1 and DSC1 are prognostic markers in anal carcinoma patients. Br. J. Cancer **106**(4), 756–762 (2012)
21. Poliseno, L.: Pseudogenes: newly discovered players in human cancer. Sci. Signal. **5**(242), re5 (2012)

Qualitative Comparison of Selected Indel Detection Methods for RNA-Seq Data

Tamara Slosarek, Milena Kraus$^{(\boxtimes)}$, Matthieu-P. Schapranow, and Erwin Boettinger

Digital Health Center, Hasso Plattner Institute,
Rudolf-Breitscheid-Str. 187, 14482 Potsdam, Germany
tamara.slosarek@student.hpi.de,
{milena.kraus,schapranow,erwin.boettinger}@hpi.de

Abstract. RNA sequencing (RNA-Seq) provides both gene expression and sequence information, which can be exploited for a joint approach to explore cell processes in general and diseases caused by genomic variants in particular. However, the identification of insertions and deletions (indels) from RNA-Seq data, which for instance play a significant role in the development, detection, and treatment of cancer, still poses a challenge. In this paper, we present a qualitative comparison of selected methods for indel detection from RNA-Seq data. More specifically, we benchmarked two promising aligners and two filter methods on simulated as well as on real RNA-Seq data. We conclude that in cases where reliable detection of indels is crucial, e.g. in a clinical setting, the usage of our pipeline setup is superior to other state-of-the-art approaches.

Keywords: RNA-Seq · Variant calling · Indels

1 Introduction

RNA-Seq is commonly applied for gene expression analysis to capture the abundance of transcripts present in the cells or tissues of interest. Aside from costlier whole-genome sequencing (WGS) and whole-exome sequencing (WES) methods, it is also possible to detect variants from RNA-Seq data as the actual sequence of the transcript is preserved [4,16]. Therefore, RNA-Seq provides a joint approach to explore cell processes in general, and diseases caused by genomic variants in particular, as it yields both expression and sequence information [19]. However, while the variant detection for single nucleotide variants (SNVs) is fairly accurate [15], the identification of indels, which for instance play a significant role in cancer development, detection, and treatment, still poses a challenge due to an increased complexity based on splicing events and allele-specific expression [22].

The main processing steps to generate genomic variants from RNA-Seq data include quality control and trimming of the raw sequencing reads, alignment of reads to a reference, filtering and preprocessing of aligned read information, and finally the variant calling step (see Fig. 1). In Sect. 2 we describe related

I. Rojas et al. (Eds.): IWBBIO 2019, LNBI 11465, pp. 166–177, 2019.
https://doi.org/10.1007/978-3-030-17938-0_16

literature in which either separate instances of the above steps were thoroughly benchmarked or full pipeline solutions were tested, with the main disadvantage of focusing only on SNVs. Current pipeline solutions are able to reach a recall in indel detection of up to 70% [22] when restricting on high coverage regions.

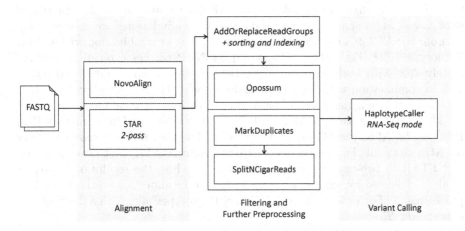

Fig. 1. Processing steps of our pipeline include alignment, filtering and further pre-processing, and variant calling. Proposed optimizations are colored green. Dashed lines indicate a comparison to the standard procedure that is colored orange. Steps in black are required in any setup. (Color figure online)

For our own approach we hypothesized that by using a splice-aware and indel-sensitive aligner together with suitable filter methods (e.g, Opossum [15]), we can improve the recall of indel detection from RNA-Seq data. In Sect. 3 we elaborate on our proposed changes of the published standard procedure. In order to test the performance, we benchmarked steps of our pipeline and their combinations on simulated data as well as on real data. In Sect. 4 we demonstrate how our approach increases the recall while maintaining precision and discuss strengths and limitations when compared to similar approaches. We conclude our findings in Sect. 5.

2 Related Work

In the following, we describe findings from aligner comparisons and filter steps, followed by a summary of current optimized pipelines for SNV and indel detection. This is done in order to decide for suitable methods and evaluation approaches on RNA-Seq data.

Baruzzo et al. (2017) extensively benchmark 14 RNA-Seq aligners on simulated data sets based on human and malaria genomes in different complexity levels. In their study NovoAlign [14] reaches the highest recall of about 80% but a comparably low precision of approximately 60% for insertions and 75% for deletions when utilizing annotations. Across all data sets, CLC [18], NovoAlign, and GNSAP [23] perform best, while STAR [5] reaches a high precision but only a mediocre recall for lower complexity levels. However, for the most complex data set, STAR reaches the fourth-best recall.

Sun et al. (2016) specifically assess the performance for indel detection of seven RNA-Seq aligners in combination with five variant callers [22]. Simulated data as well as human lung cancer data with known indel sites are used for comparison. HISAT2 [9] and GSNAP [23] perform best in combination with GATK HaplotypeCaller [17] and BCFTools [11]. GSNAP reaches a recall of approximately 70% with both variant callers, HISAT2 one of about 60%, and STAR 70% in combination with BCFTools and 60% with HaplotypeCaller. Sun et al. emphasize that alignment is the most important step for indel identification.

The Broad Institute of MIT and Harvard develops and recommends Best Practices based on their Genome Analysis Toolkit (GATK) for variant discovery on NGS data [3]. For processing RNA-Seq data, GATK Best Practices suggest STAR in two-pass mode for alignment based on the results of Engström et al. [7] followed by various filtering and score recalibration methods for preprocessing [3]. For variant discovery, GATK HaplotypeCaller with a filtering step is proposed [2].

Oikkonen et al. (2017) developed an alignment filter approach called Opossum [15] to enhance the detection of SNVs from RNA-Seq data. Different pipelines are evaluated that include TopHat2 and STAR as aligners, GATK Best Practices and Opossum as filters, and Platypus [20] and GATK HaplotypeCaller as variant callers. The Opossum filter raises both precision and recall by on average about 0.7% and 1.7% respectively compared to GATK Best Practices. However, the evaluation has not yet been conducted for indels.

In our approach, we combine findings from aligner comparisons, i.e. NovoAlign outperforms all other aligners in terms of recall in indel detection, and new filtering methods (e.g. Opossum), that until now were only used for the optimization on SNV detection, into a new pipeline setup.

3   Materials and Methods

Figure 1 depicts the general setup of a pipeline to call indels from RNA-Seq data as adapted from the GATK Best Practices. Original steps are highlighted in orange and our proposed changes in green. In the following, we first explain the choice of included tools, before we describe evaluation methods.

3.1   Choice of Included Tools

RNA-Seq reads in form of FASTQ files are first aligned to a reference sequence.

We use STAR in two-pass mode as suggested by GATK Best Practices as a baseline. Please note that a two-pass mode is comparable to using an aligner with annotations. STAR reaches high precision with low run times. Additionally, STAR is very popular in all benchmarks presented in Sect. 2, which enables us to compare our results to other studies. For our own approach we opted for NovoAlign [14], since it achieves a high recall, especially for the identification of indels [1]. In general, NovoAlign requires annotations to detect splice junctions from RNA-Seq data. However, we were not able to include them as described in the user guide [13] but were still able to run our experiments. The specific consequences of the usage of annotations will be discussed in Sect. 4. We use version 3.09.0 of NovoAlign and version 2.6.0c of STAR.

After the alignment, the mapped reads are processed by GATK AddOr-ReplaceReadGroups to ensure compatibility with succeeding tools. Filtering is applied as stated in Opossum [15] in its current revision fe8f72e provided on GitHub that improves the detection of SNVs by filtering alignment results, but has not yet been tested for indel detection. Alternatively, methods suggested by the widely used GATK Best Practices are included, while omitting steps mainly required for data with bad quality, such as indel realignment and base quality recalibration [2,12]. Finally, the adapted alignment is analyzed by GATK HaplotypeCaller [17] that also is proposed by GATK Best Practices and performs well for indel detection [22]. All tools are run with their default parameters for RNA-Seq data to assess their robustness and general applicability to unexplored data sets. The GATK version we use is 4.0.9.0.

3.2 Evaluation Strategies

In a first experiment, the performance of aligners is compared on simulated data, independent from other processing methods. We do this despite the exhaustive evaluation of Baruzzo et al., since NovoAlign without annotations was not assessed yet. Thereafter, we benchmark the full pipeline including further preprocessing and variant calling on real data from the GM12878 cell line. We focus on the performance of Opossum and GATK filters to select the methods best suitable for an optimized setup. All experiments were conducted on a virtual machine with the following specification: Eight cores of Intel(R) Xeon(R) CPU E7- 8870 @ 2.40 GHz, 48 GB RAM, and 1 TB disk space.

Comparison of Alignments with Simulated Data Sets. For the alignment evaluation, we use simulated data sets as generated with the Benchmarker for Evaluating the Effectiveness of RNA-Seq Software (BEERS). They were published by Baruzzo et al.[1] in context of their comparative studies [1] and contain data in three complexity levels. Each set holds 10^7 paired-end reads with a length of 100 bases. T1 has low variation and error rates that are expected in a human genome sequenced with Illumina (SNV rate 0.001, indel rate 0.0001, error rate 0.005), T2 and T3 increase the complexity with higher rates (T2: SNV rate 0.005, indel rate 0.002, error rate 0.01; T3: SNV rate 0.03, indel rate 0.005, error rate 0.02). Simulations were based on the human genome (hg19).

---

[1] http://bioinf.itmat.upenn.edu/BEERS/bp1/datasets.php.

The aligned sequences are compared with the true alignment given in a CIGAR file that is generated when a data set is simulated using the scripts[2] as provided by Baruzzo et al. The accuracy (ACC) is measured on read- and base-level. Moreover, the false discovery rate and false negative rate for insertions and deletions are provided on read- and junction-level, from which precision (PREC) and recall (REC) are derived.

Evaluation of Pipelines with High-Confidence Calls. In order to test on real RNA-Seq data sets, the GM12878 cell line can be used together with GIAB high-confidence variant calls [24]. We analyze an RNA-Seq data set for GM12878 that is publicly available on ENCODE [6,21] with $117,876,320$ paired-end reads (GEO accession code: GSE86658, experiment ENCSR000COQ, sample ENCBS-095RNA). As a reference, we use GIAB high-confidence variants in version 3.3.2.

To minimize run time, we decided to limit the analysis to single chromosomes, based on the distribution of indels in GIAB high-confidence variants. We choose chromosome one, 17, and 21 to represent a large, medium, and small indel abundance. Counts were obtained from variations that are present with a minimum coverage of 2X in the STAR alignment using our custom script[3]. With this low threshold we aim to reduce the number of excluded variants to a minimum while keeping a rather large amount for the evaluation.

As an evaluation tool, the GIAB consortium prompts the use of Haplotype Comparison Tools (hap.py) [10]. Two files are required as input for hap.py, a truth set, in our case the high-confidence variant calls, and a query set, which is the pipeline result. For the evaluation of RNA-Seq pipelines, the GIAB high-confidence calls need to be restricted to regions present in the particular RNA-Seq data. Separately for SNVs and indels, hap.py assesses true positives (TP), false positives (FP), false negatives (FN), and non-assessed calls (UNK) that lie outside confident call regions defined in a third input file. False negatives refer to variants that were not identified, as well as genotype or allele mismatches. From these counts, precision, recall, the F1 score, and the ratio of non-assessed calls to all called variants are calculated. Precision and recall refer to genotype matches.

4   Results and Discussion

In this section, evaluation results of the RNA-Seq aligners and pipelines defined in Sect. 3 are presented and discussed.

4.1   STAR Shows High Precision—NovoAlign Shows High Recall

Figure 2 depicts the results of the aligner comparison, exact values are given in Table 1. For the aligner comparison, we used simulated data sets in different complexity levels. NovoAlign reaches higher recall values in the range of

---

[2] https://github.com/khayer/aligner_benchmark.
[3] https://github.com/tamslo/koala/tree/master/scripts/count_insertions_and_deletions.

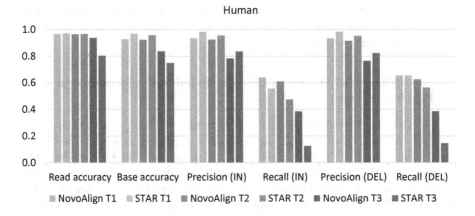

Fig. 2. Comparison of NovoAlign and STAR for simulated data based on hg19. For each measure, results for NovoAlign and STAR are presented alternately, in order of raising complexity levels.

Table 1. Results of experiments on simulated data sets with included indels collected from BEERS evaluation results. H = Human, T1-3 data sets in increasing complexity. For precision (PREC) and recall (REC) superior values of a comparison are **bold**.

Data	IN	DEL	Aligner	ACC (read)	ACC (base)	PREC (IN)	REC (IN)	PREC (DEL)	REC (DEL)
H T1	1,740	1,730	NovoAlign	0.965	0.982	0.936	**0.639**	0.933	**0.655**
			STAR	0.971	0.969	**0.983**	0.557	**0.985**	0.654
H T2	7,110	6,800	NovoAlign	0.965	0.923	0.924	**0.609**	0.915	**0.625**
			STAR	0.967	0.958	**0.956**	0.475	**0.953**	0.564
H T3	18,150	16,450	NovoAlign	0.936	0.837	0.784	**0.385**	0.764	**0.387**
			STAR	0.803	0.749	**0.836**	0.125	**0.824**	0.145

0.385–0.655 and detects both insertions and deletions on a balanced level. STAR achieves a higher precision in the range of 0.824–0.985 and shows a raised sensitivity for deletions. However, the superiority of STAR in precision is rather low with a mean (±standard deviation) of 4.7% (±0.9), while the increase in recall is rather large 13.0% (±8.9) for NovoAlign. With increasing complexity, both aligners show decreased performance, while the recall for NovoAlign declines less severely than for STAR. For lower complexities, STAR is more accurate on read- and base-level, but NovoAlign overcomes STAR for the T3 data sets.

Our results are consistent with the results of Baruzzo et al.: Compared to STAR, NovoAlign reaches a higher recall with a lower precision. However, in our study NovoAlign did not reach the 80% in recall as in Baruzzo et al., which might be due to the lack of annotations. Because STAR in two-pass mode is comparable to one-pass mode including annotations, STAR has a general advantage over NovoAlign. Despite of the annotations NovoAlign still showed better recall values and therefore confirms that NovoAlign is an indel-sensitive aligner, which will be further investigated in the pipeline comparison. However, the effect of annotations—associated with the importance of splice-awareness—for indel detection still needs further investigation.

4.2 NovoAlign and Opossum Show a Reliable Performance in a Pipeline Setup

In this section, we present the variant calling results of the pipeline setups on the GM12878 data set. We focus on our optimized approach that uses NovoAlign with Opossum, and the reference pipeline, which includes STAR together with GATK filter methods. For comparison, results for unfiltered alignments and the combination of STAR with Opossum are included. Variant calling was executed with GATK HaplotypeCaller and applied to chromosomes one, 17, and 21 to cover the range of possible amounts of indels (decreasing from high to low). The evaluation was conducted with GIAB high-confidence variant calls. All analysis steps were run against the hg38 reference because it contains new findings, which compared to hg19 improve the data analysis [8]. Figure 3 depicts pipeline results for genotype matches. The underlying metrics are listed in Table 2, together with results for a further restriction of high-confidence regions, on which we elaborate later.

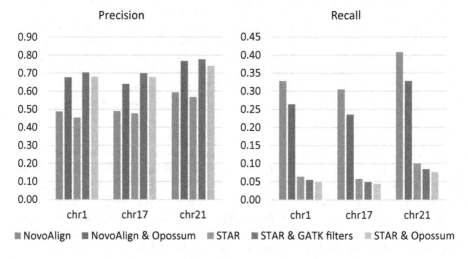

Fig. 3. Results for indel detection of different pipelines. Precision is displayed on the left, recall on the right.

In general, the filtered alignments reach higher precision and lower recall values than unfiltered approaches. In a direct comparison based on the STAR alignment, GATK filters reach slightly better results than Opossum. With regards to whole pipelines, the STAR and GATK filters pipeline on average reaches a

Table 2. Results of different pipelines. The count of actual positives can be computed by adding true positives and false negatives. The number of predicted positives is the sum of true positives, false positives, and non-assessed calls. Results for common alignment regions of NovoAlign and STAR are marked with an *.

Pipeline	Chr	TP	FN	FP	UNK	PREC	REC
NovoAlign	chr1	1,682	3,446	1,776	1,126	0.488	0.328
	chr17	637	1,452	667	482	0.490	0.305
	chr21	387	561	268	118	0.593	0.408
NovoAlign & Opossum	chr1	1,353	3,775	649	563	0.676	0.264
	chr17	493	1,596	278	240	0.640	0.236
	chr21	311	637	95	59	0.768	0.328
NovoAlign & Opossum*	chr1	1,351	3,693	515	533	0.689	0.268
	chr17	493	1,577	271	247	0.646	0.238
	chr21	311	616	88	66	0.781	0.335
STAR	chr1	1,455	21,248	1,757	669	0.453	0.064
	chr17	554	9,016	608	310	0.478	0.058
	chr21	328	2,933	252	95	0.567	0.101
STAR & GATK filters	chr1	1,251	21,452	530	510	0.703	0.055
	chr17	470	9,100	204	210	0.698	0.049
	chr21	276	2,985	80	67	0.777	0.085
STAR & GATK filters*	chr1	1,243	3,801	515	533	0.708	0.246
	chr17	466	1,604	199	219	0.702	0.225
	chr21	276	651	78	69	0.781	0.298
STAR & Opossum	chr1	1,126	21,577	529	455	0.681	0.050
	chr17	420	9,150	200	198	0.678	0.044
	chr21	249	3012	88	60	0.740	0.076

precision of 72.6% and a recall of 6.3%; our optimized NovoAlign and Opossum pipeline reaches a precision of 69.5% and a recall of 27.6%. Compared to STAR, NovoAlign identifies more true positives; the precision is slightly decreased, however, the recall is significantly higher. Nevertheless, the number of high-confidence variant calls considered for STAR and GATK filters notably exceeds the amount for NovoAlign and Opossum, on average multiplied by a factor of four. As a consequence, we conducted an additional evaluation, in which we restricted the high-confidence regions to common alignment regions of STAR and NovoAlign. The results are depicted in Fig. 4. In the further limited regions, 8,041 high-confidence variant calls are present for the selected chromosomes, opposed to 8,165 for NovoAlign alone and 35,534 for STAR. While precision and recall of the NovoAlign-based pipeline only increases marginally, the recall for STAR benefits with an increase in the range of 6.3% to 25.6%. This is due to the drastically decreased number of false negatives.

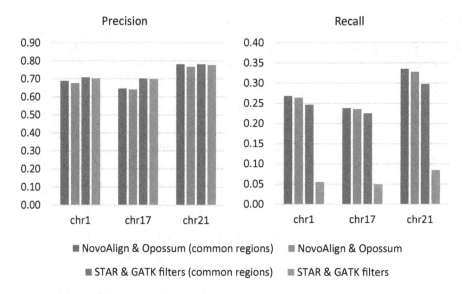

Fig. 4. Pipeline results of NovoAlign with Opossum and STAR with GATK filters, the high-confidence regions were restricted to common alignment regions. For comparison, the previous results for pipelines restricted to their alignment regions are displayed again. Precision is displayed on the left, recall on the right.

In summary, compared to the STAR and GATK filters pipeline, the recall of the NovoAlign and Opossum pipeline is higher, while the precision remains stable. As a consequence, the combination of NovoAlign and Opossum indeed outperforms the reference pipeline. In addition, Opossum and GATK filters were compared directly to each other, based on the alignment with STAR. The results show that GATK filters perform slightly better than Opossum, for both, precision and recall. The improvement in SNV detection could therefore not be reproduced for indels. For common regions of STAR and NovoAlign, the recall of the STAR and GATK filters pipeline closes up to NovoAlign and Opossum, however, does not reach it. Moreover, without another method to compare to, it is not clear, to which regions the STAR alignment needs to be restricted to increase the recall. Most considered high-confidence variants also appear in NovoAlign aligned regions, which suggests that NovoAlign identifies regions with callable insertions and deletions more reliably.

With respect to the results of Oikkonen et al., the precision and recall values of our experiments turn out rather low. This is probably due to significantly more indels in the version of GIAB high-confidence we used [24].

When it comes to real numbers and a clinical application of indel detection from RNA-Seq data, the increase in recall means a larger amount of true insertions and deletions that can, in a subsequent analysis, be considered for their importance in the studied disease, e.g., in tumours.

4.3   STAR and GATK Filters Are Faster than NovoAlign and Opossum

We assessed run time and disk usage for our experiments to comprehend requirements of the included tools. All together, the NovoAlign and Opossum pipeline takes more than three days on our machines, the STAR and GATK filters pipeline two days and a half. The index creation of STAR takes much more time than of NovoAlign, while the alignment is substantially faster. Opossum takes considerably less time than GATK filters and significantly reduces the file size, which diminishes the run time of the variant calling. With uncompressed RNA-Seq data sets, the reference genome, and files with a size of less than 1 GB ignored, the disk usage for the NovoAlign and Opossum pipeline sums up to about 173.5 GB, for STAR and GATK filters to 185 GB.

While both pipelines require a comparable amount of disk space, the STAR and GATK filters pipeline has a considerably lower run time than NovoAlign with Opossum. This is mostly due to a significantly faster alignment. However, Opossum only needs less than half of the run time of GATK filters and speeds up succeeding steps by dramatically reducing the file size.

5   Conclusion

Based on the given results, we suggest to use NovoAlign rather than STAR in alignment tasks that yield a high and reliable amount of insertions and deletions detected from RNA-Seq data. Ultimately, the aligner choice highly depends on the clinical use case; NovoAlign identifies more true variants at the cost of more false positive calls. When a specific variant yields severe health risks or additional tests can be conducted to verify variants, NovoAlign is the better choice. If certainty of called variants is more important, STAR should be used. GATK filters yield a slightly increased performance, however, Opossum is considerably faster and additionally reduces run times for subsequent steps. Therefore, we recommend Opossum, unless superior results are crucial and time is irrelevant.

In summary, we verify that the appliance of an indel-sensitive aligner in combination with suitable filter methods improves the quality of indel detection from RNA-Seq data. Our proposed pipeline with NovoAlign and Opossum achieves a higher recall than the state-of-the-art pipeline recommended by GATK Best Practices while reaching a similar precision. However, interchanging Opossum with GATK filter methods yields the potential for further improvements as well as the usage of NovoAlign with annotations.

Acknowledgement. Parts of this work were generously supported by a grant of the German Federal Ministry of Education and Research (031A427B).

References

1. Baruzzo, G., Hayer, K.E., Kim, E.J., Di Camillo, B., FitzGerald, G.A., Grant, G.R.: Simulation-based comprehensive benchmarking of RNA-seq aligners. Nat. Methods **14**(2), 135 (2017)
2. Broad Institute: Calling variants in RNAseq, January 2017. https://software. broadinstitute.org/gatk/documentation/article.php?id=3891
3. Broad Institute: Introduction to the GATK best practices, January 2018. https:// software.broadinstitute.org/gatk/best-practices
4. Chen, L.Y., et al.: RNASEQR-a streamlined and accurate RNA-seq sequence analysis program. Nucleic Acids Res. **40**(6), e42 (2011)
5. Dobin, A., Gingeras, T.R.: Mapping RNA-seq reads with star. Curr. Protoc. Bioinform. **51**(1), 11–14 (2015)
6. ENCODE Project Consortium and Others: An integrated encyclopedia of DNA elements in the human genome. Nature **489**(7414), 57 (2012)
7. Engström, P.G., et al.: Systematic evaluation of spliced alignment programs for RNA-seq data. Nat. Methods **10**(12), 1185 (2013)
8. Guo, Y., Dai, Y., Yu, H., Zhao, S., Samuels, D.C., Shyr, Y.: Improvements and impacts of GRCh38 human reference on high throughput sequencing data analysis. Genomics **109**(2), 83–90 (2017)
9. Kim, D., Langmead, B., Salzberg, S.L.: HISAT: a fast spliced aligner with low memory requirements. Nat. Methods **12**(4), 357 (2015)
10. Krusche, P., et al.: Best practices for benchmarking germline small variant calls in human genomes. bioRxiv, p. 270157 (2018)
11. Li, H.: A statistical framework for SNP calling, mutation discovery, association mapping and population genetical parameter estimation from sequencing data. Bioinformatics **27**(21), 2987–2993 (2011)
12. Li, H.: Toward better understanding of artifacts in variant calling from high-coverage samples. Bioinformatics **30**(20), 2843–2851 (2014)
13. Novocraft Technologies Sdn Bhd: RNAseq analysis: mRNA and the spliceosome. http://www.novocraft.com/documentation/novoalign-2/novoalign-user-guide/ rnaseq-analysis-mrna-and-the-spliceosome
14. Novocraft Technologies Sdn Bhd: Novoalign reference manual, March 2014. http:// www.novocraft.com/wp-content/uploads/Novocraft.pdf
15. Oikkonen, L., Lise, S.: Making the most of RNA-seq: pre-processing sequencing data with opossum for reliable SNP variant detection. Wellcome Open Res. **2**, 6 (2017)
16. Piskol, R., Ramaswami, G., Li, J.B.: Reliable identification of genomic variants from RNA-seq data. Am. J. Hum. Genet. **93**(4), 641–651 (2013)
17. Poplin, R., et al.: Scaling accurate genetic variant discovery to tens of thousands of samples. bioRxiv, p. 201178 (2017)
18. QIAGEN Bioinformatics: CLC genomics workbench. https://www. qiagenbioinformatics.com/products/clc-genomics-workbench
19. Quinn, E.M., et al.: Development of strategies for SNP detection in RNA-seq data: application to lymphoblastoid cell lines and evaluation using 1000 Genomes data. PloS One **8**(3), e58815 (2013)
20. Rimmer, A., et al.: Integrating mapping-, assembly-and haplotype-based approaches for calling variants in clinical sequencing applications. Nat. Genet. **46**(8), 912 (2014)

21. Sloan, C.A., et al.: ENCODE data at the ENCODE portal. Nucleic Acids Res. **44**(D1), D726–D732 (2015)
22. Sun, Z., Bhagwate, A., Prodduturi, N., Yang, P., Kocher, J.P.A.: Indel detection from RNA-seq data: tool evaluation and strategies for accurate detection of actionable mutations. Brief. Bioinform. **18**(6), 973–983 (2016)
23. Wu, T.D., Nacu, S.: Fast and SNP-tolerant detection of complex variants and splicing in short reads. Bioinformatics **26**(7), 873–881 (2010)
24. Zook, J., et al.: Reproducible integration of multiple sequencing datasets to form high-confidence SNP, indel, and reference calls for five human genome reference materials. bioRxiv, p. 281006 (2018)

Structural and Functional Features of Glutathione Reductase Transcripts from Olive (*Olea europaea* L.) Seeds

Elena Lima-Cabello[1] , Isabel Martínez-Beas[1],
Estefanía García-Quirós[1] , Rosario Carmona[1] ,
M. Gonzalo Claros[2] , Jose Carlos Jimenez-Lopez[1] ,
and Juan de Dios Alché[1(✉)]

[1] Estación Experimental del Zaidín. CSIC, 18008 Granada, Spain
juandedios.alche@eez.csic.es
[2] Departamento de Biología Molecular y Bioquímica, Universidad de Málaga,
Málaga, Spain

Abstract. The olive seed is a promising by product generated in the olive oil related industries, with increasing interest because of its nutritional value and potential nutraceutical properties. Knowledge concerning the antioxidant capacity of this new alimentary material is scarce. Moreover, oxidative homeostasis and signaling involved physiological processes such as development, dormancy and germination in the olive seed are also unknown. Glutathione (one of the most abundant antioxidants in plant cells), is crucial for seeds physiology, and for defense and detoxification mechanisms. The availability of glutathione in its reduced (GSH) and oxidized (GSSG) forms, the ratio of both forms (GSH/GSSG), and their concurrence in other numerous metabolic pathways is tightly regulated by numerous enzymes. Prominent among these enzymes is glutathione reductase (GR), which has been considered essential for seedling growth and development. The present work aims to increase the knowledge about the functional insights of GR in olive seeds. Searching in the olive transcriptome, at least 19 GR homologues (10 from seed and 9 from vegetative tissue) were identified and retrieved. An *in silico* analysis was carried out, which included phylogeny, 3-D modelling of the N-terminus, and the prediction of cellular localization and post-translational modifications (PTM) for these gene products. The high variability of forms detected for this enzyme in olive seeds and their susceptibility to numerous PTMs suggest a relevant role for this enzyme in redox metabolism and signalling events.

Keywords: Antioxidant · Glutathione reductase · Modelling · Olive ·
Post-translational modifications · Seed · Signalling

1 Introduction

Olive tree is the most economically important oil-producing crop in Mediterranean countries. Olive fruits are consumed only after processing, which ends up to either table olives or olive oil. Both types of products are well-known natural sources of phenolic

© Springer Nature Switzerland AG 2019
I. Rojas et al. (Eds.): IWBBIO 2019, LNBI 11465, pp. 178–191, 2019.
https://doi.org/10.1007/978-3-030-17938-0_17

and other antioxidants that exert multiple biological functions [1–3]. The exact profile and content in these antioxidants may vary depending on the varietal origin and processing procedures [4]. Along with developments in the traditional processing in the olive sector, new alternative processing procedures are appearing, including the use of de-stoning alternatives prior to olive milling. Thus, an increasing body of evidence has highlighted the potential of the olive seed as a complementary emerging material. At this regard, other potential properties of this material, such as its antioxidant capacity and the presence of antioxidant enzymes, such as superoxide dismutase, catalase, ascorbate peroxidase, glutathione reductase and others, is being explored, making olive seed itself a material of interest determining a source for novel food applications in the near future [5, 6].

The low-molecular weight tripeptide glutathione is the main antioxidant in mature and stored seeds since ascorbate increases during early seed development but decreases during maturation steps. During development, desiccation, aging, and germination, GSH protects seeds from unregulated oxidative damage that may reduce viability and vigour [7]. Sources of reactive oxygen species (ROS) may vary considerably in different stages of seed development [8]. Down-regulation of metabolism in seeds decreases the generation of ROS, and minimizes membrane damage by lipid and protein oxidation during dehydration. Such regulation of the cytosolic redox environment is vital for cell endurance, which is largely maintained by glutathione reductase (GR) –a flavo-protein oxidoreductase NAD(P)H-dependent cellular enzymatic antioxidant and an important component of ascorbate–glutathione (AsA–GSH) pathway. GR converts oxidized glutathione (GSSG) to reduced glutathione (GSH) thus helping in maintaining a high GSH/GSSG ratio under various abiotic stresses [9]. Thereby, GR helps in maintaining GSH pool and a reducing environment in the cell, which is crucial for the active functioning of proteins. Also, GSH has been suggested to be a potential regulator of epigenetic modifications, playing important roles in the regulation of genes involved in the response of seeds to a changing environment. GRs have been purified from diverse plant species. The native enzyme of most GRs is a homodimer of c. 100–120 kDa, and its subunit size ranges between 53 kDa and 59 kDa [10].

Two plant GR genes, GR1 and GR2, have been identified in several plant species, including pea, soybean, and Arabidopsis [11–14]. In Arabidopsis, GR1 encodes a chloroplastidic GR, with 49% identity to Arabidopsis GR2. Pea GR1 targets to the mitochondria a newly formed protein, while GR2 gene encodes a cytosolic form of the enzyme as a response to oxidative stress at the transcriptional level, in contrast to GR1. In the bean seed, GR activity was higher as the seeds were acquiring desiccation tolerance. Contrarily, in the recalcitrant Quercus embryonic axes and in Triticum durum seeds, GR activity remained constant in early development then abruptly declined during desiccation [15]. Previous bioinformatics studies in the olive pollen have been able to detect the existence of GR1 and GR2 genes, homologous to other GR genes described in plant species such as A. thaliana or T. cacao [16]. Prediction of the cellular location of these genes showed that GR2 expression occurs in the chloroplast, thus confirming previous studies which suggested that this isoform is involved in adequate pollen development [7].

The present study aimed to identify signs of the presence of GR transcripts in olive seeds at the transcriptomic level, and to perform an *in silico* analysis of these transcript, which includes phylogeny and 3-D modelling of the N-terminus, as well as predictions of their cellular localization and susceptibility to PTMs. Such analyses will represent the basis for future experimental determinations of the enzyme levels and activity, allowing us to determine the antioxidant capacity of these seeds. These data of great importance for determining their use as new potential food ingredients.

2 Materials and Methods

2.1 Identification of GR Transcripts in the Olive Reproductive Transcriptome

Different strategies for identification of GR full-length transcripts were applied. The searches were conducted by using GO, EC, KEGG and InterPro terms and codes definitions, as well as gene names and orthologues against the annotated seed transcriptome, available at: http://reprolive.eez.csic.es/olivodb/ [17]. BLAST (Basic Local Alignment Search Tool) (https://blast.ncbi.nlm.nih.gov/Blast.cgi) searches were carried on using homologous sequences from close species available in public databases and library resources.

2.2 *In silico* Analysis of the Sequences

Nucleotide sequences were aligned using CLUSTALX2 multiple alignment tool with default parameters [18]. Phylogenetic analyses were conducted and trees were constructed with the aid of the software Seaview [19] using the maximum likelihood (PhyML) method and implementing the most probable nucleotide substitution model (GTR) previously calculated by JmodelTest2 [20]. The branch support was estimated by bootstrap resampling with 100 replications. Sequences obtained from the olive transcriptome were translated into amino acid sequences using ExPASy program (http://web.expasy.org/translate/). The software WoLF PSORT (https://wolfpsort.hgc.jp/) was used for the prediction of protein cell localization. Prediction of serine, threonine and tyrosine phosphorylation was implemented by using Scanprosite (http://web.expasy.org/scanprosite/). Prediction of potential S-nitrosylation sites was made with GPS-SNO 1.0 [21], and TermiNator was used to predict N-terminal methionine excision, acetylation, myristoylation or palmitoylation [22]. Structure modelling for the N-terminal region of GR was performed by using the fold recognition-based Phyre2 server [23], with c4dnaA as the template. 3D reconstruction was carried out by PDB viewer (https://spdbv.vital-it.ch).

3   Results and Discussion

3.1   Retrieval and Phylogenetic Analysis of GR Sequences from Olive Seed and Olive Vegetative Transcriptomes

Table 1 describes either partial sequences or complete sequences corresponding to GR retrieved from the olive seed transcriptome (11 sequences) and the olive vegetative transcriptome (9 sequences), which are covered in the Reprolive (http://reprolive.eez. csic.es/olivodb/) database. Additional data include putative subcellular localization, organisms in which certain homology has been identified, length of the sequence, and the correspondence with homologous sequences within the recently described olive tree genome [24], as well as the percentage of identity between each pair of olive sequences.

Table 1. Output sequences identified as GRs after searching ReprOlive database including seed and a vegetative tissue transcriptome, predicted localization, and identity with the corresponding olive genome sequences. Chl.: chloroplastidial; Cyt.: cytosolic; Mit.: mitochondrial; At: *Arabidopsis thaliana*; Gm: *Glycine max*; Ps: *Pisum sativum*.

Tissue	Sequence	Sequence name in ReprOlive db	Localization/ homology	Length (pb)	Genomic Sequence	Identity
Seed	GR_Seed_1	se11_olive_027181	Chl./At	855	Oe6_s02429	100%
	GR_Seed_2	se11_olive_025545	GR	869	Oe6_s07823	100%
	GR_Seed_3	se11_olive_021387	Cyt. At	918	Oe6_s07823	100%
	GR_Seed_4	se11_olive_001379	Cyt. Ps complete	1626	Oe6_s02433	100%
	GR_Seed_5	se11_olive_001019	Cyt. Ps	1705	Oe6_s02433	100%
	GR_Seed_6	se11_olive_037697	Chl. At	769	–	0%
	GR_Seed_7	se11_olive_036877	Cyt. Os	782	Oe6_s02433	100%
	GR_Seed_8	se11_olive_034395	Chl. At	794	Oe6_s02429	100%
	GR_Seed_9	se11_olive_042727	Cyt. At	738	Oe6_s03877	100%
	GR_Seed_10	se11_olive_045783	Chl. At	721	Oe6_s02429	97,94%
	GR_Seed_11	se11_olive_047071	Cyt. Ps	714	Oe6_s00818	100%
Vegetative tissue	GR_vt_1	vg11_olive_027980	Cyt. At	556	Oe6_s07823	98,68%
	GR_vt_2	vg11_olive_024788	Cyt. Ps	539	Oe6_s02433	100%
	GR_vt_3	vg11_olive_026534	Chl. Gm.	547	Oe6_s02429	100%
	GR_vt_4	vg11_olive_019974	GR	518	Oe6_s02433	100%
	GR_vt_5	vg11_olive_037074	Chl./Mit. Ps	600	Oe6_s02429	99,57%
	GR_vt_6	vg11_olive_016841	Chl./Mit. Ps	801	Oe6_s08493	100%
	GR_vt_7	vg11_olive_007534	Cyt. Ps	435	Oe6_s03877	100%
	GR_vt_8	vg11_olive_004742	Cyt. At	397	Oe6_s03877	100%
	GR_vt_9	vg11_olive_000305	Cyt. Ps	712	Oe6_s02433	100%

The most frequently matched genomic olive sequence entries are Oe6_s02429, a chloroplastidial GR (GR2) sequence, and Oe6_s02433 identified as cytosolic (GR1). In both cases, the identity with its homologous sequences in the transcriptome is greater than 97%; GR_Seed_6 did not match with other sequences when compared with genome, but exhibited correspondence with a chloroplastidial GR of *Arabidopsis thaliana* (At3g54660.1). Searching the olive tree transcriptome, only the GR_seed_4 sequence was identified as a full-length sequence. Identity matrices were built to select these sequences with high percentage of nucleotide/nucleotide identity, respectively (not shown), considering as significant matches these surpassing 70% of ID and a E-value of significance greater than 10–12, following the same significance criteria for our complete GR_seed_4 sequence.

Two seed sequences (GR_Seed_4 and GR_Seed_5) reaching 73% of identity were selected for further analyses, as well as two vegetative tissue sequences (GR_vt_6 and GR_vt_9), and another two (GR_Seed_5 and 8) found in seed. All of them exhibited an ID higher than 97%. Figure 1 shows a phylogenetic analysis built with the GR sequences identified within the annotated transcriptome of the olive seed and the vegetative tissues, together with a wide representation of GRs identified in different taxonomical groups of interest (*Solanum lycopersicum, Vitis vinífera, Camellia sinensis, Populus euphratica,* and *Arabidopsis thaliana*). Four clusters were identified, with no clear differentiation between GR1 and GR2 isoforms. Both isoforms were grouped in the same cluster, however showing different predicted cellular localization, and maybe functional differences. Most of the seed and vegetative GR sequences from the olive transcriptome were identified as cytosolic forms with abundant number (80%) of the GR isoforms predicted in chloroplastidial cell localization.

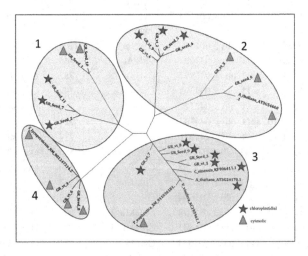

Fig. 1. Phylogenetic relationships between olive seeds GRs, their homologues in vegetative tissue, and heterologous sequences in several plant species of interest. GR_Seed, olive seed; GR_vt, olive vegetative tissue; A_thaliana, *Arabidopsis thaliana*; P_euphratica, *Populus euphratica*; V_vinifera, *Vitis vinífera*; C_sinensis, *Camellia sinensis*; S_lycopersicum, *Solanum lycopersicum.*

3.2   Predictive Analysis of Physical, Chemical and Functional Features of Olive Seed and Vegetative GR Sequences

The analysis of the sequences with the Protparam program allowed us to identify and characterize basic physical and chemical properties of the predicted proteins resulting from the translation of the GR sequences from olive, such as molecular weight, isoelectric point, and stability and aliphatic indexes. The results obtained from this analysis are shown in Table 2.

Table 2. Predictive analysis of the essential physical, chemical and functional features of GRs from olive seed and vegetative tissues. (−): basic isoelectric point; (+): acid isoelectric point.

Tissue	Sequence	Molecular weight (Da)	Isoelectric point	Stability index	Aliphatic index
Seed	GR_Seed_1	16935,71	7,74 (−)	35,55	79,75
	GR_Seed_2	16492,12	5,67 (+)	30,15	80,71
	GR_Seed_3	27985,69	8,83 (−)	28,84	92,93
	GR_Seed_4	53705,92	5,89 (+)	23,15	90,81
	GR_Seed_5	39266,03	5,46 (+)	23,96	91,64
	GR_Seed_6	5218,07	5,48 (+)	41,25 (unstable)	82,22
	GR_Seed_7	18092,35	5,00 (+)	25,10	91,14
	GR_Seed_8	12635,65	9,80 (−)	37,60	91,62
	GR_Seed_9	25908,22	8,54 (−)	33,39	89,96
	GR_Seed_10	7595,62	6,53 (+)	57,70 (unstable)	62,69
	GR_Seed_11	16634,72	8,46 (−)	15,74	90,52
Vegetative tissue	GR_vt_1	18327,52	7,66 (−)	29,02	92,20
	GR_vt_2	11740,56	5,82 (+)	26,08	102,91
	GR_vt_3	10782,50	5,25 (+)	38,73	93,56
	GR_vt_4	–	–	17,40	91,37
	GR_vt_5	5587,68	6,81 (+)	50,78 (unstable)	103,47
	GR_vt_6	5926,82	7,80 (−)	51,99 (unstable)	95,66
	GR_vt_7	3434,99	10,44 (−)	59,87 (unstable)	69,64
	GR_vt_8	13079,94	6,07 (+)	35,66	81,78
	GR_vt_9	39585,88	5,36 (+)	25,11	92,47

Sequences with a molecular weight lower than 10,000 Da are unstable, since stability index values <40 indicate low stability. The aliphatic index does not establish a clear relationship with the rest of the properties but has an average value of approximately 90 in all the sequences. The prediction of functional characteristics of proteins with the Scanprosite program indicates that GR_Seed_3, 4 and 9 sequences exhibited an active class I site belonging to the family pyridine nucleotide-disulphide oxidore-ductase (PS00076) (Table 3). This family includes flavoproteins with a pair of acti-vated cysteines responsible of transferring reduced equivalents from the cofactor FAD to the substrate, as does glutathione reductase. This domain also is also found in the vegetative tissue sequences GR_vt_1 and GR_vt_8.

Table 3. Prediction of functional characteristics of GRs from olive seed and vegetative tissues (functional domains).

Tissue	Protein	Start	End	Domain	Region
Seed	GR_seed_3	67	77	PYRIDINE_REDOX_1	GGtCVirGCVP
	GR_seed_4	67	77	PYRIDINE_REDOX_1	GGtCVirGCVP
	GR_seed_9	67	77	PYRIDINE_REDOX_1	GGtCVirGCVP
Vegetative tissue	GR_vt_1	74	84	PYRIDINE_REDOX_1	GGtCVirGCVP
	GR_vt_8	67	77	PYRIDINE_REDOX_1	GGtCVirGCVP

The "PIRIDINE_REDOX_1 domain does not appear in the rest of the GR sequences analysed. This may be indicative of functional diversity in these isoforms of the GR enzyme. In the majority of the GR sequences analysed, positions 6 and 7 of the regions containing this domain are represented by amino acids such as N-V or N-I, with the exception of the chloroplastidic GR of plants and cyanobacteria, whose amino acids are I-R [25]. This modification is present in all transcriptome sequences retrieved from ReprOlive database (Table 3), which are recognized as homologues of GR cytosolic isoforms in *Arabidopsis thaliana* and *Pisum sativum.*

3.3   *In silico* Prediction of Cellular Localization and Occurrence of PTMs

Subcellular localization was predicted by using WoLF PSORT software. Most of the sequences were predicted as cytosolic isoforms, with the exception of GR_Seed_8, GR_vt_1 and 3, which were predicted to be localised in the nucleus, GR_Seed_10 and GR_vt_4, expected as peroxisomal, and GR_vt_6, identified as chloroplastidial. Potential N-myristoylation and phosphorylation in the amino acids serine (S), threonine (T) and tyrosine (Y) was also assessed, and is described in Table 4 (this table contains modifications predicted with a high degree of certainty only).

Recent studies, have demonstrated the importance of N-myristoylation in plant viability. This lipid modification in the N-terminal residue of proteins is believed to involve nearly 2% of all plant proteins [26]. Targeting of the modified protein to a membrane (where they play crucial roles in signal transduction pathways) is one of the best determined roles for this PTM. The presence of myristoylation susceptibility in conserved sequences of the olive seed and vegetative GRs analysed here, likely accounts for a regulative role of this PTM in the activity of this enzyme, as has been described for other protein families predicted to undergo N-myristoylation in *Ara-bidopsis* like the h-type thioredoxin protein family (h-TRX) [27]. Moreover, potential *S*-nitrosylation sites were also identified in the olive seed and vegetative GRs, as described in Table 5.

Table 4. Predicted N-myristoylation & phosphorylation in GRs from olive seed/vegetative.

Tissue	Sequence	N-myristoilation	Phosphorylations		
			Tyrosine kinase	Protein kinase C	Casein kinase 2
Seed	GR_Seed_4	29 GAgsGG 34 33 GGvrAS 38 34 GVraSR 39 45 GAkvGI 50 64 GGvgGT 69 65 GVggTC 70 84 GAsfGS 89 158 GTkiSY 163 183 GQelAI 188 222 GMgaSV 227 351 GTcfAK 356 435 GAsmCG 440 449 GIavAL 454 457 GAtkAQ 462 468 GIhpSA 473	13 KpneEktqY 21	164 SaK 166 294 TgR 296 300 TkR 302 338 TdR 340 414 SgR 416 430 TdK 432 479 TmR 481 484 SrR 486	19 TqyD 22 89 SelE 92 189 TsdE 192 195 SleE 198 264 TltE 267 364 SkpD 367 472 SaaE 475
	GR_Seed_5	29 GTkiSY 34 54 GQelAI 59 79 GGyiAV 84 92 GMgaSV 97 221 GTcfAK 226 305 GAsmCG 310 319 GIavAL 324 327 GAtkAQ 332 338 GIhpSA 343		35 SaK 37 164 TgR 166 170 TkR 172 208 TdR 210 284 SgR 286 300 TdK 302 349 TmR 351 354 SrR 356	60 TsdE 63 66 SleE 69 134 TltE 137 234 SkpD 237 342 SaaE 345
Vegetative tissue	GR_vt_6	32 GTidGF 37		26 SlK 28 29 TnK 31	
	GR_vt_9	29 GTkiSY 34 54 GQelAI 59 79 GGyiAV 84 92 GMgasV 97 223 GTcfAK 228 307 GAsmCG 312 321 GIavAL 326 329 GAtkAQ 334 340 GIhpSA 345		35 SaK 37 172 TkR 174 210 TdR 212 286 SgR 288 302 TdK 304 351 TmR 353 356 SrR 358	60 TsdE 63 66 SleE 69 134 TltE 137 236 SkpD 239 344 SaaE 347

Table 5. Cys-containing peptides prone to *S*-nitrosylation in GRs from olive seed and vegetative tissue.

Tissue	Sequence	Position	Peptide
Seed	GR_Seed_2	13	VALMEGSCFAKTVFG
	GR_Seed_4	439	KVLGASMCGPDAAEI
	GR_Seed_5	309	KVLGASMCGPDAAEI
	GR_Seed_10	43	QSSWFAHCGEELTQE
	GR_Seed_11	97	KVLGASMCGPDAAEI
Vegetative tissue	GR_vt_7	17	NKDFSRNCHPLSSLR
	GR_vt_9	311	KVLGASMCGPDAAEI

The presence of a potential *S*-nitrosylation site in different sequences could be implicated in the regulation of the seed desiccation process [28]. In order to determine whether the olive GR sequences analysed here are grouped accordingly to the different properties described above (susceptibility to PTMs), a graphical superimposition of these properties was performed over the phylogenetic analysis previously described in Fig. 1 (Fig. 2).

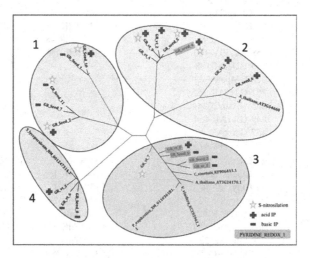

Fig. 2. Susceptibility of olive seed and vegetative sequences to different PTM superimposed over the phylogenetic analysis of the same sequences as depicted in Fig. 1.

In relation to the isoelectric point (PI), sequences with an acidic IP are grouped in cluster 2, while clusters 1, 3 and 4 are mainly composed of basic IP GR sequences. Furthermore, the sequences of cytosolic isoforms taking part of cluster 2 exhibited homology with a cytosolic GR of *Pisum sativum*, according to the olive transcriptome data, and with the Oe6_s02433 sequence from the *Olea europaea* assembly and annotation database (ReprOlive).

We have also observed that almost all sequences harbouring potential *S*-nitrosylation sites are cytosolic. Regarding the PYRIDINE_REDOX_1 functional domain sequence, most of GRs displaying it were identified as cytosolic isoforms as well, and were almost entirely grouped in cluster 3. Sequences exhibited high identity among them and even with sequences from vegetative tissues. Although two isoforms of the enzyme were distinguished, the phylogenetic analysis carried out did not cluster the two predicted isoforms predicted within specific group, thus we cannot suggest distinctive features for these GR homologous in seeds. However, by predicting the cellular location of these genes we identified that most of the GRs are putatively expressed in the cytosol. This can be explained by the scarce presence of chloroplasts in the seed at the mature stage. Noteworthy, physical and chemical analyses showed in the present work also identify the cytosolic forms of GRs as those forms with the higher stability.

3.4   2-D and 3-D Structural Modelling and Functional Assessment of Olive Seed and Vegetative GRs

The GR_Seed_4 sequence corresponding to a cytosolic GR type was chosen as a representative protein to perform the analysis of the 2-D and 3-D modelling. This sequence was identified within the olive transcriptome as a complete sequence, with a high homology with the genomic *Olea europaea* assembly and annotation database [24], where this olive genome sequence was annotated as a chloroplastidial GR. The amino acid sequence predicted for a GR1 from the olive seed transcriptome displayed a high identity with the one predicted as GR2 in a previous work [29]. Amino acids are presented in groups of different colours according to some of their characteristics (Fig. 3). Thus, A, S, T, G and P were represented in yellow (small or polar); M, I, L and V were represented in green (hydrophobic); K, R, E, N, D, H and Q were represented in red (charged), and W, Y, F and C were represented in blue (aromatic and containing cysteines). Among the amino acid sequence of the secondary structure the potentially-modified cysteine at position 439 (as predicted by the GPS-SON program), can be observed together with the highly conserved domain of interaction with the FAD cofactor (GxGxxG(x)17E) present in all members of the GR family. This domain was located at the N-terminal end and is part of the loop that connects the first beta chain with an alpha helix within the Rossmann-type folding (Fig. 4). From amino acid at position 202 onwards (and coinciding with a conserved area of the protein), a second folding of the Rossmann type can be observed, which likely corresponds to the dimerization domain of the enzyme. Moreover, a partially conserved domain (D(x) 6GxxP) [30], placed between beta chains β6 and β7, was observed.

The predicted secondary structure of the olive seed protein GR1 allowed distinguishing β1, α1 and β2 components of in the Rossmann loop folding encompassing the conserved domain GxGxxG(x)17E (which is coincident with a region of high predictive confidence). We found next a beta sheet, which did not coincide with the general model for GRs, and that was present in a zone with little confidence in the prediction. In this region, a PYRIDINE_REDOX_1 domain characteristic of this type of enzymes was also present. After this domain, an alpha helix and another beta chain (in areas of great confidence) were detected, which could coincide with the α2 and β3 components of the Rossmann loop. Next, and after another alpha helix region, 5 beta sheet areas were present, that would coincide with the β4, 5, 6, 7 and 8 represented in Fig. 4. This last alpha helix, connecting β3 and β4, is an exceptional characteristic that is present in the chloroplastidic isoforms of the enzyme [30]. Such feature, together with the presence of modified amino acids within the GGtCVirGCVP region of the PYRIDINE_REDOX_1 domain, might represent the reasons why the phylogenetic

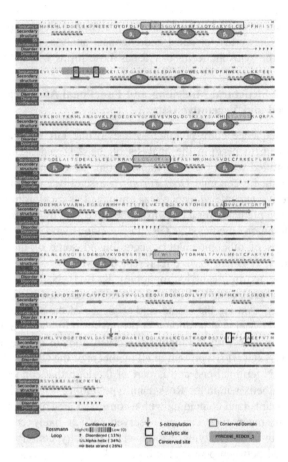

Fig. 3. Predicted secondary structure of the olive seed protein GR1. Blue arrows represent beta sheet. Green loops represent alpha loops. Grey line represents turns and loops of protein secondary structure. Confidence key is represented with a colour score (red: high/blue: low). Conserved domains in the GR are shown with a red box (most conserved domains). Conserved areas of the protein are shaded in Green. Catalytic residues of the CSA (*Catalytic Site Atlas*, EMBL-EBI) are shown with a black box. Blue circles represent Rossmann loops, and red arrows represent potentially *S*-nitrosylated sites. (Color figure online)

analysis of GR sequences from olive seed and vegetative tissues does not generates a sharp separation between cytosolic and chloropastidic forms of the GR enzyme.

After performing 3-D modelling of GR1, we were able to identify this sequence as a glutathione reductase with 93% coverage compared to the selected template (c2v6oA_; structure of thioredoxinglutathione reductase 2 from *Schistosoma mansoni*) and 100% confidence (Fig. 5). Figure 5 shows these characteristics on the 3D modelling of the predicted protein. Some protein residues are tolerated others could disrupt structure. Predictions of the most important residues are indicated with a colour scale. In red, highly conserved areas are displayed, whereas areas with low conservation are depicted in blue.

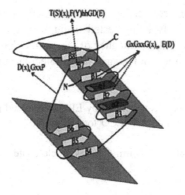

Fig. 4. Rossmann-type loop $(\beta_1\alpha_1\beta_2\alpha_2\beta_3)$ as present in the members of the GR family. (Adapted from [30])

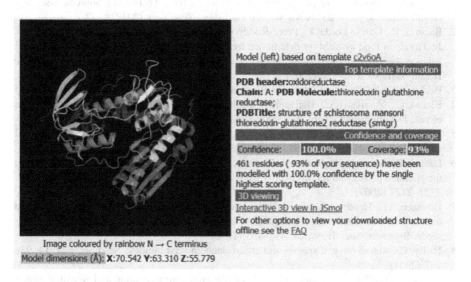

Fig. 5. Structural modelling of the olive seed GR1. (Color figure online)

4   Conclusions

The present study identifies the presence of at least four groups of GR transcripts in olive seeds and olive vegetative tissues, whose sequences were recovered from the ReprOlive database. *In silico* analysis allowed defining GRs from olive seed and vegetative tissues like highly conserved as regard to the presence of functionally-relevant motifs. However, the sequences identified display a large variety of forms, likely with cytosolic and chloroplastidic cell localization. Additional features like susceptibility to numerous post-translational modifications make these enzymes highly polymorphic and prone to differential modulation of their activity, which may represent an adaptive response to diverse scenarios, including all types of stresses. Such predictive properties are currently being tested using biochemical approaches, and the biological and biotechnological significance of the forms identified here are also being assessed like redox regulators and source for molecular tools (i.e. antibodies and probes).

Acknowledgments. This work was supported by ERDF-cofunded projects BFU2016-77243-P and RTC-2017-6654-2. EGQ thanks MINECO for FPI grant.

Conflicts of Interest. The authors confirm that this article content has no conflicts of interest.

References

1. Tsimidou, M.Z.: Virgin olive oil (VOO) and other olive tree products as sources of α-tocopherol. Updating and perspective. In: Catala, A. (ed.) Tocopherol Sources, Uses and Health Benefits, pp. 1–21. Nova Science Publisher, New York (2012)
2. Balanza, R., Garcia-Lorda, P., Perez-Rodrigo, C., Aranceta, J., Bonet, M.B., Salas-Salvado, J.: Trends in food availability determined by the Food and Agriculture Organization's food balance sheets in Mediterranean Europe in comparison with other European areas. Public Health Nutr. **10**, 168–176 (2007)
3. Boskou, D.G.: Olives and Olive Oil Bioactive Constituents, 1st ed. AOCS, Urbana (2015)
4. Blekas, G., Vassilakis, C., Harizanis, C., Tsimidou, M., Boskou, D.G.: Biophenols in table olives. J. Agric. Food Chem. **50**, 3688–3692 (2002)
5. Luaces, P., Pérez, A.G., Sanz, C.: Role of olive seed in the biogenesis of virgin olive oil aroma. J. Agric. Food Chem. **51**, 4741–4745 (2003)
6. Luaces, P., Romero, C., Gutierrez, F., Sanz, C., Pérez, A.G.: Contribution of olive seed to the phenolic profile and related quality parameters of virgin olive oil. J. Sci. Food Agric. **87**, 2721–2727 (2007)
7. Leprince, O., Hendry, G.A.F., Atherton, N.M., Walters-Vertucci, C.: Free radicals and metabolism associated with the acquisition and loss of desiccation tolerance in developing seeds. Biochem. Soc. Trans. **24**, 451–455 (1996)
8. Bailly, C.: Active oxygen species and antioxidants in seed biology. Seed Sci. Res. **14**, 93–107 (2004)
9. Kocsy, G., Galiba, G., Brunold, C.: Role of glutathione in adaptation and signaling during chilling and cold acclimation in plants. Physiol. Plantarium **113**, 158–164 (2001)
10. Schopfer, P., Plachy, C., Frahry, G.: Release of reactive oxygen intermediates (superoxide radicals, hydrogen peroxide, and hydroxyl radicals) and peroxidase in germinating radish seeds controlled by light, gibberellin, and abscisic acid. Plant Physiol. **125**, 1591–1602 (2001)
11. Gechev, T.S., Van Breusegem, F., Stone, J.M., Denev, I., Laloi, C.: Reactive oxygen species as signals that modulate plant stress responses and programmed cell death. BioEssays **28**, 1091–1101 (2006)
12. Starke, D.W., Chock, P.B., Mieyal, J.J.: Glutathione-thiyl radical scavenging and transferase properties of human glutaredoxin, (thioltransferase): potential role in redox signal transduction. J. Biol. Chem. **278**, 14607–14613 (2003)
13. Meyer, A.J., May, M.J., Fricker, M.: Quantitative *in vivo* measurement of glutathione in Arabidopsis cells. Plant J. **27**, 67–78 (2001)

14. Mullineaux, P., Creissen, G.: Glutathione reductase: regulation and role in oxidative stress. In: Scandalios, J.G. (ed.) Oxidative Stress and the Molecular Biology of Antioxidant Defense, pp. 667–713. Cold Spring Harbor, Plainview (1997)
15. Bailly, C., Benamar, A., Corbineau, F., Come, D.: Changes in malondialdehyde content and in superoxide dismutase, catalase and glutathione reductase activities in sunflower seeds as related to deterioration during accelerated ageing. Physiol. Plantarium **97**, 104–110 (1996)
16. Hendry, G.A., et al.: Free radical processes and loss of seed viability during desiccation in the recalcitrant species *Quercus robur* L. New Phytol. **122**, 273–279 (1992)
17. Carmona, R., et al.: ReprOlive: a database with linked data for the olive tree (*Olea europaea* L.) reproductive transcriptome. Front. Plant Sci. **6**, 625 (2015)
18. McWilliam, H., et al.: Analysis tool web services from the EMBL-EBI. Nucleic Acids Res. **41**, W597–W600 (2013)
19. Gouy, M., Guindon, S., Gascuel, O.: SeaView version 4: a multiplatform graphical user interface for sequence alignment and phylogenetic tree building. Mol. Biol. Evol. **27**, 221–224 (2010)
20. Darriba, D., Taboada, G.L., Doallo, R., Posada, D.: jModelTest 2: more models, new heuristics and parallel computing. Nat. Methods **9**, 772 (2012)
21. Xue, Y., et al.: GPS-SNO: computational prediction of protein S-nitrosylation sites with a modified GPS algorithm. PLoS One **5**, e11290 (2010)
22. Martinez, A., et al.: Extent of N-terminal modifications in cytosolic proteins from eukaryotes. Proteomics **8**, 2809–2831 (2008)
23. Kelley, L.A., Sternberg, M.J.: Protein structure prediction on the web: a case study using the Phyre server. Nat. Protoc. **4**, 363–371 (2009)
24. Cruz, F., et al.: Genome sequence of the olive tree *Olea europaea*. Gigascience **5**, 29 (2016)
25. Creissen, G., Edwards, E.A., Enard, C., Wellburn, A.R., Mullineaux, P.: Molecular characterization of glutathione reductase cDNAs from pea (Pisum sativum L.). Plant J. **2**(1), 129–131 (1992)
26. Traverso, J.A., Meinnel, T., Giglione, C.: Expanded impact of protein N-myristoylation in plants. Plant Sig. Behav. **3**(7), 501–502 (2008)
27. Traverso, J.A., et al.: Roles of N-terminal fatty acid acylations in membrane compartment partitioning: Arabidosis h-Type thioredoxins as a case study. Plant Cell **25**, 1056–1077 (2013)
28. Bai, X., et al.: Nitric oxide desiccation tolerance of recalcitrant Antiaris toxicaria seeds via protein S-nitrosylation and carbonylation. PLoS One **6**(6), e20174 (2011)
29. García-Quirós, E., Carmona, R., Zafra, A., Gonzalo Claros, M., Alché, J.D.: Identification and *in silico* analysis of glutathione reductase transcripts expressed in olive (*Olea europaea* L.) Pollen and Pistil. In: Rojas, I., Ortuño, F. (eds.) IWBBIO 2017. LNCS, vol. 10209, pp. 185–195. Springer, Cham (2017). https://doi.org/10.1007/978-3-319-56154-7_18
30. Dym, O., Eisenberg, D.: Sequence-structure analysis of FAD-containing proteins. Protein Sci. **10**, 1712–1728 (2001)

Bioinformatics Approaches for Analyzing Cancer Sequencing Data

Prediction of Thermophilic Proteins Using Voting Algorithm

Jing Li[1], Pengfei Zhu[1(✉)], and Quan Zou[2(✉)]

[1] College of Intelligence and Computing, Tianjin University, Tianjin, China
lijingtju@foxmail.com, zhupengfei@tju.edu.cn
[2] Institute of Fundamental and Frontier Sciences,
University of Electronic Science and Technology of China, Chengdu, China
zouquan@nclab.net

Abstract. Thermophilic proteins have widely used in food, medicine, tanning, and oil drilling. By analyzing the protein sequence, the superior structure and properties of the protein sequence are obtained, which is used to efficiently predict the protein species. In this paper, a voting algorithm was designed independently. Protein features and dimensions were extracted and reduced, respectively. Data was predicted by WEKA. Next, the voting algorithm was applied to the data obtained by the above processing. In this experiment, the highest accuracy rate of 93.03% was achieved. This experiment has at least two advantages: First, the voting algorithm was developed independently. Second, any optimization method was not used for this experiment, which prevents over-fitting. Therefore, voting is a very effective strategy for the thermal stability of proteins. The prediction data set used in this paper can be freely downloaded from http://lab.malab.cn/~lijing/thermo_data.html.

Keywords: Thermophilic proteins · Voting algorithm ·
Feature selection · Machine learning

1 Introduction

Since the extreme thermophilic microbe genome (the Methanococus jannaschii) has been published, the method of comparing genomes (proteome) has been widely used for the research of protein thermostability.

By mining the charged residues and hydrophobic residues, Bayesian rules, logic functions, neural networks, support vector machines, decision trees are used to distinguish between thermophilic proteins and non-thermophilic proteins. For data of 4684 and 653 protein sequences, 85% and 91% were obtained by neural network and 5-fold cross-validation [11]. By analyzing the distribution of neighbouring amino acids, there are dramatic differences in thermophilic and non-thermophilic proteins. A statistical method was designed for the detection of dipeptide data. 86.3%, 85.5% and 89.7% were displayed, including comparative experiments [30]. Structural information is applied to the logitboost classifiers by

© Springer Nature Switzerland AG 2019
I. Rojas et al. (Eds.): IWBBIO 2019, LNBI 11465, pp. 195–203, 2019.
https://doi.org/10.1007/978-3-030-17938-0_18

recognition of the first-class protein structure, and the principle of 5-fold cross-validation is set. Experiments show that 97% and 86.6% accuracy are captured separately. It is found that the logitboost classifier has strong generalization capacity and low demanding on the length of the protein sequence [32]. Experimental material is used in a variety of protein identification patterns, which has high degree of confidence. Among these methods, the credibility of the back propagation neural network is up to 98%. The experimental results show that the accuracy of 75% and 85% of thermophilic and non-thermophilic protein, respectively [31]. Potential models and sealed information were mined and found by Chaos game representation (CGR). The pseudo-amino acid information was calculated and extended into protein sequences, which were visualized by the CGR model. Features were extracted via CGR section and 87.92% was captured [17]. Considering the problem of mutations caused by the growth or shortening of protein sequences, this article claims that protein stability can be promoted by Support Vector Machine (SVM). Test results show that the classification accuracy rate reaches 88% [18]. In order to distinguish thermophilic proteins from non-thermophilic proteins and to deal with the stability changes of protein mutations, this paper invented a new type scoring function. Feature weights were taken into account by rewriting the random forest classifier. In the end, 97.3% accuracy was completed [13].

In this paper, a new voting program was developed. By extracting 13 features and integrating 24 classifiers, the better integrated combination was selected for voting, and relatively high accuracy was captured. The extracted features were CKSAAGP, AAC, CKSAAP, CTPC, GAAC, GTPC, GDPC, CTDC, DDE, DPC, CTDT, KSCTRIAD and TPC. Because there are too many classifiers, only voting classifiers will be explained in the following sections. Next, the dimensions of all features are cut, appropriately. WEKA was applied to preliminarily predict, and the results of preliminary prediction were used in the voting program. Ultimately, the accuracy of 93.62% and 92.8% was achieved, separately. The experiment found that data without dimension reduction has better performance.

Compared with published schemes to distinguish between thermophilic and non-thermophilic proteins, the strengths of this study are obvious.

(1) The accuracy is higher.
   The result of the vote was 93.03%
(2) The voting program was developed, independent.
   Without engineering contribution to support theory, many published papers merely describe a general method for identifying thermophilic and non-thermophilic proteins in the field of bioinformatics. In contrast, this research has corresponding engineering as the theoretical basis. In other words, professional ability of the operator is less demanding. This is crucial for the development bioinformatics [4].
(3) The data has not been optimized to prevent over-fitting.
   Sometimes, in order to get better results, optimizer will be applied to the experimental process in the field of data mining. Most of the time, data

optimization does more disadvantages than advantages. Optimization will cause many problems that cannot be ignored and the prediction effect of the model is poor [33].

2   Material and Method

2.1   Data Sources

The data source is http://lab.malab.cn/~lijing/thermo_data.html, including 915 thermophilic proteins and 793 non-thermophilic proteins. The labels of the data are positive and negative.

2.2   Feature Extraction

The features extracted are significant, which will largely affect the experimental results. The theoretical basis of the amino acids features extracted is that location information and structural composition. In the key step, 13 features were extracted, namely CKSAAGP, AAC, CKSAAP, CTPC, GAAC, GTPC, GDPC, CTDC, DDE, DPC, CTDT, KSCTRIAD. Given the limited space, feature extraction algorithms will be overly generalized and will not delve into the details.

The features of AAC algorithm are extracted based on the number of appearance. 20 different amino acids were found, respectively [3]. The DDE algorithm is based on the formation of dipeptides. After a series of reversals, the ideal mean and the ideal variance are calculated, which are used to obtain the final indicator [12]. The design theory of the CKSAAGP algorithm is the frequentness of amino acid, and the homologous eigenvalues are captured by reasoning [7]. The number of protein species is a major consideration in the TPC algorithm [9]. Due to space constraints, only feature descriptors for voting are introduced.

2.3   Max Relevance Max Distance (MRMD)

After feature extracted, the MRMD [42] is used for feature selection. Cutting the less relevant features is the primary task of MRMD [25].

2.4   Classifier Selection and Tools

In the preliminary classification of amino acids, WEKA is the main operating environment for data before and after feature selection, which is fast and efficient [20]. Besides, a large number of classifiers are built into WEKA, and 24 classifiers are screened out. The classifier for voting is discribed in the following content.

LIBSVM is widely used in machine learning and data mining, whose software packages can be used across platforms [22,24]. The goal of Simple Logistic classifier is to achieve the fitting regression effect through Logistic Boost. Through multiple iterations, the models are updated constantly. When the deviation value of the logistic regression model reduces, the update ends [23]. The random committee classifier is an extension of the random tree classifier, which is mostly used for the formation of low-level classifiers for different data sources [39]. The classification rule of the Logistic classifier is a function, which is derived from the maximum likelihood function, the activation function and the gradient descent algorithm [16]. The principle of PART is the matching of data and "decision lists". When the match reports an error [10], the default category will be called [15].

3   Experiment

In order to confirm the effectiveness of the voting algorithm, other experiments were compared. In Experiment 1, 188D was used for feature extraction of raw data (188D means 188 features were extracted from raw data, which includes 11 extraction principles of amino acid content, hydrophilicity, van der Waals force and polarity, etc.). In Experiment 2, the features were extracted utilizing IFEATURE [5] algorithm, and the WEKA and voting algorithms were used in subsequent experimental procedures. In Experiment 3, MRMD was used to select the extracted features to retain necessary features. WEKA and voting procedures were used to expect better experimental results.

3.1   Performance of Evaluation Standards

$$SN = TP/(TP + FN) \tag{1}$$

$$SP = TN/(TN + FN) \tag{2}$$

$$ACC = (TP + TN)/(TP + TN + FP + FN) \tag{3}$$

3.2   Performance of Experiments

Experiment 1. The raw data includes 915 thermophilic proteins and 793 non-thermophilic proteins. The 188D was used for feature extraction of raw data. After a series of conversions, the data results were processed into the ARFF format, which was run on WEKA (cross-validation was set to 10-fold, and 8 classifiers were selected, namely Bayesian network, Naive Bayes, Decision tree J4.8, Bagging meta learning, Logistic function, Multiclass classifier, Classification via Regression and random forest). Experiment 1 finds that the multi-class classifier and Logistic function classifier have the highest accuracy. The details are demonstrated in Table 1.

Table 1. The different classifiers performance of 188D.

Methods	AAC
Bays Net	82.50%
Random Forest	88.64%
Decision tree J4.8	81.85%
Bagging meta learning	88.06%
Logistic function	88.93%
Multiclass classifier	88.93%
Classification via Regression	86.71%
Naïve Bayes	83.43%

Data of Experiment 2. Affected by the design principle of IFEATURE, 13 features were extracted from 1708 protein sequences, which are AAC, CKSAAGP, CKSAAP, GTPC, GDPC, CTDC, DDE, DPC, CTDT, KSTRIAD, TPC, GAAC, and CTDD. Besides, many classifiers were tested on WEKA, and only Random Forest results were shown in the Table 2. The highest accuracy rate is 90.57%.

Table 2. The different features accuracy of RF.

Feature	Dimension	ACC
AAC	20	90.57%
CKSAAGP	150	79.22%
CKSAAP	2400	88.23%
CTPC	125	79.04%
GDPC	25	79.63%
CTDC	39	88.06%
CTDT	39	83.49%
DDE	400	88.47%
TPC	8000	84.66%
KSCTRIAD	343	80.91%
CTDD	195	69.67%
GAAC	5	77.22%
DPC	400	88.0%

Data of Experiment 3. The extracted features is selected by MRMD. For comparison, Table 3 shows that the accuracy after dimension reduction with Random Forest classifier on WEKA. For the purposes of comparison, the dimension information is displayed in the Table 3.

Table 3. The different features accuracy of RF after dimension reduction.

Feature	Dimension	ACC
AAC	19	90.93%
CKSAAGP	123	78.98%
CKSAAP	1501	88.23%
CTPC	113	79.04%
GDPC	23	79.63%
CTDC	35	87.7%
CTDT	38	83.49%
DDE	44	85.77%
TPC	25	79.74%
KSCTRIAD	343	80.91%
CTDD	136	68.27%
GAAC	4	76.93%
DPC	398	88.29%

3.3   Data of Voting

Lin's experiment was recurrence. Since the Jackknife took a long time, the experiment switched to 10-fold cross-validation and 92.15% accuracy was achieved. The data of Experiment 2 and Experiment 3 were used for preliminary prediction on WEKA, and a total of 24 classifiers were utilized. In this process, the information of accuracy below 80% is deleted. After all the steps are completed, a matrix of 1702 * 264 was obtained. For the comparison experiment, the data before and after the feature selection were operated like above.

3.4   Performance of the Algorithm

The voting-based program was developed independently, whose design ideas are as follows:

(1) BASE
   After careful consideration, AAC's LIBSVM information is used as a benchmark. The data source is Lin's paper, and it is general accepted to use Lin's results as a voting benchmark.
(2) Based on the information of BASE, the data that is least relevant to BASE is selected.
(3) The algorithm can directly calculate the voting composition, and the accuracy, confusion matrix, F-score and other indicators.
(4) Repeat steps (2) and (3) to achieve higher voting accuracy with fewer data as far as possible.

The data test results of Experiment 1 show that 93.03% is the best result. Not only is higher accuracy achieved, but less information is utilized. The voting combination are LIBSVM (c = 2, g = 2), Random Committee and PART of AAC, LIBSVM (default parameters) and Logistic of DDE, Simple Logistic of TPC and Multi-class classifier of CKSAAGP.

Compared with Experiment 1, the data results of Experiment 2 were relatively poor. After comprehensive consideration, 92.8% was regarded as the best performance. This result integrates information of LIBSVM (c = 2, g = 2) of AAC, LIBSVM (c = 2, g = 2), Naïve Bayes and Logistic of CKSAAP, Multi-class classifier and Simple Logistic of DPC, Logistic of CKSAAGP. It deserves special explanation that the cross-validation of all experiments was set to 10-fold.

4 Conclusion

Amino acid classification is a major problem in bioinformatics. Since the development of bioinformatics, many theories and algorithms based on amino acid classification have been proposed. Due to the limitation of generalization ability, the classification has not reached the ideal accuracy. In this paper, various factors are considered and a voting algorithm is proposed, whose execution result is the integration of LIBSVM (c = 2, g = 2), Random Committee and PART of AAC, LIBSVM (default parameters) and Logistic of DDE, Simple Logistic of TPC and Multi-class classifier of CKSAAGP. The final accuracy rate was 93.03.

As a new interdisciplinary technology in the bioinformatics field, thermophilic proteins play very important role in the study of human health. To systematically present the experimental results and improve ease of use, a server for predicting thermophilic proteins has been developed. The user only needs to input protein sequence, and the highest accuracy of voting and corresponding protein data can be obtained, automatically. On the other hand, Link prediction paradigms [40] have been applied in the prediction of disease genes [27], circular RNAs [29], miRNAs [6,8,21,37], drug side effects [35] and LncRNAs [1,34,36,38]. Also, computational intelligence such as neural networks [2,19], evolutionary algorithms [26,41] and unsupervised learning [14,28] can be applied to predict health related thermophilic proteins.

References

1. Alshahrani, M., Khan, M.A., Maddouri, O., Kinjo, A.R., Queralt-Rosinach, N., Hoehndorf, R.: Neuro-symbolic representation learning on biological knowledge graphs. Bioinformatics 33(17), 2723–2730 (2017)
2. Cabarle, F.G.C., Adorna, H.N., Jiang, M., Zeng, X.: Spiking neural P systems with scheduled synapses. IEEE Trans. Nanobiosci. 16(8), 792–801 (2017)
3. Chen, W., Ding, H., Zhou, X., Lin, H., Chou, K.-C.: iRNA(m6A)-PseDNC: identifying N6-methyladenosine sites using pseudo dinucleotide composition. Anal. Biochem. 561, 59–65 (2018)

4. Chen, W., Yang, H., Feng, P., Ding, H., Lin, H.: iDNA4mC: identifying DNA N4-methylcytosine sites based on nucleotide chemical properties. Bioinformatics **33**(22), 3518–3523 (2017)

5. Chen, Z., et al.: iFeature: a python package and web server for features extraction and selection from protein and peptide sequences. Bioinformatics **34**(14), 2499–2502 (2018)

6. Cheng, L., Hu, Y., Sun, J., Zhou, M., Jiang, Q.: DincRNA: a comprehensive web-based bioinformatics toolkit for exploring disease associations and ncRNA function. Bioinformatics **34**(11), 1953–1956 (2018)

7. Cheng, L., et al.: InfAcrOnt: calculating cross-ontology term similarities using information flow by a random walk. BMC Genom. **19**(1), 919 (2018)

8. Cheng, L., et al.: LncRNA2Target v2. 0: a comprehensive database for target genes of lncRNAs in human and mouse. Nucleic Acids Res. **47**(D1), D140–D144 (2018)

9. Cheng, L., et al.: MetSigDis: a manually curated resource for the metabolic signatures of diseases. Briefings Bioinform. **20**(1), 203–209 (2017)

10. Feng, C.-Q., et al.: iTerm-PseKNC: a sequence-based tool for predicting bacterial transcriptional terminators. Bioinformatics (2018)

11. Michael Gromiha, M., Xavier Suresh, M.: Discrimination of mesophilic and thermophilic proteins using machine learning algorithms. Proteins: Struct. Funct. Bioinform. **70**(4), 1274–1279 (2008)

12. Hu, Y., Zhao, T., Zhang, N., Zang, T., Zhang, J., Cheng, L.: Identifying diseases-related metabolites using random walk. BMC Bioinform. **19**(5), 116 (2018)

13. Li, Y., Russell Middaugh, C., Fang, J.: A novel scoring function for discriminating hyperthermophilic and mesophilic proteins with application to predicting relative thermostability of protein mutants. BMC Bioinform. **11**(1), 62 (2010)

14. Liao, Z., Li, D., Wang, X., Li, L., Zou, Q.: Cancer diagnosis through isomiR expression with machine learning method. Curr. Bioinform. **13**(1), 57–63 (2018)

15. Liu, B., Yang, F., Chou, K.-C.: 2L-piRNA: a two-layer ensemble classifier for identifying Piwi-interacting RNAs and their function. Mol. Ther.-Nucleic Acids **7**, 267–277 (2017)

16. Liu, B., Yang, F., Huang, D.-S., Chou, K.-C.: iPromoter-2L: a two-layer predictor for identifying promoters and their types by multi-window-based PseKNC. Bioinformatics **34**(1), 33–40 (2017)

17. Liu, X.-L., Lu, J.-L., Hu, X.-H.: Predicting thermophilic proteins with pseudo amino acid composition: approached from chaos game representation and principal component analysis. Protein Peptide Lett. **18**(12), 1244–1250 (2011)

18. Montanucci, L., Fariselli, P., Martelli, P.L., Casadio, R.: Predicting protein thermostability changes from sequence upon multiple mutations. Bioinformatics **24**(13), i190–i195 (2008)

19. Song, T., Rodríguez-Patón, A., Zheng, P., Zeng, X.: Spiking neural P systems with colored spikes. IEEE Trans. Cogn. Dev. Syst. **10**(4), 1106–1115 (2018)

20. Su, R., Wu, H., Xu, B., Liu, X., Wei, L.: Developing a multi-dose computational model for drug-induced hepatotoxicity prediction based on toxicogenomics data. IEEE/ACM Trans. Comput. Biol. Bioinform. (2018)

21. Tang, Y., Liu, D., Wang, Z., Wen, T., Deng, L.: A boosting approach for prediction of protein-RNA binding residues. BMC Bioinform. **18**(13), 465 (2017)

22. Wei, L., Chen, H., Su, R.: M6APred-EL: a sequence-based predictor for identifying N6-methyladenosine sites using ensemble learning. Mol. Ther.-Nucleic Acids **12**, 635–644 (2018)

23. Wei, L., Wan, S., Guo, J., Wong, K.K.L.: A novel hierarchical selective ensemble classifier with bioinformatics application. Artif. Intell. Med. **83**, 82–90 (2017)

24. Wei, L., Xing, P., Zeng, J., Chen, J.X., Su, R., Guo, F.: Improved prediction of protein-protein interactions using novel negative samples, features, and an ensemble classifier. Artif. Intell. Med. **83**, 67–74 (2017)
25. Wei, L., Zhou, C., Chen, H., Song, J., Su, R.: ACPred-FL: a sequence-based predictor using effective feature representation to improve the prediction of anti-cancer peptides. Bioinformatics **34**(23), 4007–4016 (2018)
26. Xu, H., Zeng, W., Zeng, X., Yen, G.G.: An evolutionary algorithm based on Minkowski distance for many-objective optimization. IEEE Trans. Cybern. (99), 1–12 (2018)
27. Zeng, X., Ding, N., Rodríguez-Patón, A., Zou, Q.: Probability-based collaborative filtering model for predicting gene-disease associations. BMC Med. Genom. **10**(5), 76 (2017)
28. Zeng, X., Liao, Y., Liu, Y., Zou, Q.: Prediction and validation of disease genes using hetesim scores. IEEE/ACM Trans. Comput. Biol. Bioinform. (TCBB) **14**(3), 687–695 (2017)
29. Zeng, X., Lin, W., Guo, M., Zou, Q.: A comprehensive overview and evaluation of circular RNA detection tools. PLoS Comput. Biol. **13**(6), e1005420 (2017)
30. Zhang, G., Fang, B.: Application of amino acid distribution along the sequence for discriminating mesophilic and thermophilic proteins. Process Biochem. **41**(8), 1792–1798 (2006)
31. Zhang, G., Fang, B.: Discrimination of thermophilic and mesophilic proteins via pattern recognition methods. Process Biochem. **41**(3), 552–556 (2006)
32. Zhang, G., Fang, B.: Logitboost classifier for discriminating thermophilic and mesophilic proteins. J. Biotechnol. **127**(3), 417–424 (2007)
33. Zhang, J., Feng, P., Lin, H., Chen, W.: Identifying RNA N6-methyladenosine sites in escherichia coli genome. Front. Microbiol. **9**, 955 (2018)
34. Zhang, J., Zhang, Z., Chen, Z., Deng, L.: Integrating multiple heterogeneous networks for novel LncRNA-disease association inference. IEEE/ACM Trans. Comput. Biol. Bioinform. (2017)
35. Zhang, W., Liu, X., Chen, Y., Wu, W., Wang, W., Li, X.: Feature-derived graph regularized matrix factorization for predicting drug side effects. Neurocomputing **287**, 154–162 (2018)
36. Zhang, W., Qu, Q., Zhang, Y., Wang, W.: The linear neighborhood propagation method for predicting long non-coding RNA-protein interactions. Neurocomputing **273**, 526–534 (2018)
37. Zhang, X., Zou, Q., Rodriguez-Paton, A., et al.: Meta-path methods for prioritizing candidate disease miRNAs. IEEE/ACM Trans. Comput. Biol. Bioinform (2017)
38. Zhang, Z., Zhang, J., Fan, C., Tang, Y., Deng, L.: KATZLGO: large-scale prediction of LncRNA functions by using the KATZ measure based on multiple networks. IEEE/ACM Trans. Comput. Biol. Bioinform (2017)
39. Zhu, X.-J., Feng, C.-Q., Lai, H.-Y., Chen, W., Hao, L.: Predicting protein structural classes for low-similarity sequences by evaluating different features. Knowl.-Based Syst. **163**, 787–793 (2019)
40. Zou, Q., Li, J., Song, L., Zeng, X., Wang, G.: Similarity computation strategies in the microrna-disease network: a survey. Briefings Func. Genom. **15**(1), 55–64 (2015)
41. Zou, Q., Wan, S., Zeng, X., Ma, Z.S.: Reconstructing evolutionary trees in parallel for massive sequences. BMC Syst. Biol. **11**(6), 100 (2017)
42. Zou, Q., Zeng, J., Cao, L., Ji, R.: A novel features ranking metric with application to scalable visual and bioinformatics data classification. Neurocomputing **173**, 346–354 (2016)

Classifying Breast Cancer Histopathological Images Using a Robust Artificial Neural Network Architecture

Xianli Zhang[1,2], Yinbin Zhang[3(✉)], Buyue Qian[1,2], Xiaotong Liu[1,2],
Xiaoyu Li[1,2], Xudong Wang[1,2], Changchang Yin[1,2], Xin Lv[1], Lingyun Song[1,2],
and Liang Wang[3]

[1] National Engineering Lab for Big Data Analytics, Xi'an Jiaotong University,
Xi'an 710049, Shaanxi, China
[2] School of Electronic and Information Engineering, Xi'an Jiaotong University,
Xi'an 710049, Shaanxi, China
[3] The Second Affiliated Hospital of Medical College, Xi'an Jiaotong University,
Xi'an 710004, Shaanxi, China
zhangyinbin865@sohu.com

Abstract. Pathological diagnosis is the standard for the diagnosis and identification of breast malignancies. Computer-aided diagnosis (CAD) is widely applied in pathological image analysis to help pathologists improving the accuracy, efficiency, and consistency in diagnosis. The traditional CAD methods rely on the expert domain knowledge, time-consuming feature engineering, which is insufficient to real-world systems. In recent studies, deep learning methods have been explored to improve the performance of pathological CAD. However, typical deep methods mainly suffer from the following limitations on pathological image classification. (i) The model cannot extract rich and informative features due to the shallow network structure. (ii) The commonly adopted patch-wise classification strategy makes it impossible to obtain the global features at the image level. To address the two issues, in this paper we propose to use a deep ResNet structure with Convolutional Block Attention Module (CBAM), in order to extract richer and finer features from pathological images. Moreover, we abandon the patch-wise classification strategy and perform an end-to-end training instead. The public BreakHis dataset is used to evaluate our proposed method. The results show that our model achieves a significant improvement over the baseline methods.

Keywords: Breast cancer · Histology image · Classification · ResNet · Deep learning

1 Introduction

Breast cancer is the most common malignant tumor among women worldwide. According to the data of the International Agency for Research on Cancer

© Springer Nature Switzerland AG 2019
I. Rojas et al. (Eds.): IWBBIO 2019, LNBI 11465, pp. 204–215, 2019.
https://doi.org/10.1007/978-3-030-17938-0_19

(IARC) of the World Health Organization (WHO) [1], the age-standardized incidence of Breast Cancer (43.1/100,000 people) in the World in 2012 ranked first among female cancers, accounting for 35.3% of new tumors in women and 20.8% of all cancer deaths in women.

Early detection and precise diagnosis are the keys to reduce the mortality of breast cancer. In order to distinguish malignant breast tumor from benign lesions, regular physical examination and imaging examination (mammography, B-ultrasound, and breast MRI) are wildly used, with an accuracy of 90% [2]. However, pathological diagnosis remains the golden standard of breast cancer diagnosis, which provides direct evidence for clinical treatment and prognosis evaluation. At present, immunohistochemical staining is the main method in pathological diagnosis. 80% of benign and malignant breast tumors can be diagnosed by hematoxylin-eosin (HE) staining [3]. During pathological analysis, pathologists need to repeatedly observe the histological with high power view and cytological morphological characteristics in different regions with low power microscopy field, which requires a lot of time and leads to fatigue, even experienced pathologists will lead to the deviation of diagnosis. The concordance rate of breast cancer diagnosis by different pathologists was only 75.3%. In some cases of atypical breast cancer, the concordance rate of diagnosis even dropped to 48% [4].

Recently, thanks to the rapid development of image processing and machine learning technology, Computer-aided diagnosis (CAD) has been widely used to help pathologists analyze pathological images or other medical data [5–7]. As the second optional system, CAD can help pathologists to be more efficient and objective in diagnosis. The existing CAD methods can be roughly divided into two categories. The first one is based on traditional manual feature extraction [8–11], which is criticized by data scientists because of the required very professional domain knowledge and time-consuming feature engineering. The second one is based on the recent deep learning methods [12,13], which can automatically learn useful features from the data according to the loss function, and usually have a weak dependence on domain knowledge. Existing deep learning approaches for Breast Cancer (BC) histology images classification task include cell nuclei segmentation [14,15] and the patch-wise classification [12,16]. The nuclei segmentation based techniques require professional pathologists to label the training image accurately at the cell level, which is very time-consuming. Additionally, the nuclei segmentation based methods often suffer from over-segmentation and do not perform well when some cells overlap [10]. Speak of the patch-wise classification, typical approaches are to extract some patches from the original images, and then use these patches to train a specific structure Convolutional neural network (CNN) [12,16]. However, in this way, CNN usually extracts local features around the nucleus, ignoring the overall characteristics of the entire tissue. Moreover, different patches extraction strategies also have a great impact on the final classification performance, which makes the patches-wise based approaches unstable. In addition to the disadvantage of using patches, the shallow architecture of CNN used in the above methods is not sufficient to extract more abstract and finer features from patients' breast histology images.

In order to classify the BC pathological images steadily and accurately, in this paper, we adopt an improved ResNet architecture to extract local and global features from pathological images and perform end-to-end training. ResNet [17] is a recently popular CNN architecture, it prevents vanishing gradient by using the residual connection, which allows the network architecture to be deeper to obtain richer and more abstract features. However, unlike general image classification tasks, the classification of pathological images usually requires finer feature representation. To this end, we use Convolutional Block Attention Module (CBAM) [18] to enhance the performance of the ResNet. CBAM enables networks to focus on important features and suppress unnecessary ones. Hence, the features of breast tissue extracted from each ResNet block can be adaptively refined by the CBAM.

To evaluate the effectiveness of our method, we conduct a series of experiments on the publicly available BreakHis dataset. The BreakHis dataset, a benchmark proposed by [11] for the BC histological images classification, consists of consists of 7909 breast histopathological images from 82 patients. We compare the results of our method with those reported in other state-of-the-art approaches and show that our method improves the performance by 1.8~4.2% patient-level accuracy, 5.3~8.4% image-level accuracy and 0.7~5.3% F1-score. In summary, the main contributions of this work are as follows:

- We use the deeper ResNet CNN architecture to extract richer and more abstract features of patients' BC tissue.
- We use CABM to refine the tissue features extracted from each layer of ResNet, which can improve the classification performance.
- We have significantly improved the accuracy of classification on the publicly available BreakHis Dataset.

2   Related Work

Classifying histopathological images into non-cancerous or cancerous patterns for analysis, which is the original target of the image analysis system, has been explored in automatic assistant diagnosis of cancer for more than 40 years. However, dealing with the intrinsic complexity of histopathological images was still a major challenge due to the complexity of image analysis [19]. Recently, the development of computerized systems for automatic recognition of malignant breast cancer has become an active area of research with the goal of developing decision support systems to be able to relieve the workload of pathologists [9].

A number of recent works related to breast cancer classification were carried out on small and private dataset. For example, Kowal et al. [8] report accuracy ranging from 96% to 100% on 500 images dataset for nuclei segmentation. Besides, Filipczuk et al. [9] get a performance of 98% on 737 images of fine needle biopsies. Similarly to [8] and [9], George et al. [20] use 92 images in experiment and get accuracy ranging from 76% to 94%. These datasets are usually not available to the scientific community, so their results are not comparable. Fortunately, Breast cancer histopathology database provided 7,909 (2,480 benign and 5,429 malignant samples) microscopic images of breast tumor which collected from 82 patients by surgery [11]. Recently, many researchers have conducted research based on this dataset and compared the results with each other. In [11], an evaluation of different combinations of six different feature descriptors and different classifiers is presented. The mean accuracy is range from 80% to 85% due to different image magnification factor. Sanchez-Morillo *et al.* [10] use the KAZE method [21] to extract KAZE key points from every image. Then they use the K-means for mapping the key points into vectors and use SVM as a binary classifier. Both of the above methods are based on hand-crafted pipelines of feature extraction, which is time-consuming and unstable. Luckily, Deep learning has made great breakthroughs in image classification. AlexNet is a typical deep learning model, which [22] achieves a winning top-5 test error rate of 15.3%, which is 10.9% lower than the second one who uses SIFT [23] and FVS [24]. Afterward, more research on deep learning has been absorbed. Inspired from AlexNet, [16] and [12] use a similar CNN structure to extract features from breast pathological images. They divide the original pathological image into patches, then train the model with patches and perform classification at the patch level. However, the patch-wise classification suffers from the incompleteness of feature extraction, which means that the model can only extract local features at nuclear-level, ignoring the global features of the larger organization. Additionally, AlexNet is not deep enough to extract richer and more abstract features of BC pathology images.

More recently, several pieces of research investigate some important factors of networks to enhance the performance of CNNs. ResNet [17] increases the depth of CNNs and uses residual connections to solve the problem of gradient vanishing. It shows extraordinary abilities in auto-extracting and classification when compared with AlexNet. From the perspective of refining features, CBAM [18] uses two attention mechanisms to refine the features extracted by CNN, which improve the performance of CNNs.

3   Dataset

The Breast Cancer Histopathological Image Classification (BreakHis) dataset composes of 7909 microscopic images of breast tumor tissue. They are collected from 82 patients in different magnifying factors include 40×, 100×, 200×, and 400×. The BreakHis dataset contains 2480 benign and 5429 malignant samples (700 × 460 pixels, 3-channel RGB, 8-bit depth in each channel, PNG format). Table 1 shows the class distribution of images in the Dataset. The dataset contains four histological distinct types of benign breast tumors: adenosis (A), fibroadenoma (F), phyllodes tumor (PT), and tubular adenona (TA), as shown in Fig. 1; and four malignant tumors (breast cancer): carcinoma (DC), lobular carcinoma (LC), mucinous carcinoma (MC) and papillary (PC) as shown in Fig. 2. In order to compare with the state-of-the-art methods fairly, we use the same partitions for the five-fold replications as [10–13, 16, 25]. The partitioning method can be obtained from the download page of the BreakHis dataset[1].

Table 1. Class Distribution of the images in the BreakHis Dataset

Magnification	Benign	Malignant	Total
40×	652	1370	1995
100×	644	1437	2081
200×	623	1390	2013
400×	588	1232	1820
Total	2480	5429	7909

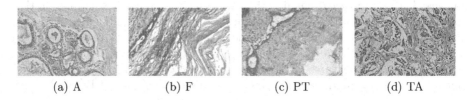

(a) A   (b) F   (c) PT   (d) TA

Fig. 1. Example of pathological images of benign breast tumors stained with HE: Adenosis (A), Fibroadenoma (F), Phyllodes Tumour (PT), Tubular Adenoma (TA). The magnification factor of the above is 100×.

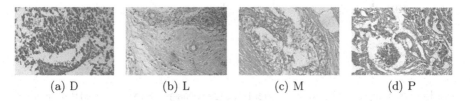

(a) D   (b) L   (c) M   (d) P

Fig. 2. Example of pathological images of malignant breast tumors stained with HE: Ductal (D), Lobular (L), Mucinous (M), Papillary (P). The magnification factor of the above is 100×.

---

[1] https://web.inf.ufpr.br/vri/databases/breast-cancer-histopathological-database-breakhis/.

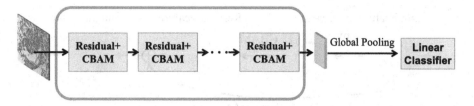

Fig. 3. An overview of the proposed model. Given a BC histology image, the model first extracts and refines the features by each residual block with CBAM (represented by blue boxes). Next, a global average-pooling is used to convert the refined features to a vector. Finally, a fully connected layer is followed as a linear classifier. (Color figure online)

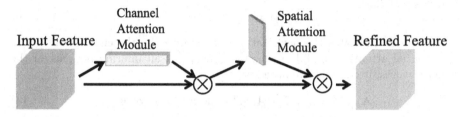

Fig. 4. Details of Convolutional Block Attention Module (CBAM).

4   Method

4.1   Overall Framework

Figure 3 shows an overview of our proposed model. There are two core ideas in our proposed methods. The first one is to adopt the ResNet [17] to extract features of BC histology images. ResNet stacks the same topological of residual blocks along with skip connections to build an extremely deep CNN architecture [18]. The skip connections are used to solve the optimization issues when the networks become deeper. Thanks to the deep network structure but easier gradient propagation, ResNet can effectively extract more abundant and abstract features of BC histology images.

The second core idea of this work is to use CBAM [18] to refine the output features of each residual block in ResNet. CBAM consists of two attention modules, which are channel attention module and spatial attention module. Figure 4 shows the details of the CBAM. The channel attention focuses on 'what' is meaningful given an input BC histology, and it can be computed as:

$$\mathbf{A}_c(\mathbf{F}) = \sigma(MLP(AvgPool(\mathbf{F})) + MLP(MaxPool(\mathbf{F}))) \tag{1}$$

where $\mathbf{F}$ indicates the output features of previous residual block, $\mathbf{A}_c(\mathbf{F}) \in \mathbb{R}^{c \times 1 \times 1}$ is the 1D channel attention map, σ indicates the sigmoid function, MLP is a shared Multi-Layer Perceptron neural network, AvgPool and MaxPool denote the global average pooling and global max pooling respectively. Then, use the

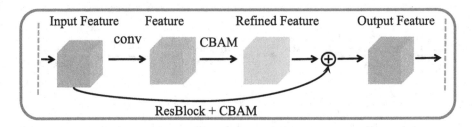

Fig. 5. The residual block with CBAM.

channel attention map $\mathbf{A}_c(\mathbf{F})$ and the input features $\mathbf{F}$, we can obtain the new features as follow:

$$\mathbf{F}' = \mathbf{A}_c(\mathbf{F}) \otimes \mathbf{F} \tag{2}$$

where the $\mathbf{F}'$ indicates the new feature map refined by the channel attention. The spatial attention module focuses on 'where' is an informative part of the given features, it can be formulated as:

$$\mathbf{A}_s(\mathbf{F}') = \sigma(f^{7 \times 7}([AvgPool(\mathbf{F}'); MaxPool(\mathbf{F}')])) \tag{3}$$

where $\mathbf{A}_s(\mathbf{F}') \in \mathbb{R}^{1 \times h \times w}$ denotes the spatial attention map, σ indicates the sigmoid function, $f^{7 \times 7}$ denotes a convolution operation with the filter size of 7 × 7. Note that the two pooling operations are performed across channels. Then we compute the output features $\mathbf{F}''$ of CBAM as:

$$\mathbf{F}'' = \mathbf{A}_s(\mathbf{F}') \otimes \mathbf{F}' \tag{4}$$

The two attention modules complement each other to make the network focus on important features and suppress unnecessary ones. Figure 5 shows the details of the residual block with CBAM. Given a BC histology image, our model first extracts and refines the features by each residual block with CBAM. After a global average-pooling layer, we send the final vector to the linear classifier and obtain the output classification probability.

4.2   Detailed Settings

There are different depths of ResNet structures to choose from, such as 18, 34, 50 and 101. In this work, we use the ResNet with 50 layers, which can be represented by ResNet-50. Then we add CBAM to each block of the ResNet-50. The original top layer of ResNet-50 is replaced by a global average-pooling layer. A dropout of 0.3 is also used after the fully connected layer for helping reduce overfitting. The loss is categorical cross-entropy and the optimizer is Stochastic Gradient Descent (SGD). We set the learning rate at 0.001. We pre-train our model on the ImageNet dataset, and then fine-tune our model on the BreakHis dataset. The BC histology images are reshaped from 700 × 460 to 512 × 336. Data augmentation strategies such as rotations and flip are also used for increasing the amount and generality of the training data.

5   Experiment

In this section, we conduct an extensive experimental for evaluation on the BreaKHis dataset. The BreaKHis dataset is divided into a training (70%) and a testing (30%) subset. Patients who are used in the training set are not in the testing set. For fair comparison with the state-of-the-art methods, we use the same partitions for the five-fold replications as [10–13, 16, 25]. For each fold, we perform five replicate experiments. Finally, we report the average and standard deviation of the total twenty five results. In the rest of this section, we will introduce the evaluation metric used in this paper, and then we report and discuss the results.

5.1   Evaluation Metric

There are three common performance measurements on the BreaKHis. The first one is recognition rate at patient level. Let $\mathbf{N}_p$ be the number of images of patient $\mathbf{p}$. For patient $\mathbf{p}$, let $\mathbf{N}_{rec}$ represent the number of images correctly classified. Then the patient score ($\mathbf{Ps}$) for patient $\mathbf{p}$ can be defined as follow:

$$\mathbf{Ps}_p = \frac{\mathbf{N}_{rec}}{\mathbf{N}_p} \tag{5}$$

If there are $\mathbf{S}$ patients in the test dataset. Then the global patient-level accuracy ($\mathbf{P}_{acc}$) can be defined as follow:

$$\mathbf{P}_{acc} = \sum_{p=1}^{\mathbf{S}} \frac{\mathbf{Ps}_p}{\mathbf{S}} \tag{6}$$

The second one is the recognition rate at image level ($\mathbf{I}_{acc}$), which can be defined as follow:

$$\mathbf{I}_{acc} = \frac{\mathbf{I}_{rec}}{\mathbf{N}_I} \tag{7}$$

where $\mathbf{I}_{rec}$ denotes the total number of correctly classified images, and $\mathbf{N}_I$ is the total number of images.

The third on is F1-score ($\mathbf{F}$), which is used to take into account both the precision $\mathbf{Pr}$ and recall $\mathbf{Re}$ of our method, they can be calculated as follows:

$$\mathbf{Pr} = \frac{\mathbf{TP}}{\mathbf{TP} + \mathbf{FN}}, \quad \mathbf{Rc} = \frac{\mathbf{TP}}{\mathbf{TP} + \mathbf{FN}}, \quad \mathbf{F} = 2\frac{\mathbf{PrRc}}{\mathbf{Pr} + \mathbf{Rc}} \tag{8}$$

where $\mathbf{TP}$ is true positive, $\mathbf{FP}$ is false positive and $\mathbf{FN}$ is false negative cases.

Besides, some other metrics for evaluating binary classification, such as Sensitivity ($\mathbf{Se}$), Specificity ($\mathbf{Sp}$), and area under the receiving operating characteristic curve ($\mathbf{AUC}$) are also reported.

5.2 Results and Discussion

Table 2 reports the results of all evaluation metric corresponding to different magnification factors. As we can see from the results, our method get the best accuracy (both patient level and image-level), AUC, Specificity, F-score at the 200× magnification factor. And it get the best Sensitivity when the magnification factor is 40×. Overall, our model seems to perform best at the magnification factor of 200×.

Table 2. Accuracy at patient-level and image-level, AUC, Sensitivity, Specificity, F-score of our method with different magnification factor

		Magnification factor			
		40×	100×	200×	400×
Patient-level	Accuracy	91.8±3.5	92.1±2.3	**92.2±3.2**	87.9±0.9
Image-level	AUC	90.2±4.4	90.8±2.2	**91.8±3.6**	88.2±1.2
	Sensitivity	**95.9±2.2**	93.6±2.4	94.7±3.4	90.4±2.0
	Specificity	81.6±7.2	88.1±4.0	**88.9±6.4**	86.6±3.2
	Accuracy	91.2±3.5	91.7±2.0	**92.6±3.1**	88.9±1.3
	F1-score	93.6±2.5	93.7±1.6	**94.1±2.3**	91.3±1.1

To prove the effectiveness of using CBAM, Table 3 list the results of ResNet-50 and ResNet-50 with CBAM in 200× magnification factor. As we can see, the ResNet-50 with CBAM achieves better results, this is attributed to CBAM's feature refinement capability.

Table 3. Results of ResNet-50 and ResNet-50+CBAM in 200× magnification factor.

		ResNet-50	ResNet-50 + CBAM
Patient-level	Accuracy	91.7±2.8	**92.2±3.2**
Image-level	AUC	91.0±3.8	**91.8±3.6**
	Sensitivity	94.5±2.2	**94.7±3.4**
	Specificity	87.5±7.1	**88.9±6.4**
	Accuracy	92.1±3.2	**92.6±3.1**
	F1-score	**94.2±1.8**	94.1±2.3

Table 4 shows the best F1-score of our method and that reported in [10–12, 16]. Note that in Table 4 we do not list the results of [13] and [25], because these work did not report F-score. As we can see, our model outperform all the approaches listed in Table 4. Specifically, when compared with the results of

Table 4. F1-score of our methods and that of [10–12,16].

Approach	Magnification factor			
	40×	100×	200×	400×
[10]	90.2	86.5	84.6	80.3
[11]	87.8	86.1	88.5	86.3
[12]	92.9	88.9	88.7	85.9
[16]	88.0	88.8	88.7	86.7
This work	**93.6**	**93.7**	**94.1**	**91.3**

other work, our method achieves 0.7%, 4.8%, 5.4%, and 4.6% improvements at 40×, 100×, 200×, and 400× magnification factors, respectively.

For a better understanding of the results in this work, in Table 5 we compare the best accuracy (both patient-level and image-level) of our method and that in [10–13,16,25]. Note that the image-level accuracy in [13] and [25] is not available. The main observation is that our deeper CNN architecture with CBAM achieved the best results when compared with traditional methods and other deep learning methods. Compared with all the results on the BreakHis dataset, we achieved 1.8%, 3.7%, 4.2%, 1.8% improvements in patient-level accuracy, and 5.3%, 6.9%, 8.4%, 7.3% improvements in image-level accuracy at 40×, 100×, 200×, and 400× magnification factors, respectively.

Table 5. Accuracy at patient-level and image-level of our method and others.

Accuracy at	Approach	Magnification factor			
		40×	100×	200×	400×
Patient-level	[10]	86.4	81.6	77.8	72.9
	[11]	83.8	82.1	85.1	82.3
	[12]	90.0	88.4	84.6	86.1
	[13]	83.0	83.1	84.6	82.1
	[16]	84.0	83.9	86.3	82.1
	[25]	87.7	85.8	88.0	84.6
	This work	**91.8**	**92.1**	**92.2**	**87.9**
Image-level	[10]	85.9	80.4	78.1	71.3
	[11]	82.8	80.7	84.2	81.2
	[12]	85.6	83.5	83.1	80.8
	[16]	84.6	84.8	84.2	81.6
	This work	**91.2**	**91.7**	**92.6**	**88.9**

6   Conclusion

In this paper, we propose a breast cancer histology images classification framework based on ResNet and Convolutional Block Attention Module (CBAM). The proposed method can effectively extract richer and finer features associated with benign and malignant tumors, thanks to the powerful deep structure of ResNet and the good feature refinement capabilities of CBAM. When compared with several state-of-the-art approaches on the publicly available BreakHis dataset, our method shows a significant improvement in accuracy (both patient-level and image-level) and F-score. Future work includes improving classification accuracy and exploring a more granular classification of pathological images.

Acknowledgment. This work was supported in part by National Key Research and Development Program of China under grant number 2018YFC130078; National Natural Science Foundation of China under No.61672420; Project of China Knowledge Center for Engineering Science and Technology; the consulting research project of Chinese academy of engineering "The Online and Offline Mixed Educational Service System for 'The Belt and Road' Training in MOOC China"; Innovative Research Group of the National Natural Science Foundation of China under No.61721002; Innovation Research Team of Ministry of Education No. IRT_17R86; CERNET Innovation Project (NGII20170101).

References

1. Lauby-Secretan, B., et al.: Breast-cancer screening - viewpoint of the IARC working group. New Engl. J. Med. **372**(24), 2353–2358 (2015). https://doi.org/10.1056/NEJMsr1504363. pMID: 26039523
2. United States Preventive Services Task Force: Screening for breast cancer: U.S. preventive services task force recommendation statement. Ann. Intern. Med. **151**(10), 716–726 (2009). https://doi.org/10.7326/0003-4819-151-10-200911170-00008
3. Fischer, A.H., Jacobson, K.A., Rose, J., Zeller, R.: Hematoxylin and eosin staining of tissue and cell sections. CSH Protocols 2008, pdb.prot4986, May 2008
4. Elmore, J.G., et al.: Diagnostic concordance among pathologists interpreting breast biopsy specimens. JAMA **313**, 1122–1132 (2015)
5. Geng, Y., et al.: An improved burden-test pipeline for identifying associations from rare germline and somatic variants. BMC Genom. **18**, 55–62 (2017)
6. Wang, J., Zhao, Z., Cao, Z., Yang, A., Zhang, J.: A probabilistic method for identifying rare variants underlying complex traits. BMC Genom. **14**(S1), S11 (2013)
7. Zhang, X., Wang, Y., Zhao, Z., Wang, J.: An efficient algorithm for sensitively detecting circular RNA from RNA-seq data. Int. J. Mol. Sci. **19**(10) (2018). http://www.mdpi.com/1422-0067/19/10/2897
8. Kowal, M., Filipczuk, P., Obuchowicz, A., Korbicz, J., Monczak, R.: Computer-aided diagnosis of breast cancer based on fine needle biopsy microscopic images. Comput. Biol. Med. **43**(10), 1563–1572 (2013)
9. Filipczuk, P., Fevens, T., Krzyzak, A., Monczak, R.: Computer-aided breast cancer diagnosis based on the analysis of cytological images of fine needle biopsies. IEEE Trans. Med. Imaging **32**(12), 2169–2178 (2013)

10. Sanchez-Morillo, D., González, J., García-Rojo, M., Ortega, J.: Classification of breast cancer histopathological images using KAZE features. In: Rojas, I., Ortuño, F. (eds.) IWBBIO 2018. LNCS, vol. 10814, pp. 276–286. Springer, Cham (2018). https://doi.org/10.1007/978-3-319-78759-6_26

11. Spanhol, F.A., Oliveira, L.S., Petitjean, C., Heutte, L.: A dataset for breast cancer histopathological image classification. IEEE Trans. Biomed. Eng. **63**(7), 1455–1462 (2016)

12. Spanhol, F.A., Oliveira, L.S., Petitjean, C., Heutte, L.: Breast cancer histopathological image classification using convolutional neural networks. In: International Joint Conference on Neural Networks (2016)

13. Bayramoglu, N., Kannala, J., Heikkila, J.: Deep learning for magnification independent breast cancer histopathology image classification. In: International Conference on Pattern Recognition, pp. 2440–2445 (2017)

14. Wang, P., Hu, X., Li, Y., Liu, Q., Zhu, X.: Automatic Cell Nuclei Segmentation and Classification of Breast Cancer Histopathology Images. Elsevier North-Holland, Inc. (2016)

15. Zhang, D., et al.: Panoptic segmentation with an end-to-end cell R-CNN for pathology image analysis. In: Frangi, A.F., Schnabel, J.A., Davatzikos, C., Alberola-López, C., Fichtinger, G. (eds.) MICCAI 2018. LNCS, vol. 11071, pp. 237–244. Springer, Cham (2018). https://doi.org/10.1007/978-3-030-00934-2_27

16. Spanhol, F.A., Oliveira, L.S., Cavalin, P.R., Petitjean, C., Heutte, L.: Deep features for breast cancer histopathological image classification. In: 2017 IEEE International Conference on Systems, Man, and Cybernetics (SMC), pp. 1868–1873, October 2017

17. He, K., Zhang, X., Ren, S., Sun, J.: Deep residual learning for image recognition. CoRR abs/1512.03385 (2015). http://arxiv.org/abs/1512.03385

18. Woo, S., Park, J., Lee, J., Kweon, I.S.: CBAM: convolutional block attention module. CoRR abs/1807.06521 (2018). http://arxiv.org/abs/1807.06521

19. Mitra, S., Shankar, B.U.: Medical image analysis for cancer management in natural computing framework. Inf. Sci. **306**, 111–131 (2015)

20. George, Y.M., Zayed, H.H., Roushdy, M.I., Bagoury, B.M.: Remote computer-aided breast cancer detection and diagnosis system based on cytological images. IEEE Syst. J. **8**(3), 949–964 (2014)

21. Alcantarilla, P.F., Bartoli, A., Davison, A.J.: KAZE features. In: Fitzgibbon, A., Lazebnik, S., Perona, P., Sato, Y., Schmid, C. (eds.) ECCV 2012. LNCS, vol. 7577, pp. 214–227. Springer, Heidelberg (2012). https://doi.org/10.1007/978-3-642-33783-3_16

22. Krizhevsky, A., Sutskever, I., Hinton, G.E.: ImageNet classification with deep convolutional neural networks. In: International Conference on Neural Information Processing Systems, pp. 1097–1105 (2012)

23. Lowe, D.G.: Distinctive image features from scale-invariant keypoints. Int. J. Comput. Vis. **60**(2), 91–110 (2004)

24. Sánchez, J., Perronnin, F.: High-dimensional signature compression for large-scale image classification. In: IEEE Conference on Computer Vision and Pattern Recognition, pp. 1665–1672 (2011)

25. Chattoraj, S., Vishwakarma, K.: Classification of histopathological breast cancer images using iterative VMD aided Zernike moments & textural signatures. CoRR abs/1801.04880 (2018). http://arxiv.org/abs/1801.04880

Spatial Attention Lesion Detection on Automated Breast Ultrasound

Feiqian Wang[1], Xiaotong Liu[2,3], Buyue Qian[2,3]([✉]), Litao Ruan[1]([✉]),
Rongjian Zhao[2,3], Changchang Yin[2,3], Na Yuan[1], Rong Wei[1], Xin Ma[2,3],
and Jishang Wei[1,2,3]

[1] The First Affiliated Hospital of Xi'an Jiaotong University,
Xi'an 710061, Shaanxi, People's Republic of China
ruanlitao@163.com
[2] National Engineering Lab for Big Data Analytics, Xi'an Jiaotong University,
Xi'an 710049, Shaanxi, China
qianbuyue@xjtu.edu.cn
[3] School of Electronic and Information Engineering, Xi'an Jiaotong University,
Xi'an 710049, Shaanxi, China

Abstract. Automated Breast Ultrasound (ABUS) is widely applied in breast screening mainly because of its non-invasive, and radiation-free nature, and the high interoperator reproducibility. Due to the complexity and high volume of data, reading ABUS images is a routine but time-consuming task for sonographers. Accordingly, the computer-aided diagnosis (CAD) has been introduced to help, in order to detect breast lesion efficiently. Traditional techniques such as watershed and fuzzy c-means did not perform satisfactorily, due to the strong underlying assumptions and complex image processing. Lately, deep learning has been explored in medical image analysis. However, it often leads to high false positive rates, which is mainly caused by its requirement of abundant training data and the lack of domain knowledge. To address these issues, we propose a novel lesion detection framework based on the U-net segmentation architecture, and explore a novel method using spatial feature map and attention skip connection. We retrospectively evaluate our model on the data of 142 patients with 305 lesions and 70 no-lesion volumes, and it significantly outperforms the comparison methods with the sensitivity of 92.1% with 1.92 false positives per volume. The promising results suggest that our proposed framework is a solid tool to assist ABUS in breast screening.

Keywords: Deep learning · Fully convolutional network ·
Automated Breast Ultrasound · Lesion detection

1 Introduction

Breast cancer is the most common malignancy in women and is a leading cause of cancer death among women worldwide [1]. The mortality rates for breast cancer

---

F. Wang and X. Liu—Equal contribution.

© Springer Nature Switzerland AG 2019
I. Rojas et al. (Eds.): IWBBIO 2019, LNBI 11465, pp. 216–227, 2019.
https://doi.org/10.1007/978-3-030-17938-0_20

in women have been declining steadily by 1.9% per year from 2003 through 2012 according to the Center for Disease Control. This observed decline may be a reflection of early detection through screening and advances in treatment [2].

Ultrasonography (US) has been used as a first-line screening and routine diagnostic tool for breast cancer in Europe since the 1980s due to its real-time, non-invasion, radiation-free, as well as easy of operation, portability, convenience, widespread availability [3]. Notably, women with dense breasts have higher risks of breast cancer than those with less dense breasts, and US may detect these dense breast lesions more readily than mammography [4]. Recently, automated breast US (ABUS) has been developed as a new promising ultrasound technique for early detection of breast lesions. The ABUS provides 3D scanning on different views of the whole breast [5]. It also has several remarkable advantages over traditional conventional handheld US (HHUS), such as higher reproducibility, less operator dependence, and less required physician time. However, three to four volumes including hundreds of slices of one breast make manual images-reading extremely time-consuming and result in significant inter-observer variation. Designed with the intention of overcoming such limitations, automated lesion detection in ABUS is highly expected to assist clinicians in facilitating the identification of breast lesions.

Nevertheless, computer-aided detection (CADe) for ABUS images remains very challenging. The low contrast of ABUS images and the blurred boundary between the lesion and non-lesion regions make it difficult to accurately detect lesions. The size and the morphology appearance of different lesions vary greatly due to deformations and variations of the intensity distribution. In addition, there is a disproportionate amount of lesions versus non-lesion regions data in the ABUS voxels, which could cause biased predictions. Recently, there have been a few studies which use CAD system for detection of breast lesions in ABUS. The models can be roughly divided into two categories. The first is based on traditional detection methods, which rely on hand-crafted pipelines of feature extraction and techniques such as watershed [6,7], thresholding [4] and clustering [8,9]. However, a hand-designed system is heavily constrained by the assumptions made during feature extraction, which might be not adequate for the detection of breast lesions, because the features are not specially designed for the medical application. Furthermore, these methods usually depend on complex image processing, which make them less generalizable to medical applications. The second is based on the deep learning which learns to directly extract the medical features. However, most of the deep learning approaches omit the spatial distribution information of the lesions, which can cause a high false positive rate [10]. For example, some normal areas in muscle layers are possibly predicted as lesions because of their shape or texture similarities with the lesions, while the lesions only appears in gland layers.

To cope with the challenges and the limitations mentioned above, we propose a new framework to segment and detect the lesions. The model is based on the standard UNet segmentation architecture with a down-sampling and upsampling path, where all the simply stacked convolutional layers are replaced with residual blocks to prevent gradient vanishing. We combine spatial information into different layers of UNet to reduce the false positive rate. Besides, the attention skip connection which combines the high-resolution information with the low-resolution information is adopted to highlight possible lesion regions and suppress unnecessary activations. Finally, considering the blurred boundaries between lesion and no-lesions, we adopt the sub-hard mining strategy in loss function, which only computes the loss of the specific samples to update the parameter of the network, in order to ensure the network learn correct information.

To validate the effectiveness of the proposed framework, we conduct extensive experiments on an ABUS dataset of 142 patients' images with 375 volumes. The experiment results demonstrate that our model outperforms all baseline techniques, and the contributions of the paper are summarized as follows:

1. We introduce the spatial information into different layers of the segmentation network, which significantly reduce the false positive rate.
2. The attention skip connection module is applied to generate soft region proposals easily and highlight low-level features for this segmentation task.
3. We propose a new sub-hard mining strategy on both positive and negative samples, which can address the problem of blurred annotations between the lesion and non-lesion regions.

2   Related Work

The detection of objects of interest in medical images is a crucial part of diagnosis and is one of the most labor-intensive for clinicians [11, 12]. During the past few years, a number of CADe approaches based on traditional machine learning methods have been developed for ABUS lesion detection. Moon et al. developed a CAD system based on a two-stage multi-scale blob analysis method, showed sensitivities of 70% with 2.7 false positives (FPs) per volume [13]. Tan et al. proposed a multi-stage system using an ensemble of neural networks to classify breast cancers with 64% the sensitivity at 1 FPs per volume [14]. To filter false positives caused by rib shadow under pectoral muscle layers or some normal tissue out of the mammary gland layer, the author excludes the slices occupied by the chestwall or the nipples. Lo et al. applied watershed segmentation to extract potential abnormalities in ABUS and reduced FPs using various quantitative features [15]. The sensitivities were 80% with 3.33 FPs per volume. In [4], the fuzzy c-means clustering method was applied to detect tumor candidates from these ABUS images. The sensitivity of the CADe system was 74.14% with 1.76 FPs per volume. Generally, those traditional methods are not robust due to the dependence on complex image processing and specific assumptions.

Convolutional neural networks (CNNs) are currently most widely used in medical image analysis, achieving state-of-the-art performances due to its ability to learn a hierarchical representation of the raw input data, without relying on hand-crafted features. The first medical object detection system using CNNs with four layers to detect nodules in x-ray images was proposed in 1995 [16].

Ciresan et al. [17] trained a network for neuronal membranes segmentation in a sliding-window setup to predict the class label of each pixel by providing a patch around that pixel as input. To avoid computational redundancy, Ronneberger et al. [18] proposed a fully convolutional network (FCN) known as UNet to increase the efficiency by training on whole images. It produces segmentation by pixelwise prediction rather than single probability distribution in the classification task for each image. A similar approach was used by Cicek et al. [19] for 3D data to get a full annotation from a sparsely annotated 3D MR volume. Milletari et al. [20] proposed an extension of the UNet layout that incorporates residual blocks and a dice loss layer, rather than the conventional cross-entropy so that to minimize the commonly used segmentation error measure directly. Milletari et al. [21], proposed a 3D-variant of U-net architecture, called V-net, performing 3D image segmentation using 3D convolutional layers with an objective function directly based on the Dice coefficient.

More recently, some important factors that can improve the performance of CNNs have been investigated. The ResNet architecture consists of residual blocks was proposed by He et al. [22], which can increase the depth of CNNs and solves the problem of gradient vanishing. The attention mechanism firstly derived from neural machine translation [23], and then is applied in image captioning which aims to highlight relevant activations for specific task [24]. To utilize those effective factors and achieve better performance, we proposed a new lesion detection system as follows.

3   Method

Our method is based on UNet. Figure 2 shows a standard UNet structure, which consists of a downsampling path and an upsampling path with a skip connection that concatenates the high-level and the low-level features. The image resolution reduces by half after max pooling along the downsampling path, while the resolution doubles via deconvolution operation along the upsampling path. Finally, a softmax layer is used to transform the result into a two-class problem. We improve UNet by replacing the simply stacked convolutional layers with residual blocks to prevent vanishing gradient. Furthermore, the attention skip connection is included to make the model pay more attention to useful spatial areas and improve the ability of localization.

3.1   Feature Maps with Spatial Information

Figure 1 shows a heat map which denotes the distribution of lesions among the ABUS images. The volumes are split along the traverse axis into 8 parts, every subfigure indicates a part of average distribution probability. The lighter the area is, the more possible it is that there are lesions in the corresponding regions. The statistical results are consistent with medical knowledge that the lesions only appear in breast gland layers rather than muscle layers or subcutaneous fat layers. Therefore, at each resolution step, the proposed network utilizes the

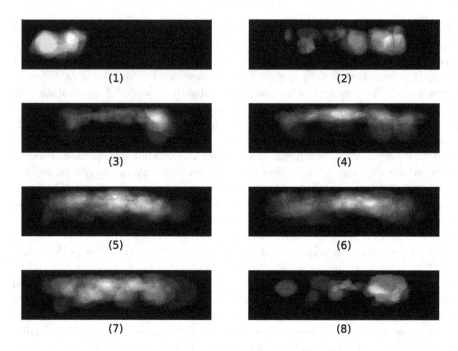

Fig. 1. The distribution probability of lesions. The lighter the area is, the higher probability there are lesions.

spatial information of ABUS images by concatenating spatial information with the corresponding feature maps. The spatial information is calculated according to the relative distance between every pixel and the nipple which is marked manually.

The calculation of spatial information is shown in the formula below:

$$m_i = (p_i - o_i)/w \tag{1}$$

$$m_j = (p_j - o_j)/h \tag{2}$$

$$m_k = (p_k - o_k)/z \tag{3}$$

where m is the three-dimensional coordinates of each pixel, n is the coordinates of the nipple, i,j,k correspond to different dimension, and w, h, z represent the width, height, and length of the ABUS volume. According to the prior spatial information, the model can learn to split the input volumes into several areas automatically.

3.2   Residual Block

At each step of the proposed network, similarly to the approach presented in [22], residual blocks are applied to prevent gradient vanishing. As is shown in Fig. 3, it contains two $3 \times 3 \times 3$ convolutional layers with stride 1, and the same

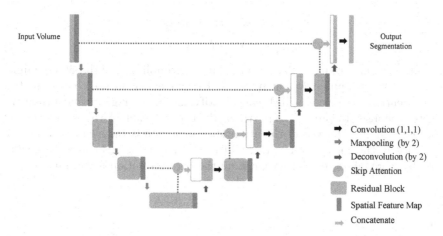

Fig. 2. The network structure of our proposed framework. At each resolution step, we concatenate a three channels spatial map with the corresponding feature map, so as to utilize the prior information to reduce false positive rate.

padding is used to ensure the output has the same size with the input. The reception field of two successive $3 \times 3 \times 3$ convolutional layers is the same with that of a $5 \times 5 \times 5$ convolutional layer but with fewer parameters to be computed. For the residual block we have:

$$Y = ReLU(W_1 * (ReLU(BN(W_2 * X)) + X) \tag{4}$$

where X is the input of the residual block, W symbolizes the convolutional operation, $ReLU$ is the non-linear function, and Y is the output. We also introduce batch normalization (BN) [25] for faster convergence before each ReLU.

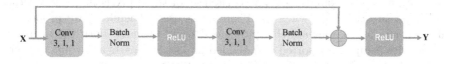

Fig. 3. The residual block.

3.3 Attention Skip Connection

As is shown in Fig. 4, the attention skip connection takes place of the simple "copy and crop" approach in the UNet, which aims to teach the network to focus on activations relevant to this segmentation task. Assuming P is the high-resolution feature map from the down-sampling path, Q is the feature map from corresponding up-sampling path, C is the computed context vector, in this attention skip connection, the shape of P is twice that of Q, we have:

$$P^{'} = W_1 * P \tag{5}$$

$$Q' = Upsample(W_2 * Q) \tag{6}$$

$$C = Sigmod(W_3 * (P' + Q')) * P \tag{7}$$

Literally, the bilinear interpolation is used for up-sampling, and the sigmoid function maps the computed weight into 0–1. Every element in the context matrix is an attention coefficient, which can identify the image regions and prune the feature responses to preserve only the activation relevant to this segmentation task. And the final output of the attention connection is the element-wise multiplication of the coefficients matrix and P, so that the model can learn how to copy the more effective information from the high-resolution feature map.

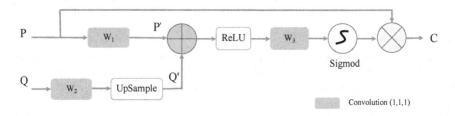

Fig. 4. Attention skip connection

3.4   Sub-hard Mining

We use the cross-entropy loss function to learn to segment the lesion regions. Since the extreme imbalance between normal regions and lesion regions, the model tends to predict most areas as normal regions. In order to obtain better performance, we adopt hard mining on the loss. However, the annotation boundaries between lesions and normal regions are usually blurred, and the hardest pixels are near the boundaries. The existences of these pixels might make the model thrash, especially when hard mining is used. Therefore, we propose a sub-hard mining to alleviate this problem. During training, both the negative and positive samples are sorted and divided into several parts according to their predicted probability respectively. For the positive samples, those have low possibility to be 1 are excluded because most of them are near the blurred boundaries, and the unclear information may confuse the model. For the negative samples, since most of them are easy to discriminate, those have a high probability to be 0 are excluded. In practice, the ratio between positive and negative samples is 1:3.

4   Experiments

4.1   Materials

All the ABUS examinations are performed with ACUSON S2000 Automated Breast Volume Scanner systems. This ABUS system acquires views of $154 \times 168 \times$

$60\,\mathrm{mm}^3$ with a 14L5BV 5–14 MHz automatically driven high-frequency and large footprint transducer. From May 2018 to December 2018, 142 female patients who were detected with breast lesions by ABUS and subsequently underwent surgery or biopsy recruited. 30 volunteer with 70 normal breast volumes were enrolled for negative control. All the patients enrolled had definite histopathological results. To ensure coverage of the entire breast two to four passes of scanning are performed at predefined locations of each breast. Two radiologists with 3 to 6 years of experience in breast imaging review the ultrasound images respectively and independently. The manual delineation of lesions are done in every slice on the axial plane of ABUS images with ITK-SNAP 3^1.

4.2 Data Preprocessing

The ABUS images have various resolutions and number of pixels in different dimensions. To normally represent the actual shape and size of lesions, we firstly normalize the pixel spacing of the ABUS volume in each direction. In our experiment, the original number of pixels in three directions is $330 * 482 * 841$, and the corresponding pixel spacings are $0.504\,\mathrm{mm/pixel}$ (traverse plane), $0.082\,\mathrm{mm/pixel}$ (longitudinal plane), $0.200\,\mathrm{mm/pixel}$ (coronal plane). After the normalization, the pixel spacings of the three directions are adjusted to $0.5\,\mathrm{mm/pixel}$ by bilinear interpolation.

4.3 Implementation Details

Our proposed framework is implemented with PyTorch and trained on NVIDIA Tesla M40 GPU. During training, in order to alleviate the computational complexity in each batch and increase the augmentation, we randomly crop the ABUS images into small patches with the size of $160 \times 80 \times 160$, 70% of which are lesion-centered. In addition, we adopt various data augmentations, e.g., contrast adjustment, rotation, cropping, flipping, especially the elastic transform [26], which are widely used in small medical image dataset to increase the data diversity. At test time, all the ABUS volumes are also split into many small patches with the same size as the training phase. Then the outputs are connected to reconstruct the whole volume prediction. The dataset is split into three parts as training, validation, and test. The model achieved the best performance on the validation set is used for testing. Adam optimizer is used to train the whole network, and the learning rate is set as 1e-4, and training is stopped when the model has similar performance on training and validation dataset.

4.4 Evaluation Metric

After the segmentation of the ABUS images, connected regions are generated from the results with probability larger than a given threshold, which is set to 0.5 in our experiments. The connected regions are taken as lesion proposals. Each

---

[1] http://www.itksnap.org/pmwiki/pmwiki.php?n=Main.HomePage.

proposal is considered as a True Positive (TP) if the detection center point is placed within the ground truth bounding box. Otherwise, it is considered to be a False Positive (FP). In this paper, the performances of lesion detection methods in ABUS images are measured in sensitivity and False Positives per volumes (FPs/volume):

$$Sensitivity = \frac{number \quad of \quad TPs}{number \quad of \quad actual \quad lesions} \tag{8}$$

$$FPs/volumes = \frac{number \quad of \quad FPs}{number \quad of \quad volumes} \tag{9}$$

Sensitivity measures how many lesions are detected, while False Positives per volume indicates how many useless candidates the system detects.

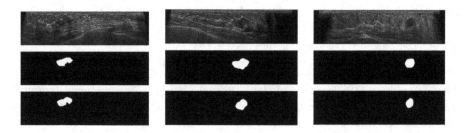

Fig. 5. Example results of lesion segmentation in ABUS. Top row: ABUS images in traverse plane. Middle row: annotated groundtruth by experienced radiologists. Bottom row: segmented probability map by the proposed attention model.

4.5 Detection Performance

We extensively compare our proposed detection framework with state of the art, including Watershed [6], UNet [18], 3D-FCN [27]. To illustrate the efficiency of the utilized spatial information and attention skip connection, we further conduct the proposed network with different versions. The four-fold cross-validation is used to ensure the reliability of results.

Figure 5 visualizes the lesion detection results by our network. By utilizing the proposed spatial attention and skip attention, our network can generate accurate cancer probability maps even when the lesions are small or some mimicry tissues exist. Table 1 lists the sensitivity and corresponding FPs per volume for different methods. As is expected, the watershed achieved worst performance with poor sensitivity and high false positive rate due to its requirements of complex image processing procedure. Compared to FCN and 3D UNet, our framework significantly improve detection sensitivity while controlled false positive per volume at 1.92. When the skip attention module is integrated into the residual network, more candidates are detected so that both the sensitivity and FPs get higher. Yet when the spatial information is included, our network learns to abandon some unreasonable candidates but keeps the right predictions. In total, our network obtained a sensitivity of 92.1% with 1.92 FPS per ABUS volume.

Table 1. Sensitivities and corresponding FPs per volume for different methods

Method	Sensitivity(%)	FPs/volume
Watershed [6]	55.63	4.25
3D-Unet [18]	81.7	2.04
3D-FCN [27]	79.51	1.85
V-Net [21]	84.33	1.91
Attention-UNet	90.36	2.77
Spatial-UNet	86.74	1.89
Spatial Attention-UNet	92.1	1.92

5 Conclusion

In this paper, we propose a fast and effective 3D fully convolutional neural network for the lesion detection in ABUS. In the proposed framework, we introduce a spatial feature map in every resolution step to combine the prior medical information with the convolution operation, thus to reduce false positive rate. Then the skip attention is used to identify the image regions and prune the irrelevant feature responses so as to preserve only the activation relevant to this segmentation task. Furthermore, the sub-hard mining loss can alleviate the influences caused by the data imbalance and blur boundary between lesion and non-lesion areas. Therefore our method generates a segmentation map with higher accuracy. Experiments show our network obtains a sensitivity of 92.1% with 1.92 FPs per ABUS volume. Our framework can provide a steady, accurate and automatic cancer detection tool for breast cancer screening by maintaining high sensitivity and low FPs.

Acknowledgment. This work was supported in part by National Key Research and Development Program of China under grant number 2016YFB1000303; National Natural Science Foundation of China under No. 61672420; Project of China Knowledge Center for Engineering Science and Technology; the consulting research project of Chinese academy of engineering "The Online and Offline Mixed Educational Service System for 'The Belt and Road' Training in MOOC China"; Innovative Research Group of the National Natural Science Foundation of China under No. 61721002; Innovation Research Team of Ministry of Education No. IRT 17R86; CERNET Innovation Project (NGII20170101).

References

1. Jacques, F., Hai-Rim, S., Freddie, B., David, F., Colin, M., Donald Maxwell, P.: Estimates of worldwide burden of cancer in 2008: GLOBOCAN 2008. Int. J. Cancer **127**(12), 2893–2917 (2010)
2. Jørgensen, K.J., et al.: Overview of guidelines on breast screening: why recommendations differ and what to do about it. Breast **31**, 261–269 (2017)

3. Fiorica, J.V.: Breast cancer screening, mammography, and other modalities. Clin. Obstet. Gynecol. **59**(4), 688–709 (2016)

4. Chiao, L., Yi-Wei, S., Chiun-Sheng, H., Ruey-Feng, C.: Computer-aided multiview tumor detection for automated whole breast ultrasound. Ultrason. Imaging **36**(1), 3 (2014)

5. Zanotel, M., et al.: Automated breast ultrasound: basic principles and emerging clinical applications. La Radiologia Medica **123**(1), 1–12 (2017)

6. Huang, Y.L., Chen, D.R.: Watershed segmentation for breast tumor in 2-D sonography. Ultrasound Med. Biol. **30**(5), 625–632 (2004)

7. Gomez, W., Leija, L.A.: Computerized lesion segmentation of breast ultrasound based on marker-controlled watershed transformation. Med. Phys. **37**(1), 82 (2010)

8. Moon, W.K., et al.: Tumor detection in automated breast ultrasound images using quantitative tissue clustering. Med. Phys. **41**(4), 042901 (2014)

9. Kuo, H.C., et al.: Segmentation of breast masses on dedicated breast computed tomography and three-dimensional breast ultrasound images. J. Med. Imaging **1**(1), 014501 (2014)

10. Drukker, K., Giger, M.L., Horsch, K., Kupinski, M.A., Vyborny, C.J., Mendelson, E.B.: Computerized lesion detection on breast ultrasound. Med. Phys. **29**(7), 1438–1446 (2002)

11. Xu, M., Zhao, Z., Zhang, X., Gao, A., Wu, S., Wang, J.: Synstable fusion: a network-based algorithm for estimating driver genes in fusion structures. Molecules **23**(8), 2055 (2018)

12. Geng, Y., et al.: An improved burden-test pipeline for identifying associations from rare germline and somatic variants. BMC Genomics **18**(7), 753 (2017)

13. Woo Kyung, M., Yi-Wei, S., Sun, B.M., Chiun-Sheng, H., Jeon-Hor, C., Ruey-Feng, C.: Computer-aided tumor detection based on multi-scale blob detection algorithm in automated breast ultrasound images. IEEE Trans. Med. Imaging **32**(7), 1191–1200 (2013)

14. Tao, T., Platel, B., Mus, R., Tabar, L., Mann, R.M., Karssemeijer, N.: Computer-aided detection of cancer in automated 3-D breast ultrasound. IEEE Trans. Med. Imaging **32**(9), 1698–1706 (2013)

15. Chung-Ming, L., et al.: Multi-dimensional tumor detection in automated whole breast ultrasound using topographic watershed. IEEE Trans. Med. Imaging **33**(7), 1503–1511 (2014)

16. Lo, S.C., Lou, S.L., Lin, J.S., Freedman, M.T., Chien, M.V., Mun, S.K.: Artificial convolution neural network techniques and applications for lung nodule detection. IEEE Trans. Med. Imaging **14**(4), 711–718 (1995)

17. Ciresan, D., Giusti, A., Gambardella, L.M., Schmidhuber, J.: Deep neural networks segment neuronal membranes in electron microscopy images. In: Advances in Neural Information Processing Systems, pp. 2843–2851 (2012)

18. Ronneberger, O., Fischer, P., Brox, T.: U-net: convolutional networks for biomedical image segmentation. In: Navab, N., Hornegger, J., Wells, W., Frangi, A. (eds.) MICCAI 2015. LNCS, vol. 9351, pp. 234–241. Springer, Heidelberg (2015). https://doi.org/10.1007/978-3-319-24574-4_28

19. Çiçek, Ö., Abdulkadir, A., Lienkamp, S.S., Brox, T., Ronneberger, O.: 3D U-Net: learning dense volumetric segmentation from sparse annotation. In: Ourselin, S., Joskowicz, L., Sabuncu, M.R., Unal, G., Wells, W. (eds.) MICCAI 2016. LNCS, vol. 9901, pp. 424–432. Springer, Cham (2016). https://doi.org/10.1007/978-3-319-46723-8_49

20. Drozdzal, M., Vorontsov, E., Chartrand, G., Kadoury, S., Pal, C.: The importance of skip connections in biomedical image segmentation. In: Carneiro, G., et al. (eds.) LABELS/DLMIA -2016. LNCS, vol. 10008, pp. 179–187. Springer, Cham (2016). https://doi.org/10.1007/978-3-319-46976-8_19
21. Milletari, F., Navab, N., Ahmadi, S.A.: V-net: fully convolutional neural networks for volumetric medical image segmentation. In: 2016 Fourth International Conference on 3D Vision (3DV), pp. 565–571. IEEE (2016)
22. He, K., Zhang, X., Ren, S., Sun, J.: Deep residual learning for image recognition. In: Proceedings of the IEEE Conference on Computer Vision and Pattern Recognition, pp. 770–778 (2016)
23. Cho, K., et al.: Learning phrase representations using rnn encoder-decoder for statistical machine translation. arXiv preprint arXiv:1406.1078 (2014)
24. Xu, K., et al.: Show, attend and tell: neural image caption generation with visual attention. In: International Conference on Machine Learning, pp. 2048–2057 (2015)
25. Ioffe, S., Szegedy, C.: Batch normalization: Accelerating deep network training by reducing internal covariate shift. arXiv preprint arXiv:1502.03167 (2015)
26. Simard, P.Y., Steinkraus, D., Platt, J.C.: Best practices for convolutional neural networks applied to visual document analysis. In: Null, p. 958. IEEE (2003)
27. Long, J., Shelhamer, E., Darrell, T.: Fully convolutional networks for semantic segmentation. CoRR abs/1411.4038 (2014)

Essential Protein Detection
from Protein-Protein Interaction Networks
Using Immune Algorithm

Xiaoqin Yang[1], Xiujuan Lei[1(✉)] ⓘ, and Jiayin Wang[2(✉)]

[1] School of Computer Science, Shaanxi Normal University,
Xi'an 710119, Shaanxi, China
xjlei@snnu.edu.cn
[2] School of Electronic and Information Engineering,
Xi'an Jiaotong University, Xi'an 710049, Shaanxi, China
wangjiayin@xjtu.edu.cn

Abstract. The prediction of essential proteins in protein-protein interaction (PPI) networks plays a pivotal part in improving the cognition of biological organisms. This study presents a novel computational technique, called EPIA, to discover essential proteins by employing immune algorithm. In EPIA, each antibody denotes a candidate essential protein set, which is initialized in a random way among all proteins in a PPI network. Then the vaccine is extracted based on the prediction results of the existing essential protein identification methods. Next, EPIA utilizes four operators, crossover, mutation, vaccination and immune selection to update the antibody population and search for the optimal candidate essential protein set. The experimental results on two species (Saccharomyces cerevisiae and Drosophila melanogaster) demonstrate that EPIA can obtain a better performance on identifying essential proteins compared to other existing methods.

Keywords: Essential protein detection · Immune algorithm · PPI network · GO annotation

1 Introduction

Essential proteins are known to be the structural and functional foundation of biological organisms. Studying essential proteins assists in revealing the molecular mechanisms and biological processes. Essential protein detection using experimental methods is time consuming and costly. With the rapid advance of high-throughput experimental technologies, a huge volume of protein-protein interaction (PPI) data can be collected. As a result, discovering essential proteins by utilizing computational methods has become a hot spot in proteomics. Hitherto, various computational approaches have been proposed to detect essential proteins, and the existing works could be divided into two categories: the network topology-based approaches and the information fusion-based approaches.

Proteins in cells interact with each other and construct a PPI network [1]. Jeong et al. proposed the centrality-lethality theory [2], which declared that those proteins at the central position of a PPI network are more apt to express essentiality. Based on this,

© Springer Nature Switzerland AG 2019
I. Rojas et al. (Eds.): IWBBIO 2019, LNBI 11465, pp. 228–239, 2019.
https://doi.org/10.1007/978-3-030-17938-0_21

a series of topological centrality measures have been applied to essential protein prediction, such as Degree Centrality (DC) [3], Betweenness Centrality (BC) [4], Closeness Centrality (CC) [5], Eigenvector Centrality (EC) [6], Information Centrality (IC) [7] and Subgraph Centrality (SC) [8]. Moreover, Edge Clustering Coefficient Centrality (NC) [9] and Local Average Connectivity (LAC) [10] are also effective computational techniques based on network topology. All these approaches score proteins according to their centralities in PPI networks and then use the ranking scores to measure protein essentiality.

To improve the accuracy of essential protein prediction, several multi-information fusion measures have been presented, among which the topological properties of PPI networks are integrated with various types of biological information, such as gene expression profiles (PeC [11] and WDC [12]), protein domains (UDoNC [13]), gene ontology (GO) terms (TEO [14]), protein complexes (UC [15] and LBCC [16]), subcellular localization (CIC [17] and SON [18]) and orthologous protein information (ION [19]). Taking into account the high false positives and false negatives contained in PPI networks, some researchers made efforts to enhance the reliability of PPI networks by combining topological and biological properties. Xiao et al. constructed a noise-filtered active protein interaction network (NF-APIN) by using gene expression data and static PPI network [20]. Luo et al. integrated the time-course gene expression data and the PPI data to build a dynamic PPI network [21]. Li et al. refined PPI network by using gene expression profiles and subcellular location information [22]. Extensive experimental results show that the integration of PPI network and some biological data can contribute to a better prediction of essential proteins in comparison to those pure topological centrality measures.

Although the above computational approaches have made some progress, there are still great room for accuracy improvement in essential protein prediction. In addition, most of the existing methods follow the rule of scoring and sorting based on a certain centrality measure, which are not always effective and reliable.

In this study, we take the process of essential protein detection as an optimization problem and propose to tackle it with a novel approach EPIA based on immune algorithm (IA) [23]. IA is a heuristic search algorithm, which is inspired by the immunity mechanism in biology. Compared to other nature inspired algorithms, IA is capable of solving optimization problems with higher stability and faster convergence rate by performing immune operator [24]. In the EPIA approach, an antibody represents a candidate essential protein set, and the vaccines are extracted on the basis of the prediction results of the existing method WDC. The update process of antibody population consists of crossover, mutation, vaccination and immune selection. When the algorithm achieves convergence, those proteins forming the antibody with the best fitness value are considered as detected essential proteins. We apply our proposed EPIA algorithm on two species (Saccharomyces cerevisiae and Drosophila melanogaster) using three different PPI networks (DIP, Krogan and HINT), and compare it with several existing classical methods including DC, SC, IC, LAC, NC, WDC, CIC, PeC, UDoNC and LBCC. The experimental results indicate that the performance of EPIA algorithm outperforms those of other competing methods.

2 Methods

2.1 Immune Algorithm

Immune Algorithm (IA) is a heuristic search algorithm that mimics the mechanism of antigen recognition and antibody reproduction in immune system [23]. IA assumes the problem that needs to be optimized as an antigen and its candidate solutions as antibodies, therefore the optimization target is searching for the antibody that has the highest affinity for the antigen [23]. IA introduces the immune operator into classical Genetic Algorithm (GA). In IA, genetic operators guarantee the diversity of antibodies; vaccination improves the fitness of antibody population; and immune selection effectively prevents the degeneration of population. The flowchart of IA is shown as Fig. 1 [24].

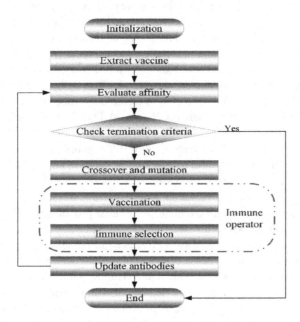

Fig. 1. The flowchart of the Immune algorithm

2.2 Initialization

In this step, initial antibody population whose number is N are generated randomly. Each antibody is a candidate set of essential proteins and consists of p different proteins. An antibody can be encoded as a p-dimensional integer set $A_i = \{a_{i1}, a_{i2}, \ldots, a_{ij}, \ldots, a_{ip}\}$ $(i = 1, 2, \ldots, N; j = 1, 2, \ldots, p)$, where each of these elements denotes the serial number of a protein in a PPI network. The affinity for the ith antibody can be calculated by Eq. (1).

$$Aff(i) = \sum_{j=1}^{p} \left(\sum_{v_k \in N_j} \left(S_{jk} \times F_{jk} \right) \right) \tag{1}$$

where S_{jk} and F_{jk} represent the structural similarity and functional similarity between protein v_j and v_k, which can be calculated by Eq. (2) [25] and Eq. (3) [26], respectively. v_k is the direct neighbor of v_j, N_j is the set of neighbors of protein v_j.

$$S_{jk} = \frac{|\Gamma(j) \cap \Gamma(k)|}{\sqrt{|\Gamma(j)| \times |\Gamma(k)|}} \tag{2}$$

where $\Gamma(j)$ is a set of the neighborhood proteins of proteins v_j plus itself, and $|\Gamma(j)|$ is the size of this set.

$$F_{jk} = \frac{|G(j) \cap G(k)|}{|G(j) \cup G(k)|} \tag{3}$$

where $G(j)$ and $G(k)$ are the GO term sets that annotate proteins v_j and v_k, respectively.

2.3 Vaccine Extraction

In the basic IA, the vaccine operation is introduced to solve the potential degradation caused by crossover and mutation. In our EPIA algorithm, the vaccine is extracted on the basis of the identification results of WDC [12], which is an effective method for predicting essential proteins. The WDC value for a protein v_j can be calculated by Eq. (4).

$$WDC(j) = \sum_{v_k \in N_j} \left(\lambda \times ECC_{jk} + (1 - \lambda) \times PCC_{jk} \right) \tag{4}$$

where N_j is the set of neighbors of protein v_j, and ECC_{jk} and PCC_{jk} represent the edge clustering coefficient and Pearson correlation coefficient of the interaction between protein v_j and v_k, respectively. The parameter λ uses the default value 0.5 set by the authors. Then we rank all the proteins in descending order according to their WDC values and select the p top-ranked proteins as vaccine. That is to say, the number of proteins contained in the vaccine is the same as that of an antibody.

2.4 The Process of Reproduction

Genetic Operation. The genetic operation includes crossover and mutation. Unlike the traditional crossover operator adopted by many evolution algorithms, which exchanges partial fragment between two selected parent individuals, in EPIA algorithm, we select a proportion of antibodies with higher affinity values as crossover templates to update those antibodies with lower affinity values. The modified crossover operator can effectively speed up the convergence. After crossover, the antibody subsequently will be further updated by mutation based on the mutation rate. Genetic operation is shown in Algorithm 1.

Algorithm 1: Genetic operation

Input: $G=(V,E)$: the PPI network; A: the antibody population; α: the proportion of crossover templates; β: the proportion of antibodies that need to be updated; P_{cross}: crossover rate; P_{mutate}: mutation rate; mp: the ratio of proteins needing to be mutated in an antibody.
Output: the set of updated antibodies by genetic operation
for each antibody $A_i \in A$
   calculate affinity score $Aff(i)$ by Eq.(1)
end
sort the antibody population on their decreasing Aff values
$Gup=\emptyset$
for each antibody $A_i \in A\_\beta$ **do** // A_i belongs to the bottom β of ranked antibodies in A
// Crossover
   select a template A_j from A_α // A_j belongs to the top α of ranked antibodies in A
   $Cset=\{v|v\in A_j, \; v\notin A_i\}$
   for each $v_k \in A_i$ **do**
     if $v_k \notin A_j$ and random()$>P_{cross}$ **do**
        replace v_k by arbitrary v_t within $Cset$
        $Cset=Cset-\{v_t\}$
     end if
   end for
//Mutation
   $mn=p*mp$
   $Mset=\{v|v\in V, \; v\notin A_i\}$
   choose mn different proteins from A_i randomly, store them in M
   for each $v_k \in M$ **do**
     if random()$>P_{mutate}$ **do**
        replace v_k by arbitrary v_t within $Mset$
       $Mset= Mset-\{v_t\}$
     end if
   end for
insert A_i into Gup
end for
Return Gup

Immune Operation. The core of the IA lies on the conformation of immune operator that is realized by means of vaccination and immune selection. The immune operation is capable of effectively improving the fitness of antibody population. In the EPIA algorithm, the vaccination is performed by injecting the vaccine extracted based on the WDC method into those antibodies which have undergone crossover and mutation operation. Furthermore, to prevent the degeneration of antibody population, the immune selection is implemented after these operations above. Immune operation is shown in Algorithm 2.

Algorithm 2: Immune operation

Input: *Gup*: the set of updated antibodies by genetic operation; *Vac*: extracted vaccine; *STD*: the standard essential protein dataset; $P_{vaccine}$: vaccination rate.
Output: the set of updated antibodies by immune operation
$Iup=\emptyset$
for each antibody $A_i \in Gup$ **do**
// Vaccination
    for each $v_k \in A_i$ **do**
        if $v_k \notin STD$ and random()$>P_{vaccine}$ **do**
            $Vset=\{v \mid v \in Vac, \; v \notin A_i\}$
            replace v_k by arbitrary v_t within $Vset$
            $Vset=Vset-\{v_t\}$
        end if
    end for
// Immune selection
    if $Aff(i)$ (new) $>Aff(i)$ (parent) **do**
        relace A_i(parent) by A_i(new)
        insert A_i(new) into Iup
    else do
        abandon A_i(new)
        insert A_i(parent) into Iup
    end if
end for
Return Iup

With all the steps above, EPIA generates a new descendant. Repeat the above process until the termination condition of algorithm is reached. The pseudocode of EPIA algorithm is shown in Algorithm 3.

Algorithm 3: EPIA algorithm

Input: $G=(V,E)$: the PPI network; N: population size; p: the number of proteins contained in an antibody.
Output: A candidate essential protein set
  Initialize an antibody population A at random
  Extract the vaccine Vac according to the WDC method
Repeat
  Gup=Genetic operation(G, A, α, β, P_{cross}, P_{mutate}, mp)
  Iup=Immune operation(Gup, Vac, STD, $P_{vaccine}$)
  Update A according Iup
  Find out the antibody with the highest affinity value
Until Termination criteria for algorithm (If the iteration number is finished or the difference of the affinity value of the optimal antibody between two successive iterations is not more than ε)
Return the optimal antibody

After the running of the algorithm EPIA, the antibody with the highest affinity value can be obtained, in which all proteins are deemed to be detected essential proteins.

3 Experiments and Discussion

3.1 Experimental Data

In this study, all the computational experiments and result analysis have been implemented on two species, including two Saccharomyces cerevisiae data, namely DIP [27] and Krogan [28], and one Drosophila melanogaster data, HINT [29]. The DIP PPIs were downloaded from DIP database (http://dip.mbi.ucla.edu/dip/). The Krogan PPIs were obtained from the BioGRID database version 3.4.142 [30]. HINT (High-quality INTeractomes) is a curated compilation of high-quality protein-protein interactions from 8 interactome resources (BioGRID, MINT, iRefWeb, DIP, IntAct, HPRD, MIPS and the PDB). After pretreatment, DIP dataset included 5093 proteins and 24743 interactions, Krogan dataset included 2674 proteins and 7075 interactions, HINT dataset contained 7285 proteins and 24436 interaction. The GO annotation data of Saccharomyces cerevisiae and Drosophila melanogaster were derived from (http://www.yeastgenome.org/download-data/curation) and the COMPARTMENTS database (April 6, 2017), respectively. The essential proteins were collected from four databases, MIPS [31], SGD [32], DEG [33] and OGEE [34]. There are 1285 essential proteins for Saccharomyces cerevisiae in total, among which 1167 and 784 essential proteins appear in DIP dataset and Krogan dataset, respectively. For Drosophila melanogaster, the number of essential proteins is 408, and 261 essential proteins can be mapped to the HINT dataset.

3.2 Comparison with Other Methods

In order to evaluate the effectiveness of our EPIA algorithm, we compare it with ten other typical methods, including five topology-based centrality measures (DC, SC, IC, LAC and NC) and five information fusion-based centrality measures (WDC, PeC, CIC, UDoNC and LBCC). The method UDoNC and LBCC are only executed on DIP dataset as mentioned in their papers [13, 16] due to the unachievable source code and relevant data corresponding to them. The comparison of histograms of different methods on three different datasets are shown in Figs. 2, 3 and 4, which intuitively visualize the proportion of essential proteins in top ranked proteins for all methods. For the three datasets, we set $N = 100$, $\alpha = 0.03$, $\beta = 0.5$, $P_{cross} = 0.5$, $P_{mutate} = 0.99$, $P_{vaccine} = 0.2$, $mp = 0.01$, and p is set as 1, 5, 10, 15, 20 and 25 percent of the quantities of proteins in a certain PPI network, respectively. As can be seen from these figures, no matter which PPI network is used, the EPIA algorithm always has an obvious superiority compared to the five topology-based centrality methods, all of which depend on the topology properties and fail to take the biological properties into account. Furthermore, EPIA also performs better than the other five information fusion-based centrality methods. Specifically, the more candidate proteins are selected, the more obvious the advantage that EPIA has in the identification of essential proteins.

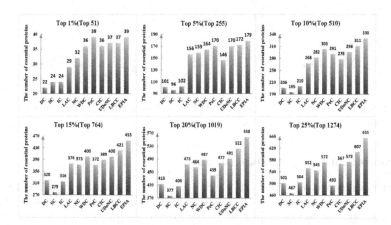

Fig. 2. Comparison of the number of essential proteins detected by EPIA and other methods on DIP dataset

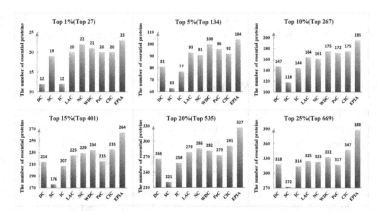

Fig. 3. Comparison of the number of essential proteins detected by EPIA and other methods on Krogan dataset

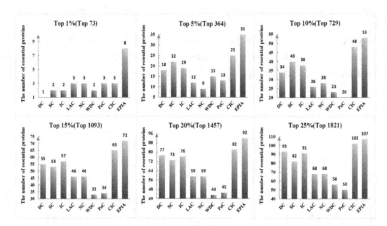

Fig. 4. Comparison of the number of essential proteins detected by EPIA and other methods on HINT dataset

3.3 Assessment Using Statistical Measures

To further evaluate the detection performance of EPIA, we make a comparison between EPIA and other methods by applying six statistical measures, including sensitivity (SN), specificity (SP), positive predictive value (PPV), negative predictive value (NPV), F-measure (F) and accuracy (ACC), which have been extensively used for the evaluation of essential protein prediction in previous studies [16]. The comparison results in terms of the six statistical indicators are shown in Table 1. It is obvious that our EPIA algorithm obtains an excellent performance on the six statistics, all of which are higher than those of any other methods on three different datasets.

Table 1. Comparison of the values of *SN, SP, PPV, NPV, F* and *ACC* for EPIA and other methods

Dataset	Methods	SN	SP	PPV	NPV	F	ACC
DIP	DC	0.4293	0.8031	0.3932	0.8256	0.4105	0.7175
	SC	0.4002	0.7944	0.3666	0.8167	0.3827	0.7041
	IC	0.4319	0.8039	0.3956	0.8264	0.4130	0.7186
	LAC	0.4730	0.8161	0.4333	0.8390	0.4523	0.7375
	NC	0.4670	0.8143	0.4278	0.8371	0.4465	0.7347
	WDC	0.4901	0.8212	0.4490	0.8442	0.4687	0.7453
	PeC	0.4225	0.8011	0.3870	0.8235	0.4040	0.7143
	CIC	0.4859	0.8199	0.4461	0.8429	0.4652	0.7434
	UDoNC	0.4910	0.8214	0.4498	0.8445	0.4695	0.7457
	LBCC	0.5201	0.8301	0.4765	0.8534	0.4973	0.7591
	EPIA	**0.5613**	**0.8423**	**0.5141**	**0.8659**	**0.5367**	**0.7779**
Krogan	DC	0.4056	0.8143	0.4753	0.7676	0.4377	0.6945
	SC	0.3469	0.7899	0.4066	0.7446	0.3744	0.6601
	IC	0.4005	0.8122	0.4694	0.7656	0.4322	0.6915
	LAC	0.4145	0.8180	0.4858	0.7711	0.4473	0.6997
	NC	0.4120	0.8169	0.4828	0.7701	0.4446	0.6982
	WDC	0.4235	0.8217	0.4963	0.7746	0.4570	0.7049
	PeC	0.4043	0.8138	0.4738	0.7671	0.4363	0.6937
	CIC	0.4426	0.8296	0.5187	0.7820	0.4776	0.7162
	EPIA	**0.4949**	**0.8513**	**0.5800**	**0.8025**	**0.5341**	**0.7468**
HINT	DC	0.3563	0.7540	0.0511	0.9693	0.0894	0.7397
	SC	0.3142	0.7524	0.0450	0.9672	0.0787	0.7367
	IC	0.3487	0.7537	0.0500	0.9689	0.0875	0.7392
	LAC	0.2605	0.7504	0.0373	0.9647	0.0653	0.7329
	NC	0.2605	0.7504	0.0373	0.9647	0.0653	0.7329
	WDC	0.2146	0.7487	0.0308	0.9625	0.0539	0.7296
	PeC	0.1916	0.7479	0.0275	0.9614	0.0481	0.7279
	CIC	0.3928	0.7553	0.0560	0.9709	0.0980	0.7422
	EPIA	**0.4100**	**0.7560**	**0.0588**	**0.9718**	**0.1028**	**0.7436**

4   Conclusions

In this study, we develop a novel EPIA algorithm that employs the immune algorithm to detect essential proteins from PPI networks. In the EPIA, each antibody denotes a candidate essential protein set and the whole algorithm consists of population initialization, vaccine extraction, genetic operation and immune operation. Finally, upon termination of the algorithm, those proteins contained in the optimal antibody are considered as the detected essential proteins. Experimental results over three datasets have demonstrated that our EPIA algorithm has the superiority compared with other methods and can detect essential proteins more effectively and accurately.

Acknowledgements. This paper is supported by the National Natural Science Foundation of China (61672334) and the Fundamental Research Funds for the Central Universities (GK201901010).

References

1. Pattin, K.A., Moore, J.H.: Role for protein-protein interaction databases in human genetics. Expert Rev. Proteomics **6**, 647–659 (2009)
2. Jeong, H., Mason, S.P., Barabási, A.L., Oltvai, Z.N.: Lethality and centrality in protein networks. Nature **411**, 41–42 (2001)
3. Vallabhajosyula, R.R., Chakravarti, D., Lutfeali, S., Ray, A., Raval, A.: Identifying hubs in protein interaction networks. PLoS ONE **4**, e5344 (2009)
4. Newman, M.E.J.: A measure of betweenness centrality based on random walks. Soc. Netw. **27**, 39–54 (2005)
5. Wuchty, S., Stadler, P.F.: Centers of complex networks. J. Theor. Biol. **223**, 45–53 (2003)
6. Bonacich, P.: Power and centrality: a family of measures. Am. J. Sociol. **92**, 1170–1182 (1987)
7. Stephenson, K., Zelen, M.: Rethinking centrality: methods and examples. Soc. Netw. **11**, 1–37 (1989)
8. Estrada, E., Rodríguez-Velázquez, J.A.: Subgraph centrality in complex networks. Phys. Rev. E Stat. Nonlinear Soft Matter Phys. **71**, 056103 (2005)
9. Wang, J., Li, M., Wang, H., Pan, Y.: Identification of essential proteins based on edge clustering coefficient. IEEE/ACM Trans. Comput. Biol. Bioinform. **9**, 1070–1080 (2012)
10. Li, M., Wang, J., Chen, X., Wang, H., Pan, Y.: A local average connectivity-based method for identifying essential proteins from the network level. Comput. Biol. Chem. **35**, 143 (2011)
11. Li, M., Zhang, H., Wang, J.X., Pan, Y.: A new essential protein discovery method based on the integration of protein-protein interaction and gene expression data. BMC Syst. Biol. **6**, 15 (2012)
12. Tang, X., Wang, J., Zhong, J., Pan, Y.: Predicting essential proteins based on weighted degree centrality. IEEE/ACM Trans. Comput. Biol. Bioinform. **11**, 407–418 (2014)
13. Peng, W., Wang, J., Cheng, Y., Lu, Y.: UDoNC: an algorithm for identifying essential proteins based on protein domains and protein-protein interaction networks. IEEE/ACM Trans. Comput. Biol. Bioinform. **12**, 276–288 (2015)
14. Zhang, W., Xu, J., Li, Y., Zou, X.: Detecting essential proteins based on network topology, gene expression data and gene ontology information. IEEE/ACM Trans. Comput. Biol. Bioinform. **15**, 109–116 (2016)
15. Li, M., Lu, Y., Niu, Z., Wu, F.X.: United complex centrality for identification of essential proteins from PPI networks. IEEE/ACM Trans. Comput. Biol. Bioinform. **14**, 370–380 (2017)
16. Qin, C., Sun, Y., Dong, Y.: A new method for identifying essential proteins based on network topology properties and protein complexes. PLoS ONE **11**, e0161042 (2016)
17. Peng, X., Wang, J., Zhong, J., Luo, J., Pan, Y.: An efficient method to identify essential proteins for different species by integrating protein subcellular localization information. In: IEEE International Conference on Bioinformatics and Biomedicine, pp. 277–280 (2015). https://doi.org/10.1109/BIBM.2015.7359693
18. Li, G., Min, L., Wang, J., Wu, J., Wu, F.X., Yi, P.: Predicting essential proteins based on subcellular localization, orthology and PPI networks. BMC Bioinform. **17**, 279 (2016)

19. Wei, P., Wang, J., Wang, W., Liu, Q., Wu, F.X., Yi, P.: Iteration method for predicting essential proteins based on orthology and protein-protein interaction networks. BMC Syst. Biol. **6**, 1–17 (2012)
20. Xiao, Q., Wang, J., Peng, X., Wu, F.X., Pan, Y.: Identifying essential proteins from active PPI networks constructed with dynamic gene expression. BMC Genom. **16**, S1 (2015)
21. Luo, J., Kuang, L.: A new method for predicting essential proteins based on dynamic network topology and complex information. Comput. Biol. Chem. **52**, 34–42 (2014)
22. Li, M., Ni, P., Chen, X., Wang, J., Wu, F., Pan, Y.: Construction of refined protein interaction network for predicting essential proteins. IEEE/ACM Trans. Comput. Biol. Bioinform. 1 (2017, early access)
23. Nakayama, T., Seno, S., Takenaka, Y., Matsuda, H.: Inference of S-system models of gene regulatory networks using immune algorithm. J. Bioinform. Comput. Biol. **09**, 75–86 (2011)
24. Jiao, L., Lei, W.: A novel genetic algorithm based on immunity. IEEE Trans. Syst. Man Cybern. Part A Syst. Hum. **30**, 552–561 (2000)
25. Mete, M., Tang, F., Xu, X., Yuruk, N.: A structural approach for finding functional modules from large biological networks. BMC Bioinform. **9**, S19–S19 (2008)
26. Andreas, S., Mario, A.: FunSimMat: a comprehensive functional similarity database. Nucleic Acids Res. **36**, D434–D439 (2008)
27. Xenarios, I., Salwínski, L., Duan, X.J., Higney, P., Kim, S.M., Eisenberg, D.: DIP, the Database of Interacting Proteins: a research tool for studying cellular networks of protein interactions. Nucleic Acids Res. **30**, 303 (2002)
28. Krogan, N.J., et al.: Global landscape of protein complexes in the yeast Saccharomyces cerevisiae. Nature **440**, 637–643 (2006)
29. Das, J., Yu, H.: HINT: high-quality protein interactomes and their applications in understanding human disease. BMC Syst. Biol. **6**(1), 92 (2012)
30. Stark, C., Breitkreutz, B.J., Reguly, T., Boucher, L., Breitkreutz, A., Tyers, M.: BioGRID: a general repository for interaction datasets. Nucleic Acids Res. **34**, 535–539 (2006)
31. Mewes, H.W., et al.: MIPS: analysis and annotation of proteins from whole genomes in 2005. Nucleic Acids Res. **34**, D169 (2006)
32. Cherry, J., et al.: SGD: Saccharomyces Genome Database. Nucleic Acids Res. **26**, 73–79 (1998)
33. Zhang, R., Lin, Y.: DEG 5.0, a database of essential genes in both prokaryotes and eukaryotes. Nucleic Acids Res. **37**, 455–458 (2009)
34. Chen, W.H., Minguez, P., Lercher, M.J., Bork, P.: OGEE: an online gene essentiality database. Nucleic Acids Res. **40**, D901–D906 (2012)

Integrating Multiple Datasets to Discover Stage-Specific Cancer Related Genes and Stage-Specific Pathways

Bolin Chen[1,2(✉)], Chaima Aouiche[1(✉)], and Xuequn Shang[1,2]

[1] School of Computer Science, Northwestern Polytechnical University,
Xi'an 710072, China
[2] Key Laboratory of Big Data Storage and Management,
Northwestern Polytechnical University,
Ministry of Industry and Information Technology, Xi'an, China
blchen@nwpu.edu.cn

Abstract. Investigating the evolution of complex diseases through different disease stages is critical for understanding the root cause of these diseases, which is fundamental for their accurate prognosis and effective treatment. There have been numerous studies that have identified many single genes, static modules and individual pathways related cancer progression, but few attempt has been developed to identify specific genes and pathways interactions related individual disease stages via data integration. To address these issues, we have proposed a general working flow, to reveal disease stages dynamics by joint analysis of multi-level datasets. Our contribution is two-fold. Firstly, we present a classical regression method to identify stage-specific cancer genes, where the gene expression and DNA methylation datasets are integrated. Secondly, we construct a pathway evolution network, which considered interactions among specific mapped pathways and their overlapped genes. Interestingly, the potential discovered biological functions from this network together with the common bridges and genes, not only help us to understand the functional evolution and dynamics of complex diseases in a more deep fashion, but also useful for clinical management to design customized drugs with more effective therapy.

Keywords: Data integration · Disease evolution · Disease genes ·
Pathological staging · Seed pathways ·
Pathway interaction sub-network

B. Chen and C. Aouiche—Equal contributors.

Electronic supplementary material The online version of this chapter (https://doi.org/10.1007/978-3-030-17938-0_22) contains supplementary material, which is available to authorized users.

I. Rojas et al. (Eds.): IWBBIO 2019, LNBI 11465, pp. 240–250, 2019.
https://doi.org/10.1007/978-3-030-17938-0_22

1 Introduction

Complex diseases, such as cancers, are kinds of evolutionary diseases [1–3], which involve successive stages from early initiation to advanced end-stages. Determining the possible biological changes associated with these stages is necessary for understanding the progression of many diseases, thereby specifying their best treatment strategy. Take the colorectal cancer for example, these stages can be classified generally into four phases based on their level of extension, lymphatic involvement and metastatic features. Specifically, stage I refers to a tumor of small size confined to the organ of origin; stage II describes the disease that has locally advanced beyond the site of origin; stage III characterize the disease that has spread to the neighboring organs, and stage IV represents distant metastatic disease. Here, cancers at early stages (stage I or II) are usually considered curable and might only need an active surveillance compared to advanced stages (stage III or IV) which might require more radical and active treatment.

Therefore, there was a critical need to characterize the dynamics associated with these stages through extraction of reliable biomarkers, which generally involve stage-specific cancer genes, dynamic modules and pathway dysregulations [4–7].

With the recent developments on biological technologies and accumulation of large amounts of omics-data, it has become more easy to investigate the dynamics of many genes and pathways involved in diseases [8–10]. Furthermore, the international consortia, such as The Cancer Genome Atlas (TCGA) [11], have generated several large-scale cancer datasets and samples with various clinical/pathological stages on, for example gene expression, DNA methylation and Copy number variation (sCNA) together with clinical data, which also provides a great opportunity to discover more robust predictive signatures for diseases. Each of these genomic data have been widely used in many contributions, to resolve many issues, but most of them focused on single genes, pathways and static networks, which cannot fully reveal the dynamics of the molecular events and patterns related to most human diseases including cancers. Since cellular functions related-cancers are mediated through complex systems interconnected by physical interactions [12,13]. In fact, these complex systems require (1) the integration of diverse types of data over the genomic, epigenetic, transcriptome and metabolome levels with interaction networks, for an unprecedented number of patients, and (2) a large number of samples with various stages, taking into account the considerable associations that can be made between their genomic profiles, clinical parameters, modules and pathways [14–20]. To might get a more comprehensive view of biology and specifically to cancer evolution or transition.

Fortunately, network/pathway analysis has been proved to be an effective way to elucidate the biological features and molecular interactions that underlie each stage of disease progression. Where each node can represents a gene or pathway and each edge corresponds to an interaction between a pair of genes/pathways. In this field, a considerable number of biological networks have been studied, but unfortunately in a static context including gene regulation networks [21], protein-protein interaction (PPI) networks [22] and disease networks [23]. In addition, the pathway interaction network has recently confirmed to be a powerful tool in interpreting how gene perturbations can lead to disease [24,25]. Specifically, across multiple stages, further shedding lights on the action mechanism on how cellular systems functions operate [26].

Notably, and in a dynamic context, extracting pathway sub-networks, taking interactions among specific pathways across specific stages. Different findings can be captured such as (1) predicting the real evolution of complex diseases, (2) understating the biological function role of the involved key genes and (3) exploring the relations between the predicted cellular functions through their common bridges and genes.

In this study, we proposed to investigate a general working flow that addresses 2 major issues concerning the staging evolution processes of complex diseases, which including: (1) the identification of stage-specific cancer related genes, (2) the extraction of stage-specific pathways and generation of their pathway interaction sub-network. The sections of this study are organized as follows: The methods and related materials are presented in Sect. 2. The experimental results are discussed in Sect. 3. The conclusion is provided in Sect. 4.

2 Materials and Methods

2.1 Data Collection and Preprocessing

Level 3 open access data for clinical and genomic datasets was accessed on 2016 from the Broad Institute of Harvard and MIT's FireBrowse (http://firebrowse. org/). These datasets include clinical data, gene expression and DNA methylation profiles for the same group of patients. The patient's samples were grouped into 4 stages based on clinical information, with a total size of 219 samples. The summarized information can be found in Table 1.

Table 1. Summary of the incorporated datasets informations from TCGA.

Data type	Platform	Samples
Gene expression	UNC-AgilentG4502A	219
DNA methylation	JHU-USC-HumanMethylation27	219
Clinical data	-	219

At the same time, pathway data were extracted from the Reactome database (http://www.reactome.org), which is an online curated resource and peer-reviewed knowledgebase of human reactions and pathways. As pathways with less genes may not have sufficient biological information, we generated a set of pathways by discarding the ones with one gene in this study. Overall, we ended up with 269 informative pathways.

To also enhance the quality of the obtained datasets more accurate, standard preprocessing steps were conducted. Further, for initially measured gene sets of both gene expression and DNA methylation profiles, which contain 17505 gene expressions and 26224 DNA methylations, we only consider the intersection of the two measured gene sets. Moreover, genes with missing values, such as NA or NULL, were removed, and methylation CpG loci that related to multiple genes were shared equally for those genes. Ultimately, a set of 12586 genes were obtained and used for subsequent regression analysis. In addition, for clinical data, where the clinical stage information for patients was available, we only adopt the "pathology_t_stage" parameter, which describes the anatomic extent of the cancer at the time of diagnosis for individual samples (t_1, t_2, t_3 and t_4). These pathological variables were converted into binary values for further analysis. For the sample selection, we count the number of patients across these pathological stages, and a total of 219 samples are obtained and considered in this study. The work flow of the whole process and analysis implemented is shown in Fig. 1.

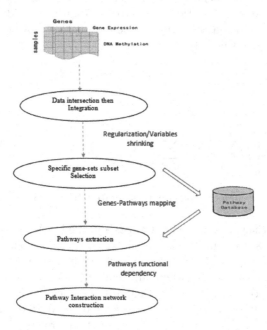

Fig. 1. Workflow of the overall defined stage-specific cancer genes and pathways in the evolution process based-pathological_staging.

2.2   Stage-Specific Related Gene Identification

The first essential step to investigate the evolution progress of complex diseases is to identify signature genes for individual stages. Various penalized methods have been proposed in this regard, and we here consider the elastic-net method.

The elastic-net [27] is a widely used classical penalized regression approach to deal with high-dimensional models with a large number of parameters. This penalization can model more than one type of omics data, execute variable selection and parameter estimation, thus reducing the computation time. Interestingly, it has already been applied to GWAS studies [28–30] as well as in the context of integrative studies [31] and clinical findings [32,33].

Consider the standard linear regression model with generally y response variables and x standardized predictors defined by:

$$B = \|Y - X\beta\|_2. \tag{1}$$

where $\beta = (\beta_1, \beta_2, \ldots, \beta_n)^T$ is the coefficient vector for all predictors.

The problem of stage-specific related gene identification is to detect a set of genes that minimize this standard objective function based on two preliminaries parameter: m samples (patients) and n features (gene expression or methylation profiles). Where, the feature matrix can be denoted as a $m \times n$ dimension matrix X. Given a m dimension label vector Y (pathology stage labels).

Since, the Elastic-net penalty used here is a convex combination of the Least Absolute Shrinkage and Selection Operator (LASSO) and ridge penalties, for the model described above, we estimate the parameters β by the following optimization problem equivalent to:

$$\widehat{B} = \|Y - X\beta\|_2 + \lambda_1|\beta| + \lambda_2\|\beta\|_2. \tag{2}$$

or

$$\widehat{B} = argmin\left\{\frac{1}{m}\sum_{i=1}^{m}\left(y_i - \sum_{j=1}^{n}x_{ij}\beta_j\right)^2 + \lambda_1\sum_{j=1}^{n}|\beta_j| + \lambda_2\sum_{j=1}^{n}\beta_j{}^2\right\}. \tag{3}$$

where λ_1, λ_2 are the penalty parameters related to LASSO and ridge penalty, respectively.

In this study, the gene expression profiles and the DNA methylation information were integrated to form the feature matrix X, and four binary stage-specific label vectors $Y_t, t = 1, 2, 3, 4$ were employed to identify disease related genes for individual stages, respectively (where an element in Y_t represents if that sample was recognized as the t_{th} pathology stage in the clinical dataset).

The objective function (3) was implemented in Matlab R2015a with the tuning parameter $\lambda_1 = \lambda_2 = 0.5$. The fitted least-squares regression coefficients were used for gene selection. Giving a pair of X and Y_t, the Matlab program calculated the fitted coefficients at around 50 times (automatically determined by Matlab). At each time, a set of signature genes could be selected, and we finally selected all the genes determined across all the times of calculation. Remarkably, 50 times produce 323 genes in this study. More details are shown in Table 2, which summarizes the times of running and the total number of selected genes at each time of calculation across the 4 pathology stages.

2.3   Stage-Specific Pathways Extraction

The stage-specific genes obtained at every pathology stage were then aligned to pathways from the Reactome pathway database. Totally, 1450 pathways were collected from this database, including 361 pathways from Pathology_t1, 376 pathways from Pathology_t2, 362 pathways from Pathology_t3 and 351 pathways from Pathology_t4.

Since the same gene can be mapped to different pathways and different pathways would have a different number of genes. At every pathology stage, only pathways with a gene set size greater than one were reserved as the study objectives owing to the fact that some pathways with considerably less genes may have insufficient biological information [34]. Hence, a total of 478 ($t_1 = 121, t_2 = 124, t_3 = 117, t_4 = 116$) pathways were collected as shown in Table 2. Finally, the duplicated pathways among 478 were filtered out, and thus only 269 extracted pathways were considered as our stage-specific pathways or seed pathways, and regarded for further analysis.

2.4   Pathway Interaction Network Construction

Once the signature genes were identified for each stage, and their specific Reactome pathways were extracted and integrated. We then pooled these pathways altogether, unify their terms and get their official annotated pathway descriptions from the database. Next, a complete pathway evolution network was constructed with each node representing a specific pathway, where one edge was laid between two pathways if they share common genes.

The pathway network would serve as a better aspect to show the dynamic evolution processes of the interested disease, since we can use the color of individual vertices to indicate their pathology stage, and the width of edges can show the overlapped score between two pathways. Here, the overlap score was calculated as follows

$$W = \frac{k^2}{p * q}. \tag{4}$$

where k is the number of the overlapped genes between a pair of pathway P_i and pathway P_j, p and q are the total numbers of genes in P_i and P_j, respectively.

Notably, to uncover the biological significance of the generated evolution network, only a sub-network of 29 pathways were considered and validated.

3   Results and Discussion

3.1   The Number of Stage-Specific Related Genes

In this contribution, we used elastic-net regression to estimate an optimal multiple linear regression of the pathological outcome on the space of genomic features. Taking the advantage of the geometric sequence of its generated models, we therefore selected all the genes that were detected by all these models. These seed genes robustly delineate early and advanced pathological stages. Table 2 summarized the overall number of genes selected at different stages. To be more specific, stage t1 has obtained 321 genes from 51 models; stage t2 has obtained 297 genes from 48 models; stage t3 has obtained 317 genes from 45 models,

and finally stage t4 has obtained 323 genes from 50 models, respectively. All of
these potential biomarker genes were well characterized the dynamics of the 4
pathological stages, due to their possible role in cancer progression.

Table 2. The overall number of genes and pathways detected. This table illustrated the
number of the generated models resulted at each pathology_stage, their total number
of genes and the corresponding aligned pathways.

Stages	Models#	Detected# of genes	# of aligned pathways
Pathology_t1	51	321	121
Pathology_t2	48	297	124
Pathology_t3	45	317	117
Pathology_t4	50	323	116

3.2   Dynamic Pathway Interaction Network Generation and Visualization

We built the informative pathway evolution network relying on the different
interaction dependencies between specific pathways related to disease-stages. It
comprised of 269 nodes and 2187 edges. Subsequently, the network was imported
into Cytoscape, an open source for visualizing molecular interaction networks
and integrated data. Different colors (green, blue, orange, red) were used to
highlight the important evolved pathways indicating their evolution through the
4 pathology stages, whereas the connections width denoted the overlapped score.
These further details are depicted in Fig. 2.

3.3   Pathway Interaction Sub-network Construction

To determine the most important and strongly connected related pathways about
cancer evolution, we extracted a minimum set of pathways and the corresponding
sub-network from the pathway evolution network. We employed two conditions
for the accurate selection: pathways with a considerable number of overlapped
genes between or within the 4 pathological stages. As a result, we ended up
with 29 seed pathways. These seed pathways are available in the supplementary material (see S1 Table) with extra informations (detected_genes_count and
official_pathways_genes_count with their 3 level names), and the sub-network is
shown in Fig. 3.

3.4   Pathways-Specific Functions to Genes Evaluation

The analysis showed that the pathological-stages specific pathways and genes
obtained in this study were successfully involved in many critical cellular functions related to cell cycle, disease, gene expression, Immune system and signal

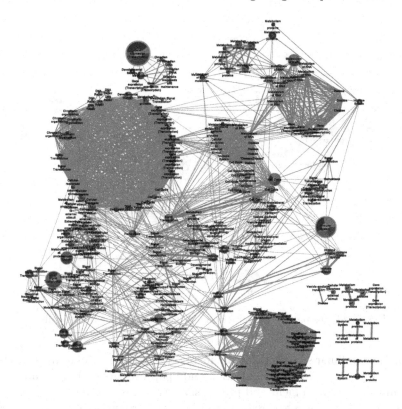

Fig. 2. Pathway interaction network (Color figure online)

transduction. More specifically, we show how the different discovered functions connected with each other and how the involved genes related stages act in these functions.

Among these functions, cell cycle was the most activated here including (1) separation of sister chromatids, (2) condensation of prophase chromosomes and (3) meiotic recombination. As it highly connected with developmental biology, chromatin organization and cellular responses to external stimuli, crossing common bridges and sharing common genes. This is in accord with the established paradigm that the dysfunction of cell cycle regulation is the primary cause of tumor development leading to DNA division and replication [35].

Moreover, another worthy mention function named Disease included Constitutive signaling by aberrant PI3K in cancer have also connected with signal transduction which in turn included RAF/MAPK cascade. Where, Phosphatidylinositol-4,5-bisphosphonate 3-kinase (PI3K) plays a crucial role in the pathogenesis of many cancer types, including colorectal cancer, and serves as a major therapeutic target in controlling cancer progression. It has the potential to identify predictive biomarkers for response to treatments, and to guide the development of targeted therapies for precision medicine. Additionally, MAPKs

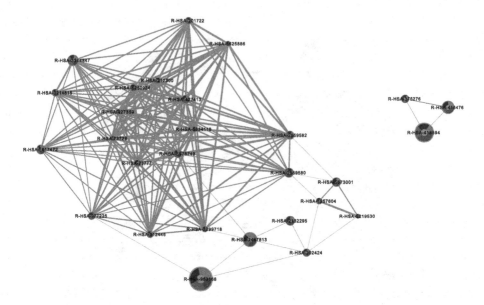

Fig. 3. Stages pathway interaction sub-network (Color figure online)

mediate intracellular signaling and are involved in diverse cellular processes that include cell proliferation, differentiation and apoptosis. As such, they are implicated in cancer development and progression. Whereas, the RAF proteins activate the MAPK pathway where inappropriate and/or persistent activation leads to abnormal differentiation, proliferation, apoptosis and cancer development [36].

4   Conclusions

The mechanism of complex diseases evolution is too complex to be revealed by only one type of genomic data, single genes and individual pathways.

There are considerable changes that happened across different disease types and individual disease samples, owing to the fact that no disease directly end up with a mortal situation. A complete understanding of these underlie changes associated especially with cancers is essential to identify potential therapeutic options and vulnerabilities.

In the present study, we tried to address these issues more deeply by using many strategies. Therefore, to get a better biological and clinical insight into the progression of cancer diseases, interestingly through pathological staging mechanism. These strategies including multi-omics data integration, specific genes identification at specific stages, pathway interaction network generation rather than gene networks, seed pathways-related stages selection, and finally specific functions interpretation.

To the end, the biological findings confirmed the efficacy of the proposed work, highlighting how specific indicators at specific pathological stages may help improve clinical research endpoints and ultimately aid in clinical utility.

Acknowledgments. This work was supported by the National Natural Science Foundation of China under [Grant No. 61602386, 61772426 and 61332014]; the Natural Science Foundation of Shaanxi Province under [Grant No. 2017JQ6008]; and the Top International University Visiting Program for Outstanding Young scholars of Northwestern Polytechnical University.

References

1. Horne, S., Chowdhury, S., Heng, H.: Stress, genomic adaptation, and the evolutionary trade-off. Front. Genet. **5**, 92 (2014)
2. Horne, S., Pollick, S., Heng, H.: Evolutionary mechanism unifies the hallmarks of cancer. Int. J. Cancer **136**, 2012–21 (2015)
3. Spiller, D.G., Wood, C.D., Rand, D.A., White, M.R.H.: Measurement of single-cell dynamics. Nature **465**(7299), 736–748 (2010)
4. Chen, L., Wang, R.S., Zhang, X.S.: Biomolecular Networks: Methods and Applications in Systems Biology, vol. 10. Wiley, Hoboken (2009)
5. Chen, L., Wang, R., Li, C., Aihara, K.: Modeling Biomolecular Networks in Cells: Structures and Dynamics. Springer, London (2010). https://doi.org/10.1007/978-1-84996-214-8
6. Lee, J., Zhao, X., Yoon, I., Lee, J., Kwon, N., Wang, Y., et al.: Integrative analysis of mutational and transcriptional profiles reveals driver mutations of metastatic breast cancers. Cell Discov. **2**, 16025 (2016)
7. Guanghui, Z., Hui, Y., Xiao, C., Jun, W., Yong, Z., Xing-Ming, Z.: CSTEA: a webserver for the cell state transition expression atlas. Nucleic Acids Res. **45**, 103–108 (2017)
8. Bosinger, S.E., Jacquelin, B., Benecke, A., Silvestri, G., Muller-Trutwin, M.: Systems biology of natural Simian immunodeficiency virus infections. Curr. Opin. HIV AIDS **7**(1), 71–78 (2012)
9. Jordan, N.V., et al.: HER2 expression identifies dynamic functional states within circulating breast cancer cells. Nature **537**(7618), 102–106 (2016)
10. Nakamura, A., Osonoi, T., Terauchi, Y.: relationship between urinary sodium excretion and pioglitazone-induced edema. J. Diab. Invest. **1**(5), 208–211 (2010)
11. Michor, F., Iwasa, Y., Nowak, M.A.: Dynamics of cancer progression. Nat. Rev. Cancer **4**(3), 197 (2004)
12. Karczewski, K.J., Snyder, M.P.: Integrative omics for health and disease. Nat. Rev. Genet. **19**(5), 299 (2018)
13. Ma, X., Sun, P.G., Zhang, Z.Y.: An integrative framework for protein interaction network and methylation data to discover epigenetic modules. IEEE/ACM Trans. Comput. Biol. Bioinform. (Early Access), 1 (2018). https://doi.org/10.1109/TCBB.2018.2831666
14. Hsu, F., Serpedin, E., Hsiao, T., Bishop, A., Dougherty, E., Chen, Y.: Reducing confounding and suppression effects in TCGA data: an integrated analysis of chemotherapy response in ovarian cancer. BMC Genomics **13**, 13 (2012)
15. Parker, J., Mullins, M., Cheang, M., Leung, S., Voduc, D., Vickery, T., et al.: Supervised risk predictor of breast cancer based on intrinsic subtypes. J. Clin. Oncol. **27**, 1160–1167 (2009)
16. Curtis, C., Shah, S., Chin, S., Turashvili, G., Rueda, O., Dunning, M., et al.: The genomic and transcriptomic architecture of 2,000 breast tumours reveals novel subgroups. Nature **486**, 346–352 (2012)
17. Kittaneh, M., Montero, A., Gluck, S.: Molecular profiling for breast cancer: a comprehensive review. Biomark. Cancer **5**, 61–70 (2013)

18. Li, A., Walling, J., Ahn, S., Kotliarov, Y., Su, Q., Quezado, M., et al.: Unsupervised analysis of transcriptomic profiles reveals six glioma subtypes. Cancer Res. **69**, 2091–2099 (2009)
19. Shen, L., Toyota, M., Kondo, Y., Lin, E., Zhang, L., Guo, Y., et al.: Integrated genetic and epigenetic analysis identifies three different subclasses of colon cancer. Proc. Nat. Acad. Sci. U.S.A. **104**, 18654–18659 (2007)
20. van't Veer, L., Dai, H., van de Vijver, M., He, Y., Hart, A., Mao, M., et al.: Gene expression profiling predicts clinical outcome of breast cancer. Nature **415**, 530–536 (2002)
21. Vaquerizas, J.M., Kummerfeld, S.K., Teichmann, S.A., Luscombe, N.M.: A census of human transcription factors: function, expression and evolution. Nat. Biotechnol. **10**, 252–263 (2009)
22. Schwikowski, B., Uetz, P., Fields, S.: A network of protein-protein interactions in yeast. Nat. Biotechnol. **18**, 1257–1261 (2010)
23. Menche, J., et al.: Uncovering disease-disease relationships through the incomplete interactome. Science **347**, 1257601 (2015)
24. Tong, A.H., Lesage, G., Bader, G.D., et al.: Global mapping of the yeast genetic interaction network: discovering gene and drug function. Science **303**(5659), 808–813 (2004)
25. Glazko, G.V., Emmert-Streib, F.: Unite and conquer: univariate and multivariate approaches for finding differentially expressed gene sets. Bioinformatics **25**(18), 2348–2354 (2009)
26. Xia, Y., Yu, H., Jansen, R., Seringhaus, M., Baxter, S., Greenbaum, D., et al.: Analyzing cellular biochemistry in terms of molecular networks. Ann. Rev. Biochem. **73**, 1051–1087 (2004)
27. Zou, H., Hastie, T.: Regularization and variable selection via the elastic net. J. Royal Stat. Soc. Ser. B **67**, 301–320 (2005)
28. Pineda, S., Milne, R.L., Calle, M.L., Rothman, N., De Maturana, E., et al.: Genetic variation in the TP53 pathway and bladder cancer risk. A comprehensive analysis. PLoS One **9**(5), e89952 (2014)
29. Cho, S., Kim, K., Kim, Y.J., Lee, J.K., Cho, Y.S., et al.: Joint identification of multiple genetic variants via elastic-net variable selection in a genome-wide association analysis. Ann. Hum. Genet. **74**, 416–428 (2010)
30. Zhou, H., Sehl, M.E., Sinsheimer, J.S., Lange, K.: Association screening of common and rare genetic variants by penalized regression. Bioinformatics **26**, 2375–2382 (2010)
31. Mankoo, P.K., Shen, R., Schultz, N., Levine, D.A., Sander, C.: Time to recurrence and survival in serous ovarian tumors predicted from integrated genomic profiles. PLoS One **6**, e24709 (2011)
32. Lee, H., Flaherty, P., Ji, H.: Systematic genomic identification of colorectal cancer genes delineating advanced from early clinical stage and metastasis. BMC Med. Genomics **6**, 54 (2013)
33. Lee, H., Palm, J., Grimes, S., Ji, H.: The cancer genome atlas clinical explorer: a web and mobile interface for identifying clinical-genomic driver associations. Genome Med. **7**, 112 (2015)
34. Ahn, T., Lee, E., Huh, N., Park, T.: Personalized identification of altered pathways in cancer using accumulated normal tissue data. Bioinformatics **30**, i422–i429 (2014)
35. Perez, R., Wu, N., Klipfel, A.A., Beart Jr., R.W.: A better cell cycle target for gene therapy of colorectal cancer: cyclin G. J. Gastrointest. Surg. **7**, 884–889 (2003)
36. Maurer, G., Tarkowski, B., Baccarini, M.: Raf kinases in cancer-roles and therapeutic opportunities. Oncogene **30**(32), 3477–3488 (2011)

Integrated Detection of Copy Number Variation Based on the Assembly of NGS and 3GS Data

Feng Gao[1], Liwei Gao[2], and JingYang Gao[1(✉)]

[1] College of Information Science and Technology,
Beijing University of Chemical Technology, Beijing 100029, China
gaojy@mail.buct.edu.cn
[2] Department of Radiation Oncology,
China-Japan Friendship Hospital, Beijing, China

Abstract. The genomic coverage of copy number variations (CNVs) ranges from 5% to 10%, which is one of the essential pathogenic factors of human diseases. The detection of large CNVs is still defective. However, the read length of the third-generation sequencing (3GS) data is longer than that of the next-generation sequencing (NGS) data, which can theoretically solve the defect that the long variation can't be detected. However, due to the low accuracy of the 3GS data, it is difficult to apply in practice. To a large extent, it is a supplement to the NGS data research. To solve these problems, we developed a new mutation detection tool named AssCNV23 in this paper. Firstly, this tool corrects the 3GS data to solve the problem of high error rate, and then combines the results of a variety of mutation detection tools to improve the accuracy of the initial mutation set and to solve the detection bias of a single detection tool. At the same time, the high-quality 3GS data was introduced by AssCNV23 to guide the NGS data to assemble, and then detects the CNV after getting enough length data. Finally, to improve the detection efficiency, the tool generates images containing the sequence depth information based on the read depth strategy and uses the convolutional neural network to detect the existing CNVs. The experimental results show that AssCNV23 guarantees a high level of breakpoint accuracy and performs well in identifying large variation. Compared with other tools, the deep learning model has advantages in accuracy and sensitivity, and Matthew correlation coefficient (MCC) performs well in various experiments. This algorithm is relatively reliable.

Keywords: NGS · 3GS · Assembly · Integrated detection · AssCNV23

1 Introduction

Copy Number Variation (CNV) is an integral part of Structural Variations (SVs). It refers to submicroscopic variations in the size of the genomic fragment, ranging from kb to Mb, including single-segment deletion and duplication, and complex SVs derived from each other [1, 2]. The coverage of CNV on the genome reaches 5% to 10%, which is much higher than that of single nucleotide variation (SNV). Being one of the most

© Springer Nature Switzerland AG 2019
I. Rojas et al. (Eds.): IWBBIO 2019, LNBI 11465, pp. 251–260, 2019.
https://doi.org/10.1007/978-3-030-17938-0_23

critical pathogenic factors of human diseases, CNV has been confirmed to be associated with many complex disorders: neonatal seizures are associated with intellectual development and the whole-gene duplication of SCN2A and SCN3A [3, 4]. Moreover, CNV is one of the most important types of variations in the cancer genome. CNV on GSTM1 increases the incidence of bladder cancer [5], and CNV affects susceptibility genes for breast and ovarian cancers [3]. Because of the association of CNV and diseases, the accurate detection for CNV within the whole genome is of great significance.

Now the most commonly used sequencing technology is the next-generation sequencing (NGS) technology, also known as high-throughput sequencing technology. Since its appearance in 2004, after continuous technological improvement [6] and other sequencing technology development, it has far exceeded the Sanger sequencing technology regarding cost reduction and throughput enhancement [7], and completely changed DNA sequencing. But the length of the read that NGS technology produced is short, which makes genomic assembly and related genetic structural function analysis difficult. Theoretical studies have shown that reducing the length of reads from 1,000 bp to 100 bp may result in a six-fold or more reduction in gene fragment continuity [8]. The third-generation sequencing (3GS) technology, represented by PacBio, uses Single Molecule Real Time (SMRT) DNA Sequencing technology [9] to solve the problems in NGS technology. The emergence of long read provides new ideas for solving the complex repetitive sequence problems in NGS data and the problem of assembling the whole genome. The advantage of the 3GS data is that it generates long read, far exceeding that of the Sanger sequencing technology and the next-generation technology. However, the 3GS technology still has defects, the average accuracy of reads is only 82%–85% [10]. This high error rate makes the mapping between the various reads more complicated.

2 Methods

Given the shortcomings of the existing sequencing errors of the 3GS data, combined with the advantages of the NGS data, we propose a method to correct the 3GS data using the NGS high-quality data. And CNV detection of the NGS data using multiple detection tools can compensate for the detection bias of single tool and obtain high-quality mutation breakpoints. The read where the mutation breakpoint is located is screened out from the NGS data, and it is assembled under the guidance of the 3GS data, taking advantage of the long reading length of the 3GS data. The results after assembly are analyzed to find out the variation of copy number. Combined with the read depth strategy, the number of copies of each site is estimated, and the pictures that contain the number of copies are generated for deep learning training. We have integrated the above methods and compiled a set of tools for detecting CNVs, named AssCNV23. Its workflow mainly includes the following four parts: the first step is to correct the third generation sequencing data to obtain high-quality third generation data. In the second step, multi-tool ensemble detection is carried out for the NGS data to get highly reliable mutation site information and select read sets with mutation sites. In the third step, under the guidance of the high-quality 3GS data with long read, read collections with long CNV can be obtained by assembling the selected reads. In the fourth step, using the read depth strategy to count the number of copies and generate pictures for training to get a model to detect CNVs.

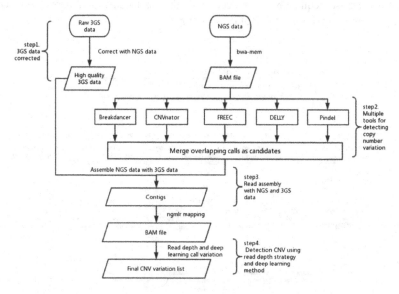

Fig. 1. An overview of the AssCNV23. AssCNV23 takes both advantages of NGS data and 3GS data, using assembly-based and read depth strategy to improve the detection accuracy and sensitivity.

2.1 3GS Data Corrected

The 3GS data were adjusted using the developed tool Cor3GS. The schematic diagram of the operation process is shown in Fig. 1. The specific method is as follows:

(a) Data format conversion module: remove the low-quality read and the adapter read of the original 3GS data, and then converted into *.fasta* format data and saved and indexed as a reference genome for the NGS data.

(b) DNA mutation detection module: using the *bwa-mem* algorithm mapping the NGS data to the 3GS data, and get the BAM file. According to the sequence quality information in the BAM file, remove the low-quality sequence, and then perform the sequence deduplication. Since there is a large number of base mismatches near Indel, it is easy to be considered as the wrong base generated by the sequencing error in the calibration process. So it is necessary to perform Indel Realigner on the de-duplicated file to minimize the error rate. It is necessary to re-correct the base quality value of the reads in the BAM file, so that the quality value of the reads in the final output BAM file is close to the actual value. Using DNAseq to obtain the variation information between the NGS data and the 3GS data, which is the location of the sequencing error existing in the 3GS data.

(c) Base correction module on mapping region: extract the name and the mutation site and the base sequence of this site in 3GS data and the base sequence of this site in NGS data, these four columns data are stored. Each sequence is replaced according to the third and second base sequences of each variant site, and the third generation base sequence is replaced with the next-generation base sequence to complete the correction of the 3GS data (Fig. 2).

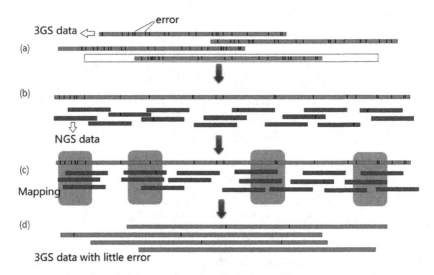

Fig. 2. Cor3GS flow chart. The workflow includes three modules. (a) Data format conversion module. (b) DNA mutation detection module. (c) Base correction module on mapping region

2.2   Multiple Tools for Detecting Copy Number Variation

Experiments were performed using the currently favorite tools for detecting CNV, include BreakDancer, CNVnator, FREEC, DELLY Pindel. For their detection results, only DEL and DUP with a length greater than 1000 bp were analyzed. Calculate the accuracy and sensitivity of the test results. Set two thresholds e = 0.4 and m = 2000, indicating that the deviations of the two detected breakpoints with the benchmark data did not exceed 2000 bp and 40% of the variable length; such breakpoints were considered as true positive. Under this set of thresholds, if at least three tools have a sensitivity higher than 0.45 and an accuracy greater than 0.4, their detection results are combined into candidate variation breakpoints. Otherwise, the program will divide the genome by a fixed length to obtain the variant breakpoint candidates.

2.3   Read Assembly with NGS and 3GS Data

After getting the local breakpoint candidates, the reads in the range were extracted from the BAM file, and the high-quality reads were filtered to perform the subsequent sequence assembly. The high-quality filtered reads are assembled using 3GS data, as follows:

At first, according to the quality information of the partial base sequence reads in the selected NGS data, delete the low-quality base reads, generate new partial base reads. According to the k parameter input by the user, cut the new reads into *k-mer*, store the *k-mers* and their number of occurrences in the hash table, and then using the *k-mers* constructs the *de* Bruijn graph.

Secondly, the edges of the *de* Bruijn graph without multiple exits or entries are compressed, merged into one edge, and a compressed *de* Bruijn graph is generated and get the read multiplicity of the compressed edge in the compressed *de* Bruijn graph

Lastly, repost the 3GS data to the NGS data, and complete the assembly by disassembling and compressing the *de* Bruijn graph, generating the contigs (Fig. 3).

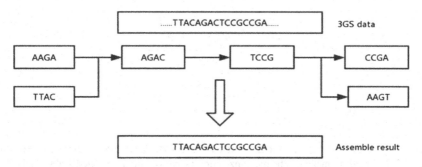

Fig. 3. Repeat sequence selection diagram. The edges in the De Bruijn graph constructed by NGS data may have multiple entries and exits, 3GS reads are used to determine which entry and exit are correct.

2.4   Detection CNV Using Read Depth Strategy and Deep Learning Method

Mapping the contigs generated by the assembly to the reference genome using *ngmlr*, a BAM file that containing the mapping result of the long reads is created, and then the CNV mutation detection is performed using the BAM file in combination with the read depth strategy and deep learning method.

The average local coverage can be obtained by summing the coverage of each site within the local breakpoints and then dividing by the local length. The base coverage at a site can be obtained using the depth command of Samtools. We considered 200 bp as a window, and 20 bp was overlapped between two adjacent windows to calculate the average coverage depth. If the average coverage depth of a series of connected windows was close, connected these continuous windows into one window, and then obtained the new average coverage. In this way, the local range was divided into some windows with a tremendous difference in average coverage. The Qualimap [11] was used to obtain the mean coverage information about the original BAM file, so that these windows can be tagged as normal, deletion, or duplication.

Then convert BAM file to pileup file using samtools. The circular binary segmentation (CBS) algorithm was used to delineate segments by copy number and identify significant change-points. And the flowing formula was applied to generate the file to create pictures to conduct deep learning (Fig. 4).

$$LRR_I = 2 * log_{10}\left(\frac{CN_i}{2}\right)$$

The relationship between LRR and CN as followed.

CN	1	2	3	4	5	6
LRR	−0.6	0	0.35	0.6	0.8	0.95

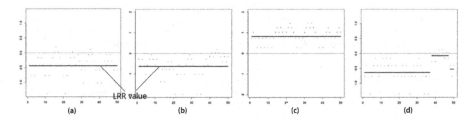

(a)   (b)   (c)   (d)

Fig. 4. The red line can be considered as LRR value. (a) The LRR value is more than −0.6 and close to 0, this kind of image can be recognized don't have variation. (b) The LRR value is close to −0.6, this kind of picture can be considered as deletion. (c) The LRR value is between 0.8 and 0.95, this can be judged to have duplication, but can't determine how many repetitions had occurred. (d) The LRR value in the picture changed. The reason for the diversification is that it comes to the breakpoint. (Color figure online)

The architecture of AssCNV23's network contains four convolutional layers intersected with three max pooling layers, three fully connected layers. Details are presented in Table 1.

Table 1. CNN architecture. I: input. C: convolutional layer. P: max pooling layer. NN: fully connected neural network layer. ReLU: rectified linear function f(x) = max(0, x)

Layer	Type	Activation	Feature maps	Filter size	Output size
0					100 × 100
1	C	ReLU	96	11 × 11	90 × 90
2	P		96	3 × 3	30 × 30
3	C	ReLU	256	4 × 4	26 × 26
4	P		256	3 × 3	9 × 9
5	C	ReLU	384	3 × 3	7 × 7
6	C	ReLU	256	3 × 3	5 × 5
7	P		256	3 × 3	2 × 2
8	NN	ReLU	1	2304	1 × 2304
8	NN	Softmax	1	3	1 × 3
9	NN	Softmax	1	3	1

3 Results

To simplify the complexity of the problem, only deletion and repetition are considered as sub-types of CNV, and the length range of them are more than 1000 bp. The detected results were compared to the benchmark data, and the performance of each tool was evaluated by Precision (Pre), Sensitivity (Sen), F1-score (F1) and Matthews correlation coefficient (MCC). MCC represents the reliability of the results of the algorithm. Its value range is $[-1, +1]$. When FP and FN are all zero, MCC is 1, indicating that the results of classification are completely correct. When TP and TN are all zero, MCC value is -1, indicating that the results of classification are completely wrong.

The precision is the correct ratio of the detected result, and the sensitivity is the correct ratio of the benchmark data. F1-score is a comprehensive indicator of precision and sensitivity. TP indicates true positive, FP indicates false positive, FN indicates false negative. And the formulas are:

$$\text{Precision} = \frac{TP}{TP + FP} \tag{1}$$

$$\text{Sensitivity} = \frac{TP}{TP + FN} \tag{2}$$

$$\text{F1 score} = \frac{2 * Precision * Sensitivity}{Precision + Sensitivity} \tag{3}$$

$$\text{MCC} = \frac{TP * TN - FP * FN}{\sqrt{(TP + FP)(TP + FN)(TN + FP)(TN + FN)}} \tag{4}$$

3.1 Generate Simulated Data

To simplify the problem and verify the feasibility of the method, we first experimented with the simulated data. We utilized chromosome 11 to generate NGS data with a depth of 20X using ART_illumina [12]. The read length was set to 150 bp. To avoid interference from other mutations, only copy number variation is introduced, and SinC [13] is used to simulate the generation of copy number variation of NGS data generated by ART_illumina. The variable length range is set to three parts, which are 1000 bp–5000 bp, 5000 bp–10000 bp and 10000 bp or more. Using the pbsim [14] tool to generate simulated 3GS PacBio data using chromosome 11, the coverage depth was 30X, the shortest read length was 2000, the longest read length was 10,000, and the sequencing error was 15%.

3.2 Calling CNV with Simulated Data

Count the number of CNVs of different lengths in the 20X simulation data, as shown in Table 2:

Table 2. Simulation data variation statistics

Read length	Insert size length	Insert size deviation	Read coverage	CNV length	DEL count	DUP count
150	200	10	20X	1k–5k	18975	5114
150	200	10	20X	5k–10k	7260	1933
150	200	10	20X	10k–	749	207

BreakDancer, CNVnator, DELLY, Pindel, FREEC and AssCNV23 were used to detect CNV of simulated data using default parameters. The results are summarized in Figs. 5 and 6.

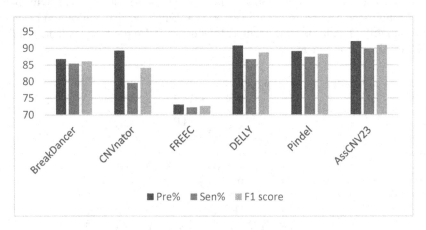

Fig. 5. Tools comparison of DEL detection on simulated data

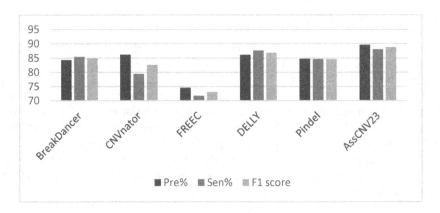

Fig. 6. Tools comparison of DUP detection on simulated data

According to the chart analysis, BreakDancer, DELLY, Pindel and AssCNV23 have better and stable performance on the simulation data. The detection accuracy and sensitivity of AssCNV23 are the highest among all tools. Concerning F1-score, AssCNV23 outperforms the other four tools. The precision of AssCNV23 is also higher than the others, while AssCNV23's MCC value is close to the best one. Because of the high value of MCC, the algorithm is relatively reliable.

Next, the F1 scores of CNV variations of different lengths are counted separately, as summarized in Table 3.

Table 3. Statistics of **F1 scores** for different lengths of CNV variation detected by different tools

Tools		BreakDancer	CNVnator	FREEC	DELLY	Pindel	AssCNV23
1000 bp–5000 bp	DEL	93.78	89.81	73.11	96.87	97.55	**98.39**
	DUP	90.82	87.14	73.57	94.67	91.13	**95.27**
5000 bp–10000 bp	DEL	81.98	81.10	73.08	84.97	83.01	**87.71**
	DUP	82.08	79.74	73.73	83.89	80.88	**85.24**
10000 bp+	DEL	82.43	81.25	71.63	84.09	84.17	**87.27**
	DUP	81.69	81.05	72.12	82.05	82.02	**86.99**

The analysis of CNVs of different lengths is carried out. With the increase of the range of variation, the detection accuracy and sensitivity of all tools show a down-ward trend. Among them, FREEC has the best stability, but its detection accuracy and sensitivity have been low. DELLY, Pindel and AssCNV23 have high precision and sensitivity in the short variation, and the precision and sensitivity decrease when the variable length increases, but still higher than FREEC, and AssCNV23 maintains a more senior level among the three tools, thus showing good stability. The reason is that 3GS data are introduced into AssCNV23 to participate in the assembling process, which increases the length of the sequenced fragments, so that more long variations can be detected in the detection than other tools.

In the whole simulation data experiment, the detection of different copy lengths of the two types of copy number variants (DEL and DUP) respectively showed the performance of different detection tools, which provides a reference of the high coverage data experiment and real data experiment.

4 Conclusion

CNV is an important category of structural variation. In this paper, a simple method of 3GS data correction is proposed, and a CNV detection method AssCNV23 based on the combination of NGS data and 3GS data is proposed. Combined with the initial detection results of BreakDancer, CNVnator, FREEC, DELLY, Pindel and other detection tools, the reliability of mutation breakpoints is ensured. At the same time, under the guidance of high-quality 3GS data, NGS reads with breakpoints is partially assembled. The transformation of sequence from short to long has been completed, and the detection length of mutation has been improved. The structural variation of more than 10 000 bp can be detected. Finally, based on the read depth strategy and the cyclic binary segmentation algorithm, the character data is transformed into an image containing the depth information of sequencing, and a deep learning method is introduced to train a model that can be applied to other data, which is helpful to the subsequent detection process.

Compared with other tools, AssCNV23 can detect more copy number variations and reduce the number of false positives. This makes AssCNV23 more accurate and sensitive than other tools. AssCNV23 is not only limited to using the tools mentioned in this paper to obtain the initial mutation set, but also can be combined with more tools to further improve the accuracy of mutation breakpoints.

Acknowledgment. Project supported by Beijing Natural Science Foundation (5182018) and the Fundamental Research Funds for the Central Universities & Research projects on biomedical transformation of China-Japan Friendship Hospital (PYBZ1834).

References

1. Ye, K., Wang, J., Jayasinghe, R., et al.: Systematic discovery of complex indels in human cancers. Nat. Med. **22**(1), 97–104 (2016)
2. Redon, R., Ishikawa, S., Fitch, K.R., et al.: Global variation in copy number in the human genome. Nature **444**(7118), 444–454 (2006)
3. Yu, G., et al.: An improved burden-test pipeline for identifying associations from rare germline and somatic variants. BMC Genom. **18**(Suppl 7:753), 55–62 (2017)
4. Thuresson, A.C., Van Buggenhout, G., Sheth, F., et al.: Whole gene duplication of SCN2A and SCN3A is associated with neonatal seizures and a normal intellectual development. Clin. Genet. **91**(1), 106–110 (2017)
5. Lu, C., Xie, M., Wendl, M., Wang, J., McLellan, M., Leiserson, M., et al.: Patterns and functional implications of rare germline variants across 12 cancer types. Nat. Commun. **6**, Article no. 10086 (2015)
6. Bentley, D.: Whole-genome re-sequencing. Curr. Opin. Genet. Dev. **16**, 545–552 (2006)
7. Sanger, F., Nicklen, S., Coulson, A.: DNA sequencing with chain-terminating inhibitors. PNAS **74**, 5463–5467 (1977)
8. Kingsford, C., Schatz, M., Pop, M.: Assembly complexity of prokaryotic genomes using short reads. BMC Bioinf. **11**, 21 (2010)
9. Chin, C.S., et al.: The origin of the Haitian cholera outbreak strain. N. Engl. J. Med. **364**, 33–42 (2011)
10. Rasko, D.A., et al.: Origins of the E. coli strain causing an outbreak of Hemolytic–Uremic syndrome in Germany. N. Engl. J. Med. 365, 709–717 2011
11. Garcíaalcalde, F., Okonechnikov, K., Carbonell, J., et al.: Qualimap: evaluating next-generation sequence alignment data. Bioinformatics **28**(20), 2678 (2012)
12. Huang, W., Li, L., Myers, J.R., Marth, G.T.: ART: a next-generation sequencing read simulator. Bioinformatics **28**(4), 593–594 (2012)
13. Pattnaik, S., Gupta, S., Rao, A.A., et al.: SInC: an accurate and fast error-model based simulator for SNPs, Indels and CNVs coupled with a read generator for short-read sequence data. BMC Bioinf. **15**(1), 40 (2014)
14. Ono, Y., Asai, K., Hamada, M.: PBSIM: PacBio reads simulator–toward accurate genome assembly. Bioinformatics **29**(1), 119–121 (2013)

Protein Remote Homology Detection Based on Profiles

Qing Liao, Mingyue Guo, and Bin Liu[✉]

School of Computer Science and Technology, Harbin Institute of Technology,
Shenzhen 518055, Guangdong, China
liaoqing@hit.edu.cn, 362717784@qq.com,
bliu@insun.hit.edu.cn

Abstract. As a most important task in protein sequence analysis, protein remote homology detection has been extensively studied for decades. Currently, the profile-based methods show the state-of-the-art performance. Position-Specific Frequency Matrix (PSFM) is a widely used profile. The reason is that this profile contains evolutionary information, which is critical for protein sequence analysis. However, there exists noise information in the profiles introduced by the amino acids with low frequencies, which are not likely to occur in the corresponding sequence positions during evolutionary process. In this study, we propose one method to remove the noise information in the PSFM by removing the amino acids with low frequencies and two a profile can be generated, called Top frequency profile (TFP). Autocross covariance (ACC) transformation is performed on the profile to convert them into fixed length feature vectors. Combined with Support Vector Machines (SVMs), the predictor is constructed. Evaluated on a benchmark dataset, experimental results show that the proposed method outperforms other state-of-the-art predictors for protein remote homology detection, indicating that the proposed method is useful tools for protein sequence analysis. Because the profiles generated from multiple sequence alignments are important for protein structure and function prediction, the TFP will has many potential applications.

Keywords: Protein remote homology detection · Top Frequency Profile (TFP)

1 Introduction

Protein remote homology detection is important approaches for inferring the structures and functions of proteins [1]. In this regard, some computational methods have been proposed, which can be divided into three categories [2], including alignment methods, discriminative methods, and ranking methods. Among these methods, the discriminative methods achieved the-state-of-the-art performance, and have been widely used in this field, which require fixed-length feature vectors as inputs [3, 4]. Methods based on Support Vector Machines (SVMs) [5, 6] are the top performing methods due to the advantages of the kernel tricks. Their performance mainly depends on how to accurately represent protein sequences as feature vectors. The profile-based representation is one of the most efficient approaches for extracting the features of proteins. Profiles are

I. Rojas et al. (Eds.): IWBBIO 2019, LNBI 11465, pp. 261–268, 2019.
https://doi.org/10.1007/978-3-030-17938-0_24

calculated based on the Multiple Sequence Alignments (MSAs) [7–9], which contains the evolutionary information. However, it is not an easy task to convert the profile into fixed length vector since it is a matrix with different length. Unfortunately, almost all the machine learning and algorithms require fixed length feature vectors as inputs. In this regard, some powerful vectorization methods have been proposed, such as top-n-gram [10], autocross-covariance (ACC) transformation [11], secondary structure features [12, 13], etc.

Although these methods did great contributions to the development of this important field, there exist several disadvantages: (1) Although profiles contain the evolutionary information, noise information also exists, for example, in Position Specific Frequency Matrix (PSFM), the amino acids with low frequencies are unlikely to occur in the corresponding sequence positions during evolutionary process, which will prevent the predictive performance improvement of the predictors if these amino acids are considered; (2) During the vectorization process, the sequence-order effects of residues in proteins cannot be efficiently incorporated. However, as discussed in some recent studies [14], this information is critical for extracting features with high discriminative power.

To overcome these disadvantages in the field, in this study, we propose a new profile by removing the noise information in the PSFM, called Top Frequency Profile (TFP). A method considering both the global and local sequence-order effects of proteins (Autocross-covariance (ACC) transformation [11]) is then performed on the profiles. Experimental results on a widely used benchmark dataset showed that the proposed approach outperforms other existing methods. Therefore, the proposed method would be useful tool for protein sequence analysis.

2 Materials and Methods

2.1 Benchmark Datasets

We employ a widely used benchmark datasets constructed based on SCOP to evaluate the performance of various methods for protein remote homology detection [15]. All the protein sequences are extracted from Astral database [16] and the sequence identities of any pair of proteins in all these benchmark are no more than 95%.

The superfamily benchmark dataset [17] (for remote homology detection task) contains 4352 proteins from 54 different families.

2.2 Profiles

For protein remote homology detection, the evolutionary information in the multiple sequence alignments is useful for improving the predictive performance. In this regard, two profiles are used in this study, and their detailed information will be introduced in the followings.

Position Specific Frequency Matrix (PSFM). Position Specific Frequency Matrix (PSFM) is a widely used profile, which is generated by running PSI-BLAST [18] to search against a non-redundant protein NCBI's nrdb90 database [19], and the parameters of PSI-BLAST are set as default except that the number of iterative is set as 10. PSFM can be represented as:

$$\text{PSFM} = \begin{bmatrix} m_{1,1} & m_{1,2} & \cdots & m_{1,L} \\ m_{2,1} & m_{2,2} & \cdots & m_{2,L} \\ \vdots & \vdots & \vdots & \vdots \\ m_{20,1} & m_{20,2} & \cdots & m_{20,L} \end{bmatrix} \tag{1}$$

where L is the length of the protein; 20 represents the total number of standard amino acids. The element $m_{i,j}$ ($0 \leq m_{i,j} \leq 1$) in this matrix reflects the frequency of amino acid i occurring in column j ($j = 1, 2, ..., L$).

In order to remove the noise in PSFM, here we are to propose a new profile: Top frequency profile (TFP). This profile reduces the noise information in PSFM by removing the amino acids with low frequencies, which will descript in more details in the following sections.

Top Frequency Profile (TFP). Top Frequency Profile (TFP) removes the noise information in the PSFM by only considering the most frequent amino acids during evolutionary process, because previous studies show that these amino acids are critical for protein remote homology detection, and amino acids with low frequency values are unlikely to occur in the specific sequence position during evolutionary process [10]. In this method, for each column in PSFM (Eq. 1), the top N most frequent amino acids are considered as important residues in this sequence position during the evolutionary process, the frequency values of all the other (20-N) standard amino acids were set as 0.

2.3 Matrix Transformation Methods

Evolutionary information in profiles is useful for protein remote homology detection. However, how to efficiently extract this information is a difficult task due to the different lengths of the profiles [14]. Therefore, powerful feature extraction techniques are highly required. Here we employ the Autocross covariance (ACC) transformation [11] to extract the evolutionary information from the three profiles: PSFM and TFP. The detailed information of this method will be introduced in the following sections.

Autocross-Covariance (ACC). Autocross covariance (ACC) transformation [11] is able to transform profiles into fixed length vectors. ACC method is composed of two approaches: AC and cross-covariance (CC).

AC measures the same property's correlation of two amino acids separated by a distance of d along the protein sequence, which can be calculated by [11]:

$$\text{AC}(i,d) = \sum_{j=1}^{L-d} (m_{i,j} - \bar{m}_i)(m_{i,j+d} - \bar{m}_i)/(L - d) \tag{2}$$

where i represents the i-th row in PSFM (Eq. 1); j represents the j-th column in PSFM; L is the length of the protein sequence; $m_{i,j}$ represents the element in PSFM; $\bar{m}_i$ is the average value of all the elements in the i-th row in PSFM. For each distance d ($d = 1, 2, ..., D$, where D is the maximum distance), AC will generate 20 different features based on the 20 properties (number of standard amino acids), and therefore, the dimension of the resulting feature vector is 20 * D.

CC measures any two different properties' correlation between two amino acids separated by a distance d along the protein [11]:

$$CC(i_1, i_2, d) = \sum_{j=1}^{L-d} (m_{i_1,j} - \bar{m}_{i_1})(m_{i_2,j+d} - \bar{m}_{i_2})/(L - d) \tag{3}$$

where i_1 and i_2 represent the i_1-th and i_2-th rows in PSFM, respectively; $m_{i_1,j}$ represents the element in PSFM; $\bar{m}_{i_1}$ and $\bar{m}_{i_2}$ are the average values of the elements in the i_1-th and i_2-th rows in PSFM, respectively. Since the number of asymmetric combinations of any two different properties is 380, the dimension of the resulting feature vector is 380 * D.

In this study, each protein sequence is represented as a feature vector by ACC, which is the combination of the feature vectors generated by AC and CC, whose dimension is 400 * D. For more implementation of ACC, please refer to [11].

2.4   Construction of SVM Classifiers

The Support Vector Machine (SVM) is employed as the classifier to construct the predictor, which has been successfully applied in many areas. Three profiles (PSFM and TFP) of proteins are first generated, and then the three matrix transformation methods are performed on these profiles so as to convert them into fixed length feature vectors. Finally they are inputted into SVM for training and testing. In this study, the publicly available Gist SVM package (http://www.chibi.ubc.ca/gist/) is used as the implementation of the SVM algorithm with default parameter setting. Finally, nine predictors are constructed based on the three profiles (PSFM and TFP), and one matrix transformation method (ACC).

2.5   Evaluation Method of Performance

In this study, the average ROC score and ROC50 score [20] are used to evaluate the performance of various methods. An ROC score is a quantification of performance by the normalized area under a curve that plots true positives against false positives for different classification thresholds. The ROC50 score is a specialization of ROC score, which is the area under the ROC curve up to the first 50 false positives. For separation of positive samples from negative ones, the score of better performance is closer to 1, whereas a score around 0.5 means that the results are randomly predicted.

3   Results and Analysis

In this study, we propose the TFP method to remove the noise in the PSFM. The ACC is then performed on the profiles, and finally, two predictors are constructed for protein remote homology, including PSFM-ACC and TFP-ACC.

3.1   The Impact of Parameters on ACC-Based Predictors

As introduced in the method section, N is a parameters in TFP. In this section, we will investigate its impact on the performance of the TFP-ACC.

Parameter N in TFP approach is a threshold to remove the noise information. In this approach, only the top N most frequent amino acids are considered in each column of PSFM (Eq. 1), and the values of all the other elements are set as zero. The average ROC50 scores of TFP-ACC predictors with different N values on the benchmark dataset are shown in Fig. 1, from which we can see that the parameter N has little impact on its performance. In this study, the value of N is set as 3 considering both the computational cost and performance.

Based on the above analysis, we come to a conclusion that the parameter of N has little impact on the performance of the corresponding predictors.

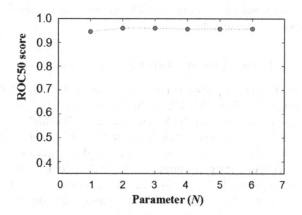

Fig. 1. ROC50 scores of TFP-ACC with different N values on SCOP benchmark dataset.

3.2   Performance Comparison Among the Three ACC-Based Predictors

It has been proofed that the noise information can be reduced by removing the elements with low frequency values in profiles [10]. In this regard, we propose the TFP to extract the evolutionary information and reduce the noise information. Here, we are to explore if these methods can improve the performance of protein remote homology detection. The pairwise comparisons between TFP-ACC and PSFM-ACC based on ROC50 scores on the benchmark dataset are shown in Fig. 2. In these two figures, we can see that most of the points fall in the area below the diagonal line, indicating that for protein remote homology detection, the methods labelled in x-axis (TFP-ACC) out-performs the methods labelled in the y-axis (PSFM-ACC). These results further confirmed that the proposed approach TFP are useful for protein remote homology detection, and removing the amino acids with low frequency values is an efficient way to reduce the noise in PSFM.

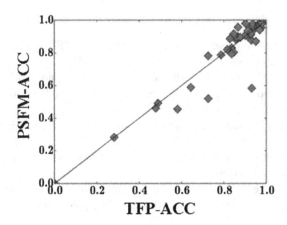

Fig. 2. The pairwise comparisons between TFP-ACC and PSFM-ACC based on ROC50 scores on SCOP superfamily benchmark dataset.

3.3   Performance Comparison with Other Existing Methods

The performance of the nine proposed methods are compared with 9 state-of-the-art methods, including SVM-Bprofile [21], SVM-Top-n-gram [10], SVM-pattern [22], SVM-motif [22], PSI-BLAST [18], SVM-pairwise [23], GPkernel [24], LSTM [24], SVM-LA [23]. Table 1 summarizes their performance for protein remote homology detection. From this table we can see that the proposed two methods all outperform other predictors. Generally, the performance of the TFP-based method outperforms the PSFM-based method, which is fully consistent with previous study [10]. In contrast, the proposed TFP approach is able to efficiently remove the amino acids with low frequency values, and therefore, performance improvement can be achieved.

Table 1. Performance comparison on SCOP superfamily benchmark dataset.

Methods	ROC	ROC50	Source
PSFM-ACC ($LG = 6$)	0.954	0.853	This study
TFP-ACC ($N = 3$, $LG = 6$)	0.960	0.861	This study
SVM-Bprofile ($Ph = 0.13$)	0.903	0.681	[21]
SVM-Top-n-gram	0.933	0.763	[10]
SVM-Pattern	0.835	0.589	[22]
SVM-Motif	0.814	0.616	[22]
PSI-BLAST	0.675	0.330	[22]
SVM-Pairwise	0.896	0.464	[23]
GPkernel	0.899	–	[24]
SVM-LA	0.925	0.649	[23]
LSTM	0.932	0.652	[24]

4   Conclusion

In this study, we propose two predictors by combining three profiles (PSFM and TFP) with ACC. Experimental results show that these two predictors outperform other existing methods, and the TFP is able to efficiently improve the predictive performance for protein remote homology detection by reducing the noise information in PSFM. The profiles, such as PSFM, and PSSM have been widely used in the field of protein sequence analysis, especially for the studies of protein structure and function, and therefore, the TFP will have many potential applications.

Acknowledgements. This work was supported by the Shenzhen Overseas High Level Talents Innovation Foundation (Grant No. KQJSCX20170327161949608) and National Natural Science Foundation of China (Grant No. 61702134).

References

1. Liu, B., et al.: Combining evolutionary information extracted from frequency profiles with sequence-based kernels for protein remote homology detection. Bioinformatics **30**, 472–479 (2014)
2. Chen, J., Guo, M., Wang, X., Liu, B.: A comprehensive review and comparison of different computational methods for protein remote homology detection. Brief. Bioinform. **9**, 231–244 (2018)
3. Zhao, X., Zou, Q., Liu, B., Liu, X.: Exploratory predicting protein folding model with random forest and hybrid features. Curr. Proteomics **11**, 289–299 (2014)
4. Wei, L., Zou, Q.: Recent progresses in machine learning-based methods for protein fold recognition. Int. J. Mol. Sci. **17**, 2118 (2016)
5. Leslie, C.S., Eskin, E., Noble, W.S.: Pacific Symposium on Biocomputing, vol. 7, pp. 566–575. World Scientific (2002)
6. Li, D., Ju, Y., Zou, Q.: Protein folds prediction with hierarchical structured SVM. Curr. Proteomics **13**, 79–85 (2016)
7. Gribskov, M., McLachlan, A.D., Eisenberg, D.: Profile analysis: detection of distantly related proteins. Proc. Natl. Acad. Sci. **84**, 4355–4358 (1987)
8. Zou, Q., Hu, Q., Guo, M., Wang, G.: HAlign: fast multiple similar DNA/RNA sequence alignment based on the centre star strategy. Bioinformatics **31**, 2475–2481 (2015)
9. Li, S., Chen, J., Liu, B.: Protein remote homology detection based on bidirectional long short-term memory. BMC Bioinform. **18**, 443 (2017)
10. Liu, B., Wang, X., Lin, L., Dong, Q., Wang, X.: A discriminative method for protein remote homology detection and fold recognition combining Top-n-grams and latent semantic analysis. BMC Bioinform. **9**, 510 (2008)
11. Dong, Q., Zhou, S., Guan, J.: A new taxonomy-based protein fold recognition approach based on autocross-covariance transformation. Bioinformatics **25**, 2655–2662 (2009)
12. Wei, L., Liao, M., Gao, X., Zou, Q.: Enhanced protein fold prediction method through a novel feature extraction technique. IEEE Trans. Nanobiosci. **14**, 649–659 (2015)
13. Wei, L., Liao, M., Gao, X., Zou, Q.: An improved protein structural classes prediction method by incorporating both sequence and structure information. IEEE Trans. Nanobiosci. **14**, 339–349 (2015)

14. Liu, B., Liu, F., Wang, X., Chen, J., Fang, L., Chou, K.-C.: Pse-in-One: a web server for generating various modes of pseudo components of DNA, RNA, and protein sequences. Nucleic Acids Res. **43**, W65–W71 (2015)
15. Rangwala, H., Karypis, G.: Profile-based direct kernels for remote homology detection and fold recognition. Bioinformatics **21**, 4239–4247 (2005)
16. Brenner, S.E., Koehl, P., Levitt, M.: The ASTRAL compendium for protein structure and sequence analysis. Nucleic Acids Res. **28**, 254–256 (2000)
17. Liao, L., Noble, W.S.: Combining pairwise sequence similarity and support vector machines for detecting remote protein evolutionary and structural relationships. J. Comput. Biol. **10**, 857–868 (2003)
18. Altschul, S.F., et al.: Gapped BLAST and PSI-BLAST: a new generation of protein database search programs. Nucleic Acids Res. **25**, 3389–3402 (1997)
19. Holm, L., Sander, C.: Removing near-neighbour redundancy from large protein sequence collections. Bioinformatics **14**, 423–429 (1998)
20. Liu, B., Jiang, S., Zou, Q.: HITS-PR-HHblits: protein remote homology detection by combining PageRank and Hyperlink-Induced Topic Search. Brief. Bioinform. https://doi.org/10.1093/bib/bby104
21. Dong, Q., Lin, L., Wang, X.: Protein remote homology detection based on binary profiles. In: Hochreiter, S., Wagner, R. (eds.) BIRD 2007. LNCS, vol. 4414, pp. 212–223. Springer, Heidelberg (2007). https://doi.org/10.1007/978-3-540-71233-6_17
22. Yu, X., Cao, J., Cai, Y., Shi, T., Li, Y.: Predicting rRNA-, RNA-, and DNA-binding proteins from primary structure with support vector machines. J. Theor. Biol. **240**, 175–184 (2006)
23. Saigo, H., Vert, J.P., Ueda, N., Akutsu, T.: Protein homology detection using string alignment kernels. Bioinformatics **20**, 1682–1689 (2004)
24. Hochreiter, S., Heusel, M., Obermayer, K.: Fast model-based protein homology detection without alignment. Bioinformatics **23**, 1728–1736 (2007)

Next Generation Sequencing and Sequence Analysis

Reads in NGS Are Distributed over a Sequence Very Inhomogeneously

Michael Sadovsky[1,2]([✉]), Victory Kobets[2], Georgy Khodos[2], Dmitry Kuzmin[3], and Vadim Sharov[3]

[1] Institute of Computational Modelling of SB RAS,
Akademgorodok, 660036 Krasnoyarsk, Russia
`msad@icm.krasn.ru`
[2] Institute of Fundamental Biology and Biotechnology, Siberian Federal University,
Svobodny prosp., 79, 660049 Krasnoyarsk, Russia
`victory.kobets@gmail.com, kalcifer@list.ru`
[3] Institute of Space Research and Computer Sciences, Siberian Federal University,
Kirenskogo str., 26, 660074 Krasnoyarsk, Russia
`{dkuzmin,vsharov}@sfu-kras.ru`
`http://icm.krasn.ru`

Abstract. Distribution of read starts over a sequences genetic entity is studied. Key question was whether the starts are distributed uniformly and homogeneously along a sequence, or there exist some spots of the increased local density of the starts. To answer the question, 15 bacterial genomes have been studied. It was found that some genomes exhibit extremely far distribution pattern, from an homogeneity, while others show lower level of the inhomogeneity. The inhomogeneity level was determined through the Kullback-Leibler distance between the real string distribution, and that one bearing the most probable continuations of the shorter strings.

Keywords: Order · Digitalization · Entropy · Mutual entropy · Equilibrium

1 Introduction

Currently, the sequencing technologies are growing up rapidly. These technologies are both smart and complex, thus challenging researchers to figure out the issues resulted from biology, and those resulted from the technology details. These latter may be quite complicated and not obvious, at the first glance. A variety and abundance of the problems ranging from biological issues (so called "wet protocol") to computational and ever hard mathematical (i.e. assembling and the uniqueness of that latter) points hardly could be just outlined, not speaking about a comprehensive analysis. Here we focus on the specific problem that becomes acute due to the progress in sequencing and processing of genetic data.

© Springer Nature Switzerland AG 2019
I. Rojas et al. (Eds.): IWBBIO 2019, LNBI 11465, pp. 271–282, 2019.
https://doi.org/10.1007/978-3-030-17938-0_25

There are many tools and pipelines to assemble the read ensemble into a set of contigs, scaffolds, and further on. All of them are based on de Bruijn graph methodology [1–4]. Regardless the specific details of the assembling algorithm, all of them have one key idea standing behind the approach: the starts of reads obtained by a sequencer from a genetic entity are supposed to be distributed (almost) uniformly and homogeneously along the sequenced genetic sequence. Our paper aims to prove (or disprove) the validity of this supposition, through a simulation of read generation.

Coverage (local coverage, to be exact) $H_L(n)$ is the most common index of a quality of sequencing. It is defined as a number of unique reads covering a given nucleotide; here L stands for the length of reads (they are supposed to be of equal length, for simplicity). Evidently, this index is not expected to the same, for different fragments of a genetic sequence under consideration, that is why the local index should be introduced [3,4]. Obviously

$$\overline{H} = N^{-1} \sum_{n=1}^{N} H_L(n) \tag{1}$$

is the average (over a genome) cover index. A quality of sequencing output could be characterized with two figures: the former is the average cover (1), and the latter is its variance (or standard deviation) determined over a genome.

Indeed, that is a common place that the figure of the standard deviation of (1) is small, and a sequence is covered rather homogeneously by reads. Such homogeneity is not observed, in reality: as a rule, local cover is extremely inhomogeneous. Of course, the up-to-date algorithms and software platforms are able to process such inhomogeneous data flows, while it takes significantly greater time and resources. Reciprocally, the assembling quality becomes doubtful, not speaking about an uniqueness.

Here we aim to simulate a sequencer operation, in order to model the distribution of read starts over a sequence. Also, we study the patterns of real distribution and compare them to simulation ensembles, in order to find out the rules standing behind the distribution. Such rules are of great value for evaluation of an assembling quality, for any genome entity, and any sequencing machine and pipeline.

2   Study of the Real Distribution of Start Points of Reads Along a Genome

To begin with, we have studied the distribution of the real read starts along a genome sequence. To do that, we downloaded the assembled genomes and the reads ensemble. Then, we mapped the reads back over the genome, and fixed the positions of the starts of the reads. Mapping has been carried out with Bowtie 2 software. Two output files were developed, due to the mapping: the former was $\{0,1\}$-sequence of the length N (here N is the length of a genome under consideration), and the latter was the sequence of integers m_j of the length N, $1 \leq j \leq N$, where m_j was the number of reads (of various lengths) starting at the j-th position.

Consider firstly a binary sequence obtained from mapping of reads over a genome. The key question here is whether zeros and ones are following in some (statistically revealed) manner, or they run randomly, with no order or pattern in their interlocation. Two approaches here should be explored:

1. Supposing the sequence of zeros and ones follows some probabilistic law, fit the sequence with some proper distribution function and identify the parameters of the distribution for further analysis;
2. Considering the sequence of zeros and ones as a symbol one, convert it into a series of frequency dictionaries $\{W_q\}$ of increased thickness q, $1 \leq q \leq q^*$ and figure out the most unexpected strings of the length q derived from the frequencies of the strings of the length $l < q$.

Here we follow the second approach that is completely similar to that one used to study the statistical properties of nucleotide sequences [5–11].

3 Simulation of Start Points of Reads: Theoretical Background

Let now describe the approach to study the statistical properties of the start points distribution in more detail. A digitalization described above converts a genome sequence into a symbol one, with two types of alphabet: the former is binary one $\{0, 1\}$, and the latter consists of M symbols, where M is the maximal number of reads starting at the same point, in a genome.

As soon as a genome is converted into a symbol sequence, it must be transformed into a series of frequency dictionaries $\{W_q\}$ of increasing thickness q. The thickness q is the length of words (strings) comprising a dictionary. More exactly, let q be the length of window that identifies a fragment in a sequence. Frequency dictionary W_q is the list of all the words (strings) observed within a sequence, so that each word ω in a dictionary is supplied with its frequency. The frequency

$$f_\omega = \frac{n_\omega}{N}, \tag{2}$$

where n_ω is the number of copies of a word ω, and N is the length of a sequence; to make the definition (2) feasible, one must connect a sequence into a ring, see details in [5–11]. Such closure results in appearance of $q - 1$ phantom words in a dictionary, while we neglect them.

Consider now the series

$$W_1, W_2, W_3, \ldots, W_{q-1}, W_q$$

of the frequency dictionaries in more detail. The key question here is the relation between the dictionaries observed in this series. Actually, a "downward" transfer (i.e., the transfer from W_j to W_{j-1} dictionary) is obvious: to do it, one must sum

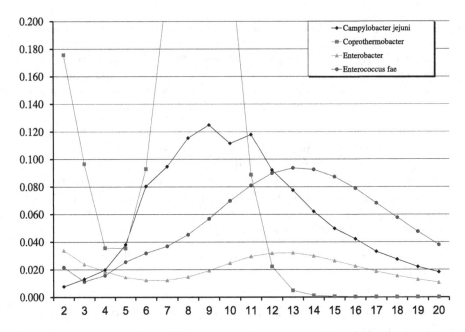

Fig. 1. Information capacity (5), vertical axis, determined for the symbol sequence representing the distribution of starts, with respect to the number of the starts observed in each nucleotide. Horizontal axis represents the thickness q.

up the frequencies of all the words differing in the first (or in the last) symbol[1]. The "upward" transfer $W_j \mapsto W_{j+1}$ is less evident.

Indeed, in general such transfer yields a family of dictionaries $\{W_{j+1}\}$, instead of a single one. Of course, the family contains the real frequency dictionary W_{j+1}, while there is no way to identify it. Simultaneously, there exists the specific frequency dictionary $\widetilde{W}_{j+1}$ in this family that comprise the most expected continuations of the words of the length j into the words of the length $j + 1$. This specific dictionary (let's call it *reconstructed* one) exhibits the maximal entropy, among others comprising the family.

This extremal principle, together with the linear constraints of the "downward" transfer in a series of frequency dictionaries, yields the expected frequency explicitly:

$$\widetilde{f}_{\nu_1 \nu_2 ... \nu_{q-1} \nu_q} = \frac{f_{\nu_1 \nu_2 ... \nu_{q-2} \nu_{q-1}} \times f_{\nu_2 \nu_3 ... \nu_{q-1} \nu_q}}{f_{\nu_2 \nu_3 ... \nu_{q-2} \nu_{q-1}}}; \qquad (3)$$

here we derive $\widetilde{f}(\omega_q)$ from $f(\omega_{q-1})$, see details in [5–11]. Finally, one must compare the real frequency dictionary W_q to that one bearing the most expected continuations: $\widetilde{W}_q$. To do that, the specific entropy of real frequency dictionary W_q against the reconstructed one must be calculated:

---

[1] The equality of these two sums stands behind the connection of a sequence into a ring.

Table 1. Information capacity (5) for the genomes *Acinetobacter baumannii* (1), *Clostridium autoethanogenum* DSM 10061 (2), *E. coli* K12 (3), *E. coli* O157 (4), *Saccharopolyspora erythraea* (7), *Stanieria spp.* NIES-3757 (8), *Staphilococcus aureus* NCTC 8325 (9), *Yersinia pseudotuberculosis* YPIII (10) with respect to the number of starts in each nucleotide.

q	1	2	3	4	7	8	9	10
2	0.006855	0.000151	0.017722	0.000194	0.001249	0.000003	0.006578	0.001581
3	0.007378	0.000131	0.010428	0.000194	0.001073	0.000004	0.008799	0.000969
4	0.009004	0.000318	0.006923	0.000620	0.001005	0.000006	0.012815	0.000998
5	0.009519	0.000899	0.004901	0.002075	0.001076	0.000008	0.013922	0.001346
6	0.009734	0.002810	0.005316	0.006755	0.001513	0.000012	0.015945	0.002269
7	0.010228	0.007856	0.005875	0.018833	0.001909	0.000018	0.020776	0.004497
8	0.011036	0.018508	0.005971	0.044770	0.001709	0.000024	0.031730	0.009182
9	0.013226	0.038805	0.005699	0.089915	0.001422	0.000042	0.050902	0.017498
10	0.017014	0.071194	0.005139	0.151169	0.001490	0.000052	0.078325	0.029912
11	0.022700	0.111794	0.004541	0.210797	0.001954	0.000066	0.109040	0.046115
12	0.030096	0.153165	0.003988	0.242604	0.002791	0.000077	0.134495	0.064515
13	0.038545	0.181569	0.003534	0.227371	0.003967	0.000105	0.148101	0.081751
14	0.047060	0.184315	0.003184	0.174954	0.005500	0.000132	0.141853	0.095160
15	0.054321	0.160808	0.002907	0.110638	0.007324	0.000151	0.120265	0.101762
16	0.059905	0.120647	0.002617	0.058855	0.009580	0.000176	0.089713	0.101649
17	0.062156	0.077580	0.002541	0.027123	0.012024	0.000208	0.059527	0.095004
18	0.060816	0.043650	0.002395	0.011079	0.014533	0.000228	0.035781	0.083791
19	0.057186	0.021745	0.002449	0.004268	0.017319	0.000249	0.019894	0.070384
20	0.050751	0.009841	0.002410	0.001649	0.020015	0.000303	0.010248	0.056214
N	4335793	4352205	4641652	5498578	8212805	5319768	2821361	4689441
Dth	98.40	198.67	203.79	234.70	102.74	108.72	188.51	84.03
σ	58.05	25.57	20.63	35.93	41.07	18.86	74.93	22.87

$$\overline{S}\left[\widetilde{W}_q|W_q\right] = \sum_{\omega\in\Omega} f_\omega \cdot \ln\left(\frac{f_\omega}{\widetilde{f}_\omega}\right). \tag{4}$$

Keeping in mind the expression (3), one gets

$$\overline{S}_q\left[\widetilde{W}_q|W_q\right] = 2S_{q-1} - S_q - S_{q-2}; \qquad \overline{S}_q\left[\widetilde{W}_2|W_2\right] = 2S_1 - S_2. \tag{5}$$

More details on these formulae could be found in [6–8].

Table 2. Information capacity (5) for the genomes *Acinetobacter baumannii* (1), *Clostridium autoethanogenum* DSM 10061 (2), *E. coli* K12 (3), *E. coli* O157 (4), *Saccharopolyspora erythraea* (7), *Stanieria spp.* NIES-3757 (8), *Staphilococcus aureus* NCTC 8325 (9), *Yersinia pseudotuberculosis* YPIII (10) for binary genome representation.

q	1	2	3	4	7	8	9	10
2	0.000016	0.171241	0.016968	0.000051	0.032699	0.000071	0.000002	0.000690
3	0.000005	0.092698	0.009659	0.000044	0.022948	0.000067	0.000002	0.000358
4	0.000039	0.027254	0.005648	0.000078	0.017313	0.000188	0.000002	0.000296
5	0.000017	0.006960	0.002577	0.000107	0.012300	0.000296	0.000002	0.000282
6	0.000027	0.002292	0.001462	0.000147	0.008861	0.000321	0.000003	0.000252
7	0.000045	0.001216	0.000894	0.000209	0.006493	0.000287	0.000004	0.000241
8	0.000039	0.000672	0.000522	0.000241	0.004674	0.000261	0.000004	0.000224
9	0.000033	0.000516	0.000390	0.000240	0.003582	0.000209	0.000007	0.000245
10	0.000096	0.000508	0.000331	0.000246	0.002880	0.000201	0.000012	0.000256
11	0.000082	0.000485	0.000296	0.000220	0.002359	0.000165	0.000017	0.000268
12	0.000139	0.000743	0.000341	0.000238	0.001891	0.000180	0.000024	0.000323
13	0.000280	0.001145	0.000431	0.000333	0.001753	0.000216	0.000034	0.000438
14	0.000509	0.001630	0.000674	0.000481	0.001693	0.000337	0.000046	0.000638
15	0.000989	0.002188	0.001256	0.000825	0.001897	0.000604	0.000057	0.001061
16	0.001938	0.002616	0.002020	0.001567	0.002697	0.001074	0.000067	0.001930
17	0.003840	0.003202	0.002553	0.003088	0.004363	0.001666	0.000082	0.003718
18	0.007741	0.003807	0.002690	0.006113	0.008357	0.002521	0.000097	0.007337
19	0.016023	0.004396	0.002607	0.012768	0.015387	0.003796	0.000115	0.015557
20	0.035304	0.004948	0.002472	0.028037	0.021780	0.005396	0.000145	0.034443

4   Results

We examined 16 bacterial genomes that meet the criteria; namely, we need the genome that

(1) sequenced by Illumina technology;
(2) are duly assembled and annotated, and
(3) there is a set of original reads available for the further analysis.

There are few non-bacterial genomes meeting these criteria; besides, a genome consisting of several chromosomes poses some other technical and essential problems, so we kept ourselves within the prokaryotic genomes, twelve entities in total.

Table 1 shows the data on information capacity (5) obtained for eight bacterial genomes. The genomes gathered in the table have rather smooth and similar pattern of the information capacity behaviour; the upper part of the table shows the data obtained for the digitalization with number of starts taken into account.

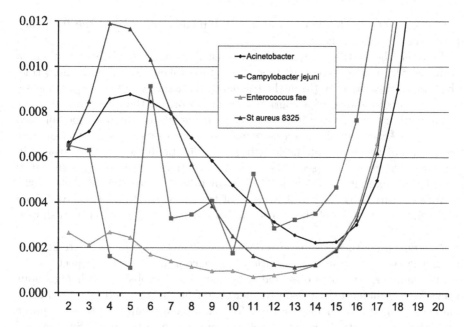

Fig. 2. Information capacity (5) determined for the binary symbol sequence representing the distribution of starts.

The lower part of that former shows similar data for binary digitalization (i.e. regardless the number of starts occurred in a nucleotide). Also, this table lengths of the bacterial genomes (denoted N), total genome cover depth (Dth), and the standard deviation for the set of local cover indices (1). All these figures are shown in the bottom of the table.

Figure 1 shows the patterns observed for four bacterial genomes; these genomes exhibit quite variable behaviour and rough pattern of the information capacity variation with the frequency dictionary thickness q growth. Similarly, Table 2 and Fig. 2 show the figures and pattern, respectively, for the same set of bacterial genomes, while transformed into a binary sequence each. *Coprothermobacter proteolyticus* genome yields a tremendous growth of information capacity (5) (see Fig. 1) with maximum figure of 0.404645, for non-binary digitalization. Reciprocally, the pattern of information capacity (5) observed for binary digitalization of the genome has four local minima; probably, these two observations make an evidence of the increased complexity of the reads starts distribution along a sequence.

Let now concentrate on Figs. 1 and 2. They show the behavioural patterns of information capacity (5), for four bacterial genomes each. First of all, all the curves are bell-shaped and it results from the finite sampling effect: an abundance of a frequency dictionary W_q grows exponentially, as q grows linearly. Hence, the greatest majority of the strings occur in a single copy, as q becomes great enough. Moreover, there exists specific figure q^* that yields no word occurred in two or more copies, at all; this figure makes a redundancy measure of the frequency dictionary of this thickness [11].

Figure 1 shows the pattern for the distribution of reads starts along a sequence, with respect to the number of the starts taken place in each nucleotide. The information capacity (5) of the frequency dictionaries of various thickness q reflects a predictability of a continuation of a word of the given length $q - 1$ into a word of the length q; if $\overline{S}_q \approx 0$, then all the words (the frequency of each word, to be exact) of the length q could be quite exactly predicted from the frequencies of the words of the length $q - 1$. The predictability goes worse, in general, as $\overline{S}_q$ grows up (see details in Sect. 5). Hence, the genomes shown in these figures exhibit quite low level of predictability of the distribution of the number of starts observed in a window of the given length q, as derived from the frequency ensemble of the starts numbers distribution observed in a shorter window.

Comparison of these two figures reveals significant smoothness in predictability of the starts numbers distribution, when counting it with respect to the specific numbers of starts observed in a nucleotide; probably, such behaviour comes from combinatorial reasons rather than from biology. Indeed, the specific numbers of starts taken into account for a dictionary implementation enlarge the alphabet capacity, thus cutting-off the tail of the distribution. Such cut-off manifests in a smoother pattern of the curve (5). Reciprocally, a multimodality of the distributions shown in these figures is of great interest. An occurrence of two (or more) local minima (and maxima, reciprocally) means an existence of some meso-scale structuredness in the starts distribution. The patterns shown in Figs. 1 and 2 differ in digitalization implemented for a study of the distribution of the reads starts numbers: the former shows the distribution with respect to the number of starts observed in a nucleotide, while the letter represents just the fact of a start, regardless to the specific number of reads starting in a nucleotide. There are only two common genomes in these Figs: *Campylobacter jejuni* and *Enterococcus faecalis* OG1RF; other genomes are different. It means that predictability of the strings representing the distribution of starts number is sensitive to digitalization version. In such capacity, those two genomes mentioned above exhibit the highest level of unpredictability in the starts numbers distribution along a genome.

Another interesting question concerns the variation of the number of starts to be observed in the same nucleotide, in different bacteria. Table 3 shows these data, for nine bacterial genomes. The table contains a union of the records for those genomes; blank cells in this Table mean that there was not a nucleotide with such number of starts, in the genome. Definitely, the greatest majority of nucleotides yields no start of a read; we shall not consider them. At a glance, the number nucleotides with multiple starts decreases, as that latter grows up (see Table 3). Here *E. coli* K12 genome completely falls out of the common pattern: it shows permanent and consistent non-monotony in the number of starts distribution. Moreover, it looks like a kind of a cycle of the length 2; some reasons of such behaviour are discussed below (see Sect. 5).

Table 3. *Acinetobacter baumannii* (1), *Clostridium autoethanogenum* DSM 10061 (2), *Saccharopolyspora erythraea* (3), *Staphilococcus aureus* NCTC 8325 (4), *Stanieria spp.* NIES-3757 (5), *Yersinia pseudotuberculosis* YPIII (6), *E. coli* K12 (7), *E. coli* O157 (8), *Enterobacter cloacae* (9).

n_s	1	2	3	4	5	6	7	8	9
1	835013	1421481	1449801	736737	213175	1387796	34206	1687399	298955
2	228570	595816	190999	319300	13853	429958	66101	907832	221129
3	47061	195018	21748	105585	488	103255	8093	391783	49898
4	9859	55149	3205	33897	27	21449	19111	152246	27531
5	1788	14236	837	9888	1	4103	2899	54214	6285
6	463	3457	274	3041		815	5695	18568	3314
7	129	812	99	868		182	1094	6256	866
8	52	201	29	311		70	1748	1940	519
9	22	34	6	96		39	378	728	197
10	9	11	2	48		22	546	213	123
11	6	1	1	12		13	152	72	66
12	3	1		8		12	169	22	40
13	4			4		11	44	11	17
14				2			62	5	23
15	1	1				2	28	1	12
16						1	25	1	11
17				1		1	5	1	8
18						2	2		1
19							6		2
20							2		
21							1		
22							1		2
23			1						2
24									1
25									1
26									2
27							1		
28							1		1
29							1		1
32							1		
35								1	
95				1					
116				1					

The table shows significant variation in the maximal number of starts found in a nucleotide; probably, this fact results from the peculiarities of sequencing procedure and may represent a quality of the sequencing rather than the biological issues. Extremely variable highest number of the starts (95 and 116 observed for *Staphylococcus aureus*) supports indirectly this idea. In general, the number of nucleotides giving the increasing starts number in a genome follows an exponential law: indeed, one may calculate the ratio of the numbers in two subsequent cells in Table 3 and find them to be quite proximal.

5   Discussion

The distribution of read starts along a nucleotide sequence is studied. This question is rather acute, since numerous inhomogeneities in this distribution may bring problems in assembling, annotation and further analysis of genetic entities. We explore the generalized approach to reveal some inhomogeneities in the starts distribution similar to [5–11]. Here a genome is considered as a symbol sequence, and we refrain from implementation of any biological knowledge "till the end"; in other words, we seek for the highly unexpected sites in the symbol sequences and the procedure is free from any biological knowledge. As soon as the sites are found, their biological role is studied. Basically, the hypothesis is that the sites tend to be located non-randomly, with a sounding preference to some biologically charged loci. It was found that the sites are distributed along a genome very non-randomly; whether the sites are located in the biologically important parts of a genome, still awaits for the answer.

The results provided above definitely show that the distribution of start points over a genetic entity is rather far from any equilibrium, or homogenous one. Any experimentalist knows that sequencing may skip some (rather extended) areas in a genetic sequence; the reasons of such distortion may follow both from biological issues of a matter, and from peculiarities of the sequencing technology. Here we tried to answer the question towards the character of this inhomogeneity in starts distribution.

To begin with, it should be said that the results shown above are biased. The problem may arise from the structure of reads ensemble. Indeed, we used the assembled genome, and the reads used to do it. The point is that the reads are obtained from both strands of DNA, while we used the leading one to align them. We used BowTie 2 to map the reads, and some of them might be mapped at the leading strand, while the have been sequenced from the ladder one. Thus, it might increase, to some extent, the number of observed starts (both unique, and multiple ones). The pattern of the number of starts distribution observed for *E. coli* K12 genome (see Table 3) proves indirectly this assumption. Hence, we plan to reconsider the starting points pattern with respect to the detailed analysis of the reads from the point of view of their strand origin.

To reveal the structuredness in the strings containing the nucleotides with various numbers of starts of reads, we used the idea of information capacity (3–5); this is an averaged measure telling on the distribution character in general, but nothing could be understood on individual level. To enhance the analysis, an idea of information valuable words [5–10] could be implemented. The idea is based on the detail analysis of (4) definition: if $\widetilde{f}_\omega \approx f_\omega$, then the corresponding term in the sum (4) is close to zero. On the contrary, the greatest contribution into the sum (4) is provided by the terms with the greatest deviation of $\tilde{f}_\omega$ from f_ω. Such words are claimed **information valuable** ones.

So, the idea of further analysis is as following:

(1) Count the expected frequency $\tilde{f}_\omega$ for each $\omega \in W_q$;
(2) Identify those with the deviation of $\tilde{f}_\omega$ from f_ω exceeding some given level α;
(3) Match all such information valuable words over the genome, and check it against the annotation.

The hypothesis is that such words would match some peculiar sites, within a genome.

Another very important issue that falls beyond the scope of this paper, while is expected to be done soon is the approximation of the distribution of starts points located along a genome sequence with a number of various patterns, among them are Poisson distribution, LaPlace distribution, geometric distribution, negative binomial one, and many others. The idea is to fit the observed data best of all, with some specific distribution, so that some biologically sounding results might be retrieved from this fitting. In particular, the patterns shown in Figs. 1 and 2 support the hypothesis towards the feasible simulation of those distributions by Markov chains of the order 5 to 7, and around.

All these data and observations would be used for further simulation studies of the sequencing procedures implemented in various machines. Such simulation is of great value for better understanding of the details of assembling, annotation and comparison of sequenced genetic entities.

Acknowledgement. This study was supported in part by RFBR grant 18-29-13044/18.

References

1. Van Dijk, E.L., Auger, H., Jaszczyszyn, Y., Thermes, C.: Ten years of next-generation sequencing technology. Trends Genet. **30**(9), 418–426 (2014)
2. Li, H., Homer, N.: A survey of sequence alignment algorithms for next-generation sequencing. Brief. Bioinform. **11**(5), 473–483 (2010)
3. Buermans, H., den Dunnen, J.: Next generation sequencing technology: advances and applications. Biochimica et Biophysica Acta (BBA)—Mol. Basis Dis. **1842**(10), 1932–1941 (2014)
4. Conesa, A., et al.: A survey of best practices for RNA-seq data analysis. Genome Biol. **17**(1), 13 (2016)
5. Sadovsky, M.G.: Information capacity of nucleotide sequences and its applications. Bull. Math. Biol. **68**(4), 785–806 (2006)
6. Sadovsky, M.G.: Comparison of real frequencies of strings vs. the expected ones reveals the information capacity of macromoleculae. J. Biol. Phys. **29**(1), 23–38 (2003)
7. Sadovsky, M.G., Putintseva, J.A., Shchepanovsky, A.S.: Genes, information and sense: complexity and knowledge retrieval. Theory Biosci. **127**(2), 69–78 (2008)

8. Sadovsky, M.G.: Information capacity of symbol sequences. Open Syst. Inf. Dyn. **9**(01), 37–49 (2002)
9. Borovikov, I., Sadovsky, M.G.: Sliding window analysis of binary n-grams relative information for financial time series. In: Center for Advanced Signal and Image Sciences (CASIS) at LLNL 18th Annual Workshop, p. 1 (2014)
10. Sadovsky, M., Nikitina, X.: Strong inhomogeneity in triplet distribution alongside a genome. In: Ortuño, F., Rojas, I. (eds.) IWBBIO 2015. LNCS, vol. 9044, pp. 248–255. Springer, Cham (2015). https://doi.org/10.1007/978-3-319-16480-9_25
11. Bugaenko, N.N., Gorban, A.N., Sadovsky, M.G.: Maximum entropy method in analysis of genetic text and measurement of its information content. Open Syst. Inf. Dyn. **5**(3), 265–278 (1998)

Differential Expression Analysis of ZIKV Infected Human RNA Sequence Reveals Potential Genetic Biomarkers

Almas Jabeen[1], Nadeem Ahmad[1], and Khalid Raza[2(✉)]

[1] Department of Biosciences, Jamia Millia Islamia, New Delhi 110025, India
[2] Department of Computer Science, Jamia Millia Islamia,
New Delhi 110025, India
kraza@jmi.ac.in

Abstract. Zika virus (ZIKV) infection is considered to be an emerging viral outbreak due to its link to diseases like microcephaly, Guillain-Barre Syndrome in human which is an alarming concern. In this study, we implemented our reproducible RNA-seq analysis pipeline to quantify RNA-seq data in terms of transcripts, and gained common expression results from intersection of three differential expression identification tools. This uncovered significant DEGs of high consensus, significant DEGs of moderate consensus, significant DEGs of low consensus. Moreover, the highly significant DEGs provided us with six DEGs which are transcription factors, which may be involved in the altered biological process somehow. The presented study provides researchers with highly reproducible pipeline for viral studies as well as the novel computational findings for the transcription factors (TFs) involved in ZIKV infection which could enable the researchers to develop new therapeutic strategies to tackle the infection.

Keywords: Zika virus · RNA-seq · Biomarkers · DEGs

1 Introduction

The WHO reveals several types of diseases and infections outbreak all over the world of which majority were concerned with viral infections. Few of these are Ebola Virus Disease [1], Middle East Respiratory Syndrome, H5N1 influenza infection and ZIKV infection [2]. ZIKV is a flavivirus transmitted by Aedes mosquitoes [3]. The minor infection results in low-grade fever, myalgia, maculopapular rashes and in severe cases, adults suffer neurological and congenital structural defects [4]. It may also cause congenital malformations in pregnant women, and newborns with microcephaly [5]. ZIKV infection also affects peripheral nervous systems (PNS) and central nervous system (CNS), causing transcriptional dysregulation which results in cell death [6].

Infectious response mechanism of virus is supported by various findings. Xia et al. [7] showed that ZIKV fixes the mutations in NS1 gene that enhances mosquito infection and increases its ability to dodge immune response. Rolfe et al. [8] compared the alterations in transcript expressions of the RNA-seq data for ZIKV-infected hNPCs to CMV-infected iPSC-derived hNPCs which revealed several pathways correlated

© Springer Nature Switzerland AG 2019
I. Rojas et al. (Eds.): IWBBIO 2019, LNBI 11465, pp. 283–294, 2019.
https://doi.org/10.1007/978-3-030-17938-0_26

with ZIKV infection. They also reported top 30 upregulated and downregulated genes related to ZIKV pathogenesis. On the basis of Ontologic, Phylogenetic and Pathway analysis, Moni et al. [9] concluded that ZIKV infection highly resembles Dengue fever. They reported that 929 genes were dysregulated, 47 were highly expressed in ZIKV infection as well as dengue, whereas less than 15 transcripts were significant in ZIKV as well as other flavivirus infections which were involved in the experiment. The emerging threat of ZIKV outbreak laid urgent need of developing preventive vaccines and treatments for infected patients. For developing potent vaccines, antiviral drugs and other treatments against ZIKV infection, RNA-seq finds its great importance in terms of availability of data and analysis tools.

The Next Generation Sequencing (NGS) technologies like RNA-seq (RNA sequencing) find their application in the diagnostic virology (discovery, characterization and detection of viruses), antiviral drug and vaccine development, analysis of host-virus interaction, study of viral spread [2]. Due to being cost effective and having an improved turnaround time, NGS methods specially RNA sequencing methods along with other analytical and clinical tests validation methods, can serve as essential diagnostic for viral spread [10]. The research in RNA-seq includes studying the altered pathway during infection or disease, gene expression changes (differential expression analysis) [11]. Though RNA-seq analysis is considered as the standard expression profiling methodology, still easy, open and standard pipelines for performing this task by non-expert research community with different background is a major challenge. This study provides researchers as well as the non-experts with an easy implementable RNA-seq analysis pipeline for DEGs identification. Mostly, the research interest for differential studies lies in the comparison of the transcription result under different experimental conditions, therefore the RNA-seq studies can be categorized into Differential Gene Expression (DGE) studies, Differential Transcript Expression (DTE) studies and Differential Transcript/exon Usage (DTU) studies, where comparisons are made between conditions on the basis of transcriptional measure by each gene, each of the transcript measure, usage of transcript/exon, respectively [12].

In this study, we performed DGE analysis of ZIKV exposed patient using publically available RNA-seq data from the NGSfor DEGs identification. The identified significant consensus of differentially expressed transcripts (DETs) from this pipeline revealed the corresponding DEGs that may have key roles in significant biological processes and functional pathways related to the disease; hence this would enable us to search for putative vaccines and therapeutic strategies against ZIKV infections.

2 Methodology

In this pipeline the RNA-seq dataset of ZIKV infected human induced pluripotent stem cells (hiPSCs) were retrieved from NCBI. Then after read quality check by FastQC [13], the data was preprocessed using Trimmomatic [14]. Reads were then mapped to reference human genome (hg38) using aligner named Bowtie2 [15]. The alignment result was further subjected for quantification of expression values using HTSeq-count [16]. These counts further served as input for transcript normalization. Bioconductor package of 'R' language [17, 18] which provides several statistical tools for normalization such as edgeR [19, 20], DESeq2 [21]. In this study, we implemented these statistical tools for normalization and differential expression identification. Additionally, Cuffdiff [22, 23, 24], a program provided by Cufflinks tool, was also used in this experimental study. In this research, we grouped significant DEGs with high consensus, significant DEGs with moderate consensus, and significant DEGs with low consensus derived from their respective differentially expressed transcripts (DETs) which were identified using these three tools in a consensus manner. The complete pipeline for our work is shown in Fig. 1.

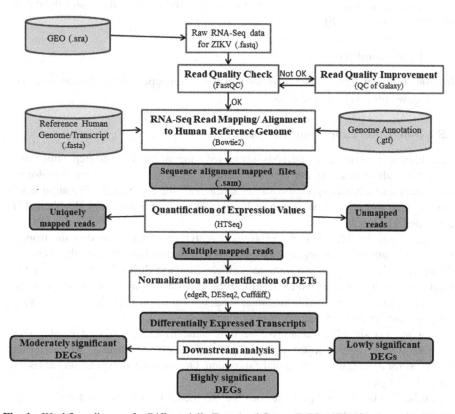

Fig. 1. Workflow diagram for Differentially Expressed Genes (DEGs) identification using ZIKV infection RNA-Seq data from the NGS.

2.1 Raw Data Collection

The dataset with NCBI-GEO accession GSE78711for ZIKV infected hiPSCs-derived cortical neural progenitor cells (hNPCs) were retrieved [25]. The data consist of eight sample files of which four were from control experiments (Mock-infected samples) and other four from virus treated experiments (ZIKV-infected samples).

2.2   Data Preprocessing

All the samples were first converted to fastq format and then read quality check using FastQC tool was implemented. We used Trimmomatic 0.36 tool for adapter trimming in order to remove the adapters from data. In order to segregate rRNAs from the sequence, we used SortMeRNA 2.1 tool [26].

2.3   Read Mapping

We used Bowtie2 aligning tool for mapping the pre-processed reads against the human reference genome (hg38).

2.4   Read Counting

In order to count the number of reads which were mapped to the human reference genome, we used HTSeq-count tool.

2.5   Differential Expression Analysis

The differential expression analysis starts with the normalization step which is a method to adjust read counts between samples in such a way to get a uniformly distributed normalized expression values throughout the experiment. We applied following three tools which were run at benchmark value for Bejamini-Hochberg [27] controlled FDR as 0.05.

edgeR: edgeR [19, 20] is the expression analysis tool which models the mapped read count data using a negative binomial (NB) model. It moderates the estimated dispersion calculated for each gene to a single common dispersion estimate, or to a local dispersion estimate, which results from genes with similar expression weight calculated using a weighted conditional likelihood method [21]. It is a measure of assessing the inter-library variation of that gene.

For the classic edgeR analysis, we took gene transcripts IDs of eight sample libraries which were grouped in two. After dispersion estimation, we performed exactTest for determining differential expression. On normalized expression values, we applied the tool.

DESeq2: The DESeq2 package uses the NB model in order to test the differential expression. It estimates the shrinkage according to the data distribution, and then adjusts the logarithmic fold changes to improvise the result stability and its interpretation [21].

For analysis through DESeq2 package, we input the read-counts of all eight sample libraries in form of matrix, and also specified sample condition i.e. whether the samples are control-treated or virus-treated. It firstly estimated size factors, and then calculated gene-wise dispersion. It finally fitted the model and tested for differential expression.

Cuffdiff: Cuffdiff [22, 23, 24] is the Cufflink transcript assembly package used to identify significant changes in transcript expression. It models the variance in groups of samples which lies beyond the expected variance calculated by Poisson model. It tests for observed logfold change in its expression against the null hypothesis of no change.

The normalization process in Cuffdiff is performed by classic-fpkm, geometric mode, quartile mode. Cuffdiff needs count files in.bam format, therefore we first converted .sam alignment file to .bam files using Samtools. After sorting it, we ran Cuffdiff command.

2.6   Consensus Approach to DEGs

We assume that A is the set of differentially expressed transcripts (DETs) identified by tool edgeR under the specified benchmark cutoffs and filters of FDR < 0.05, FDR-adjusted p value or q value < 0.05 and log2FC > |2|, B is the set of DETs identified by tool DESeq2 and C is the set of DETs identified by Cuffdiff tool. Using a consensus approach, significant transcripts with high consensus (DETs$_{High}$) in terms of differential expression can be defined as Eq. (1).

$$DETs_{High} = (A \cap B \cap C) \tag{1}$$

Similarly, using majority voting rule, significant differentially expressed transcriptswith moderate consensus (DETs$_{Moderate}$) can be defined as Eq. (2),

$$DETs_{Moderate} = (A \cap B) \cup (A \cap C) \cup (B \cap C) \tag{2}$$

And significant differentially expressed transcripts with low consensus (DETs$_{Low}$) can be defined as Eq. (3),

$$DETs_{Low} = (A \cup B \cup C) \tag{3}$$

With the help of open source browser developed by University of California, Santa Cruz called UCSC Genome browser, we found the official gene names/symbols for corresponding Refseq IDs (Transcript IDs). Then we extracted out the unique genes (DEGs) in each category of DETs. Next, we identified which of the highly significant DEGs were TFs or target, as we know TFs may be involved in disease pathways. This was done by manual search from transcription databases (TRRUST [28] and Tf2DNA [29]).The identified TFs were further validated for their relevance with ZIKV through literature searches.

3   Results and Discussion

The complete protocol (depicted in Fig. 1) was implemented and performed on a workstation with 8 GB RAM, multicore processors under Ubuntu 18.04.1 LTS operating system.

3.1   Data Preprocessing

On read quality check of raw data by FastQC, we found that four out of eight samples had quality-score (Phred Score) above 20, which were considered to be good whereas remaining four samples below 20 were considered to poor quality reads. When poor quality reads subjected for adaptor trimming; only 2–3% of the total reads got trimmed. All types of rRNAs were separated from the samples.

3.2   Read Mapping

After read mapping to the human genome it was observed that the overall alignment rate of all the eight samples was more than 99%, which implies to be very well aligned.

3.3   Read Counting and Normalization

As a standard assumption the number of reads mapped (read count) to a gene/transcript is considered to be the proxy of its expression. For data analysis, read count data from HTSeq-count further needs to be normalized by total fragment count in order to make counts comparable across the experiments. edgeR, DESeq2 and Cuffdiff were used for this task, which firstly transformed read count data into a continuous distribution. They used NB model to estimate dispersion parameter for each transcript. This dispersion parameter gave a measure of the degree of inter-library variation of particular transcript between the samples. Estimation of common dispersion provided the idea of overall variability across the dataset.

3.4   Differential Expression Analysis

When all the eight samples grouped in two were subjected for differential analysis using edgeR tool, it was found that normalization factors calculated for each sample is close to 1 which signified that all the eight libraries are similar in composition. The input estimated common dispersion before estimating tagwise dispersions in order to proceed differential expression analysis.

Firstly, Biological coefficient of variation (BCV) was applied to this input data. BCV is mathematical square root of common dispersion estimated using NB model. With an increase in the number of read counts, the BCV remains unaffected, though a decrease in technical CV can be observed. Therefore, the accurate BCV estimation is crucial for differential expression analysis studies in RNA-seq experiments. The BCV calculated from the experiment was found to be 34%. Since, higher the BCV measure, lower will be the number of differentially expressed genes/transcripts detected, therefore it is assumed that the test detected higher number of DETs. Moreover, in the Fig. 2 (a), a common dispersion (red line on BCV plot) lied between 0.2–0.4, hence considered to have detected higher number of DETs.

For Multidimensional Scaling (MDS), the input was provided inform of a distance matrix where values represented distances between the pairs of objects. MDS plot represents the relationship between different groups of samples and can be affected by high BCVs. MDS plots show distances between the samples in terms of BCV spread in two dimensions of the plot (dim 1 and dim 2) as shown in Fig. 2(b).

Dimension 1 (dim1) separates the control samples from the Virus-treated samples which signified the possibility of detecting higher number of DETs. This plot can be observed in the form of an unsupervised clustering.

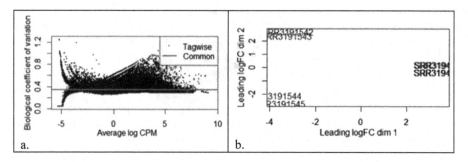

Fig. 2. (a) Plot for Biological coefficient of variation (BCV) depicts common dispersion to lie between 0.2–0.4. (b) MDS plot of various samples. (Color figure online)

After fitting the NB models and estimation of dispersions, we proceeded with tests determining the differential expression. The tagwise exactTest was applied. P-values were calculated by combining over all sums of counts that have a probability less than the probability under the null hypothesis of the observed sum of counts. The test performed at FDR < 0.05, provided us with output result with all 153726 transcripts arranged in tabular form with information regarding their geneID, logFC, logCPM, P-value (that is the default FDR-adjusted P-value also called q value). It was observed that 683 genes were down-regulated, 152668 genes were not differentially expressed and 375 were up-regulated.

The smear plot of tagwise log-fold changes (logFC) against logCPM is shown in Fig. 3. The differentially expressed tags are highlighted in the plot and the horizontal blue lines show 4-fold changes.

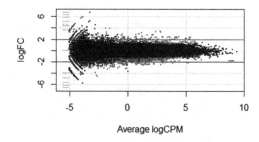

Fig. 3. Smear plot of dataset analogous to an MA-plot as for microarray data.

For the consensus method in DEGs identification, we also applied DESeq2 tool from R Bioconductor which gave the differential expression result along with normalized count data for all eight samples with 153726 transcripts. The dispersion plot of the normalized read count from DESeq2 tool is shown in Fig. 4.

We also performed a differential analysis of count data using Cuffdiff program. The processed count files from HTSeq tool when subjected to Cuffdiff tool, gave expression estimates of 153620 identified DETs in terms of FPKM and read counts.

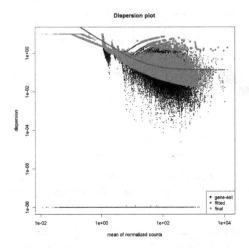

Fig. 4. Dispersion plot for mean of normalized read counts by DESeq2 tool.

Since results from these tools contains large number of transcripts which were differentially expressed, when the filtering criteria was employed, we found the best identified transcripts for each of the three tools. Using majority voting rule, we identified a total of 28 DETs as significant DETs with a high consensus, 248 DETs as significant DETs with a moderate consensus and 654 as significant DETs with a low consensus, as per Eqs. (1), (2) and (3), respectively. Further, 27 corresponding significant unique DEGs with a high consensus were identified, along with 76 DEGs with a moderate consensus and 270 significant DEGs with a low consensus (Fig. 5).

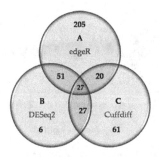

Fig. 5. Differentially Expressed Genes (DEGs) identified by consensus.

Out of these 27 DEGS with a high consensus, 6 were identified as Transcription Factors (TFs), which are DDIT3, CEBPB, TRIB3, XBP1, KLF15 and JDP2 accordingly. To validate the association of these identified DEGs in ZIKV infection, we performed an exhaustive literature search and found to be associated except for JDP2 gene. Furthermore, DDIT3 [8, 30, 31, 32], CEBPB [8], TRIB3 [30], XBP1 [8, 33, 34, 35, 36, 37] were both experimentally and computationally validated, and KLF15 [8] was computationally validated.

The information related to these identified TFs as high consensus DEGs in association to ZIKV infectionare listed in Table 1. It was observed that all the identified genes were upregulated. Flavivirus (including ZIKV) translation and RNA replication produces Endoplasmic Reticulum (ER) stress in cell causing ER. Most of these TFs are found to be actively involved in Protein Kinase R (PKR)-like ER Kinase (PERK) branch of ER-Unfolded Protein Response (UPR). CEBPB also induces the expression

Table 1. Functions of six identified significant DEGs with high consensus

RefSeq ID	Official gene symbol	Gene name	log2FC			Association to ZIKV infection	PMID/Reference
			edgeR	DESeq2	Cuffdiff		
NM_001195056.1	DDIT3	DNA-damage-inducible transcript 3	2.328	2.332	2.269	Involved in the PERK branch ER-UPR due to ER stress triggered by flavivirus translation and RNA replication. It is also a substrate for Nonsense-mediated mRNA decay (NMD) involved in cell cycle arrest and induction of apoptosis	27293547 [8], 29451494 [30], 30401782 [31], 30228241 [32]
NM_001285878.1	CEBPB	CCAAT enhancer binding protein beta	2.739	2.833	3.182	Induces the expression of inflammatory and ER stress response factors	27293547 [8]
NM_001301188.1	TRIB3	Tribbles pseudokinase 3	2.325	2.342	2.088	Involved in PERK branch of ER-UPR due to ER stress triggered by flavivirus translation and RNA replication	29451494 [30]
NM_005080.3	XBP1	X-box binding protein 1	2.297	2.301	2.468	Indicator of UPR activation in ZIKV infection as IRE1-XBP1 and ATF6 pathways of UPR in neural cells gets activated	27293547 [8], 28190239 [33], 30670030 [34], 29321318 [35], 28592527 [36], 29976926 [37]
NM_014079.3	KLF15	Kruppel-like factor 15	2.046	2.061	2.348	Upregulated in ZIKV infection	27293547 [8]
XM_005247400.3			2.022	2.041	2.178		
XM_017020973.1	JDP2	Jun dimerization protein 2	2.098	2.115	2.431	–	–

of inflammatory and ER stress response factors. UPR is also involved in IRE1-XBP1 and ATF6 pathways of neural cells when infected with ZIKV. However, these identified DEGs need further experimental validation.

4   Conclusion

The outbreak of viral disease such as ZIKV has brought the attention of computational biologists and bioinformaticians to perform the differential expression analysis of ZIKV infected patients to understand transcriptomic changes in the body that helps in designing better diagnostic tools, therapeutics and treatments. In this study, we computationally analyzed the RNA-seq data of ZIKV infected patients to identify DEGs through an easy reproducible pipeline. We report six most significant TFs using a consensus of three tools (edgeR, DESeq2, Cuffdiff) as differentially expressed in ZIKV infected patient. These genes are validated using an exhaustive literature recapitulation for its reliability. In the future, we look forward to perform gene regulation and pathway analysis, GO enrichment analysis and topological analysis for these DEGs in order to further validate their role in ZIKV infection.

Acknowledgement. The author A. Jabeen acknowledges Maulana Azad National Fellowship-Junior Research Fellowship (JRF) received from the UGC, Government of India.

References

1. Imran, M., Khan, A., Ansari, A.R., Shah, S.T.H.: Modeling transmission dynamics of Ebola virus disease. Int. J. Biomath. **10**(04), 1750057 (2017)
2. Jabeen, A., Ahmad, N., Raza, K.: Machine learning-based state-of-the-art methods for the classification of RNA-seq data. In: Dey, N., Ashour, A.S., Borra, S. (eds.) Classification in BioApps. LNCVB, vol. 26, pp. 133–172. Springer, Cham (2018). https://doi.org/10.1007/978-3-319-65981-7_6
3. Tiwari, S.K., Dang, J., Qin, Y., Lichinchi, G., Bansal, V., Rana, T.M.: Zika virus infection reprograms global transcription of host cells to allow sustained infection. Emerg. Microbes Infect. **6**(4), e24 (2017)
4. Agrawal, R., Oo, H.H., Balne, P.K., Ng, L., Tong, L., Leo, Y.S.: Zika virus and the eye. Ocul. Immunol. Inflamm. **26**(5), 654–659 (2018)
5. Shi, Y., Gao, G.F.: Structural biology of the Zika virus. Trends Biochem. Sci. **42**(6), 443–456 (2017)
6. Oh, Y., et al.: Zika virus directly infects peripheral neurons and induces cell death. Nat. Neurosci. **20**(9), 1209 (2017)
7. Xia, H., et al.: An evolutionary NS1 mutation enhances Zika virus evasion of host interferon induction. Nat. Commun. **9**(1), 414 (2018)
8. Rolfe, A.J., Bosco, D.B., Wang, J., Nowakowski, R.S., Fan, J., Ren, Y.: Bioinformatic analysis reveals the expression of unique transcriptomic signatures in Zika virus infected human neural stem cells. Cell Biosci. **6**(1), 42 (2016)
9. Moni, M.A., Lio, P.: Genetic profiling and comorbidities of Zika infection. J. Infect. Dis. **216**(6), 703–712 (2017)

10. Barzon, L., Lavezzo, E., Costanzi, G., Franchin, E., Toppo, S., Palù, G.: Next-generation sequencing technologies in diagnostic virology. J. Clin. Virol. **58**(2), 346–350 (2013)
11. Raza, K., Ahmad, S.: Recent advancement in next-generation sequencing techniques and its computational analysis. Int. J. Bioinf. Res. Appl. Inderscience (in Press)
12. Soneson, C., Love, M.I., Robinson, M.D.: Differential analyses for RNA-seq: transcript-level estimates improve gene-level inferences. F1000Research, vol. 4, p. 152 (2015)
13. Andrews, S.: FastQC: a quality control tool for high throughput sequence data (2010)
14. Bolger, A.M., Lohse, M., Usadel, B.: Trimmomatic: a flexible trimmer for Illumina sequence data. Bioinformatics **30**(15), 2114–2120 (2014)
15. Langmead, B., Salzberg, S.L.: Fast gapped-read alignment with Bowtie 2. Nat. Methods **9** (4), 357 (2012)
16. Anders, S., Pyl, P.T., Huber, W.: HTSeq—a Python framework to work with high-throughput sequencing data. Bioinformatics **31**(2), 166–169 (2015)
17. Huber, W., et al.: Orchestrating high-throughput genomic analysis with Bioconductor. Nat. Methods **12**(2), 115 (2015)
18. Gentleman, R.C., et al.: Bioconductor: open software development for computational biology and bioinformatics. Genome Biol. **5**(10), R80 (2004)
19. Robinson, M.D., McCarthy, D.J., Smyth, G.K.: edgeR: a Bioconductor package for differential expression analysis of digital gene expression data. Bioinformatics **26**(1), 139–140 (2010)
20. McCarthy, D.J., Chen, Y., Smyth, G.K.: Differential expression analysis of multifactor RNA-Seq experiments with respect to biological variation. Nucleic Acids Res. **40**(10), 4288–4297 (2012)
21. Love, M.I., Huber, W., Anders, S.: Moderated estimation of fold change and dispersion for RNA-seq data with DESeq2. Genome Biol. **15**(12), 550 (2014)
22. Trapnell, C., et al.: Transcript assembly and quantification by RNA-Seq reveals unannotated transcripts and isoform switching during cell differentiation. Nat. Biotechnol. **28**(5), 511 (2010)
23. Trapnell, C., Hendrickson, D.G., Sauvageau, M., Goff, L., Rinn, J.L., Pachter, L.: Differential analysis of gene regulation at transcript resolution with RNA-seq. Nat. Biotechnol. **31**(1), 46 (2013)
24. Roberts, A., Trapnell, C., Donaghey, J., Rinn, J.L., Pachter, L.: Improving RNA-Seq expression estimates by correcting for fragment bias. Genomebiology **12**(3), R22 (2011)
25. Tang, H., et al.: Zika virus infects human cortical neural progenitors and attenuates their growth. Cell Stem Cell **18**(5), 587–590 (2016)
26. Kopylova, E., Noé, L., Touzet, H.: SortMeRNA: fast and accurate filtering of ribosomal RNAs in metatranscriptomic data. Bioinformatics **28**(24), 3211–3217 (2012)
27. Benjamini, Y., Hochberg, Y.: Controlling the false discovery rate: a practical and powerful approach to multiple testing. J. Roy. Stat. Soc.: Ser. B (Methodol.) **57**(1), 289–300 (1995)
28. Han, H., et al.: TRRUST v2: an expanded reference database of human and mouse transcriptional regulatory interactions. Nucleic Acids Res. **46**(D1), D380–D386 (2017)
29. Pujato, M., Kieken, F., Skiles, A.A., Tapinos, N., Fiser, A.: Prediction of DNA binding motifs from 3D models of transcription factors; identifying TLX3 regulated genes. Nucleic Acids Res. **42**(22), 13500–13512 (2014)
30. Zanini, F., Pu, S.Y., Bekerman, E., Einav, S., Quake, S.R.: Single-cell transcriptional dynamics of flavivirus infection. Elife **7**, e32942 (2018)
31. Fontaine, K.A., et al.: The cellular NMD pathway restricts Zika virus infection and is targeted by the viral capsid protein. mBio, 9, e02126-18 (2018)
32. Chen, Q., et al.: Treatment of human glioblastoma with a live attenuated Zika virus vaccine candidate. MBio **9**(5), e01683-18 (2018)

33. Walter, L.T., et al.: Evaluation of possible consequences of Zika virus infection in the developing nervous system. Mol. Neurobiol. **55**(2), 1620–1629 (2018)
34. Zhao, D., Yang, J., et al.: The unfolded protein response induced by Tembusu virus infection. BMC Vet. Res. **15**(1), 34 (2019)
35. Panayiotou, C., et al.: Viperin restricts Zika virus and tick-borne encephalitis virus replication by targeting NS3 for proteasomal degradation. J. Virol. JVI-02054 (2018)
36. Hou, S., et al.: Zika virus hijacks stress granule proteins and modulates the host stress response. J. Virol. JVI-00474 (2017)
37. Volpi, V.G., Pagani, I., Ghezzi, S., Iannacone, M., D'Antonio, M., Vicenzi, E.: Zika virus replication in dorsal root ganglia explants from interferon receptor1 knockout mice causes myelin degeneration. Sci. Rep. **8**(1), 10166 (2018)

Identification of Immunoglobulin Gene Usage in Immune Repertoires Sequenced by Nanopore Technology

Roberto Ahumada-García[1], Jorge González-Puelma[2],
Diego Álvarez-Saravia[2], Ricardo J. Barrientos[1],
Roberto Uribe-Paredes[3], Xaviera A. López-Cortés[1],
and Marcelo A. Navarrete[2]([⊠])

[1] Department of Computing and Industries, Universidad Católica del Maule,
Talca, Chile
robertoahumadagarcia@gmail.com,
ricardo.j.barrientos@gmail.com,
xaviera.lopez.c@gmail.com

[2] School of Medicine, University of Magallanes, Punta Arenas, Chile
{jorge.gonzalez, diego.alvarez,
marcelo.navarrete}@umag.cl

[3] Computer Engineering Department, University of Magallanes,
Punta Arenas, Chile
roberto.uribe@umag.cl

Abstract. The immunoglobulin receptor represents a central molecule in acquired immunity. The complete set of immunoglobulins present in an individual is known as immunological repertoire. The identification of this repertoire is particularly relevant in immunology and cancer research and diagnostics. In a seminal work we provided a proof of concept of the novel ARTISAN-PCR amplification method, we adapted this technology for sequencing using Nanopore technology. This approach may represent a faster, more portable and cost-effective alternative to current methods. In this study we present the pipeline for the analysis of immunological repertoires obtained by this approach. This paper shows the performance of immune repertoires sequenced by Nanopore technology, using measures of error, coverage and gene usage identification.

In the bioinformatic methodology used in this study, first, Albacore Base calling software, was used to translate the electrical signal of Nanopore to DNA bases. Subsequently, the sequons, introduced during amplification, were aligned using bl2seq from Blast. Finally, selected reads were mapped using IMGT/HighV-QUEST and IgBlast.

Our results demonstrate the feasibility of immune repertoire sequencing by Nanopore technology, obtaining higher depth than PacBio sequencing and better coverage than pair-end based technologies. However, the high rate of systematic errors indicates the need of improvements in the analysis pipeline, sequencing chemistry and/or molecular amplification.

Keywords: Immunoglobulin · Sequencing · Data analysis · Nanopore · Pipeline

© Springer Nature Switzerland AG 2019
I. Rojas et al. (Eds.): IWBBIO 2019, LNBI 11465, pp. 295–306, 2019.
https://doi.org/10.1007/978-3-030-17938-0_27

1   Introduction

Immunoglobulins are central proteins of the adaptive immune system since they fulfill a fundamental role of defense against foreign agents. These proteins are synthesized for B Cells and are constituted by two identical heavy chains and two identical light chains, joined by disulfide bridges. Immunoglobulins that can be found forming part of their membrane or being secreted outside of the cell [1, 2]. Its formation is the result of two main phenomena: recombination of 3 gene segments (known as V (variable), D (diversity) and J (junction)) and point mutations (Somatic hypermutation) [3]. Both phenomena generate differences between each rearrangement of immunoglobulin producing a repertoire close to 10^{11} per individual [4].

The sequencing and identification of the nucleic acids that codify for variable region of these proteins has diverse applications in research, diagnosis and treatment of cancer and other diseases of the immune system [5, 6].

However, the sequencing and analysis of the immune repertoire represents a unique challenge for molecular biology and bioinformatics due to the intrinsic high variability attained to recombinant gene segments and somatic hypermutation. Accurate repertoire measurements requires unbiased PCR amplification, high depth, and full read coverage. We have previously overcame the primer binding bias of standard multiplex PCR [7] by a novel ARTISAN-PCR strategy (Anchoring reverse transcription immunoglobulin sequence and amplification by nested PCR) [8]. After amplification repertoire are subjected to high throughput sequencing. Whereas PacBio sequencing provides the required coverage (app. 1 Kb) the depth remains limited (10 K–100 K reads). On the other hand, Illumina pair-end sequencing provides excellent depth, albeit with insufficient coverage (app. 600 bp for Miseq 300 PE) [9]. Nanopore sequencing technology potentially provides high coverage and depth at a lower cost, therefore we tested the feasibility of immune repertoire sequencing of indexed amplicons generated by ARTISAN-PCR. These long-read technologies are very promising, but their error rates are higher than other current sequencing methods and require computational-based corrections and/or additional bioinformatics preprocessing before they can be valuable [10].

This work presents a pipeline for analysis of immune repertoire sequenced by Nanopore technology, showing performance measurements as error rate, percentage of substitutions and indels, coverage and identification of V(D)J genes of immunoglobulin.

2   Materials and Methods

2.1   Immunoglobulin Amplification, Library Preparation and Sequencing

In this study, 5 samples from healthy individuals were analyzed. All volunteers provided informed consent and the study as it was approved by the ethics review board of the University of Magallanes (registry 1180882).

RNA was extracted from peripheral blood mononuclear cells obtained by gradient centrifugation. Anchored cDNA was synthesized and amplified by ARTISAN-PCR as previously described [8].

A total of 5 amplicons representing each antibody chain: IgA, IgG, IgM, IgKappa and IgLambda were individually indexed and the library was processed using the 1D2 Sequencing Kit (R9.5) following the manufacturer protocol.

Each amplicon has the following structure:

| 5′ Index | Anchor | V(D)J Rearrangement | Constant Chain | End Adaptor 3′ |

Fig. 1. Amplicons structure.

In the amplicons structure the Index identifies patient chain. The Constant Chain identifies Immunoglobulin chain (IgA, IgG, IgM, IgKappa and IgLambda).

The library with 25 indexed amplicons was then sequenced during 24 h in a MinION Nanopore Sequencing Device.

2.2 Basecalling and Qscore Stratification

The first bioinformatics step was to obtain the reads from the Nanopore electrical signal by ONT Albacore Sequencing Pipeline Software (version 2.3.3). In this step it was used the AXON server of the Universidad Católica del Maule, operating system CentOS Linux 7.4.1708 (core). This server has a hard disk of 1.8 Tera Bytes. 32 Gigabytes of Memory and has 2 processors Intel (R) Xeon (R) CPU E5-2623 v4 @ 2.60 GHZ with 4 cores each processor. This step took 70 h to work out the calculation.

Then sequencing quality was assessed by the internal Qscore (Qs) provided by Albacore and the reads were classified according to this score.

2.3 Demultiplexing of Indexed Amplicons

The second step of data processing corresponds to aligning the reads generated by Minion Nanopore to their respective sample indexes (Index, Anchor, Constant Chain, and End Adaptor) by using a custom script based on a Smith-Waterman heuristic algorithm using Bl2seq de BLAST software (Basic Local Alignment Search Tool) [11].

Consecutively, matching analysis was applied using the Levenshtein distance written in the C programming language. In this case the characters or bases of DNA that change are the product of insertions, deletions or substitutions of these. It is considered as a match when two sequences are identical, the mismatch is considered when at least one insertion, deletion or substitution. The reads that have up to 6 mismatches (6 differences between bases) contain motifs that align to their respective sample indexes. Alignments to indexes with a greater difference were discarded for subsequent analysis steps.

The general script for the demultiplexing of indexed amplicons was written in the R programming language [12].

2.4   Error Rate and Coverage

Each amplicon contains adapters and extensions located in 5' and 3' (Index, Anchor, Constant Chain, and End adaptor) that contain known DNA sequences, which are used to estimate sequencing error rate, substitution profiles, indel profiles, and coverage. The error rate constitutes the percentage of mismatched bases in the alignment and that can be subdivided into substitution, insertion and deletion rates, Eq. (1) [13].

The errors correspond to the total sum between substitutions, insertions and deletions in the motives. The alignment length is the sum of the number of eligible sequences (sequences that have up to 6 mismatch for the motif) multiplied by the length of the motif.

$$
\begin{aligned}
error\ rate &= \frac{mismatches + \sum(length(insertions \in read)) + \sum(length(deletions \in read))}{matches + mismatches + \sum(length(insertions \in read)) + \sum(length(deletions \in read))} \\
&= \frac{errors}{alignment\ length} * 100\%
\end{aligned}
$$

$$(1)$$

The error rate was estimated according to the criteria of stage 1.3 demultiplexing of indexed amplicons by using all motifs of the detected reads. The error rate was calculated as Eq. (2).

$$
error\ rate = \frac{(errors_{Index} + errors_{Anchor} + errors_{Constant\ Chain} + errors_{End\ Adaptor})}{(alignment_{Index} + alignment_{Anchor} + alignment_{Constant\ Chain} + alignment_{End\ Adaptor})} \quad (2)
$$

Reads were selected according to their internal Qscore (Qscore greater than 6) for subsequent analysis. Then, the reads were separated in forward and reverse complement to analysis of substitutions and indels. For these analysis, the sample indexes that are recognized in each read were considered. The united motifs of a read form a construct that eliminates the first 4 theoretical bases of Index and the last 4 bases of End Adapter (these amplicons have low quality in their end, according to preliminary analysis using FastQC tools [14]).

The profiles of each substitution analyzed were: $C > A$, $C > G$, $C > T$, $T > A$, $T > C$, $T > G$, $G > T$, $G > C$, $G > A$, $A > T$, $A > G$ and $A > C$. In addition, all substitution were examined by incorporating information from the contiguous bases 5' and 3' for each mutated base generating 192 possible substitutions types (12 types of substitution $*$ 4 types of 5' base $*$ 4 types of 3' base). For each of the 192 possible substitutions types the percentage is obtained with respect to the total of substitutions. The procedure described above is similar for calculating the percentages of the indels.

Then, the coverage of the reads was analyzed respect to the structure of the amplicons (Fig. 1). The amplicons coverage is shown as the number of motifs that cover an area of the amplicons (does not correspond to coverage of known reference bases). In this case the presence of the motifs was quantified (Index, Anchor, Constant Chain, and End Adapter). Which are the motives of the reads is quantified, based on the criteria of stage 1.3 demultiplexing of indexed amplicons. In this stage, the presence of the V(D)J rearrangement motif was also quantified using Igblast software and Change-o tools for gene mapping [15].

2.5 Immunoglobulin Identification

Reads with Qscore greater than 6 and demultiplexing according to stage 1.3 were subsequently aligned using IMGT HighV-Quest and IgBlast. Then, sequences alignment using IMGT HighV-Quest were analyzed with the BcRep R package [16]. In case of sequences alignment using IgBlast were then analyzed with Change-o tools [15].

3 Results

3.1 Sequencing Yield, Quality, and Error Rate

From the library of 25 indexed amplicons 1,450,507 reads were identified. The frequency distribution of read quality and error rate is showed in Table 1.

The proportion of reads with a Qscore equal or greater than 7, considered with acceptable quality, was 30.9% (n = 447,693) and the percentage of lower quality reads was 69.1% (n = 1,002,814). Due to the high proportion of low quality reads we performed a stratified analysis in order to establish whether lower quality reads contain useful sequencing data. With this information it is decided to analyze reads with a Qscore equal or greater than 6 (n = 914,069). It is observed that the reads with a Qscore greater than 7 have an error rate of up to 10%, in the case of reads with higher Qscore the error decreases by one percentage point.

Table 1. Numbers of reads according to Qscore.

Group	Numbers of reads	Error rate
$1 \leq$ Qscore < 2	1,296	$2.8 \ 10^{-1}$
$2 \leq$ Qscore < 3	10,305	$1.8 \ 10^{-1}$
$3 \leq$ Qscore < 4	55,390	$1.4 \ 10^{-1}$
$4 \leq$ Qscore < 5	158,765	$1.4 \ 10^{-1}$
$5 \leq$ Qscore < 6	310,682	$1.3 \ 10^{-1}$
$6 \leq$ Qscore < 7	466,376	$1.2 \ 10^{-1}$
$7 \leq$ Qscore < 8	332,726	$1 \ 10^{-1}$
$8 \leq$ Qscore < 9	100,646	$9 \ 10^{-2}$
$9 \leq$ Qscore	14,321	$8 \ 10^{-2}$
Total	**1,450,507**	

In Fig. 2 is possible to see the reads that were demultiplexing for each type of immunoglobulin chain (IgA, IgG, IgM, IgKappa and IgLambda) and patient chain (DCR, JGP, LAH, MAN and MVC). Reads with Qscore between 5 and 9 were analyzed. 251,044 were identified with immunoglobulin chain and 317,395 reads were detected with patient chain. Besides, the amount of identified sequences show the depth and coverage of sequencing, allowing this technology as a greater platform in comparison to others commonly used for the sequencing of immunoglobulins [9].

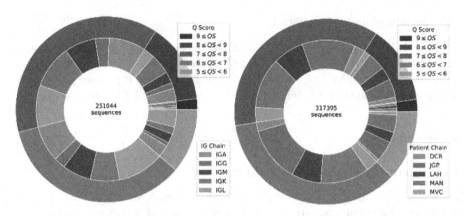

Fig. 2. Immunoglobulins (Ig) and Patient Chains identified. Sequences with a Qscore between 5 and 9 were analyzed.

The reads with a quality over 6, demultiplexing, were separated as forward and reverse complement. The forward reads was 45% (n = 273,215) and the percentage of reverse complement reads was 55% (n = 331,998). The substitutions, insertions and deletions of the reads were calculated separately for each type of sequence. Figure 3A and B shows the most important substitutions for forward and reverse complement reads. In this case the two most important substitutions G > A are shown (where G is the real base, being replaced by A) and A > G (where A is the real base, being replaced by G). It should be mentioned that in this case it is true that both for reads forward and reverse have as a maximum percentage the substitutions G > A and A > G. In the case of substitutions in the forward reads (Fig. 3A), it is observed that the profile CGC with the highest percentage (more than 20% of the substitutions in the forward reads), where G is replaced by A, and between two equal bases (between two C) Nanopore errors occur as expected. The second profile of the forward reads that has the highest percentage is AAC (more than 10%), where the central A is replaced by G, it is observed that when there are two equal bases, in this case A, the second base cannot recognize it and it is replaced.

In the case of substitutions of the reverse complement reads (Fig. 3B) it is observed that the highest substitution profile is TAC (greater than 15%), where A is replaced by G. The second most important profile corresponds to AGC with almost 15% of the substitutions, where G is replaced by A which is contiguous to the base that we wish to infer.

In summary, it is observed that in the case of forwards reads there are greater errors when two bases are equal in the profile. In the case of the reverse complement reads the substitutions change the base resembling a contiguous base.

The other substitution profiles are not shown in this document because of their space and also they have very low percentages, all of which are less than 5%.

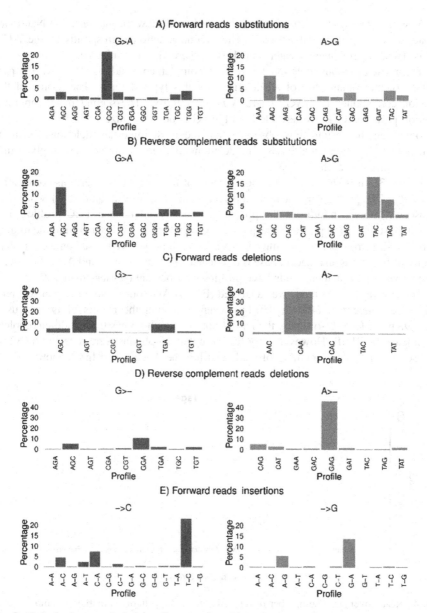

Fig. 3. Substitutions and indels. Bar graphs show the most important percentages of substitutions and indels.

Figure 3C and D, shows the most important deletion profiles to forward and reverse complement reads. It is observed that the deletion of A is the most important for the two types of reads. The most important deletions for the forward reads are those of A and G. The most important deletion for the forward reads is A> - with a percentage of almost 40% of the total deletions for this type of reads. The profile corresponds to

CAA, where the A base is not inferred by the neural network that uses the Nanopore Albacore software's neural network. The second deletion corresponds to the AGT profile (with a percentage greater than 15%), where G is not inferred.

The most important deletion for reverse complement reads is A> - with a percentage of more than 45% of all deletions for this type of reads. The profile of the deletion corresponds to GAG, where A is not inferred. The second most important deletion profile is GGA, where the bold type G is not recognized.

Based on the results it is observed that the most important deletions occur in profiles that have two equal bases. The deletions of C and T are not shown given that the percentages of each are less than 5% for all.

In Fig. 3E, it is shown that the most important forward reads insertions correspond to C and G. The profile with the highest percentage is T-C (greater than 25%), where a C is inserted into the middle of the profile (mistakenly inferring a TCC profile). The second most important profile corresponds to G-A where a G is inserted into the middle of the profile (erroneously inferring a GGA profile). It is observed based on these results that in the insertions an adjacent base is duplicated. The inserts of T and A, as the inserts of the reverse complement reads have profiles with percentages less than 10%.

The coverage of the reads was analyzed (Fig. 4). We found that all sequences have V(D)J Rearrangement (400 to 880 bp) when analyzing the results of IgBlast used MakeDb.py of Change-o (using the partial option: which allows to include incomplete alignments to V(D)J). However, very few complete amplicons were detected 6,682. It is observed that a large number of reads that has the Anchor and Index motif.

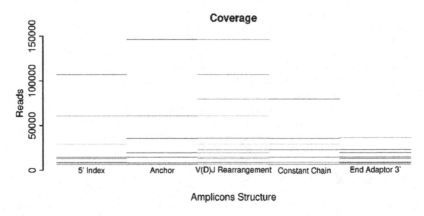

Fig. 4. Read coverage. The number of reads that have the patterns is marked in color. (Color figure online)

3.2   Nanopore Sequencing Allows V(D)J Gene Usage Identification

Demultiplexed reads were aligned to the immunoglobulin databases HighV-Quest and IgBlast. The results we have obtained using IgBlast can be better recognized incomplete immunoglobulins compared to HighV-Quest, however HighV-Quest has more tools used in the analysis of the sequence as BcRep R package that are presented below.

The overall V, D, and J gene segments usage for heavy and light chains is depicted in Fig. 5. The frequency distribution is consistent with the previously reported usage in repertoires obtained in healthy individuals.

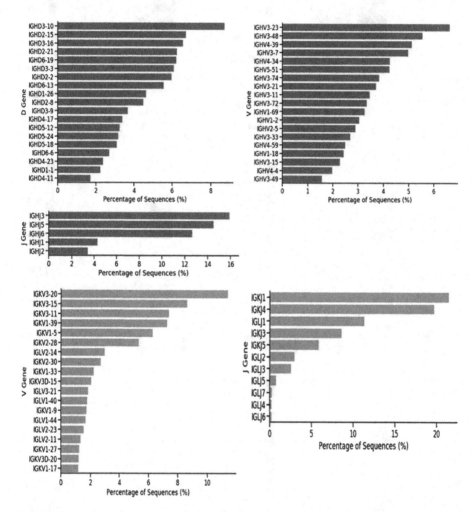

Fig. 5. Percentage of variable region from V, D, and J segments associated with families. Bar graphs in upper panels correspond to immunoglobulin heavy chains (blue bars) and graphs in the lower section to light chains (orange bar). (Color figure online)

Figure 6 represents the combinatorial gene segments usage according to each individual index for demultiplexed samples. As expected for healthy individuals the gene segments usage and combinations correlates.

304   R. Ahumada-García et al.

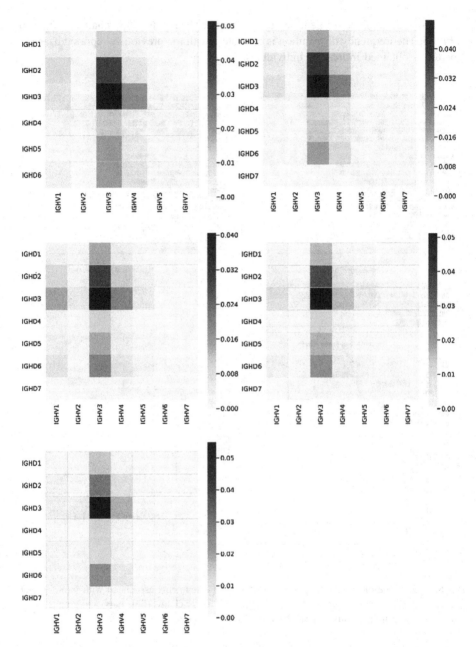

Fig. 6. Heavy chain gene segments usage and V-D recombination in demultiplexed samples. Each panel represents one individual.

4   Conclusions

This work corresponds to the first description for feasibility to amplifiying immune repertoire by ARTISAN-PCR protocol sequenced by nanopores.

The bioinformatic analysis obtained from the pipeline previously described allow us the use of gene information for the assembly of the immunoglobulin variable region, providing relevant information, for example, for the prognosis of various pathologies. This method provides full rearrangement coverage and sequencing depth that could potentially reach 10^6 sequences per flow cell.

In our previous experiments using PacBio we were able to identify an average of 50,000 reads, whereas with this novel approach we rescued more than 300,000 reads with adequate quality.

Nevertheless, this technology still faces important challenges such as high error rate [17, 18]. In our experiments error rate was as high as 10% for reads with a Qscore between 7 and 8, selection of reads with higher Qscore only decreased this error rate by 1 percentual point. Although this error rates allows accurate identification of V(D)J family usage, it is not sufficient for other analysis such as precise hypermutation rate measurement. However, the principal disadvantage of this platform is the accuracy, apparently driven by systematic errors [19]. Since most of the error rate could be associated to predictable patterns, it is envisioned that bioinformatics tools could be readily developed [10].

Acknowledgments. This project was funded by grants ESR-MAG1895 to RUP and Fondecyt#1180882 to MAN.

Author Contributions. RAG performed bioinformatics analysis, JGP performed molecular biology experiments, DAS provided concept analysis and performed bioinformatics analysis, XLC, RB and RUP provided concept analysis, MAN designed and conceptualized the project and experiments. RAG, MAN, XLC and JGP wrote the paper.

References

1. Kipnis, J.: Multifaceted interactions between adaptive immunity and the central nervous system. Science **353**, 766–771 (2016)
2. Labrecque, N., Cermakian, N.: Circadian clocks in the immune system. J. Biol. Rhythms **30**, 277–290 (2015)
3. Rajewsky, K.: Clonal selection and learning in the antibody system. Nature **381**, 751–758 (1996)
4. Greiff, V., Miho, E., Menzel, U., Reddy, S.T.: Bioinformatic and statistical analysis of adaptive immune repertoires. Trends Immunol. **36**, 738–749 (2015)
5. Navarrete, M.A., Bertinetti-Lapatki, C., Michelfelder, I., Veelken, H.: Usage of standardized antigen-presenting cells improves ELISpot performance for complex protein antigens. J. Immunol. Methods **391**, 146–153 (2013)
6. van Bergen, C.A.M., et al.: Selective graft-versus-leukemia depends on magnitude and diversity of the alloreactive T cell response. J. Clin. Invest. **127**, 517–529 (2017)

7. Koning, M.T., Nteleah, V., Veelken, H., Navarrete, M.A.: Template-switching anchored polymerase chain reaction reliably amplifies functional lambda light chain transcripts of malignant lymphoma. Leuk. Lymphoma **55**, 1212–1214 (2014)
8. Koning, M.T., et al.: ARTISAN PCR: rapid identification of full-length immunoglobulin rearrangements without primer binding bias. Br. J. Haematol. **178**, 983–986 (2017)
9. Chaudhary, N., Wesemann, D.R.: Analyzing immunoglobulin repertoires. Front. Immunol. **9** (462), 5–6 (2018)
10. Warren, R.L., et al.: LINKS: scalable, alignment-free scaffolding of draft genomes with long reads. GigaScience **4**, 35 (2015)
11. Tatusova, T.A., Madden, T.L.: BLAST 2 sequences, a new tool for comparing protein and nucleotide sequences. FEMS Microbiol. Lett. **174**, 247–250 (2006)
12. Salas, Christian: ¿Por qué comprar un programa estadístico si existe R? Ecol. Austral **18**, 223–231 (2008)
13. Rang, F.J., Kloosterman, W.P., de Ridder, J.: From squiggle to basepair: computational approaches for improving nanopore sequencing read accuracy. Genome Biol. **19**, 90 (2018)
14. Brown, J., Pirrung, M., McCue, L.A.: FQC dashboard: integrates FastQC results into a web-based, interactive, and extensible FASTQ quality control tool. Bioinformatics **33**, 3137–3139 (2017)
15. Gupta, N.T., Vander Heiden, J.A., Uduman, M., Gadala-Maria, D., Yaari, G., Kleinstein, S. H.: Change-O: a toolkit for analyzing large-scale B cell immunoglobulin repertoire sequencing data. Bioinformatics **31**, 3356–3358 (2015)
16. Bischof, J., Ibrahim, S.M.: bcRep: R package for comprehensive analysis of B cell receptor repertoire data. PLoS One **11**, e0161569 (2016)
17. Deamer, D., Akeson, M., Branton, D.: Three decades of nanopore sequencing. Nat. Biotechnol. **34**, 518–524 (2016)
18. Lu, H., Giordano, F., Ning, Z.: Oxford Nanopore MinION sequencing and genome assembly. Genomics Proteomics Bioinform. **14**, 265–279 (2016)
19. Laehnemann, D., Borkhardt, A., McHardy, A.C.: Denoising DNA deep sequencing data—high-throughput sequencing errors and their correction. Brief. Bioinform. **17**, 154–179 (2016)

Flexible and Efficient Algorithms for Abelian Matching in Genome Sequence

Simone Faro[1]([✉]) and Arianna Pavone[2]

[1] Dipartimento di Matematica e Informatica, Università di Catania,
Viale Andrea Doria 6, 95125 Catania, Italy
`faro@dmi.unict.it`
[2] Dipartimento di Scienze Cognitive, Università di Messina,
Via Concezione 6, 98122 Messina, Italy
`apavone@unime.it`

Abstract. Approximate matching in strings is a fundamental and challenging problem in computer science and in computational biology, and increasingly fast algorithms are highly demanded in many applications including text processing and DNA sequence analysis. Recently efficient solutions to specific approximate matching problems on genomic sequences have been designed using a filtering technique, based on the general abelian matching problem, which firstly locates the set of all candidate matching positions and then perform an additional verification test on the collected positions.

The *abelian pattern matching problem* consists in finding all substrings of a text which are permutations of a given pattern. In this paper we present a new class of algorithms based on a new efficient fingerprint computation approach, called *Heap-Counting*, which turns out to be fast, flexible and easy to be implemented. We prove that, when applied for searching short patterns on a DNA sequence, our solutions have a linear worst case time complexity. In addition we present an experimental evaluation which shows that our newly presented algorithms are among the most efficient and flexible solutions in practice for the abelian matching problem in DNA sequences.

Keywords: Approximate string matching ·
Abelian matching jumbled matching · Experimental algorithms

1 Introduction

Given a pattern x and a text y, the *abelian pattern matching* problem [10] (also known as *jumbled matching* [6,13] or *permutation matching* problem) is a well known special case of the approximate string matching problem and consists in finding all substrings of y, whose characters have the same multiplicities as in x, so that they could be converted into the input pattern just by permuting their characters.

© Springer Nature Switzerland AG 2019
I. Rojas et al. (Eds.): IWBBIO 2019, LNBI 11465, pp. 307–318, 2019.
https://doi.org/10.1007/978-3-030-17938-0_28

Example 1. For instance, assume that $y = ccgatacgcattgac$ is a text of length 15 and $x = accgta$ is a pattern of length 6, then x has two abelian occurrences in y, at positions 1 and 4, respectively, since both substrings $cgatac$ and $tacgca$ are permutations of the pattern.

This problem naturally finds applications in many areas, such as string alignment [3], SNP discovery [4], and also in the interpretation of mass spectrometry data [5]. We refer to the recent paper by Ghuman and Tarhio [13] for a detailed and broad list of applications of the abelian pattern matching problem.

More interestingly related with scope of this paper abelian matching finds application in the field of approximate string matching in computational biology, where algorithms for abelian matching are used as a filtering technique [2], usually referred to as *counting filter*, to speed up complex combinatorial searching problems. The basic idea is that in many approximation problems a substring of the text which is an occurrence of a given pattern, under a specific distance function, is also a permutation of it. For instance, the counting filter technique has been used solutions to the approximate string matching problem allowing for mismatches [16], differences [18], inversions [7] and translocations [15].

In this paper we are interested in the *online* version of the problem which assumes that the input pattern and text are given together for a single instant query, so that no preprocessing of the text is possible. Although its worst-case time complexity is well known to be $O(n)$, in the last few years much work has been made in order to speed up the performances of abelian matching algorithms in practice, and some very efficient algorithms have been presented, tuned for specific settings of the problem [9,13].

Specifically we present two algorithms based on a new efficient fingerprint computation approach, called *Heap-Counting*, which turns out to be fast, flexible and ease to be implemented, especially for the case of DNA sequences. The first algorithm is designed using a prefix based approach, while the second one uses a suffix based approach. We prove that both of them have a linear worst case time complexity.

In addition we present also two fast variants of the above algorithms, obtained by relaxing some algorithmic constraints, which, despite their quadratic worst case time complexity, turn out to be faster in some specific practical cases.

From our experimental results it turns out that our newly presented algorithms are among the most efficient and flexible solutions for the abelian matching problem in genomic sequences.

The paper is organized as follows. After introducing in Sect. 2 the relevant notations and describing in Sect. 3 the related literature, we present in Sect. 4 two new solutions of the online abelian pattern matching problem in strings, based on the Heap-Counting approach, and prove their correctness and their linear worst case time complexity. Then, in Sect. 5, we present a detailed experimental evaluation of the new presented algorithms, comparing them against the most effective solutions known in literature.

2   Basic Notions

Before entering into details we recall some basic notions and introduce some useful notations.

We represent a string x of length $|x| = m > 0$ as a finite array $x[0 .. m-1]$ of characters from a finite alphabet Σ of size σ. Thus, $x[i]$ will denote the $(i+1)$-st character of x, for $0 \leq i < m$, whereas $x[i .. j]$ will denote the substring of x contained between the $(i + 1)$-st and the $(j + 1)$-st characters of x.

Given a string x of length m, the occurrence function of x, $\rho_x : \Sigma \to \{0, \ldots, m\}$, associates each character of the alphabet with its number of occurrences in x. Formally, for each $c \in \Sigma$, we have:

$$\rho_x(c) = \big| \{ i : x[i] = c \} \big|.$$

The *Parikh vector* [1,20] of x (denoted by pv_x and also known as *compomer* [5], *permutation pattern* [11], and *abelian pattern* [10]) is the vector of the multiplicities of the characters in x. More precisely, for each $c \in \Sigma$, we have

$$pv_x[c] = \big| \{ i : 0 \leq i < m \text{ and } x[i] = c \} \big|.$$

In the following, the Parikh vector of the substring $x[i .. i+h-1]$ of x, of length h and starting at position i, will be denoted by $pv_{x(i,h)}$.

The procedure for computing the Parikh vector pv_x of a string x of length m needs an initialization of the vector which takes $O(\sigma)$ time, and an inspection of all characters of x which takes $O(m)$ time. Thus the Parikh vector can be computed in $O(m + \sigma)$ time.

In terms of Parikh vectors, the abelian pattern matching problem can be formally expressed as the problem of finding the set $\Gamma_{x,y}$ of positions in y, defined as

$$\Gamma_{x,y} = \big\{ s : 0 \leq s \leq n - m \text{ and } pv_{y(s,m)} = pv_x \big\}.$$

3   Previous Results

For a pattern x of length m and a text y of length n over an alphabet Σ of size σ, the *online abelian pattern matching problem* can be solved in $\mathcal{O}(n)$ time and $\mathcal{O}(\sigma)$ space by using a naïve *prefix based approach* [10,16,18,19], which slides a window of size m over the text while updating in constant time the corresponding Parikh vector.

Specifically, for each position $0 \leq s < n - m$, and character $c \in \Sigma$, we have

$$pv_{y(s+1,m)}[c] = pv_{y(s,m)}[c] - \big| \{c\} \cap \{y[s]\} \big| + \big| \{c\} \cap \{y[s+m]\} \big|,$$

so that the vector $pv_{y(s+1,m)}$ can be computed from the vector $pv_{y(s,m)}$ by incrementing the value of $pv_{y(s,m)}[y[s + m]]$ and by decrementing the value of $pv_{y(s,m)}[y[s]]$. Thus, the test "$pv_{y(s+1,m)} = p_x$" can be easily performed in constant time. This is done by maintaining an error value e such that

$$e = \sum_{c \in \Sigma} \big| pv_x[c] - pv_{y(s,m)}[c] \big|, \, for \, 0 \leq s < n - m.$$

At the beginning of the algorithm the value of e is set to $\sum_{c \in \Sigma} |pv_x[c] - pv_{y(0,m)}[c]|$. Thus, when it becomes 0 an occurrence is reported.

A *suffix-based* approach to the problem has been presented in [10], as an adaptation of the Horspool string matching algorithm [17] to abelian pattern matching problem. Rather than reading the characters of the window from left to right, characters are read from right to left. Every time the reading restarts in a new window, starting at position s, the Parikh vector is initialized by setting $pv_{y(s,m)}[c] = 0$, for all $c \in \Sigma$. Then, during each attempt, as soon as a frequency overflow occurs, the reading phase is stopped and a new alignment is attempted by sliding the window to the right. An occurrence is reported when the whole window is inspected without reporting any frequency overflow. The resulting algorithm has an $\mathcal{O}(n(\sigma + m))$ worst-case time complexity but performs well in many practical cases, especially for long patterns and large alphabets.

Successive solutions to the problem tried to speed-up both the prefix-based and suffix-based approaches described above by applying techniques of algorithm engineering, where experimental evaluations play an important role.

In [8] Cantone and Faro presented the Bit-parallel Abelian Matcher (BAM), which applies bit-parallelism to enhance the suffix-based approach for the abelian pattern matching problem. It turns out to be very fast in practical cases. It has been also enhanced in [9] by reading 2-grams (BAM2), obtaining best results in most practical cases.

However, although the adaptive width of bit fields makes possible to handle longer patterns than a fixed width, the packing approach used in BAM can be applied only in the case where the whole Parikh vector fits into a single computer word. This makes the algorithm particularly suitable for small alphabets or short patterns, but not useful in the case of long patterns and large alphabets.

In [9], an attempt to adapt such strategy in the case of long patterns have been presented. The authors proposed the BAM-shared algorithm (BAMs for short) which uses a kind of alphabet reduction in order to make the bit-vector fit into a single computer word. This implies that the algorithm works using a filtering approach and, in case of a match, the candidate occurrence should be verified.

In [9] the authors presented also a simple and efficient bit-parallel suffix-based approach. Instead of packing the Parikh vector of a string, their algorithm, named Exact Backward for Large alphabets (EBL for short) maintains a bit vector B of size σ where, for each $c \in \Sigma$, $B[c]$ is set to 1 if the character c occurs in the pattern, and is set to 0 otherwise.

More recently, in [13], Ghuman and Tarhio enhanced the EBL suffix-based solution by using SIMD (Single Instruction Multiple Data) instructions. Their solution, named Equal-Any algorithm (EA for short), uses a SIMD load instruction for reading, at each iteration, the whole window of the text in one fell swoop and storing it in a computer word w. Experimental results show that such solution is 30% faster than previous algorithms for short English patterns. However, despite this results, it works only when the length of the pattern is less or equal to 16.

Some efficient variants of the prefix based approach have been also presented in the last few years. Among them in [15] Grabowsky *et al.* presented a more efficient prefix-based approach, which uses less branch conditions. In [9] Chhabra *et al.* presented a prefix-based solution named Exact Forward for Small alphabets (EFS for short) which applies the same packing strategy adopted by the BAM algorithm to the prefix-based approach, obtaining very competitive results in the case of short alphabets.

To delve into the problem we refer to a detailed analysis of the abelian pattern matching problem and of its solutions presented in [10] and in [14].

4   The Heap-Counting Approach

Let x and y be strings of length m and n, respectively, over a common alphabet Σ of size σ. As described above previous solutions for the abelian pattern matching problem maintain in constant time the symmetric difference, e, of the multisets of the characters occurring in the current text window and of those occurring in the pattern, respectively. Thus, when $e = 0$, a match is reported. Alternatively, they use a packed representation of the Parikh vector, where some kind of overflow sentinel is implemented in order to take track that the frequency of a given character has exceeded its corresponding value in the Parikh vector of the pattern. The aim is to perform the initialization of the Parikh vector in constant time and to perform vector updates in a very fast way.

Instead of using a structured representation of the Parikh vector of a string, fitting in a single computer word, our approach tries to map the multisets of our interest into natural numbers, using a heap-mapping function h that allows for very fast updates.

Specifically we suppose to have a function $h : \Sigma \rightarrow \mathbf{N}$ (the *heap-function*), that maps each character c of the alphabet Σ to a natural number, $h(c)$ indeed. Then, we assume that the multiset of a given string $w \in \Sigma^*$, of length m, can be univocally associated to a natural number, $h(w)$, using the following relation:

$$h(w) = \sum_{i=0}^{m-1} h(w[i]) \tag{1}$$

The value $h(w)$ is called the *heap-value* of the string w. In this context a abelian match is found at position s of the text when the heap-value associated to the window starting at position s is equal to the heap-value of the pattern. This approach, when applicable, leads to two main advantages: the multisets of the characters occurring in string can be represented by a single numeric value, fitting in a single computer word; modifications and updates of such multisets can be done by means of simple integer additions.

Our heap-counting approach is based on the following elementary fact:

Let $\Sigma = \{c_0, c_1, \ldots, c_{\sigma-1}\}$ be an alphabet of size $|\Sigma| = \sigma$, let $m > 1$ be an integer, and let $h : \Sigma \rightarrow \mathbb{N}$ be the mapping $h(c_i) = m^i$, for $i = 0, \ldots, \sigma - 1$.

```
COMPUTE-HEAP-MAPPING(x, m)
1.    for each c ∈ Σ do h[c] ← 0
2.    j ← 1
3.    for i ← 1 to m do
4.         if h(x[i]) = 0 then
5.              h(x[i]) ← j
6.              j ← j × m
7.    return h
```

```
HEAP-COUNTING-ABELIAN-MATCHING(x, m, y, n)
1.    h ←COMPUTE-HEAP-MAPPING(x, m)
2.    δ ← γ ← 0
3.    for i ← 0 to m − 1 do
4.         δ ← δ + h(x[i])
5.         γ_0 ← γ_0 + h(y[i])
6.    if γ_0 = δ then OUTPUT(0)
7.    for s ← 1 to n − m do
8.         γ_s ← γ_{s−1} + h(y[s + m − 1]) − h(y[s − 1])
9.         if γ_s = δ OUTPUT(s)
```

Fig. 1. The pseudocode of the HEAP-COUNTING-ABELIAN-MATCHING for the online exact abelian matching problem, implemented using a prefix-based approach.

Then for any two distinct k-multicombinations (i.e., k-combinations with repetitions) φ_1 and φ_2 from the set Σ, with $1 \le k \le m$, we have

$$\sum_{c \in \varphi_1} h(c) \neq \sum_{c \in \varphi_2} h(c). \tag{2}$$

Example 2. Let $x = agcga$ be an input pattern of length 5 over the alphabet $\Sigma = \{a, c, g, t\}$ of size 4. Based on Lemma 4, the heap-function $h : \Sigma \to \mathbb{N}$, is defined as $h(a) = 1, h(c) = 5, h(g) = 25$, and $h(t) = 125$. The heap-value of x is then $h(x) = h(a) + h(g) + h(c) + h(g) + h(a) = 57$.

Let $\diamond$ be a character such that $\diamond \notin \Sigma$ and let $\Sigma_x \subseteq \Sigma$ be the set of all (and only) the characters occurring in x. We indicate with σ_x the size of the alphabet Σ_x. Plainly we have $\sigma_x \le \min\{\sigma, m\}$, thus we can think to this transformation as a kind of alphabet reduction.

We define the *reduced text* $\bar{y}$, over Σ_x, as a version of the text y where each character $y[i]$, not included in Σ_x, is replaced with the special character $\diamond \notin \Sigma$. Since, in general, $\sigma_x < \sigma$ (especially in the case of short patterns), to process the reduced version of the text, instead of its original version, allows the heap function to be computed on a smaller domain, reducing therefore the size of the heap-values associated to any given string.

Example 3. Let $x = agcac$ be an input DNA sequence (the pattern) of length 5 and let $y = agtcagaccatcagata$ be a text of length 17, both over the alphabet $\Sigma = \{a, g, c, t\}$ of size 4. We have that $\Sigma_x = \{a, c, d\}$. Thus the reduced version of y is $\bar{y} = $ ag◇c◇gacca◇caga◇◇

It can be proved that such alphabet reduction does not influence the output of any abelian pattern matching algorithm.

We are now ready to describe in details the two new algorithms based on the heap-counting approach. The algorithm described in Sect. 4.1 implements the heap-counting approach i a prefix-based algorithm, while the algorithm described in Sect. 4.2 uses a suffix-based mechanism.

4.1   A Prefix-Based Algorithm

Inspired by Lemma 4, the new algorithm precomputes the set Σ_x and the function $h : \Sigma_x \to \mathbb{N}$, defined as $h(c_i) = m^i$, for $i = 0, \ldots, \sigma_x - 1$, where m is the length of the pattern and σ_x is the size of Σ_x.

Figure 1 shows the HEAP-ABELIAN-MATCHING algorithm and its the auxiliary procedure.

During the preprocessing phase (lines 1–7) the algorithm precomputes the heap-mapping function h (line 1) by means of procedure COMPUTE-HEAP-MAPPING. Such procedure computes the mapping table over the alphabet $\Sigma_x \cup \{\diamond\}$ associating the value 1 with all characters not occurring in x, i.e. we set $h(\diamond) = 1$.

The heap values $\delta = h(x)$ and $\gamma_0 = h(y[0..m-1])$ are then precomputed in lines 2–5, Likewise, during the searching phase (lines 8–10), the heap value $\gamma_i = h(y[i..i+m-1])$ is computed for each window $y[i..i+m-1]$ of the text t, with $0 < i \leq n - m$. Specifically, starting from the heap value γ_{s-1}, the algorithm computes the heap value γ_s by using the relation $\gamma_s = \gamma_{s-1} - h(y[s-1]) + h(y[s+m-1])$ (line 8). Of course, in practical implementations of the algorithm it is possible to maintain a single value γ, corresponding to the heap value of the current window of the text.

The set $\Gamma_{x,y}$ of all occurrences in the text is then

$$\Gamma_{x,y} = \big\{i \mid 0 \leq i \leq n - m \text{ and } \gamma_i = \delta\big\}$$

In order to compute the space and time complexity of the algorithm, it can be easily observed that the computation of the mapping h requires $O(m + \sigma)$-time and -space. Moreover observe that $\gamma_s = \gamma_{s-1} - h(y[s-1]) + h(y[s+m-1])$, so that γ_s can be computed in constant time from γ_{s-1}. Thus the set $\Gamma_{x,y}$ can be computed in $\mathcal{O}(n)$ worst case time.   ∎

From a practical point of view it is understood that for an architecture, say, at 64 bits, all operations will take place modulo 2^{64}. Thus, when m^{σ_x+1} exceeds 2^{64} we could have some collisions in the set of the heap values and an additional verification procedure should be run every time an occurrence is found. However it has been observed experimentally that, also in this specific cases, the collision problem for the heap function h is negligible.

4.2   A Suffix-Based Algorithm

In this section we extend the idea introduced in the previous section and present a backward version of the prefix-based algorithm described above which turns out to be more efficient in the case of long patterns or large alphabets. It shows a sub-linear behaviour in practice, while maintains the same worst case time complexity.

Figure 2 shows the pseudocode of the new algorithm, called BACKWARD-HEAP-ABELIAN-MATCHING, and its the auxiliary procedure.

COMPUTE-MEMBERSHIP-MAP(x, m)
1.   for each $c \in \Sigma$ do $b(c) \leftarrow$ False
2.   for $i \leftarrow 1$ to m do $b(x[i]) \leftarrow$ True
3.   return b

BACKWARD-HEAP-COUNTING-ABELIAN-MATCHING(x, m, y, n)
1. $h \leftarrow$ COMPUTE-HEAP-MAPPING(x, m)
2. $b \leftarrow$ COMPUTE-MEMBERSHIP-MAPPING(x, m)
3. $\delta \leftarrow 0$
4.   for $i \leftarrow 1$ to m do $\delta \leftarrow \delta + h(x[i])$
5. $y \leftarrow y.x$
6. $s \leftarrow 0$
7.   while (True) do
8. $\gamma \leftarrow -\delta$
9. $j \leftarrow m - 1$
10.   while ($j \geq 0$) do
11.   if $(b(y[s + j]))$ then
12. $\gamma \leftarrow \gamma + h(y[s + j])$
13. $j \leftarrow j - 1$
14.   else
15. $\gamma \leftarrow -\delta$
16. $s \leftarrow s + j + 1$
17. $j \leftarrow m - 1$
18.   do
19.   if ($\gamma = 0$) then
20.   if ($s \leq n - m$) then OUTPUT(s)
21.   else return
22.   if $(b(y[s + m])$ =False) then break
23. $\gamma \leftarrow \gamma - h(y[s]) + h(y[s + m])$
24. $s \leftarrow s + 1$
25.   while (True)
26. $s \leftarrow s + m + 1$

Fig. 2. The pseudocode of the BACKWARD-HEAP-COUNTING-ABELIAN-MATCHING for the online exact abelian matching problem, implemented using a suffix-based approach.

During the preprocessing phase (lines 1–6) the algorithm precomputes the heap-mapping function h (line 1) and the membership function b (line 2). We use procedure COMPUTE-HEAP-MAPPING, and procedure COMPUTE-MEMBERSHIP-MAPPING, respectively.

The heap value $\delta = h(x)$ of the pattern x is then precomputed in lines 3–4. A copy of the pattern is then concatenated at the end of the pattern (line 5), as a sentinel, in order to avoid the window of the text to shift over the last position of the text.

The main cycle of the searching phase (line 7) is executed until the value of s becomes greater than $n - m$ (line 20). An iteration of the main cycle is divided into two additional cycles. The first cycle of line 10 performs a backward scanning of the current window of the text and stops when the whole window has been scanned or a character not occurring in Σ_x is encountered. The second cycle of line 18, starting from the heap value of the current window of the text, computes at each iteration the heap value of the next window in constant space using a forward scan. The second cycle stops when a character not occurring in Σ_x is encountered.

It can be proved that the algorithm BACKWARD-HEAP-COUNTING-ABELIAN-MATCHING computes all abelian occurrences of x in y with $O(\sigma + m + n)$-time and $O(\sigma + m)$-space complexity in the worst case.

4.3   Relaxed Filtering Variants

A simpler implementation of the above presented algorithms can be obtained by relaxing the heap-counting approach presented at the beginning of this section, in order to speed-up the computation of the heap values of a string and, as a consequence, to spud-up the searching phase of the algorithm.

Specifically we propose to use the natural predisposition of the characters of an alphabet to be treated as integer numbers. For instance, in many practical applications, input strings can be handled as sequences of ASCII characters. In such applications, characters can just be seen as the 8-bit integers corresponding to their ASCII code.

In this context, if we indicate with ASCII(c), the ASCII code of a character $c \in \Sigma$, we can set $h(c) = $ ASCII(c). Thus the heap value of a string can be simply computed as the sum of the ASCII codes of its characters.

As a consequence the resulting algorithms works as a filtering algorithm. Indeed, when an occurrence is found we are not sure that the substring of the text which perform a match is a real permutation of the pattern. This implies that an additional verification phase is run for each candidate occurrences.

Plainly the resulting algorithms have an $O(\sigma + nm)$ worst case time complexity, since a verification procedure could be run for each position of the text.

5   Experimental Results

We report in this section the results of an extensive experimentation of the newly presented algorithms against the most efficient solutions known in literature for the online abelian pattern matching problem. In particular we have compared 11 algorithms divided in three groups: prefix-based, suffix-based and SIMD based algorithms. Specifically we compared the following 5 prefix based algorithms: the Window-Abelian-Matching (WM) [10,16,18,19]; the Grabowsky-Faro-Giaquinta (GFG) [15]; the Exact Forward form Small alphabets (EFS) [9]; the Heap-Counting-Abelian-Matching (HCAM) described in Sect. 4.1; the Heap-Filtering-Abelian-Matching (HFAM) described in Sect. 4.3.

We compared also the following 5 suffix based algorithms: the Backward-Window-Abelian-Matching (BWM) [10]; the Bit-parallel Abelian Matching algorithm (BAM) using 2-grams [8,9]; the Exact Backward for Large alphabets (EBL) [9]; the Backward-Heap-Counting-Abelian-Matching (BHCAM) described in Sect. 4.2; the Backward-Heap-Filtering-Abelian-Matching (BHFAM) described in Sect. 4.3.

In addition We compared also the Equal Any (EA), an efficient prefix based solution [13] implemented using SIMD instructions.

All algorithms have been implemented in C, and have been tested using the SMART tool [12] and executed locally on a MacBook Pro with 4 Cores, a 2 GHz Intel Core i7 processor, 16 GB RAM 1600 MHz DDR3, 256 KB of L2 Cache and 6 MB of Cache L3.[1] Comparisons have been performed in terms of running times, including any preprocessing time.

For our tests we used a genome sequence provided by the research tool SMART, available online for download (for additional details about the sequences, see the details of the SMART tool [12]).

In the experimental evaluation, patterns of length m were randomly extracted from the sequences, with m ranging over the set of values $\{2^i \mid 1 \leq i \leq 8\}$. Thus at least one occurrence is reported for each algorithm execution. In all cases, the mean over the running times (expressed in hundredths of seconds) of 1000 runs has been reported.

Table 1 summarises the running times of our evaluations. The table is divided into four blocks. The first block presents the results relative to prefix based solutions, the second block presents the results for the suffix based algorithms, while the third block presents the results for the algorithm based on SIMD instructions. The newly presented algorithms have been marked with a star ($\star$) symbol. Best results among the two sets of algorithms have been bold-faced to ease their localization, while the overall best results have been also underlined. In addition we included in the last block the speedup (in percentage) obtained by our best newly presented algorithm against the best running time obtained by previous algorithms: positive percentages denote running times worsening, whereas negative values denote performance improvements. Percentages representing performance improvements have been bold-faced.

---

[1] The SMART tool is available online at https://smart-tool.github.io/smart/.

Table 1. Experimental results on a genome sequence.

	m	2	4	8	16	32	64
PREFIX BASED	WM	13.63	13.02	13.09	13.04	13.04	13.06
	GFG	20.47	20.19	20.42	20.41	20.47	20.30
	EFS	8.26	8.31	8.35	8.34	8.36	–
	HCAM ⋆	**6.97**	**6.86**	**6.92**	**6.93**	**6.91**	**6.89**
	HFAM ⋆	20.14	11.86	9.18	7.87	7.47	7.09
SUFFIX BASED	BWM	65.59	46.81	33.47	25.79	22.94	20.67
	BAM	**10.62**	**10.96**	13.20	11.87	10.69	9.85
	EBL	29.95	27.44	55.69	119.26	227.14	–
	BHCAM ⋆	26.82	18.03	**9.98**	**7.74**	**7.55**	**7.32**
	BHFAM ⋆	37.21	21.72	11.50	9.35	9.09	8.88
	Speed-up	+11.87%	+11.40%	+10.76%	+10.55%	**−12.12%**	**−29.16%**
SIMD	EA	**4.01**	**4.03**	**4.56**	**4.67**	–	–

Consider first the case of small alphabets, and specifically abelian string matching on strings over an alphabet of size $\sigma \leq 4$ (Table 1). From experimental results it turns out that prefix based solutions are more flexible and efficient than suffix based algorithms. This is because the shift advancements performed by suffix based solutions do not compensate the number of character inspections performed during each iteration. Thus, while prefix based algorithms maintain a linear behaviour which do not depend on the pattern length, suffix based solutions shown an increasing trend (or a slightly decreasing trend), while the length of the pattern increases, but with a very low performances on average. Specifically the HCAM algorithm obtains the best results only for $m \geq 16$, where it is approximately 10% slower than the EA algorithm, in the case of short patterns. However it remains always the best solution if compered with all other standard algorithm, with a gain from 11% to 35%. Among the suffix based solutions the BHCAM algorithm still remains the best choice in most cases, with a less sensible variance if compared with the HCAM algorithm.

6   Conclusions

In this paper we have introduced the heap-counting approach for the abelian pattern matching problem in strings and we have presented two new algorithms based on a prefix-based approach (HCAM) and on a suffix-based approach (BHCAM), respectively. We also presented two variants of these algorithms, based on a relaxed version of the heap-counting approach: the HFAM and BHFAM algorithms. From our experimental results it turns out that our approaches obtain good results when used for searching text over small alphabets, as the case of DNA sequences. The resulting algorithms turns out to be among the most effective in practical cases.

References

1. Amir, A., Apostolico, A., Landau, G.M., Satta, G.: Efficient text fingerprinting via Parikh mapping. J. Discrete Algorithms **1**(56), 409–421 (2003)
2. Baeza-Yates, R.A., Navarro, G.: New and faster filters for multiple approximate string matching. Random Struct. Algorithms **20**(1), 23–49 (2002)
3. Benson, G.: Composition alignment. In: Benson, G., Page, R.D.M. (eds.) WABI 2003. LNCS, vol. 2812, pp. 447–461. Springer, Heidelberg (2003). https://doi.org/10.1007/978-3-540-39763-2_32
4. Böcker, S.: Simulating multiplexed SNP discovery rates using base-specific cleavage and mass spectrometry. Bioinformatics **23**(2), 5–12 (2007). https://doi.org/10.1093/bioinformatics/btl291
5. Böcker, S.: Sequencing from compomers: using mass spectrometry for DNA de novo sequencing of 200+ nt. J. Comput. Biol. **11**(6), 1110–1134 (2004)
6. Burcsi, P., Cicalese, F., Fici, G., Lipták, Z.: Algorithms for jumbled pattern matching in strings. Int. J. Found. Comput. Sci. **23**(2), 357–374 (2012)
7. Cantone, D., Cristofaro, S., Faro, S.: Efficient matching of biological sequences allowing for non-overlapping inversions. In: Giancarlo, R., Manzini, G. (eds.) CPM 2011. LNCS, vol. 6661, pp. 364–375. Springer, Heidelberg (2011). https://doi.org/10.1007/978-3-642-21458-5_31
8. Cantone, D., Faro, S.: Efficient online Abelian pattern matching in strings by simulating reactive multi-automata. In: Holub, J., Zdarek, J. (eds.) Proceedings of the PSC 2014, pp. 30–42 (2014)
9. Chhabra, T., Ghuman, S.S., Tarhio, J.: Tuning algorithms for jumbled matching. In: Holub, J., Zdarek, J. (eds.) Proceedings of the PSC 2015, pp. 57–66 (2015)
10. Ejaz, E.: Abelian pattern matching in strings, Ph.D. thesis, Dortmund University of Technology (2010). http://d-nb.info/1007019956
11. Eres, R., Landau, G.M., Parida, L.: Permutation pattern discovery in biosequences. J. Comput. Biol. **11**(6), 1050–1060 (2004)
12. Faro, S., Lecroq, T., Borzì, S., Di Mauro, S., Maggio, A.: The string matching algorithms research tool. In: Proceeding of Stringology, pp. 99–111 (2016)
13. Ghuman, S.S., Tarhio, J.: Jumbled matching with SIMD. In: Holub, J., Zdarek, J. (eds.) Proceeding of the PSC 2016, pp. 114–124 (2016)
14. Ghuman, S.S.: Improved online algorithms for jumbled matching. Doctoral Dissertation 242/2017, Aalto University publication series, Aalto University, School of Science, Department of Computer Science (2017)
15. Grabowski, S., Faro, S., Giaquinta, E.: String matching with inversions and translocations in linear average time (most of the time). Inf. Process. Lett. **111**(11), 516–520 (2011)
16. Grossi, R., Luccio, F.: Simple and efficient string matching with k mismatches. Inf. Process. Lett. **33**(3), 113–120 (1989)
17. Horspool, R.N.: Practical fast searching in strings. Softw. Pract. Exp. **10**(6), 501–506 (1980)
18. Jokinen, P., Tarhio, J., Ukkonen, E.: A comparison of approximate string matching algorithms. Softw. Pract. Exp. **26**(12), 1439–1458 (1996)
19. Navarro, G.: Multiple approximate string matching by counting. In: Baeza-Yates, R. (ed.) 1997 Proceeding of the 4th South American Workshop on String Processing, pp. 125–139 (1997)
20. Salomaa, A.: Counting (scattered) subwords. Bull. EATCS **81**, 165–179 (2003)

Analysis of Gene Regulatory Networks Inferred from ChIP-seq Data

Eirini Stamoulakatou$^{(\boxtimes)}$, Carlo Piccardi, and Marco Masseroli

Dipartimento di Elettronica, Informazione e Bioingegneria,
Politecnico di Milano, 20133 Milan, Italy
{eirini.stamoulakatou,carlo.piccardi,marco.masseroli}@springer.com

Abstract. Computational network biology aims to understand cell behavior through complex network analysis. The Chromatin Immuno-Precipitation sequencing (ChIP-seq) technique allows interrogating the physical binding interactions between proteins and DNA using Next-Generation Sequencing. Taking advantage of this technique, in this study we propose a computational framework to analyze gene regulatory networks built from ChIP-seq data. We focus on two different cell lines: GM12878, a normal lymphoblastoid cell line, and K562, an immortalised myelogenous leukemia cell line. In the proposed framework, we preprocessed the data, derived network relationships in the data, analyzed their network properties, and identified differences between the two cell lines through network comparison analysis. Throughout our analysis, we identified known cancer genes and other genes that may play important roles in chronic myelogenous leukemia.

Keywords: Biomolecular networks · Transcription factors · ChIP-seq · Next-Generation Sequencing · Cancer · Bioinformatics

1 Introduction

In biological sciences, network analysis is becoming one of the main tools to study complex systems. Networks used to represent the regulation of gene expression are known as Gene Regulatory Networks (GRNs) [1]. In network biology, particularly in disease/cancer research, comparisons are often performed on GRNs [2] and DNA co-methylation networks [3], obtained from the gene expression and DNA methylation profiles of healthy and disease tissues.

Here, we focus on normal and cancer GRNs that, differently from other works, we inferred from Chromatin Immuno-Precipitation sequencing (ChIP-seq) data. ChIP-seq is a Next-Generation Sequencing (NGS) technique designed to study, map and understand protein-DNA interactions on a genome-wide scale. It provides measurements of epigenetic (transcription factor and histone) regulation of genes, retaining all the advantages of the NGS technology thanks to its coverage, high resolution and cost-effectiveness. Our goal is to study the relationship between gene-related epigenetic factors and genes in a normal vs. disease case,

© Springer Nature Switzerland AG 2019
I. Rojas et al. (Eds.): IWBBIO 2019, LNBI 11465, pp. 319–331, 2019.
https://doi.org/10.1007/978-3-030-17938-0_29

possibly leading towards the discovery of novel molecular diagnostic and prognostic signatures. Particularly, we focused on two immortalized human cell lines, K562 and GM12878; they are both from blood tissue, the first one (K562) from chronic myelogenous leukemia, whereas the second one (GM12878) from normal lymphoblastoid cells.

A major contribution of this work is the study of the relation between epigenetic transcription factors and human protein-coding genes in K562 and GM12878 cell lines in the view of complex network comparison. This was possible by defining relationships between transcription factors and protein-coding genes to create gene regulatory networks. Another major aspect of this study is the creation of a computational framework with appropriate network comparison methods, according to our network characteristics, to extract differences and similarities of the compared networks. The defined comparison models are fully "data-driven", as they do not take into consideration any form of prior biological knowledge. Finally, using our analytic framework, we highlighted behaviours directly emerging from the data, drawing insights that could drive further biological investigations.

2   Used Data Sets

Among the numerous publicly accessible available genomic databases, we chose the following two: the ENCyclopedia Of DNA Elements (ENCODE) and GENCODE [4]; the former one as source for the NGS experimental data, the second one for the gene annotations we used. GENCODE genomic samples are organized as General Feature Format (GTF) text files, whose structure is described in [4]. Each of their lines refers to a genomic feature annotation and is made up of several tab-separated fields. The first eight fields are standard GTF fields that convey information about the feature chromosome, annotation source, feature type, start and stop genomic coordinates, score, strand, and genomic phase. The ninth field is actually a sequence of key-value pairs made up of further information about the feature.

Biosamples involved in the sequencing experiments generating our considered data came from two immortalized cell lines, namely K562 and GM12878. These two cell lines are among the most investigated ones in the ENCODE project [5], being the object of a large number of sequencing experiments from research labs all over the world, each identifying thousands of epigenetic events through the whole genome. Both cell lines belong to human blood tissue, in particular: K562 cell line consists in a chronic myeloid leukemia (CML) cell line [6], GM12878 cell line is made up of lymphoblastoid cells [7].

3   Analysis Framework

Networks provide a theoretical framework that allows a convenient conceptual representation of interrelations among a large number of elements. Furthermore, they usually allow framing questions about the behavior of the underlying represented system, by applying well-established analyses on the network representing the considered data. Here, we focus on cell line specific gene regulatory networks, where source nodes represent genes encoding transcription factors (TFs), whereas target nodes are any genes. A link exists between a source TF encoding gene and a target gene if the encoded TF binds the target gene promoter; the links are weighted, and the weight represents the power of the binding.

We propose a network analysis framework to characterize commonalities and differences in behavior across normal GM12878 cells and cancerous K562 cells, using ChIP-seq datasets. We evaluate if some genes display extreme behaviors, and whether or not such behaviors highlight aspects of the underlying biology. The proposed framework includes the following steps: (1) High quality data extraction from NGS and genomic annotation datasets, through the GenoMetric Query Language (GMQL); (2) Transformation of the extracted metadata and genomic region data to adjacency matrixes, representing the most valuable information and the data relationships extracted; (3) Numeric characterization of each network structure through 8 topological measures; (4) Application of comparison methodologies to identify the most common and different gene connections.

3.1 Data Acquisition and Preprocessing

For the data acquisition and preprocessing, we chose GMQL [8] as the most suitable tool. GMQL is a high-level declarative query language, specifically designed for genomic data retrieval and processing. The GMQL portal[1] publicly provides reasonably high computational and storage capabilities and, moreover, it hosts up-to-date GENCODE and ENCODE data, among others. This last aspect allowed us to just write a GMQL query to perform the complete extraction and filtering of the genes' epigenetic status data described below, without the need to download the related data files from the GENCODE and ENCODE public repositories and write specific programs to extract the relevant data. In the following paragraphs we describe the usage of GMQL to filter and extract the highest quality epigenetic status data from ENCODE.

The goal of the defined GMQL query is to map transcription factors of the two cell lines on each gene promoter region. Thus, the first step is the selection of the transcription factors and the promoter regions. The ENCODE consortium has defined and implemented a system of 'audits', i.e., flags meant to give additional, yet essential, quality information about the provided experimental data to the research community. To extract high-quality data, we did not consider all the experiment data files labeled with at least one of the following audits: *extremely low read depth*, *extremely low read length*, or *insufficient read depth*. Furthermore, to consider only data from higly reproducible NGS ChIP-seq experiments, we selected only the called peak data files labeled as *conservative IDR threshold peaks*. Finally, in the case of more replicate data files from the same

---

[1] http://www.gmql.eu/.

transcription factor targeting experiments, we chose to only consider one data file for each transcription factor, the one with the largest number of called peaks. By choosing the peak set with the highest cardinality, we retain a larger amount of information, still being confident of its reasonably good quality thanks to the foregoing audit-based and reproducibility-aware filtering performed.

In our study we are exclusively interested in promoter regions of human known protein-coding genes, i.e., genomic regions around the starting position of a gene transcript. Therefore, an important aspect is to consider the right position along the human genome of each transcript of all genes of interest. The process of identifying and designating locations of individual genes and transcripts on the DNA sequence is called genome annotation. One of the most important active projects about human genome annotation is GENCODE.[2] Thus, for the promoter region extraction we chose GENCODE repository annotations, specifically the GENCODE v24 release version and the annotation type transcript; so, an annotation file for transcript isoforms was selected, reporting all the transcript start sites (TSSs) of each human gene. In order to build the promoter regions from the transcript isoforms, we used the typical $-2k/+1k$ base interval around their first base. All the selected transcription factor binding regions are then mapped to the considered gene promoters. As a gene can have more than one promoters, we selected for every TF only the gene with the highest signal value. The dataset created by the performed GMQL mapping operation provides a matrix-like structured outcome, ideal for subsequent data analysis. In particular, we created such a dataset/matrix for each considered cell line, where the matrix rows represent transcription factors, columns represent genes, and each matrix cell contains a value that represents the maximum binding signal of a TF in a gene promoter. To create the gene regulatory network from the above data, we finally considered each TF as representing its encoding gene, thus obtaining a gene adjacency matrix for each cell line.

3.2   Gene Regulatory Network Analysis

A primary aspect in gene regulatory networks is to capture the interactions between molecular entities from high-throughput data. The GRNs that we constructed are weighted directed networks, where nodes represent genes and links between nodes exist solely if the regulatory element, a transcription factor encoded by a source gene, binds a target gene promoter.

The problem of detecting significant dissimilarities in paired biological networks is different from popular graph theory problems, like graph isomorphism or subgraph matching, for which various graph matching and graph similarity algorithms exist and have been also applied on biological networks [9,10]. Several approaches to compare gene regulatory networks constructed from healthy and disease samples have been developed [11,12]. The majority of them focuses on the comparison of the entire networks, using statistics that describe network global properties [13]; but these statistics are not sensitive enough to detect smaller,

---

[2] https://www.gencodegenes.org/.

yet important, differences. On the other hand, there are numerous alignment-based methods that compare networks using the properties of the individual nodes, e.g., local similarity [14]. The aim of these methods is to identify matching nodes, and use these nodes to identify exact subnetwork matches. These approaches are computational intensive, as exact graph matching is NP-hard. In addition, alignment-free comparison methods exist, which have been used to identify evolutionary relationships [15]. These methods are based on the fact that differences in network structure is essential, as structural properties of biological networks can bring important biological insights, such as determining the relationships between protein functions from protein interaction network topology. To achieve network structure comparison, they count the occurrences of subgraph shapes in the local neighbourhoods of all nodes in a network [16].

Our created networks have a peculiar structure, mainly due to the fact that ChIP-seq experiment data exist only for a limited set of TFs; thus, in our GRNs the number of source nodes (TFs) is much lower (about 100) than the number of the target nodes (human protein-coding genes, about 19,000). This makes difficult to directly apply reliably the methodologies mentioned above. On the other hand, motifs and modules have long been identified as important components of biological networks [3]; thus, we focused on looking for strongly connected components (SCCs) in each considered network, and on evaluating the one-step ego-nets in each SCC. So, we avoid comparing the entire networks, and concentrate on their most informative nodes. The one-step ego-net of a node/gene g is the (sub)network consisting of all the nodes within one edge distance from g, also including all the edges between those nodes. For directed graphs, as in our case, a node g ego-net contains the g "out" neighborhood, i.e., in our case the genes where g points to and their connections. To analyse the ego-nets of each SCC of the two networks under comparison, we applied standard approaches such as pairwise (on matching nodes) metrics to quantify similarity based on network properties, discover specific features, and detect anomalous nodes/genes.

The state-of-the-art offers a well-established set of graph metrics for complex networks. The most important metrics for a detailed analysis of a weighted directed network have been previously described in [2]; they are used in the current study and here summarized. The *degree* of a node is the total number of edges incident to it. Thus, the average value of the network degrees, measured over all network nodes, is called the *average degree* of the network, as we handle directed graph we computed in and out degrees. For the *total weight*, we sum the weights of all the edges of the graph. The *diameter* of a network is the maximal distance between any pair of nodes in the network. The *modularity* measures to what extent the network is structured in communities. It takes values between 0 and 1; a higher modularity means a stronger division between well-delimited communities, i.e., subnetwotks with large internal edge density but weakly connected each other, while a lower modularity means that no such subnetwork exist. The metric that quantifies the degree correlation, i.e. to what extent nodes with large degree are connected to nodes with large degree, is called *assortativity*. The network's heterogeneity can be measured by the *degree distri-*

bution entropy. As *principal eigenvalue* we denote the largest eigenvalue of the weighted adjacency matrix of the network. For each node/gene, the *connectivity* is defined as the sum of the connection strengths with the other nodes/genes of the network.

Our proposed analysis method compares not the networks themselves, but instead the ensemble of all gene neighborhoods (ego-nets) in each SCC of the networks, through a pairwise approach. This idea of using the content of subgraphs to build a comparison method between networks arises from the fact that modules are important biological network components.

The statistical comparison measures we used in our method are the following:

- The *cosine similarity* (*CS*), a measure of similarity between two vectors: it expresses the cosine of the angle between them, not from the perspective of magnitude, but from that of orientation. The resulting similarity between the two vectors ranges from -1, meaning exactly opposite directions, to 1, meaning exactly the same direction, with 0 usually indicating independence. This measure is applied in our context by building a vector with elements consisting of each metric of interest measured on the graph.
- The *Jaccard index* (*J*), a statistic used for comparing the similarity of sample sets. It measures similarity between finite sample sets, defined as the size of the intersection divided by the size of the union of the sample sets. The Jaccard index always gives a value between 0, which means no similarity, and 1, for identical sets. In our study we used the Jaccard similarity pairwise for each matching node of the SCCs of the two compared networks. For each node we built a set of its ego-net edges, with the edges being represented as an object (source node, target node) since the networks are directed. This measure gave us the percentage of similarity between the matched genes based on their interactions with the other genes, and ranges from 0 to 100.
- The *fidelity metric* ϕ, another network similarity measure, computed following the approach proposed in [17]. It is a statistical formula that generates a single value to summarize the similarity between two sets of properties/topological features (which characterize two entities of the same nature).

Additionally, in our network comparison analysis we included the identification of patterns for neighborhoods (ego-nets) of the normal and cancer networks, and the report of deviations, if any, as proposed in [18]. The detection of outliers is intimately related with the pattern discovery: only if the majority of nodes closely obey to a pattern, we can then confidently consider as outliers the few nodes that deviate. In order to detect the patterns and the outliers of the SCCs of the normal and cancer networks, we selected and grouped the topological features of the ego-nets into pairs, where we expect to find patterns of normal behavior and point out anomalies that deviate from the patterns. All methods presented here were implemented using Python programming language and its pyGMQL [28] and Networkx [29] packages.

4 Results

Here, we present and discuss the results obtained for our considered normal and cancer cell line networks, using two distinct network analysis approaches: *single-network analysis* and *differential network analysis*, which answer different questions. In our context, the single-network analysis aims at identifying both the key genes (i.e., hub genes) and the similarities in the binding behavior of the TFs present in a given data set. Conversely, the differential network analysis aims to uncover similarities and differences in the TFs of the two data sets. More specifically, using feature vectors with the aforementioned statistical measures, we evaluated the similarity of the TFs present in both data sets, and also we identified common behavior trends and outlier nodes for the two cell lines.

4.1 Single Network Analysis

The two weighted directed networks constructed, one for the normal GM12878 and one for the cancer K562 cell line, were individually analyzed. Both resulted having a single giant strongly connected component (SCC), with only TF encoding genes (source nodes), and a single out-component, i.e., a set whose nodes are reachable with a directed path from the SCC, with about 90% of the network nodes, including a few TF encoding genes not in the SCC. Table 1 reports the topological feature values measured for the two networks and their SCCs.

Curiously, in both networks the most important (hub) nodes, identified using the page-rank algorithm [19], were mitochondrial genes. The TFs with largest degree were identified using the reverse page-rank algorithm (applying page-rank to the networks obtained by reversing the directions of all links). For the cancer network they were ATF7, RBFOX2, ATF1, NFIC, NRF1, PKNOX1, RFX1, VEZF1 and L3MBTL2, whereas for the normal network they were IKZF1, ELF1, FOXK2, PKNOX1, ZNF143 and BHLHE40. IKZF1 is a leukemia tumor suppressor associated with chromatin remodeling, with also increasing evidence that IKZF1 loss also affects signaling pathways that modulate therapy response [20]. Also ELF1 is a key transcription factor in the regulation of genes involved

Table 1. Topological feature values for the two networks and their SCCs.

Features	K562	GM12878	K562-SCC	GM12878-SCC
Nodes	18,732	18,732	230	111
Isolated nodes	2,312	4,305	-	-
Source nodes	238	115	230	111
Edges	923,025	481,704	20,320	5,556
Average degree in/out	56.261	33.384	88.343	55.051
Assortativity	−0.054	−0.043	−0.021	−0.011
Diameter	4	4	4	3
Modularity	0.29	0.34	0.27	0.33

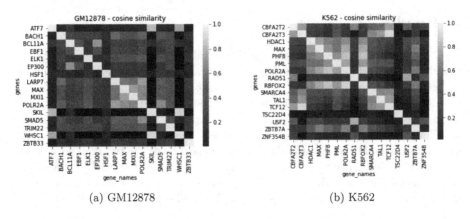

(a) GM12878   (b) K562

Fig. 1. Heatmaps showing the cosine similarity between TFs in the two cell lines.

in hematopoiesis [21]. PKNOX1 is a Hox co-factor, whose function alteration is directly linked to hematopoiesis and leukemia. ZNF143 is also involved in leukemia development [22].

Using the cosine similarity function pairwise, we identified the TFs with similar behavior in each network, i.e., that bind the same genes with similar strength. Figure 1 reports the cosine similarity heatmaps we created for some of such TFs; values closer to 1 show greater similarity. The heatmaps clearly show some clusters with high similarity in each network. For the GM12878 one, LARP7, MAX, MXI1 and POLR2A were the TFs with greatest similarity; conversely, ATF7, SKIL and WHSC1 had totally different bindings with respect to the other TFs. In the K562 cancer network, HDAC1, MAX, PHF8, PML, RBFOX2 and POLR2A created a cluster of similarity, and SMARCA4, TAL1 and TCF12 another one. The first cluster TFs resulted enriched in the *Homo sapiens transcriptional misregulation in cancer* KEGG pathway. TSC22D4 and ZNF354B resulted the TFs with the greatest dissimilarities to the others.

Table 2. Topological features for ego-nets of normal and cancer cell line SCCs.

Features	GM12878		K562	
	Average	Range (min; max)	Average	Range (min; max)
Nodes	33	(2; 67)	34	(4; 67)
Edges	798	(1; 2,153)	855	(8; 2,287)
Average degree in/out	18	(0.5; 32)	19.5	(2; 34)
Total weight	215,000	(21; 532,000)	273,000	(316; 618,000)
Density	0.633	(0.471; 1.500)	0.600	(0.511; 1.150)
Degree entropy	3.121	(0.630; 4.101)	3.330	(1.307; 4.105)
Assortativity	−0.212	(−0.500; −0.011)	−0.188	(−0.370; −0.060)
Principal eigenvalue	9,869	(0; 14,899)	29,290	(69; 39,138)
Connectivity	15,872	(1,002; 126,000)	18,389	(1,578; 164,000)

4.2   Differential Network Analysis

For the network comparison analysis we focused on the single SCC in each of the two networks, considering only the TFs whose data were available for both cell lines, i.e., 68 TFs. The average, minimum and maximum values of the ego-net features extracted for such TFs are reported in Table 2; no relevant differences in the global features were found between the two cell lines.

The obtained global results led us to apply the comparison methods at local level in order to highlight differences, if existing. As a first approximation, we simply checked which were the most different TFs between the normal and cancer cell lines. Using the Jaccard, cosine and fidelity similarity measures, we computed pairwise similarity scores for every pair of TFs. Despite the global topological features showed relatively similar values in both cell lines, at local level we discovered interesting dissimilarities (data not shown). To further explore the topological differences among the two cell lines, we characterized the structure of the ego-net extracted for each TF using the same 9 standard measures for network topology as in Table 2. These measures capture important characteristics of a network structure, which in part determines its functionality. In particular, we sought to detect the structural heterogeneity among TFs. For each ego-net of a TF, we created a feature vector with these feature values, which we used for pairwise cosine and fidelity similarity between each pair of TF/ego-nets. The cosine similarity, however, proved to be not a good metric, as all results were close to 1 (identical). In addition, we applied Jaccard similarity using as input the TF/ego-nets edges, this metric demonstrated to be a good method. Most different TFs found, according Jaccard similarity, are in Table 3.

All these TFs, except of BACH1, appeared to have greater activity in K562 than in GM12878 cell line (data not shown). Interestingly, CTBP1 appeared to bind strongly in the cancer cell line, but it had only a bond in the normal cell line data. An explanation of this behavior may be that, according to KEGG, CTBP1 is a leukemia cancer gene. In the same context, ZBTB33 and CEBPB are

Table 3. Similarity values, from three different statistical measures of the most different TFs in the compared SCCs according to Jaccard similarity.

Transcription factors	$Jaccard$ (%)	ϕ	CS
CTBP1	0.0	0.56	0.99
ZBTB33	0.22	0.58	0.99
CEBPB	0.37	0.57	0.99
NR2C2	0.41	0.55	0.99
KDM1A	0.52	0.52	0.97
BACH1	0.80	0.17	0.99
BCLAF1	1.19	0.34	0.96

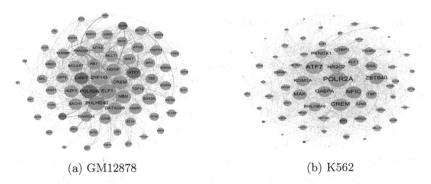

(a) GM12878   (b) K562

Fig. 2. Graphical representation of the compared SCCs. Colors denote the community [26]; node size is according to node degree.

responsible for cancer-driven myelopoiesis, which promotes cancer progression [23,24]. KDM1A plays an important role in hematopoiesis and was identified as a dependency factor in leukemia stem cell populations [25]. BCLAF1 is in the 6q23.3 cytogenetic location, a genomic region that has been reported to exhibit a high frequency of deletions in tumors such as lymphomas and leukemias. The relation of NR2C2 and BACH1 functions to cancer progression remains unclear.

We also performed pathway enrichment analysis of the communities we identified in the SCCs (Fig. 2) using the Louvain algorithm [26]. In the two largest communities in K562, which include 70% of the SCC nodes, the enriched KEGG pathways were the *Homo sapiens p53 signaling pathway* and *chronic myeloid leukemia* pathway. In the largest community of the GM12878 SCC, it was the *MAPK signaling pathway*; according to [27], the activation of this pathway is essential for the antileukemic effects of dasatinib, a target therapy used to treat certain cases of chronic myeloid leukemia.

Finally, using the approach of [18] we tried to identify TFs with significant anomalous behavior in the two SCCs. Using the number of nodes and edges, we were able to detect if the ego-nets of the TFs had a star or connected (complete) shape, i.e., minimal or maximal density. Upper diagrams in Fig. 3 show that, in both cell lines, all TFs created almost complete ego-nets, except CTBP1 that bound only one TF gene. The total weight and the number of edges detected TFs with considerable higher total edge weight compared to the number of edges in their ego-net. As shown in Fig. 3 (lower diagrams), PKNOX1, ZBTB40, NR2C1/2, FOXK2 and BCLAF1 bound with stronger connections other TF genes in both cell lines. Interesting result from this analysis is that the number of nodes and the number of edges of the ego-nets as well as the number of edges and the total weight follow power-law, as we can observe from the linear function in log-log scale.

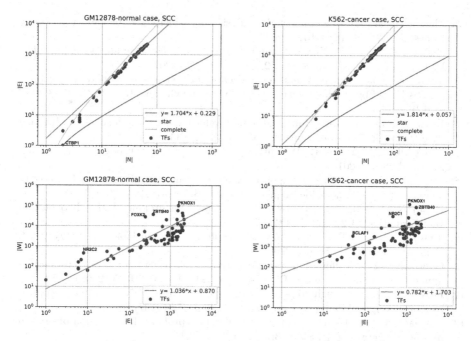

Fig. 3. (a) Ego-net edge count ($|E|$) vs. node count ($|N|$). Red line: linear function fit on median values; blue line: (N-1) function, star graphs, whereas orange line is the N(N-1) function, complete graphs, where n is the number of nodes. (b) Total weight ($|W|$) vs. total count ($|E|$) of edges in the ego-nets for all nodes. (Color figure online)

5   Conclusions

In this manuscript we have shown how to build gene regulatory networks from ChIP-seq data, and how to evaluate them individually or comparatively when built from a normal and a cancer cell line. Through our analysis, we explored the characteristics of the two compared cell lines and identified differences in their transcription factor functions. As a future work, we will explore further the biological meaning of our results trying to evaluate them using gene expression data and we will extend our analysis to more cell lines.

Acknowledgments. This work has been supported by the ERC Advanced Grant 693174 "Data Driven Genomic Computing" (GeCo).

References

1. Rodríguez-Caso, C., et al.: On the basic computational structure of gene regulatory networks. Mol. Biosyst. 5(12), 1617–1629 (2009)
2. Zhang, B., et al.: A general framework for weighted gene co-expression network analysis. Stat. Appl. Genet. Mol. Biol. 4, Article no. 17 (2005)

3. Zhu, X., et al.: Getting connected: analysis and principles of biological networks. Genes Dev. **21**(9), 1010–1024 (2007)
4. Harrow, J., et al.: GENCODE: the reference human genome annotation for The ENCODE Project. Genome Res. **22**(9), 1760–1774 (2012)
5. Wang, J., et al.: Sequence features and chromatin structure around the genomic regions bound by 119 human transcription factors. Genome Res. **22**(9), 1798–1812 (2012)
6. Lozzio, C., Lozzio, B.B.: Human chronic myelogenous leukemia cell-line with positive Philadelphia chromosome. Blood **45**(3), 321–334 (1975)
7. ENCODE Cell Types (2018). https://genome.ucsc.edu/encode/cellTypes.html. Accessed 28 Dec 2018
8. Masseroli, M., et al.: Processing of big heterogeneous genomic datasets for tertiary analysis of Next Generation Sequencing data. Bioinformatics (2018, in press)
9. Przulj, N., et al.: Biological network comparison using graphlet degree distribution. Bioinformatics **23**(2), 177–183 (2007)
10. Yang, Q., Sze, S.: Path matching and graph matching in biological networks. J. Comput. Biol. **14**(1), 56–67 (2007)
11. Choi, J.K., et al.: Differential co-expression analysis using microarray data and its application to human cancer. Bioinformatics **21**(24), 4348–4355 (2005)
12. Fuller, T.F., et al.: Weighted gene co-expression network analysis strategies applied to mouse weight. Mamm. Genome **18**(6–7), 463–472 (2007)
13. Ratmann, O., et al.: From evidence to inference: probing the evolution of protein interaction networks. HFSP J. **3**(5), 290–306 (2009)
14. Phan, H.T., et al.: PINALOG: a novel approach to align protein interaction networks-implications for complex detection and function prediction. Bioinformatics **28**(9), 1239–1245 (2012)
15. Liu, X., et al.: New powerful statistics for alignment-free sequence comparison under a pattern transfer model. Theor. Biol. **284**(1), 106–116 (2011)
16. Waqar, A., et al.: Alignment-free protein interaction network comparison. Bioinformatics **30**(17), 430–437 (2014)
17. Topirceanu, A., et al.: Statistical fidelity: a tool to quantify the similarity between multi-variable entities with application in complex networks. Int. J. Comput. Math. **94**(9), 1787–1805 (2016)
18. Akoglu, L., McGlohon, M., Faloutsos, C.: oddball: Spotting anomalies in weighted graphs. In: Zaki, M.J., Yu, J.X., Ravindran, B., Pudi, V. (eds.) PAKDD 2010. LNCS (LNAI), vol. 6119, pp. 410–421. Springer, Heidelberg (2010). https://doi.org/10.1007/978-3-642-13672-6_40
19. Page, L., et al.: The PageRank citation ranking: bringing order to the web. In: 7th International World Wide Web Conference, pp. 161–172 (1999)
20. Marke, R., et al.: The many faces of IKZF1 in B-cell precursor acute lymphoblastic leukemia. Haematologica **103**, 565–574 (2018)
21. Larsen, S., et al.: The hematopoietic regulator, ELF-1, enhances the transcriptional response to Interferon-β of the OAS1 anti-viral gene. Sci. Rep. **5**, 17497 (2015)
22. Ngondo-Mbongo, R.P., et al.: Modulation of gene expression via overlapping binding sites exerted by ZNF143, Notch1 and THAP11. Nucleic Acids Res. **41**(7), 4000–4014 (2013)
23. Koh, D.I., et al.: KAISO, a critical regulator of p53-mediated transcription of CDKN1A and apoptotic genes. Proc. Natl. Acad. Sci. USA **111**(42), 15078–15083 (2014)
24. Hirai, H., et al.: Non-steady-state hematopoiesis regulated by the C/EBPβ transcription factor. Cancer Sci. **106**(7), 797–802 (2015)

25. McGrath, J.P., et al.: Pharmacological inhibition of the histone lysine demethy-lase KDM1A suppresses the growth of multiple acute myeloid leukemia subtypes. Cancer Res. **76**(7), 1975–1988 (2016)
26. Lambiotte, R., et al.: Fast unfolding of communities in large networks. J. Stat. Mech. **2008**, P10008 (2008)
27. Dumka, D., et al.: Activation of the p38 Map kinase pathway is essential for the antileukemic effects of dasatinib. Leuk. Lymphoma **50**(12), 2017–2029 (2009)
28. Nanni, L., et al.: Exploring genomic datasets: from batch to interactive and back. In: Proceedings ExploreDB, pp. 1–6 (2018)
29. Hagberg, A.A., et al.: Exploring network structure, dynamics, and function using NetworkX. In: Proceedings 7th Python in Science Conference, pp. 11–15 (2008)

Structural Bioinformatics and Function

Function vs. Taxonomy: The Case of Fungi Mitochondria ATP Synthase Genes

Michael Sadovsky[1,2]([✉]), Victory Fedotovskaya[2], Anna Kolesnikova[2,3], Tatiana Shpagina[2], and Yulia Putintseva[2]

[1] Institute of Computational Modelling of SB RAS,
Akademgorodok, 660036 Krasnoyarsk, Russia
`msad@icm.krasn.ru`
[2] Institute of Fundamental Biology and Biotechnology, Siberian Federal University,
Svobodny prosp., 79, 660049 Krasnoyarsk, Russia
`viktoriia.fedotovskaia@gmail.com`, `kolesnikova.denovo@gmail.com`,
`shpagusa@mail.ru`, `yaputintseva@mail.ru`
[3] Laboratory of Genomics and Biotechnology, Federal Research Center RAS,
Krasnoyarsk, Russia
`http://icm.krasn.ru`

Abstract. We studied the relations between triplet composition of the family of mitochondrial *atp*6, *atp*8 and *atp*9 genes, their function, and taxonomy of the bearers. The points in 64-dimensional metric space corresponding to genes have been clustered. It was found the points are separated into three clusters corresponding to those genes. 223 mitochondrial genomes have been enrolled into the database.

Keywords: Order · Clustering · *K*-means · Elastic map · Stability · Evolution

1 Introduction

The problem of the interrelation of structure of nucleotide sequences, functions encoded in them, and taxonomy of their bearers still challenges researchers. A rapid growth of sequenced genetic data supports a progress in this problem. Yet, it is far from a completion, and the basic reason standing behind is the complexity of the phenomenon under consideration. Besides, one should keep in mind that the details of the problem statement may affect seriously both the answer, and the problem itself. In particular, one should define what is function, structure, and taxonomy, to get an exact, unambiguous and comprehensive answer on the question.

Here we try to reveal the interrelation and contribution of each entity, i.e. *structure*, *function* and *taxonomy* into their interplay and phenomenae observed in nature. Evidently, the answer depends strongly on the exact notion of what *structure* is, first of all. Luckily, the notion of a *function* is significantly less arguable, as well as the notion of *taxonomy*. The point is that the diversity and

I. Rojas et al. (Eds.): IWBBIO 2019, LNBI 11465, pp. 335–345, 2019.
https://doi.org/10.1007/978-3-030-17938-0_30

abundance of structure identified in nucleotide sequences is great enough (see, e.g. [1–6]), and those structure are quite different and may not be reduce one to another.

Everywhere below, **structure** is a frequency dictionary of triplets developed over some nucleotide sequence. We shall consider two types of triplet dictionaries, to be exact; they differ in the reading frame shift t. A triplet frequency dictionary W_3 is the set of all triplets $\omega_1 = \mathsf{AAA}$ to $\omega_{64} = \mathsf{TTT}$ together with their frequency

$$f_\omega = \frac{n_\omega}{N}. \tag{1}$$

Here n_ω is the number of specific triplet ω observed over a sequence, and the reading frame (of the length 3) moves along a sequence with the step $t = 1$. Triplet frequency $\overline{W}_3$ is developed in the way similar to W_3, but for $t = 3$. The definition (1) must be changed then for

$$f_\omega = \frac{n_\omega}{M}, \tag{2}$$

where M is the total number of triplets counted within a sequence; obviously, M is three times less than N, for $\overline{W}_3$. Such frequency dictionaries have been used to reveal the relation between structure and taxonomy, see [7–9] for details. Further, we stipulate that there are no other symbols in genetic matter, but $\aleph = \{\mathsf{A}, \mathsf{C}, \mathsf{G}, \mathsf{T}\}$.

We studied 223 mitochondrial genomes of five fungal division: *Basidiomycota* (24 entries), *Ascomycota* (185 entries), *Blastocladiomycota* (2 entries), *Chytridiomycota* (6 entries) and *Zygomycota* (6 entries) were downloaded from NCBI GenBank. To reveal the interplay between all three issues mentioned above, we used the genes *atp6*, *atp8* and *atp9* belonging to ATP synthase genes family. The primary function of mitochondria is a production of energy via oxidative phosphorylation. In general, they encode 14 conserved protein-coding electron transport and respiratory chain complexes genes (*atp6*, *atp8*, *atp9*, *cob*, *cox1*, *cox2*, *cox3*, *nad1*, *nad2*, *nad3*, *nad4*, *nad4L* and *nad6*) and have no difference in function [10–12]. Using CLC Genomic Workbench v.10 we retrieved the annotations and the sequences of three standard mitochondrial protein encoding genes involved into the oxidative phosphorylation (these are *atp6*, *atp8*, *atp9*). Next, the sequence for each gene has been prepared in two versions:

(1) *gene* is a sequence containing exons and introns as it is presented in a genome, and
(2) *CDS (coding DNA sequence)* is a sequence free from introns, in fact, it corresponds to a mature RNA ready for protein translation.

Besides, ATP synthase genes are quite often used for phylogeny implementation [13–15].

As soon, as all the genes are isolated (in two versions each), the sequences have been transformed into the frequency dictionary W_3 or $\overline{W}_3$, respectively, with *ad hoc* software. Next, due to *VidaExpert*[1] freeware the distribution of

---

[1] http://bioinfo-out.curie.fr/projects/vidaexpert/.

the points corresponding to genetic entities was analyzed. Transformation of sequences into frequency dictionaries allows to implement powerful and efficient tools of up-to-date statistical analysis and multidimensional data visualization.

1.1 Clustering Techniques

Clustering is the key tool of this research; we have used K-means and elastic map technique. K-means is well known and exhaustively described method of clustering, hence we shall not describe the method here in detail (see [16] for more details).

Also, the elastic map technique has been used to cluster and analyze data distribution, in triplets frequency space. Since this method is quite new, we describe it here in few details. To start, one must find out the first and the second principal components, and develop a plane over them (as on axes); next, each data point must be projected on this plane. Secondly, each data point must be connected to its projection with a mathematical spring. That latter has infinite expansibility and the elasticity coefficient remains permanent, for any expansion. Thirdly, figure out the minimal square comprising all the projections, and change it with the elastic membrane. That latter is supposed to be homogeneous, so that it may bend and expand. Next, release the system to reach the minimum of the total deformation energy. The elastic membrane would transform into a jammed surface, and this is the two-dimensional manifold approximating the data set. Fourthly, redefine each point on the jammed surface through the orthogonal projection. Finally, cut-off all the springs, so that the jammed surface comes back to a plane. That is the elastic map representing the cluster structuredness, if any, in the data set [17–19].

To identify clusters, we used the local density of points. That latter is defined as following. Supply each point of an elastic map (in so called inner coordinates, when the jammed surface is already flattened) with a bell-shaped function, e.g.

$$f(r) = \mathcal{A} \cdot \exp\left\{\frac{(r - r_j)^2}{\sigma^2}\right\}. \tag{3}$$

Here r_j is the coordinate vector of j-th point, and σ is an adjusting parameter (that is a specific width of the bell-shaped function). Then the sum function

$$F(r) = \mathcal{A} \cdot \sum_{j=1}^{N} \exp\left\{\frac{(r - r_j)^2}{\sigma^2}\right\}, \tag{4}$$

is calculated; the function $F(r)$ is then shown in elastic map.

2 Genes Distribution

We start from the clustering obtained due to elastic map technique and then consider the structuredness provided by K-means classification.

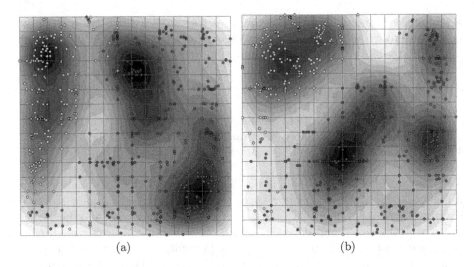

(a)   (b)

Fig. 1. Distribution of *atp* genes over elastic map. *atp*6 is in red, *atp*8 is in green and *atp*9 is in yellow; left is for W_3 and right is for $\overline{W}_3$ dictionaries. (Color figure online)

2.1 Elastic Map Clustering

Figure 1 shows the distribution of the genes on the soft elastic map (with 16 × 16 grid) and elasticity coefficients defined by default. Also, this figure shows the local density function, in grey scale. Figure 1(a) shows the distribution of genes in triplet frequency space obtained for W_3 dictionaries (developed for gene sequences with introns and exons). Evidently, there are three distinct clusters in this figure. Surprisingly, the clusters are gene specific, with a high accuracy: the left cluster gathers mainly *atp*9 genes (shown in yellow), the right cluster gathers mainly *atp*6 genes (shown in red) and the central one gathers mainly *atp*8 genes (shown in green). Or course, there are some "escapees": the genes that occupy an opportunistic cluster. The key point here is that this distribution is obtained for gene sequence (i.e. those with introns), and the reading frame shift $t = 1$.

Figure 1(b) shows similar distribution obtained for $\overline{W}_3$ dictionaries, with $t = 3$. In fact, these transformation into a triplet frequency dictionary completely corresponds to protein translation. There is no surprise in improved clustering observed for these dictionaries: the number of "escapees" goes down here.

The cluster structure shown in Fig. 1 is doubtless. Local density visualization technique makes it unambiguous. The clusters (in inner coordinates) are obviously isolated from each other. Coloring used to identify the peculiar gene type also unambiguously proves very high coherence of a cluster identified through triplet frequencies, and the gene type occupation. Of course, there are very few exceptions in the occupation: some genes join an opportunistic cluster. Nonetheless, the greatest majority of specific genes (say, *atp*6) tend to occupy the cluster that is identified through the triplet statistics, not with a functional role of a gene.

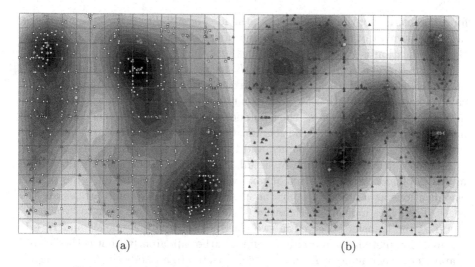

Fig. 2. Distribution of species over elastic map. *Candida* spp. are in red, *Saccharomyces* spp. are in blue and *Fusarium* spp. are in green; left is for W_3 and right is for $\overline{W}_3$ dictionaries. (Color figure online)

Originally, the basic goal of our paper is to compare the impact of each entity from a triad *structure – function – taxonomy* on their common interplay pattern. To reveal the impact of each entity into this interplay, we checked the distribution of species in the patterns obtained through the clustering of frequency dictionaries W_3 and $\overline{W}_3$, respectively. Figure 2 shows such distribution, for three most abundant genera of fungi: *Candida* spp., *Fusarium* spp. and *Saccharomyces* spp. Again, Fig. 2(a) shows the distribution of W_3 triplet frequency dictionaries, and Fig. 2(b) shows the distribution of $\overline{W}_3$. Evidently, these three most abundant species are spread among the clusters rather equally; one may expect that other species are spread in similar manner, with obvious constraint coming from the finite (and small) number of some species comprising a genus.

2.2 K-means and Structure-Function Interplay

In Sect. 2.1, a direct evidence of the prevalence of function over the taxonomy is shown, for ATP synthase genes family of fungi mitochondria. Here we consider and analyze the structuredness obtained in the set of point corresponding to triplet frequency dictionaries due to K-means.

For each database (i.e. that one with W_3 dictionaries, and that one with $\overline{W}_3$ dictionaries) a classification through K-means has been developed; we implemented the classification for $K = 2$, $K = 3$, $K = 4$ and $K = 5$. Two issues must be kept in mind here: the former is stability of classification, and the latter is separability of classes. The first problem is immanent for K-means. Since a classification starts from a random allocation of the points into K classes, then there is no guarantee of the identity of the final configuration: it might change, for different runs of the procedure. So, the idea of stability is to check whether a desirable number of runs converge to the same configuration, or not. If yes, then the classification is stable, otherwise it is unstable.

Unlike the elastic map technique, K-means does not yield the "natural" number of classes[2]; a researcher has to fix it at his own. On the other hand, one can develop a classification for various number of classes, and trace the transfer of elements of the classes, as they number grow up. This transfer is of special interest. We have developed the classifications for $2 \leq K \leq 5$ with K-means, for two issues: the former is taxonomy, and the latter is function.

Let us consider this point in detail. First, we consider the results of K-means implementation in terms of taxonomy. Figure 3 shows this series of four classifications as a layered graph; the classes are the nodes, and arrows are the edges. The edges indicate the transfer of the elements from a class to "younger" one (i.e., the transfer observed in two classifications with K-means and $K + 1$-means). Complete layered graph is defined rather apparently: that is the layered graph where each node in K-th layer is connected to all nodes in $K + 1$-th layer. In such capacity, the graph shown in Fig. 3 is almost complete: it has 25 edges, while the complete one must have 38 ones. In other words, the graph shown in this figure is far from a tree.

Figure 4 shows the graph observed for genes distribution. At the first glance, it looks pretty similar to that one shown in Fig. 3: it also has 22 edges (cf. to 28 in the graph shown in Fig. 3). Basic difference of that former consists in the abundances of objects (genes, in our case) and their preferences when transferred from node to node. This point is outlined with bold colored arrows connecting the specific nodes that comprise the genes with high predominance. The subgraph comprising the nodes and edges with high predominance of the genes makes a tree.

The composition of species in the graph shown in Fig. 3 is rather uniform: one can found any family in each node of the graph; in other words, the species tend to distribute themselves over the classes almost equally, so that no order or structuredness might be found. The pattern shown in Fig. 4 is drastically different; first of all, there are two isolated leaves in the graph. These are the classes with unique incident edge each, observed for $K = 5$; the former comprises $atp8$ genes (169 entries), and the latter comprises $atp9$ genes (86 entries). It should be stressed that $atp8$ genes differ, to some extent, from other ones, in this pattern: they are always comprised into a single cluster, for all $2 \leq K \leq 5$. The clusters enlisting $atp8$ genes also contain $atp6$ genes, for $2 \leq K \leq 4$. Evidently, the genes $atp6$ exhibits quite similar behaviour to $atp8$ genes: they tend to occupy the same cluster and split into two separated clusters only for $K = 5$, when $atp8$ comprise the isolated leaf in the graph, and $atp6$ comprise the cluster slightly deteriorated with other genes (14 entries of $atp8$ and 5 entries of $atp9$).

The family of $atp9$ genes is the only one tending to occupy a separate cluster, regardless the clustering technique used to identify the clusters. Such solidity

---

[2] An advanced version of K-means yields the maximal number of distinguishable classes, see Sect. 3.

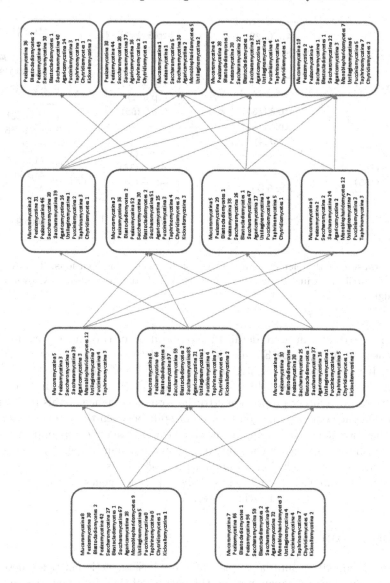

Fig. 3. Transfer of the elements in a series of K-means classifications, $2 \leq K \leq 5$, for species distribution under classification with W_3 dictionaries.

remains even for a splitting of a cluster, as K exceeds 3. Indeed, a single cluster comprising[3] the genes of $atp9$ family splits into two clusters of approximately equal abundance, for $K = 4$ and $K = 5$. Still, there is no significant presence of the genes of other families in these two individual clusters.

---

[3] With respect to a minor deterioration by other genes.

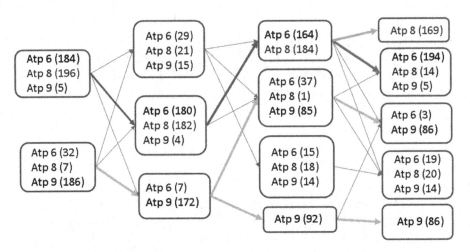

Fig. 4. Transfer of the elements in a series of K-means classifications, $2 \leq K \leq 5$, for genes distribution under classification with W_3 dictionaries developed over CDS. (Color figure online)

Let now consider in more detail the composition of two clusters comprising *atp*9 genes, as $K = 4$ (see Fig. 4). We examined the composition of these two clusters. It was found that all five divisions are splitted out between these two clusters almost homogeneously: the abundances of each specific division in a cluster is approximately proportional to the total abundance of this division in dataset. On the contrary, the genera are not split between these two clusters: it means that two genera belonging to the same division may occupy opportunistic clusters while the species belonging to a genus are mainly found in the same cluster, with a single exclusion. *Candida santjacobensis* is the only species found in the cluster comprising *atp*9 genes, only. All other genes of this genus (32 species) occupy the opportunistic cluster.

3 Discussion

Here we examined the mutual impact of three basic genetic entities (these are *structure*, *function* and *taxonomy*) on the pattern of their interplay. To do that, we created a database comprising the ATP synthase genes of fungal mitochondria, namely, *atp*6 genes. The genes were then converted into triplet frequency dictionaries, so that each gene is now represented as a point in 64-dimensional Euclidean space where the triplet frequencies are the coordinates. Then we checked whether an inner structuredness could be found in the set of such points, and the answer was positive. We have found that there exist three clusters identified with non-linear statistics (called elastic map technique); besides, other type of structuredness has been found through the implementation of linear classification technique (K-means).

At the next step, we examined all the clusters in terms of

(i) species composition, and

(ii) genes composition of each cluster.

The composition of the clusters has been checked regardless the identification technique used to figure them out.

Strong prevalence of gene (i.e. structure) in the cluster formation has been found, for all the clusters developed due to various clustering techniques; see Figs. 1 through 4. Such predominance is not self-evident, in advance. For example, paper [7] unambiguously proves the strong prevalence of taxonomy over the function, when studied over the entire mitochondrial genomes. One may expect that genes are stronger that taxonomy, while it is not evident in advance, for sure. The predominance of genes impact proves the superiority of function over taxonomy, in pattern formation within the triad *structure – function – taxonomy*. Nonetheless, this is not an ultimate proof; there are few questions to be answered to get a final evidence of the predominance mentioned above; let them list here:

(i) class (or clusters) distinguishability,

(ii) implementation of other metrics that Euclidean one,

(iii) stability of classification obtained with K-means, and

(iv) indexing of a database used to reveal the interplay, in terms of various taxa occurrence.

All these questions are rather technical than essential. Meanwhile, there are some more questions with hard biology standing behind.

3.1 CDS, $\overline{W}_3$ and Dimension Reduction

Previously, we presented structuredness observed in fungi mitochondrial ATP synthase genes through K-means implementation to classify triplet frequency dictionaries $\overline{W}_3$ developed over CDS of those genes. In fact, CDS is equivalent to mature RNA ready for translation; reciprocally, $\overline{W}_3$ frequency dictionary is the dictionary of the codons, not just common triplets, i.e. it contains the triplet occupying the positions corresponding to the reading frame at the translation process.

This fact allows to classify or cluster the genes in other space, with less dimension. Indeed, one can easily change the codon frequencies into the frequencies of corresponding amino acid residues. Since $\overline{W}_3$ comprises the codons, not the triplets, then the frequency of an amino acid residue is just the sum of the frequencies of all synonymous codons. This apparent and clear transformation results in the change of 64-dimensional Euclidean space for 21-dimensional one, where the frequencies of amino acid residues (plus *Stop* signal) are the coordinates.

3.2 Gene Family Selection

We have carried out the study of the mutual interplay of taxonomy, function and structure on the basis of ATP synthase genes of fungi mitochondria. Meanwhile, it may take sense to extend the set of genes incorporated into a study: in particular, the oxidative phosphorylation involves the proteins encoded with some other genes, but *atp* family. Thus, an inclusion of the genes (*nad1*, *nad2*, *nad3*, *nad4*, *nad4L*, *nad5*, *nad6*, *cob*, *cox1*, *cox2*, *cox3* and *cob*) both all together in isolated groups may bring a lot of new knowledge towards the relation within the triad *gene structure, function* and *taxonomy*.

3.3   Genome Selection

Similar reasoning as that one discussed in the above subsection addresses the choice of genomes to be considered for the analysis of the interplay in triad *structure – function – taxonomy*. There is no guarantee that the pattern with high prevalence of structure over taxonomy is observed always, regardless a genetic matter taken into consideration. Obviously, mitochondrion genomes are very good object for such kind of study: they are extremely homogeneous in the function encoded in it, the have a single chromosome, and are very well studied. Anyway, a universality of the observation done over these genomes must be verified through the examination of other genomes. Chloroplasts seem to be the second to none, in such capacity, for the same reasons: a single chromosome, perfect conservation of functions, good quality of sequencing and annotation.

Yet, a study of organella genomes may not be an ultimate proof of the pattern presented above. Some other genetic system must be involved into consideration, to approve it. All these issues fall beyond the scope of this paper and should be done in due time.

Acknowledgement. We are thankful to Reviewer whose remarks made the paper apparently better.

References

1. Molla, M., Delcher, A., Sunyaev, S., Cantor, C., Kasif, S.: Triplet repeat length bias and variation in the human transcriptome. Proc. Nat. Acad. Sci. **106**(40), 17095–17100 (2009)
2. Provata, A., Nicolis, C., Nicolis, G.: DNA viewed as an out-of-equilibrium structure. Phys. Rev. E **89**, 052105 (2014)
3. Qin, L., et al.: Survey and analysis of simple sequence repeats (SSRs) present in the genomes of plant viroids. FEBS Open Bio **4**(1), 185–189 (2014)
4. Moghaddasi, H., Khalifeh, K., Darooneh, A.H.: Distinguishing functional DNA words; a method for measuring clustering levels. Sci. Rep. **7**, 41543 (2017)
5. Bank, C., Hietpas, R.T., Jensen, J.D., Bolon, D.N.: A systematic survey of an intragenic epistatic landscape. Mol. Biol. Evol. **32**(1), 229–238 (2015)
6. Albrecht-Buehler, G.: Fractal genome sequences. Gene **498**(1), 20–27 (2012)
7. Sadovsky, M., Putintseva, Y., Chernyshova, A., Fedotova, V.: Genome structure of organelles strongly relates to taxonomy of bearers. In: Ortuño, F., Rojas, I. (eds.) IWBBIO 2015. LNCS, vol. 9043, pp. 481–490. Springer, Cham (2015). https://doi.org/10.1007/978-3-319-16483-0_47

8. Sadovsky, M., Putintseva, Y., Birukov, V., Novikova, S., Krutovsky, K.: *De Novo* assembly and cluster analysis of Siberian Larch transcriptome and genome. In: Ortuño, F., Rojas, I. (eds.) IWBBIO 2016. LNCS, vol. 9656, pp. 455–464. Springer, Cham (2016). https://doi.org/10.1007/978-3-319-31744-1_41

9. Sadovsky, M., Putintseva, Y., Zajtseva, N.: System biology of mitochondrion genomes. In: Qian, P.Y., Nghiem, S.V. (eds.) The Third International Conference on Bioinformatics, Biocomputational Systems and Biotechnologies, pp. 61–66, Venice/Mestre (2011)

10. Basse, C.W.: Mitochondrial inheritance in fungi. Curr. Opin. Microbiol. **13**(6), 712–719 (2010)

11. Gray, M.W., Burger, G., Lang, B.F.: Mitochondrial evolution. Science **283**(5407), 1476–1481 (1999)

12. Bullerwell, C.E., Lang, B.F.: Fungal evolution: the case of the vanishing mitochondrion. Curr. Opin. Microbiol. **8**(4), 362–369 (2005)

13. Esser, C., et al.: A genome phylogeny for mitochondria among α-proteobacteria and a predominantly eubacterial ancestry of yeast nuclear genes. Mol. Biol. Evol. **21**(9), 1643–1660 (2004)

14. Davoodian, N., et al.: A global view of *Gyroporus*: molecular phylogenetics, diversity patterns, and new species. Mycologia **110**, 985–995 (2018)

15. Nadimi, M., Daubois, L., Hijri, M.: Mitochondrial comparative genomics and phylogenetic signal assessment of mtDNA among arbuscular mycorrhizal fungi. Mol. Phylogenet. Evol. **98**, 74–83 (2016)

16. Fukunaga, K.: Introduction to Statistical Pattern Recognition. Academic Press, London (1990)

17. Gorban, A.N., Zinovyev, A.Y.: Fast and user-friendly non-linear principal manifold learning by method of elastic maps. In: 2015 IEEE International Conference on Data Science and Advanced Analytics, DSAA 2015, Campus des Cordeliers, Paris, France, 19–21 October 2015, pp. 1–9 (2015)

18. Gorban, A.N., Zinovyev, A.: Principal manifolds and graphs in practice: from molecular biology to dynamical systems. Int. J. Neural Syst. **20**(03), 219–232 (2010). PMID: 20556849

19. Gorban, A.N., Zinovyev, A.Y.: Principal manifolds for data visualisation and dimension reduction. In: Gorban, A.N., Kégl, B., Wünsch, D., Zinovyev, A.Y. (eds.) LNCSE, vol. 58, 2nd edn, pp. 153–176. Springer, Berlin, Heidelberg, New York (2007). https://doi.org/10.1007/978-3-540-73750-6

Non-Coding Regions of Chloroplast Genomes Exhibit a Structuredness of Five Types

Michael Sadovsky[1,2](✉), Maria Senashova[1], Inna Gorban[2], and Vladimir Gustov[2]

[1] Institute of Computational Modelling of SB RAS,
Akademgorodok 660036, Krasnoyarsk, Russia
{msad,msen}@icm.krasn.ru
[2] Institute of Fundamental Biology and Biotechnology,
Siberian Federal University,
Svobodny prosp., 79, 660049 Krasnoyarsk, Russia
inn.gorban@gmail.com, v.gustows@mail.ru
http://icm.krasn.ru

Abstract. We studied the statistical properties of non-coding regions of chloroplast genomes of 391 plants. To do that, each non-coding region has been tiled with a set of overlapping fragments of the same length, and those fragments were transformed into triplet frequency dictionaries. The dictionaries were clustered in 64-dimensional Euclidean space. Five types of the distributions were identified: ball, ball with tail, ball with two tails, lens with tail, and lens with two tails. Besides, the multi-genome distribution has been studied: there are ten species performing an isolated and distant cluster; surprisingly, there is no immediate and simple relation in taxonomy composition of these clusters.

Keywords: Order · Probability · Triplet · Symmetry · Projection · Clustering

1 Introduction

Non-coding regions in DNA sequences have been supposed to be a kind of an evolutionary junk; currently, it is a well knows fact that such regions play essential role in gene regulation, and in the genetic information processing, in general [1–6]. The role of non-coding regions is not absolutely clear yet, and a lot could be found behind them. The non-coding regions are found elsewhere, in a genome of any taxonomy level, including organelle genomes. Here we studied the non-coding regions of chloroplast genomes, following the way present in [7–12].

Previously, a seven-cluster pattern claiming to be a universal one in bacterial genomes has been reported and very elegant theory explaining the observed patterns was proposed [7,8,11]. Later, we have expanded the approach for chloroplast genomes [12,13]. Here se present some preliminary results of a study of

© Springer Nature Switzerland AG 2019
I. Rojas et al. (Eds.): IWBBIO 2019, LNBI 11465, pp. 346–355, 2019.
https://doi.org/10.1007/978-3-030-17938-0_31

statistical properties of non-coding regions of chloroplast genomes carried out under the methodology described above [11–13].

In papers [7–12] the difference in triplet composition determined for coding and non-coding regions has been established. Let now introduce more exact definitions and notions for further analysis. Consider a symbol sequence $\mathfrak{T}$ from four letter alphabet $\aleph = \{A, C, G, T\}$ corresponding to a (chloroplast) genome stipulating that $\mathfrak{T}$ has no other symbols but those indicated above. The sequences have been downloaded from NCBI bank (391 entities). Each sequence has been tiled with the set of intersecting fragments of the length L; the fragments located in a sequence with the step t. Next, for each genome every fragments were transformed into a triplet frequency dictionary W_3 (see Sect. 2). The transformation changed a fragment with a point in 64-dimensional Euclidean space, and the cluster structuredness of the points has been revealed and studied.

We aimed to check whether the fragments of each specific genome form a pattern where each separate genome is clustered more or less separately. Speaking in advance, the hypothesis both holds true, and it does not. More specifically, the triplet frequency dictionaries may not be separated by various clustering techniques; on the other hand, labeling each fragment with species reveals a non-random distribution of the points in Euclidean space. Moreover, an individual distribution of the fragments in the space reveals five types of the distribution. A study of a common distribution exhibits extremely unusual behaviour of ten genomes that form a kind of clearly and evidently separated dense cluster located very far from the main body of the points of other genomes.

2   Frequency Dictionaries

391 chloroplast genomes have been retrieved from NCBI bank. Each genome has been tiled with a set of (intersecting) fragments of the length $L = 603$ symbols; the fragments moved along a sequence with the step $t = 11$. It should be noticed that the length L is divisible by 3, but the step t is not; this choice of the parameters of tiling is not accidental. The idea standing behind this pattern of the tiling is described in detail in [8, 11–13].

Next, each fragment was marked with the number of central nucleotide of that former. Following the annotation of a genome, we selected the fragments completely falling into non-coding regions. No overlaps to a coding region has been permitted. Then each fragment has been transformed into a triplet frequency dictionary. Formally, a triplet frequency dictionary W_3 could be defined ambiguously, in dependence on the reading frame shift. Indeed, let $\omega = \nu_1\nu_2\nu_3$ be a triplet, i.e. three symbols in $\mathfrak{T}$ standing next each other. Locate the frame identifying a triplet at the very beginning of $\mathfrak{T}$; move then the frame along $\mathfrak{T}$ with the step t and count all the triplets occurred within $\mathfrak{T}$. Counting the number of copies n_ω of each triplet ω, one gets the finite dictionary $W_{(3,t)}$. Changing then the number of copies for their frequency

$$f_\omega = \frac{n_\omega}{N}, \qquad \text{where} \qquad N = \sum_{\omega=\text{AAA}}^{\text{TTT}} n_\omega, \qquad (1)$$

one gets the frequency dictionary $W_{(3,t)}$. Obviously, one may use a frequency dictionary determined for an arbitrary t; we shall use the frequency dictionaries $W_{(3,3)}$ type.

2.1 Clustering

We used the freely distributed software *VidaExpert*[1] to analyze and visualize the distribution of the non-coding regions of genomes, both individually, and in a group. To do that, an ensemble of the fragments covering the non-coding regions corresponding to a genome has been arranged into a data base, and the distribution of the triplet frequency dictionaries has been studied, in the space of principal components of the ensemble. Also, a set of ensembles was arranged into a joint data base, with the same analysis technique applied for visualization.

3   Results

We examined 391 chloroplast genomes trying to identify a pattern of the tripe frequency dictionaries distribution, in the principal components space. Here present some preliminary results towards the patterns yielded by non-coding regions

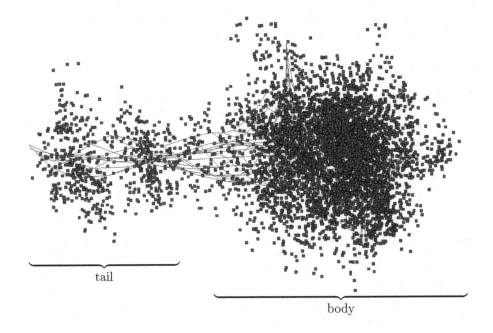

Fig. 1. Barley *Hordeum vulgare* subsp. *Spontaneum* chloroplast genome fragments distribution.

---

[1] http://bioinfo-out.curie.fr/projects/vidaexpert/.

of chloroplast genomes, in the triplet frequency space. Subsection. 3.1 presents the results concerning the shape of the distribution observed over individual genomes, and Subsect. 3.2 presents similar results on the pattern observed for a mutual distribution of many genomes.

Let us also explain the terms *profile* and *above* used below to identify various projections. All figures provided below show the points distribution; that latter is a two-dimensional projection of a three-dimensional projection from 64-dimensional Euclidean space of triplet frequencies. All the figures present the distributions in three principal components (corresponding to the greatest, next and the third eigenvalue of the covariance matrix). *Profile* view means that the first principle components is located in the plane of a figure and directed from left to right; the second principal component here is also located in the plane, and directed from bottom to up. For *above* view the first principal components is located in the same way, but the second one orthogonal to the figure plane so that is looks out from the figure plane.

(a)   (b)

Fig. 2. *Ricinus communis* chloroplast genome exhibits the lens with a tail structure. Left is *profile* view, and right is *above* view.

3.1   Individual Genome Clustering

Figure 1 shows a pattern to explain some terms used below. We classify the patterns in terms of *body* and *tail*: the patterns differ in the number of tails, and in the shape of a body. An examination of 391 genomes yielded five classes. These classes are:

(1) *Ball.* This is the pattern exhibiting no peculiar structuredness, the genome of *Erodium chrysanthum* is the typical representor (see Fig. 3(a)). This pattern differs from other ones due to a similitude of the distribution seen in various projections: any projection yields a ball. There are 7 genomes exhibiting this pattern.

(2) *Ball and tail.* This is the pattern where the main body (*ball* is supplied with a clearly detectable other cluster (called *tail*); see Fig. 1 for details. The most surprising thing is that this tail looks like a (quite thick) ring, or torus; this is very unusual pattern, so the feasibility of minimum approximating

manifold must be provided properly [14]. There are 209 genomes exhibiting this pattern.

(3) *Ball and two tails.* This is the pattern resembling the previous one, while tail comprises two rings, not a single one. In such capacity, it might be called "scissors". There are 49 genomes exhibiting this pattern.

(4) *Lens and tail.* This pattern looks like a ball with tail (see Fig. 2(a)), in one projection, but in contrary to that former, it looks like a lens, or a ball segment, in other projection (see Fig. 2(b)). There are 45 genomes exhibiting this pattern.

(5) *Lens and two tails.* This pattern is similar to previous one, while it exhibits two tails, not a single one. There are 81 genomes exhibiting this pattern; see Fig. 4 for details.

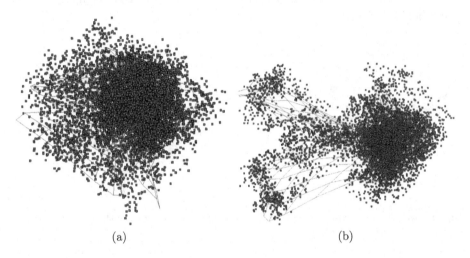

(a)   (b)

Fig. 3. *Erodium chrysanthum* chloroplast genome exhibits a ball-shaped structure (left), and *Liriodendron tulipifera* chloroplast genome exhibits a structure of ball with two tails (right).

Figures 1, 2 and 3 show all the structures observed in the family of 391 chloroplast genomes. The first question here arises whether those structures correlate to taxonomy of the genomes, or not. It should be noticed that the number of genomes exhibiting peculiar structure differs quite strongly, see the list of the structure above. In this Table, T means the total number of species in a division, L_1 (L_2, respectively) are the numbers of species within a division with *lens with tail* (*lens with two tails*, respectively) structures, B is the number

Table 1. Divisions distribution over the structure types; see text for details.

Division	T	L_1	L_2	B	B_2	B_1
Anthocerotophyta	1	0	0	0	1	0
Bryophyta	2	0	0	0	0	2
Marchantiophyta	3	0	0	0	3	0
Tracheophyta	385	45	81	7	45	207
Total	391	45	81	7	49	209

of species with *ball* structure, and B_2 (B_1, respectively) is the number of species within a division with *ball with tail* (*ball with two tails*, respectively) structure.

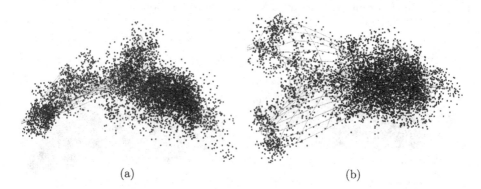

(a)   (b)

Fig. 4. *Lupinus luteus* chloroplast genome exhibits the lens with two tails structure. Left is *profile* view, and right is *above* view.

Table 1 shows the distribution of taxa at the division level. It should be said that the taxonomy composition of the divisions is quite biased: there are 6 or less species in three divisions; one hardly may expect to retrieve the taxonomy relation to a structure type over these division, due to a finite sampling effect. For *Tracheophyta* division is rather abundant and the distribution looks very far from a uniform one; besides, no other simple random distribution law might be fitted with these data (see Table 1).

3.2   Intergenomic Clustering

Previously, wonderful structuredness in bacterial genomes [7,8,11] has been reported. The structuredness manifests in clustering of considerable short fragments of a genome converted into triplet frequency arranged in seven clustering pattern, where six clusters represent coding regions of a genome, with respect to a reading frame shift, and the seventh one gathers fragments from non-coding regions. Later, this approach has been applied to a study of chloroplast genomes [12,13] and similar multi-cluster pattern has been found. The difference between bacteria and chloroplasts consists in different number of clusters observed in a pattern: bacteria genomes yield seven clusters, as maximum, while chloroplast ones yield up to eight clusters.

The structures mentioned above comprise the fragments identified both for coding and non-coding regions. In such capacity, the question arises whether one can reveal a relation between triplet composition, and taxonomy (for instance) of the genome bearers, in case of the comparison of a sufficiently abundant ensemble of genomes. Both for chloroplasts [10], and bacteria [9] the answer is positive: taxonomy may be traced in the system of clusters developed through K-means or other clustering techniques. The success of those researches has been provided mainly by implementation of the entire genome into consideration, namely, coding and non-coding regions. So, the question arises whether similar relation between structure (namely, triplet composition) and taxonomy of the bearers, if non-coding regions are taken into consideration, only.

(a)   (b)

Fig. 5. Simultaneous distribution of triplet dictionaries of non-coding regions of chloroplast genomes of five species.

Here we answer this question: yes, there is relation between taxonomy and triplet composition of the genome part comprising non-coding regions, solely. Figure 5 shows the simultaneous distribution of the fragments of non-coding regions converted into triplet frequency dictionaries of several species; to do it, we merged several data bases developed for individual genomes, into a single one and analyzed it. Different colors label different species; the cloud of the point belonging to the same species tend to form quite dense cluster, while these latter may not be separated with any unsupervised clustering technique. Figure 5(a) shows the view from above, and Fig. 5(b) shows the profile view of the distribution.

(a)   (b)

Fig. 6. Ten genomes forming a distinct and outlying clusters; Fig. 6(a) shows the profile view, and Fig. 6(b) shows from bottom view.

3.3   Nine Mysterious Genomes

Nine genomes exhibit mysterious clustering behaviour: these are *Psilotum nudum*, AC AP004638, *Oryza sativa* Indica Group, AC AY522329, *Oryza sativa* Japonica Group, AC AY522330, *Panax ginseng*, AC AY582139, *Huperzia lucidula*, AC AY660566, *Helianthus annuus*, AC DQ383815, *Jasminum nudiflorum*, AC DQ673255, *Piper cenocladum*, AC DQ887677, *Pelargonium × hortorum*, AC DQ897681. These genomes form the distinct, apparent and clearly identified cluster that is located unexpectedly far from the main body formed by the other genomes. Figure 6 shows this clustering pattern. We have examined the behaviour of all these ten genomes, both separately and individually. It means that we checked the clustering structure formed by those genomes when combined with various number of other "normal" genomes. "Normal" genomes form separately the cluster looking rather uniformly, from outer point of view. Those ten "escapees" also form the cluster that looks very uniformly from outer point of view. Meanwhile, together they exhibit the pattern where two clusters are evidently split and isolated one from other.

It should be said that the set of "normal" genomes is quite abundant: it comprises 381 genomes. Thus, we checked the separate cluster occurrence, for various less abundant subsets of "normal" genomes comprising up to 20 genomes and "escapees". It has been found that the "escapees" form the separated cluster in any combination of these latter, when compared to "normal" genomes.

4   Discussion

In papers [7,8,11] an approach to reveal a structuredness in bacterial genomes based on the comparison of frequency dictionaries $W_{(3,3)}$ of the fragments of a genome is presented; our results show that chloroplasts behave in other way. The always cluster in two coinciding triangles. The vertices of that latter correspond to phases of a reading frame shift and comprise the fragments with identical reading frame shift figure (reminder value). Moreover, unlike in [7,8,11], the chloroplast genomes exhibit a mirror symmetry.

Another important issue is that GC-content does not determine the positioning of the clusters, unlike for bacterial genomes. The pattern observed for bacterial genomes (triangle vs. hexagon) with central body comprising the non-coding regions of a genome is determined by GC-content. Both for bacteria [7,8,11] and chloroplasts [12,13], the fragments corresponding to non-coding regions of a genome always occupy the central part of a pattern; thus, the question arises towards a fine structure of those non-coding regions expressed in terms of statistical properties (and clustering) of the fragments falling purely into the non-coding regions. Here we present some preliminary results answering this question.

We analyzed non-coding regions separately from coding ones. First of all, the structuredness observed in non-coding regions differs significantly from that one observed over the whole genome. The patterns yielded by non-coding regions are more diffusive, in comparison to those observed for whole genome. Probably, the key difference consists in the lack of discernibility of the fragments belonging to different species with unsupervised statistically based clustering technique. An inverse holds true: tracing the fragments belonging to the same species, one may see they comprise a dense and apparent cluster, if the distribution of the fragments belonging to different species is developed simultaneously.

Nonetheless, for each individual species the distribution of the fragments yields a specific pattern. We have identified five types of the distribution: *ball*, *ball with tail*, *ball with two tails*, *lens with tail* and *lens with two tails*. The structure called *ball with two tails* is the most surprising one: it ay not be approximated with good accuracy with a two-dimensional manifold of genus 0 (say, with a part of a plane, or hemisphere). On the contrary, the best starting manifold to approximate the pattern is a two-dimensional manifold of genus 2, i.e. a square with two holes in it.

Thus, we have proven an existence of a structuredness in the non-coding regions of chloroplast genomes; moreover, some relation to taxonomy of the bearers of the genomes may be traced. All the results show one can find a lot standing behind the simple statistical properties of non-coding regions of a genome, while more detailed study falls beyond the scope of this paper.

Acknowledgement. This study was supported by a research grant # 14.Y26.31.0004 from the Government of the Russian Federation.

References

1. Andolfatto, P.: Adaptive evolution of non-coding DNA in Drosophila. Nature **437**(7062), 1149 (2005)
2. Shabalina, S.A., Spiridonov, N.A.: The mammalian transcriptome and the function of non-coding DNA sequences. Genome Biol. **5**(4), 105 (2004)
3. Mercer, T.R., Dinger, M.E., Mattick, J.S.: Long non-coding RNAs: insights into functions. Nat. Rev. Genet. **10**(3), 155 (2009)
4. Kelchner, S.A.: The evolution of non-coding chloroplast DNA and its application in plant systematics. Ann. Mo. Bot. Gard. **87**(4), 482–498 (2000)
5. Guttman, M., Rinn, J.L.: Modular regulatory principles of large non-coding RNAs. Nature **482**(7385), 339 (2012)
6. Mattick, J.S., Makunin, I.V.: Non-coding RNA. Hum. Mol. Genet. **15**(suppl_1), R17–R29 (2006)
7. Gorban, A.N., Zinovyev, A.Y.: The mystery of two straight lines in bacterial genome statistics. Bull. Math. Biol. **69**(7), 2429–2442 (2007)
8. Gorban, A., Popova, T., Zinovyev, A.: Codon usage trajectories and 7-cluster structure of 143 complete bacterial genomic sequences. Phys. A: Stat. Mech. Appl. **353**, 365–387 (2005)
9. Gorban, A.N., Popova, T.G., Sadovsky, M.G.: Classification of symbol sequences over their frequency dictionaries: towards the connection between structure and natural taxonomy. Open Syst. Inf. Dyn. **7**(1), 1–17 (2000)
10. Sadovsky, M., Putintseva, Y., Chernyshova, A., Fedotova, V.: Genome structure of organelles strongly relates to taxonomy of bearers. In: Ortuño, F., Rojas, I. (eds.) Bioinformatics and Biomedical Engineering. Lecture Notes in Computer Science, pp. 481–490. Springer International Publishing, Cham (2015). https://doi.org/10.1007/978-3-319-16483-0_47

11. Gorban, A.N., Popova, T.G., Zinovyev, A.Y.: Seven clusters in genomic triplet distributions. Silico Biol. **3**(4), 471–482 (2003)
12. Sadovsky, M., Senashova, M., Malyshev, A.: Chloroplast genomes exhibit eight-cluster structuredness and mirror symmetry. In: Rojas, I., Ortuño, F. (eds.) Bioinformatics and Biomedical Engineering. Lecture Notes in Computer Science, pp. 186–196. Springer International Publishing, Cham (2018). https://doi.org/10.1007/978-3-319-78723-7_16
13. Sadovsky, M.G., Senashova, M.Y., Putintseva, Y.A.: Chapter 2. In: Chloroplasts and Cytoplasm: Structure and Functions, pp. 25–95. Nova Science Publishers, Inc. (2018)
14. Sadovsky, M.G., Ostylovsky, A.N.: How to detect topology of a manifold to approximate multidimensional data. In: Applied Methods of Statistical Analysis. Nonparametric Methods in Cybernetics and System Analysis, pp. 204–210, Novosibirsk, NSTU, NSTU PLC (2017)

Characteristics of Protein Fold Space Exhibits Close Dependence on Domain Usage

Michael T. Zimmermann[1], Fadi Towfic[2], Robert L. Jernigan[3],
and Andrzej Kloczkowski[4,5(✉)]

[1] Clinical and Translational Science Institute, Medical College of Wisconsin,
Milwaukee, WI 53223, USA
[2] Celgene Corporation, Summit, NJ 07901, USA
[3] Department of Biochemistry, Biophysics, and Molecular Biology,
Iowa State University, Ames, USA
[4] Battelle Center for Mathematical Medicine,
The Research Institute at Nationwide Children's Hospital,
Columbus, OH 43205, USA
Andrzej.Kloczkowski@nationwidechildrens.org
[5] Department of Pediatrics, The Ohio State University,
Columbus, OH 43205, USA

Abstract. With the growth of the PDB and simultaneous slowing of the discovery of new protein folds, we may be able to answer the question of how discrete protein fold space is. Studies by Skolnick et al. (PNAS, 106, 15690, 2009) have concluded that it is in fact continuous. In the present work we extend our initial observation (PNAS, 106(51) E137, 2009) that this conclusion depends upon the resolution with which structures are considered, making the determination of what resolution is most useful of importance. We utilize graph theoretical approaches to investigate the connectedness of the protein structure universe, showing that the modularity of protein domain architecture is of fundamental importance for future improvements in structure matching, impacting our understanding of protein domain evolution and modification. We show that state-of-the-art structure superimposition algorithms are unable to distinguish between conformational and topological variation. This work is not only important for our understanding of the discreteness of protein fold space, but informs the more critical question of what precisely should be spatially aligned in structure superimposition. The metric-dependence is also investigated leading to the conclusion that fold usage in homology reduced datasets is very similar to usage across all of PDB and should not be ignored in large scale studies of protein structure similarity.

Keywords: Protein structure · Protein structural alignment ·
Protein fold space · Graph theory · Protein structure universe · Protein families ·
Protein folds

© Springer Nature Switzerland AG 2019
I. Rojas et al. (Eds.): IWBBIO 2019, LNBI 11465, pp. 356–369, 2019.
https://doi.org/10.1007/978-3-030-17938-0_32

1   Introduction

The three dimensional structures of proteins are often grouped into hierarchical classifications in order to facilitate our understanding of their relationships with each other. Using this concept, one can envision a "fold space" for protein structures where a fold is defined as a specific spacial arrangement of secondary structures. These folds of single domain proteins have been classified by a number of structural ontologies including CATH [1] and SCOP [2] that group most known protein structures based upon combinations of sequence homology, structural topology, and function. Pfam [3] is another well used resource, but focuses more on functional classification, rather than structural (though the two are often related). Presently, structural classifications such as CATH and SCOP still rely heavily upon expert manual curation. The prevailing view concerning protein fold space is that it is comprised of a finite number of discrete folds that are described by these structural ontologies. Recent updates have yielded increased coverage of the diverse types of folds that proteins can assume [4], with a noticeable saturation being reached. The results of such efforts, largely driven by structural genomics initiatives [6], may imply that we are reaching full enumeration of the single domain folds [5]. As such, one of the interesting and important implications that arise from these works is that fold space is discrete and not continuous.

Work by Skolnick *et al.* [7] challenges this view concerning the discreteness of protein structure space using a graph theory approach to analyzing the topological relatedness of protein structures. By considering a large representative set of structures, and a graph based on pair-wise structural relatedness judged by TM-score [8] of 5906 protein chains with low homology from the PDB [9], it was shown that the average shortest path in this network is seven. In the graph, a node represents each PDB file and edges are placed between them whenever pair-wise TM-score is greater than 0.4. We believe that this may not necessarily be informative about protein structure space [10], but instead is likely a general network property since the same result can be obtained in a simpler way using the approximation of Watts and Strogatz [11] for random small world graphs. Multiple questions still arise such as the metric-dependence of this conclusion, if state-of-the-art structure matching algorithms can distinguish topological diversity from conformational, and the overall role of domain architecture. In this work we seek a more detailed understanding of the properties of fold space graphs and their implications for our perceptive on protein structure relatedness. Our main contributions are as follows (1) Graphs generated based on various TM score cutoffs show a high degree of modularity, however (2) we show that the TM algorithm is not well suited for distinguishing topology from conformation based on our comparative analysis (utilizing TM align) of reverse transcriptase (RT) structures gathered from Pfam to manually curated categories in CATH. Thus (3) we explored structure space using a domain-based comparison utilizing CATH and SCOP categories. Our comparison showed that there exists one dominant, modular cluster with some discontinuities in structure space outside of the larger cluster. Thus, we conclude that the continuity (or discreteness) of protein structure fold space depends highly on the resolution one is willing to impose for distinguishing folds.

Modularity, graph partitioning efficiency, and community detection are three terms that refer to roughly the same concept; determining if there exist regions of a network with high connectivity of nodes within individual clusters, but relatively low connectivity between different clusters. For our application to fold space, this translates into groups of structures that are close structural matches to each other within a group, but not to members of other groups. Therefore, for community structure to be prevalent there must be groups of structures that are closely related, but few structures that are simultaneously similar to members of a different group. High modularity combined with a relatively large number of clusters would point to a discrete fold space. Low modularity or a high modularity with very few clusters would point toward a continuous view. Various metrics to evaluate the community structure in graphs have been developed including the modularity score of Newman and Girvan [12] that we apply here. The logic behind community structure and graph clustering to explain the small average shortest path is the following: Consider a cluster A that is well connected. That is, for every node n_i in A, any other node n_j in A is reachable, on average, via a greater number of shorter paths compared to another node, n_x that is a member of a different cluster B. This means that any neighbor of any node in A is quickly reachable from any node in the cluster. Strong community architecture does exist in fold space graphs and further analysis is performed by employing the Markov Cluster (MCL) Algorithm [13, 14]. If a large number of well-connected clusters exist in the graph and relatively few edges connect them, then either the dataset is not a complete representation of fold space or the space is not continuous.

Many methods to determine the relatedness of proteins and protein structures have developed. These are dominated by sequence algorithms because sequence data is abundant and sequence-based algorithms are computationally efficient and fairly intuitive. One such scheme is VAST, Vector Alignment Search Tool [15], which incorporates statistical significance thresholds and estimation of interactions chosen by chance. The widespread use of PSI-BLAST [16] and similar string algorithms in structure classifications like CATH and SCOP are further examples. Matches based on sequence homology represent a conservative subset of similar proteins due to the fact that the inverse folding problem, determining how many sequences can assume a given 3D shape (fold), is unsolved in general. Many cases exist where sequences with little to no homology assume nearly identical folds; i.e. Ubiquitin (1UBI) and SUMO (1WM2) have 15% sequence identity, but fold to practically similar structures differing only by 1.5 Å C^{α} RMSD. Many structure alignment procedures exist that are widely used in structural biology. In this work, we will primarily use TM-align [8], which has been shown to give excellent alignments and used for template detection in I-TASSER [17], currently ranked among the best performing 3D structure prediction servers.

Another fundamental question that needs to be addressed is exactly what should be compared? Proteins with different numbers of amino acids are, mathematically, objects with different conformational dimensions; therefore, we commonly simplify the problem to finding the best superimposition of two structures. Interestingly, there may be patterns in other mathematical spaces that simplify the analysis of structures, such as the relation between spectral dimension (related to energy transfer efficiency) and fractal dimension (related to packing density) in protein structures [18, 19]. However, the details of the structure can influence the energy transfer (allostery) pathways within the structure [20] or their interactions [21]. A much coarser view could consider proteins as approximate globules – amorphous 3D blobs whose surfaces are semi-molten [22] and have mostly polar character but with some internal nonpolar groups exposed to water. We have just described two very different views on protein structures where the details can highly bias an analysis toward a specific conclusion. In the first case, it is intuitive that relatively few structures will be similar whereas in the second case, many proteins could resemble each other's shape. Thus, the resolution with which we consider structures will affect our conclusion about the discreteness of fold space. In this work, we investigate the structure superimposition using TM-align and domain similarity. Our application of TM-align procedure is similar to Skolnick *et al.* [7], except that we consider various thresholds instead of a single, fixed TM score cutoff. We also collect a representative from each known fold type and apply the same graph analysis to this smaller dataset. These representatives have been deemed by expert manual curation to symbolize distinct fold types and thus represent a best case scenario for concluding that fold space is discrete. The results of this analysis are utilized to interpret data obtained by using the complete protein dataset. For domain similarity, we analyze fold space independent of any structure-based comparison by connecting nodes if the proteins they represent share a common CATH or SCOP annotation. Annotations were taken from CATH at the Topology level and from SCOP at the Fold level. Such an analysis provides an impartial baseline for how any structure similarity metric that seeks to approximate CATH or SCOP-level fold similarity will perform on the dataset.

Defining the entities that are compared: complete PDB files, PDB chains (individual polypeptides), or single domains is an important problem. Much effort has been applied to developing methods for computational domain prediction. Early contributions such as FSSP using Dali [23, 24] have been very influential, while newer algorithms such as DomNet [25] show increased refinement and agreement with manual curation. However, in this study we will focus on the manual curation levels of CATH and SCOP. If whole PDB files or chains are used, there will be cases where the peptide chain folds to two or more domains. These structures can act as cluster-linkers in the fold space graph since one domain may have a significant score with structures in one cluster, while the other domain will have high structural relation to a different cluster. Alternatively, a single domain could require the interaction of more than one polypeptide. Such proteins complicate the relationship between sequence-homology reduced datasets and fold usage. Considering the size of a protein may also be important since a small protein is more likely to possess a topology that is some subset of a larger protein.

2   Results

For any approach that relies on graph theory, understanding the structure of the graphs used is necessary. Figure 1 shows us that, for any TM-score threshold, there exist a relatively small number of nodes possessing a high degree of connectivity. We have investigated these hub nodes and draw two conclusions. Some of them are hubs because they are among the smallest proteins in the dataset with approximately 50 residues. It is more likely for a small protein to be a topologically similar subset of a

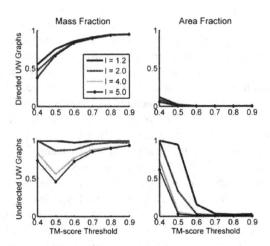

Fig. 1. Metrics evaluating MCL clustering on TM-score graphs. UW stands for uniformly weighted meaning that if an edge exists in the graph, we assign it a weight of one. Area Fraction is defined by Eq. 2 and relates to cluster size. Mass Fraction is the fraction of total edge weight that is captured within clusters and is formally defined in Eq. 3. Including the edge weights does not impact these metrics (see Tables 1 and 2).

larger protein than for two proteins of equal size to match. Others have high connectivity because they have multiple domains. Each domain can individually have a significant alignment with other structures, which inflates the connectivity relative to single domain chains.

In our previous work [10], the relationship between average shortest path computed using the WattsStrogatz approximation [11] and the TM-score threshold for retaining edges in the graph was investigated. We found that the average shortest path is less than seven for cutoffs below 0.75. Stricter cutoffs result in large areas of the graph becoming disconnected. With increasing TM-score, the edge set gets sparser, approaching a cardinality of zero. This is shown in Fig. 2 where the number of nodes with no edges increases as the TM-score threshold increases.

Since TM-scores are numerical, defined on the interval from zero to one, and are not symmetric, pairwise scores can easily be interpreted as a directed graph where we use TM-scores as edge weights. In the MCL algorithm edge weight is the probability of a random walk traversing along a given edge. We construct unweighted graphs by assigning all edges a weight of one and undirected graphs by linking nodes (with or without edge weights) based on the larger of their two TM-scores. From Supplemental Tables 1 and 2, it is evident that the edge treatment makes a minimal impact upon MCL clustering.

Since the sequence-structure relationship is not fully understood, sequence-homology reduced datasets are not necessarily the same as topology reduced datasets. The effect on graph behavior of a topologically reduced dataset is of interest for comparison to the homology reduced dataset. Parameter choices that yield expected results in the topologically reduced dataset will help us to better interpret the meaning of clusters for the homology reduced set. For this reason, we also compare distinct

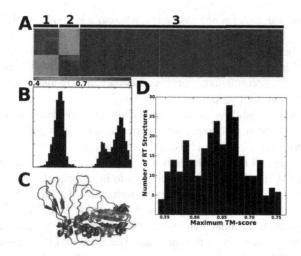

Fig. 2. (A) Heat map of TM-scores between 283 reverse transcriptase (RT) structures and 1233 topology representatives from CATH v3.3. Rows are arranged in the same order as the columns of Sects. 1 and 2. Sections 1 and 2 correspond to two orientations of the RT "fingers," while Sect. 3 is the TM score between the topology representatives and the RT structures. (B) Histogram of the TM-scores within the set of 283 reverse transcriptase structures. The set can easily be split into structures that are related to each other at a TM-score of greater or less than 0.7. No pairwise scores are below 0.4. Low scores correspond to the two finger domains being in different positions, while the higher scores correspond to the two fingers in roughly the same orientation. The high scoring population can be somewhat thought of as two groups; one where the fingers are in a more closed conformation, and one where they are both extended. (C) We show a representative of the lower TM-score population; 1RW3 aligned to 1JLA with a TM-score of 0.56. The view shown highlights the different finger positions that are characteristic of the lower scoring group. (D) Histogram of maximum TM-score between each reverse transcriptase domain and the topology representatives from CATH (max for each row of Sect. 3). Each reverse transcriptase domain has a TM-score between itself and a topology representative of at least 0.53, but none are higher than 0.76. There are 277 topologies matched to the 283 reverse transcriptase structures. Thus, large TM-scores, while relatively sparse, are not because of any single (or even a small set) of reverse transcriptase like topologic representatives.

topologies to each other by gathering 1233 CATH version 3.3 topology representatives; a collection of manually curated topology representatives that span all of PDB, performing the same procedure. Interestingly, this dataset of distinct topology representatives exhibits a high modularity, indicative of community structure (Supplemental Table 2). We calculate a modularity score defined in [12] by comparing the number of edges within clusters to the number of edges that are linking clusters to each other. At low TM thresholds (0.4), the graph exhibits high connectivity (57550 edges) and the majority of the nodes included in the largest cluster.

The extent of community structure is less than for the PDB300 dataset (as judged by F_{Mass} and F_{Area} – see Methods), but remains high. The MCL inflation parameter determines granularity of the clustering with a low inflation yielding few large clusters and high inflation producing many small clusters. Even for the high inflation value of 5, the largest cluster still contains 855 structures, whereas a low value of 1.2 retains 1217. At a TM score cutoff of 0.6 we find that MCL consistently distinguishes many of the topologies from each other (only 334 edges between the 1233 nodes). Thus, these graphs may either be modular because they are significantly related (pointing to structure space being continuous) or because they are mutually distantly related (pointing to a discrete fold space).

Conformational variability is also an important consideration for comparing topologies. Are our structure comparison metrics able to distinguish between conformational variation and topological? To address this question, we will compare 283 reverse transcriptase (RT) structures gathered from Pfam [3] family PF00078 to each other and to the CATH topology representatives to investigate the ability for structure comparison metrics to distinguish between conformational (within the RT family) and topological (between RT and fold representatives) differences. RT structure is often described by analogy to a human hand where the active site is in the center of the palm and the fingers and thumb "grip" the substrate. The Pfam family set used corresponds to two fingers and the palm, thus containing sequence (average sequence identity of 67%) and conformational variants. A TM-score above 0.4 is regarded as a significant topological relationship. We find that all members of the reverse transcriptase family have TM-scores above this threshold, but there is significant diversity of scores within the family (Fig. 2). Roughly half of the pairwise comparisons are between 0.4 and 0.7 corresponding to different finger conformations (generalized from visualizing 100 randomly chosen pairs from this group). Higher scoring pairs are characterized by the structures having the same general finger conformation. The subgroup at about 0.82 has a higher representation of a more extended finger conformation (again from visualization of 100 randomly chosen pairs; data not shown). A representative pair is shown in Fig. 2. Further, each reverse transcriptase domain has a TM-score between itself and a topology representative of at least 0.53, but none are higher than 0.76. Therefore, all RT structures have a significant structure alignment to a topology representative. One might expect that because all RT structures share a common fold, one topology would be the best match to most of the RT structures. However, matching each of the 283 RTs to its highest scoring topology yields 277 different topologies. Thus, large TM-scores, while relatively sparse, are not because of any single (or even a small set) of RT- like topologic representatives. Further, TM (and likely any rigid superimposition algorithm) is, in general, unable to distinguish between conformational and topological variation. Methods like Fr-TM-align [26] or FATCAT [27] that are capable of accounting for flexibility of the biomolecule may perform better in this specific test, but fast and accurate methods for incorporating flexibility in structure matching are still being improved. Current structure comparison algorithms have difficulty in distinguishing between conformational and topological differences.

Metrics similar to Silhouettes [28] have also been generated (not shown). These are basically average path length from a node to any other node within a given cluster compared to the average path length from a node to every node that is not in that cluster. Evidence of the high number of connections within each cluster exists in that the average out-of-cluster path is only slightly longer than the average within-cluster path.

A critical point of the above analysis hinges on the efficacy of the TM score algorithm in quantifying the fold space of proteins. Thus, it is reasonable to ask: are these investigations into protein fold space dependent upon the metric used? We have already shown that the state-of-the-art structure comparison method has difficulty in distinguishing topological and conformational differences, but can we explore fold space independent from structure superimposition? One way is to make a graph where each protein chain is represented by a node and nodes are connected by edges if the two proteins share a common fold. Common folds are determined by a shared CATH topology or SCOP fold using CATH version 3.4 and SCOP version 1.75. In the PDB300 dataset 90% of the protein chains are annotated by at least one of these ontologies, while all of the PISCES proteins are annotated (see Methods for dataset details). Unannotated nodes are neglected in the following analysis. Using the same graph analysis procedure, we find that this domain based graph also has a very high degree of connectivity and modularity. See Table 1 for details. We again find that there exists one dominant cluster. It has been shown that MCL usually generates a dominant cluster and for some applications modifications that generate a more even granularity are preferred [29]. However, using these approaches would be equivalent to assuming fold space is discrete. Another explanation for the dominant cluster is the imbalance in topology usage. Table 2 summarizes the usage of the ten most used topologies across all of CATH, PDB300, and the PISCES dataset. Seven of the ten most used annotations across all of CATH are also in the top ten most used topologies in the datasets used here. Further, if we sort the topology classes by their use across all of CATH, and compare with the topology use in each of our datasets, PDB300 and PISCES have a correlation coefficient with CATH of 0.94 and 0.93, respectively. Thus, the relative distribution of domain types is similar in these datasets compared to the whole PDB. We conclude that the reason for the observed shortest paths in TM-score based graphs is the modularity of proteins and the bias in topology usage. Protein structures exhibit variation upon themes – stable domains develop and are embellished upon for further modification of function.

Viksna and Gilbert [30] proposed a new method of assessment of domain evolution by measuring the rate of certain kinds of structural changes that can lead to novel fold development. Birzele *et al.* [31] find fascinating evidence that alternative splicing plays a role in protein structure evolution by developing transitional structures between fold types. Fong and colleagues [32] emphasize the modularity and importance of domain fusion events in the evolution of protein domains. Meier *et al.* [33] suggest a link between conformational flexibility and domain evolution where the native state ensemble can partially occupy at least two intermediate fold types and the relative population of each may be influenced by single amino acid mutations. The results presented here combined with these studies point to the importance of considering protein folds more rigorously in structure matching. It is not only important for our understanding of the discreteness of protein fold space, but informs the more critical question of what precisely should be spatially aligned in structure superimposition.

3 Discussion

We have shown that graphs generated either from TM scores or domain annotation show a high degree of community structure (modularity). TM-scores alone are not able to fully distinguish manually curated topology representatives from each other at the same threshold levels that have been used to analyze fold space across the PDB. This is partly due to the effect of conformation on TM-score. It is shown here that conformational variability within a set of reverse transcriptase structures can lead to very different conclusions about which CATH topology is a closest representative. It is important to realize that since we do not fully understand the relationship between protein sequence and structure, homology reduced datasets may not be topology reduced. This has been shown by analyzing graphs generated by connecting nodes if they share any common CATH topology or SCOP fold and showing that they have similar modularity and graph structure compared to graphs based on structure super-imposition (see Fig. 3). It is possible that improving coarse-grained representations like TOPS strings [34, 35] will be useful in the future for handling the multi-resolution complexities of structure comparison.

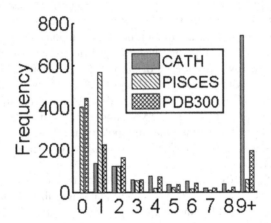

Fig. 3. Frequency of CATH topology usage in three datasets. Using CATH version 3.4, we consider the 1282 topology IDs, counting the number of times a structure in each dataset is annotated with each topology ID. For both of the datasets used in this study, we find that many topologies are not used at all (abscissa value of 0), and that relatively few topologies have a high rate of use. In the CATH database, most topologies have a high rate of use. Interestingly, the PISCES dataset is more topologically diverse than the PDB300 dataset.

Classifying protein tertiary structures into a discrete set of domains is useful in that it helps our conceptual understanding of protein structures, aids in reducing the possible outcomes for sequence based folding procedures, adds to our understanding of the structure-function relationship, as well as many other applications. Whether protein fold space is continuous or discrete depends upon the resolution with which it is viewed. We believe the more fundamental observation is the usage of topology types.

4 Methods

Datasets
We use the PDB300 dataset from [7], which consists of 5906 protein chains of lengths between 40 and 300 residues sharing less than 35% sequence identity. Here, the word "chain" refers to a single polypeptide. It is worth noting that numerous individual protein chains in this dataset contain more than one domain (topology or fold) as defined by CATH or SCOP. Also, the PDB has been updated since this dataset was gathered; 3 structures have become obsolete and were not superseded by a new ID, while 41 have been updated to new IDs. For the purpose of comparison, we continue to use the original version of each PDB ID when employing TM-align.

Domain centric datasets are constructed in two ways. The first is to cut the CATH hierarchy at the topology level resulting in 1233 or 1282 representative domains for version 3.3 and 3.4 respectively. The second begins by using the PISCES server [36] to gather a representative set of chains that are of better than 2.5Å resolution, less than 20% mutual sequence identity, and a crystallographic R-factor of less than 0.25. This dataset contains 4750 PDB chains. We then use CATH to identify individual domains at the Topology level within this set.

The final dataset used is Pfam family PF00078, corresponding to reverse transcriptase (RT). At the time of data download, 283 members with full 3D coordinates were available in the PDB. These structures were downloaded and the subset of points agreeing with the Pfam family definition was retained.

Protein Structure Evaluation Metrics
In this paper, the TM-score defined in [8], is used to analyze the structural similarity of protein structures. This metric is interesting in that it is not symmetric; TM (A,B) does not necessarily equal TM (B,A) particularly when proteins A and B are of different lengths. The TM-score is defined as:

$$TM - score = \text{Max} \left[\frac{1}{L_{\text{Target}}} \sum_{i}^{L_{ali}} \frac{1}{1 + \left(\frac{d_i}{d_0 \left(L_{\text{Target}} \right)} \right)^2} \right] \tag{1}$$

Where L_{ali} is the length of the alignment, L_{Target} is the sequence length of the target structure, d_i is the Euclidean distance between aligned points, and d_0 is a normalization factor based on L_{Target}.

Analysis of Structural Classification
A number of structural classification schemes exist including the CATH database (1), SCOP (2), and PFAM (3). Both CATH and SCOP are hierarchical in nature and utilize a combination of homology, topology, and biochemical function to organize protein structures. The first level of CATH and SCOP classification consists of 4 classes, binning structures into predominantly α-helix or β-sheet content, presence of both, or lack of secondary structure elements. The second level of SCOP, as well as the second and third levels of CATH, is based on overall secondary structure orientation. These

levels are manually curated and place proteins into general categories like beta-barrel and two-layer sandwich. The third level of SCOP takes into account the topology and function of a given protein to decide how related they are evolutionarily. All subsequent levels in both classifications are decided by sequence identity or, in some cases, other sequence based scoring schemes. Pfam families are generated by manual functional curation, multiple sequence alignments, and Hidden Markov Models and come in two varieties: Pfam-A for only manually curated entries and Pfam-B where automated methods are also used to extend the sequence space covered by classification. All three of these databases rely on sequence homology and biochemical functions to group proteins into fold types rather than directly comparing quantitatively the topology of the biomolecules.

Graph Construction

We define a graph based on TM-score as $G_t = \{E, V\}$ where $e_i \in E$ is an edge in G_t if it connects two vertices $a \in V$ and $b \in V$ and $TM(a,b) > t$. Each PDB chain in the dataset is represented by a single node. The TM-score threshold t is initially set at 0.4 as in [7], but values up to 0.9 are also considered to further analyze the graph structure. Graphs are either undirected or directed. To make the directed graphs we consider $t = max(t_1,t_2)$ where $TM(a,b) = t_1$ and $TM(b,a) = t_2$.

Cluster Generation and Comparison

To investigate the community structure of graphs we first employ the Markov Cluster Algorithm (MCL) [13, 14]. In this procedure, graphs are clustered based on random walks that simulate flow along the graph's edges. Nodes that are well connected will exhibit more flow between them than nodes with few connections; the probability of selecting an edge to walk along within the cluster is higher than choosing an edge that leads you out of the cluster (provided the probability of selecting any edge at random is uniform). As the algorithm progresses, nodes that share high amounts of flow (many common walks) are grouped together into clusters.

MCL has evaluation protocols to explore the relatedness of clustering with different parameters. In MCL each edge has a weight. Here we use uniformly weighted (UW) graphs or we use the TM-score as the edge weight. Defining cluster size as the number of nodes within a cluster, MCL computes the Area Fraction (F_{Area}) defined by Eq. 2. This metric gives an indication for the size of clusters as many small clusters or isolated nodes will result in a low F_{Area}. The Mass Fraction (F_{Mass}) is the sum of all edge weights within clusters and is shown in Eq. 3 where w_i is the edge weight of edge i such that edge i is in Cluster c.

$$F_{Area} = \frac{\sum clusterSize^2}{N(N-1)} \qquad (2)$$

$$F_{Mass} = \sum_{c=1}^{|c|} \sum_{i=1}^{|E|} w_i \ s.t. w_i \in C_c \qquad (3)$$

Having F_{Area} close to zero implies that the graph has been clustered into many small clusters, while a value of one implies that all nodes occupy one cluster.

Possessing F_{Mass} close to one indicates that clusters are tightly connected with relatively few edges connecting them. How the algorithm treats the length of a walk (number of edges traversed) is very important to the process and is controlled by a parameter called Inflation, I. Penalizing longer walks produces a large number of small clusters. Allowing longer walks generates fewer, but larger, clusters. It is informative to compare results across multiple inflation values to better understand the organization of the graph.

Table 1. Clustering metrics for graphs of common CATH or SCOP annotations

Dataset	Inflation	Mod_5	Mod_{10}	Mod_{all}	Eff	F_{Mass}	F_{Area}	#C	Max	Avg
PDB300	1.2	.98	.98	.99	.21	.98	.58	137	3617	34.8
	2.0	.89	.90	.92	.33	.91	.25	183	2155	26.1
	4.0	.75	.78	.82	.49	.81	.10	325	1429	14.7
	6.0	.51	.54	.59	.49	.71	.07	450	1179	10.6
	8.0	.37	.40	.45	.49	.66	.05	519	1006	9.2
	12.0	.26	.28	.33	.47	.62	.04	580	931	8.2
PISCES	1.2	.96	.97	.97	.34	.97	.27	138	864	15.7
	2.0	.89	.91	.94	.59	.93	.09	174	465	12.4
	4.0	.78	.82	.86	.66	.88	.06	236	357	9.2
	6.0	.61	.64	.68	.65	.82	.04	289	303	7.5
	8.0	.59	.62	.65	.64	.81	.04	302	296	7.2
	12.0	.55	.57	.61	.64	.80	.04	316	281	6.9

Table 2. Top 10 CATH topology usage

CATH				PDB300			PISCES		
ID	Usage	Architecture	Topology	ID	Usage	C_r	ID	Usage	C_r
3.40.50	19229	3-Layer(aba) sandwich	Rossmann fold	3.40.50	1339	1	3.40.50	379	1
2.60.40	13806	Sandwich	Immunoglobulin-like	2.60.40	678	2	2.60.40	132	2
3.20.20	6106	Alpha-beta barrel	TIM Barrel	1.10.10	384	11	3.20.20	118	3
3.30.70	4236	2-layer Sandwich	Alpha-Beta Plaits	3.30.70	334	4	3.30.70	101	4
2.40.10	3954	Beta barrel	Thrombin	2.60.120	312	6	2.60.120	96	6
2.60.120	3433	Sandwich	Jelly Rolls	2.40.50	268	7	1.10.10	81	11
2.40.50	2244	Beta barrel	OB fold	1.20.5	263	15	1.20.5	65	15
3.30.200	2012	2-layer sandwich	Phosphorylase Kinase	3.20.20	250	3	1.10.287	57	19
1.10.510	1992	Orthogonal bundle	Phosphotransferase	2.40.10	206	5	2.40.50	49	7
1.10.490	1983	Orthogonal bundle	Globin-like	2.30.30	205	16	1.20.120	44	29

Column titles are: Inflation for the MCL parameter that determines granularity of the clustering, Mod_5 - modularity using the 5 largest clusters, Eff - efficiency of the clustering, F_{Mass} and F_{Area} are given in Eqs. 2 and 3, #C - number of clusters, Max - number of nodes in the largest cluster, Avg - average number of nodes across all clusters.

Usage is the number of protein chains that are annotated with the given CATH topology ID. C_r is the rank of this topology ID in CATH.

Acknowledgements. AK and RLJ acknowledge support from the National Science Foundation (DBI 1661391) and from National Institutes of Health (R01GM127701 and R01GM127701-01S1).

References

1. Cuff, A.L., et al.: Nucleic Acids Res. **37**, D310–D314 (2009)
2. Murzin, A.G., Brenner, S.E., Hubbard, T., Chothia, C.: J. Mol. Biol. **247**, 536–540 (1995)
3. Finn, R.D., et al.: Nucleic Acids Res **36**, D281–D288 (2008)
4. Cuff, A.L., et al.: Nucleic Acids Res. **39**, D420–D426 (2011)
5. Zhang, Y., Hubner, I.A., Arakaki, A.K., Shakhnovich, E., Skolnick, J.: Proc. Natl. Acad. Sci. U.S.A. **103**, 2605–2610 (2006)
6. Grabowski, M., Joachimiak, A., Otwinowski, Z., Minor, W.: Curr. Opin. Struct. Biol. **17**, 347–353 (2007)
7. Skolnick, J., Arakaki, A.K., Lee, S.Y., Brylinski, M.: Proc. Natl. Acad. Sci. U.S.A. **106**, 15690–15695 (2009)
8. Zhang, Y., Skolnick, J.: Nucleic Acids Res. **33**, 2302–2309 (2005)
9. Berman, H.M., et al.: Nucleic Acids Res. **28**, 235–242 (2000)
10. Zimmermann, M., Towfic, F., Jernigan, R.L., Kloczkowski, A.: Proc. Natl. Acad. Sci. U. S. A **106**, E137 (2009)
11. Watts, D.J., Strogatz, S.H.: Nature **393**, 440–442 (1998)
12. Newman, M.E., Girvan, M.: Phys. Rev. E **69**, 026113 (2004)
13. Van Dongen, S.: Technical Report INS-R0010. National Research Institute for Mathematics and Computer Science in the Netherlands (2000)
14. Van Dongen, S.: Ph.D. Thesis, Univ Utrecht, The Netherlands (2000)
15. Gibrat, J.F., Madej, T., Bryant, S.H.: Curr. Opin. Struct. Biol. **6**, 377–385 (1996)
16. Altschul, S.F., et al.: Nucleic Acids Res. **25**, 3389–3402 (1997)
17. Zhang, Y.: BMC Bioinf. **9**, 40 (2008)
18. de Leeuw, M., Reuveni, S., Klafter, J., Granek, R.: PLoS One **4**, e7296 (2009)
19. Reuveni, S., Granek, R., Klafter, J.: Proc. Natl. Acad. Sci. U.S.A. **107**, 13696–13700 (2010)
20. Lee, J., et al.: Science **322**, 438–442 (2008)
21. Guntas, G., Purbeck, C., Kuhlman, B.: Proc. Natl. Acad. Sci. U.S.A. **107**, 19296–19301 (2010)
22. Zhou, Y., Vitkup, D., Karplus, M.: J. Mol. Biol. **285**, 1371–1375 (1999)
23. Holm, L., Sander, C.: Nucleic Acids Res. **25**, 231–234 (1997)
24. Holm, L., Sander, C.: Nucleic Acids Res. **26**, 316–319 (1998)
25. Yoo, P.D., Sikder, A.R., Taheri, J., Zhou, B.B., Zomaya, A.Y.: IEEE Trans. Nanobiosci. **7**, 172–181 (2008)
26. Pandit, S.B., Skolnick, J.: BMC Bioinf. **9**, 531 (2008)
27. Ye, Y., Godzik, A.: Bioinformatics **19**(Suppl 2), ii246–ii255 (2003)

28. Horimoto, K., Toh, H.: Bioinformatics **17**, 1143–1151 (2001)
29. Satuluri, V., Parthasarathy, S., Ucar, D.: Proceedings of the First ACM International Conference on Bioinformatics and Computational Biology, BCB 2010, pp. 247–256 (2010). https://dl.acm.org/citation.cfm?doid=1854776.1854812
30. Viksna, J., Gilbert, D.: Bioinformatics **23**, 832–841 (2007)
31. Birzele, F., Csaba, G., Zimmer, R.: Nucleic Acids Res. **36**, 550–558 (2008)
32. Fong, J.H., Geer, L.Y., Panchenko, A.R., Bryant, S.H.: J. Mol. Biol. **366**, 307–315 (2007)
33. Meier, S., et al.: Curr. Biol. **17**, 173–178 (2007)
34. Gilbert, D., Westhead, D., Nagano, N., Thornton, J.: Bioinformatics **15**, 317–326 (1999)
35. Torrance, G.M., Gilbert, D.R., Michalopoulos, I., Westhead, D.W.: Bioinformatics **21**, 2537–2538 (2005)
36. Wang, G., Dunbrack Jr., R.L.: Bioinformatics **19**, 1589–1591 (2003)

Triplet Frequencies Implementation in Total Transcriptome Analysis

Michael Sadovsky[1,2(✉)], Tatiana Guseva[2], and Vladislav Biriukov[2]

[1] Institute of Computational Modelling of SB RAS,
Akademgorodok 660036, Krasnoyarsk, Russia
msad@icm.krasn.ru
[2] Institute of Fundamental Biology and Biotechnology,
Siberian Federal University, Svobodny prosp., 79, 660049 Krasnoyarsk, Russia
dianema2010@mail.ru, vladislav.v.biriukov@gmail.com
http://icm.krasn.ru

Abstract. We studied the structuredness in total transcriptome of Siberian larch. To do that, the contigs from total transcriptome has been labeled with the reads comprising the tissue specific transcriptomes, and the distribution of the contigs from the total transcriptome has been developed with respect to the mutual entropy of the frequencies of occurrence of reads from tissue specific transcriptomes. It was found that a number of contigs contain comparable amounts of reads from different tissues, so the chimeric transcripts to be extremely abundant. On the contrary, the transcripts with high tissue specificity do not yield a reliable clustering revealing the tissue specificity. This fact makes usage of total transcriptome for the purposes of differential expression arguable.

Keywords: Order · Probability · Triplet · Symmetry · Projection · Clustering

1 Introduction

Transcriptome is a set of all the symbol sequences from $\aleph = \{A, C, G, T\}$ alphabet corresponding to the entire ensemble of RNA moleculae (of mRNA moleculae) found in a cell (or in a sample). In a genome deciphering, transcriptome sequencing, assembling and annotation goes ahead. The point is that one may not be sure a transcriptome is stable, in terms of the composition of the sequences mentioned above. Indeed, the set definitely depends on a tissue, on a development stage, on a life cycle stage, and many other factors.

Stipulating a stability of a genome in an organism, one may expect that various tissues exhibit different expression of genes; this is a common place for multicellular organisms, and may take place in unicellular ones, if different stages of a life cycle are considered. Such difference is claimed *differential expression*. This latter is essential in a study of various physiological processes run in an

© Springer Nature Switzerland AG 2019
I. Rojas et al. (Eds.): IWBBIO 2019, LNBI 11465, pp. 370–378, 2019.
https://doi.org/10.1007/978-3-030-17938-0_33

organism, and may tell a researcher a lot concerning some peculiarities in functioning of biochemical and genetic networks.

Total transcriptome is the ensemble of all RNA (or mRNA) sequences gathered regardless their origin, through a bulky source sampled from an organism, or a tissue, etc. Since some genes in specific tissues, or cells may be suppressed or yield lowered expression due to some other reasons, one may expect that total transcriptome make a useful tool for assembling of all the genes observed in a sample, if assembled totally. Here we checked this idea on the total transcriptome of *Larix sibirica* Ledeb.

So, the goal of the study was to compare the efficiency of a "help and support" in specific transcriptome assembling, through the implementation of the total one. To do that, we have sequenced, filtered and cleaned the reads, for four specific tissues: needles, cambium, shoot, and seedling. These four specific transcriptome have been assembled; simultaneously, a total set of reads has been obtained through merging of all four specific ensembles into a single one. Then the assembling of the (total) transcriptome has been carried out. Finally, we tried to compare the total transcriptome with four specific ones to see whether some improvement in assembling "bottle neck" transcripts in specific transcriptomes takes place, or not; speaking in advance, we found greater losses than profits, in such approach.

2   Materials and Methods

Sequencing of *L. sibirica* Ledeb. total transcriptome was carried out in Laboouratory of forest genomics of Siberian federal university. Four groups of tissue specific read ensembles have been obtained separately: needles, cambium, shoot and seedling. Also, later we merged all the reads ensembles into a single one, and assembled the total transcriptome.

Real transcriptomes (both tissue specific, and for the total one) comprise the contigs of various lengths. Some figures characterizing the specific (as well, as the total one) transcriptomes are shown in Table 1; the table presents the figures for the longest contig ($L_{\max}$), average length of transcripts ($\langle L \rangle$), and total abundance of contigs in a transcriptome (M). All transcripts were longer 200 b. p.

For the proposes of the clustering and analysis of transcriptomes, we selected the subsets of contigs, in each specific transcriptome (including the total one). We took into the subsets sufficiently long contigs, only. The idea standing behind such selection is following: shorter contigs would yield rather abundant subsets of points (in 64-dimensional space) that are in local quasi-equilibrium: in other words, too many short contigs would have zero frequency of some triplets. Moreover, a greater number of triplets would be presented in a single copy, in a number of such shorter contigs, thus yielding a kind of quasi-equilibrium over the subspace determined by these triplets.

To avoid the above mentioned effect, we have eliminated shorter contigs. We comprise sufficiently long contigs, to carry out clustering and visualization of the data. Table 1 shows the figures used to select the contigs involved into analysis: L_d is the cut-off length of the contigs, in each specific transcriptome. That former means that we selected the contigs longer than L_d; M_d figures show the abundances of the sets of selected longer contigs.

To gain the total transcriptome, the reads ensembles obtained for each specific tissue have been merged into a single ensemble, and assembling has been carried out [1,2]. Common idea in total transcriptome implementation is to enforce the coverage level of the genes expressed in various tissues, thus improving assembling of *de novo* sequence. Not discussing here an efficiency (quite arguable, frankly speaking), we just stress that a total transcriptome still is a good first step, in any genome deciphering being a kind of *mean filed* approximation.

2.1 Frequency Dictionaries

To analyze statistical properties of transcriptomes, we used a conversion of them into frequency dictionaries; in particular, we focused on triplet frequency dictionaries, only. Formally, a triplet frequency dictionary is the list of all triplets $\omega = \nu_1\nu_2\nu_3$ observed in a sequence $\mathfrak{T}$. This is the triplet frequency dictionary $W_{(3,1)}$. More generally, let t be the step of a move of the reading frame (of the length 3) identifying a triplet ω. Then the frequency dictionary $W_{(3,t)}$ is the list of triplets identified in $\mathfrak{T}$, if the reading frame moves along $\mathfrak{T}$ with the step t. Definitely, one gets t different triplet frequency dictionaries here: there are t different starting positions of the first location of the reading frame.

Further, we shall focus on the dictionaries $W_{(3,1)}$ and $W_{(3,3)}$. In such capacity, there could be 3 triplet frequency dictionaries of $W_{(3,3)}$ type. The analysis of statistical properties of transcriptome provided here is based on the fact that three different frequency dictionaries $W_{(3,3)}$ determined over coding part of a genome differ seriously from similar dictionaries determined over non-coding ones [3–6]. This difference stands behind the analysis.

We did not derive all three versions of triplet frequency dictionaries of $W_{(3,3)}$ type for the transcripts; instead, we developed the clustering of triplet frequency dictionaries expecting them to gather into the clusters corresponding to the phase (i.e. reading frame shift figure $t = \{0; 1; 2\}$) and strand embedment (leading vs. ladder).

2.2 Clustering and Visualization

We used freely distributed software *ViDaExpert* by Andrew Zinovyev (bioinfo.curie.fr) for visualization data. Also, K-means clustering technique [7] has been applied, to prove a structuredness in transcriptome data. To retrieve a structure pattern in transcriptome (any of them, enlisted above), each contig was converted into frequency dictionary $W_{(3,1)}$. Everywhere further we shall denote it as W_3; to distinguish different dictionaries, we shall use an upper index in square brackets: $W_3^{[j]}$, so that $f_\omega^{[j]} \in W_3^{[j]}$. Here $f_\omega^{[j]}$ is the frequency of a triplet ω. Well known Euclidean metrics

$$\rho\left(W_3^{[1]}, W_3^{[2]}\right) = \sqrt{\sum_{\omega=AAA}^{TTT} \left(f_\omega^{[1]} - f_\omega^{[2]}\right)^2} \qquad (1)$$

has been used to determine a distance between two triplet frequency dictionaries $W_3^{[1]}$ and $W_3^{[2]}$, for clustering and visualization purposes.

Using *ViDaExpert* software, we considered the distribution of points corresponding to frequency dictionaries in three-dimensional projection; the choice of axes for the projection was carried out automatically, since we observed the distribution in three principal components (the first one, the second one, and the third one), mainly, not in triplets.

To prove (or disprove) visually observed clustering, we used K-means, provided by the same software. The choice of K was determined by the stability of clustering: we always started from $K = 2$ and stopped at $K^\star$ where clustering became unstable. Besides, we also used elastic map technique, for the purposes of visualization, mainly. Detailed description of that methodology could be found in [8–12].

2.3 Chargaff's Parity Discrepancy

Chargaff's parity rules stipulate several fundamental properties of nucleotide sequences describing a kind of symmetry in them. We used these rules to analyze the observed cluster patterns, in transcriptomes. Tot begin with, Chargaff's substitution rule stipulates that in double stranded DNA molecule nucleotide A always opposes to nucleotide T, and vice versa. Same is true for the couple of nucleotides C $\Leftrightarrow$ G.

The first Chargaff's parity rule stipulates that the number of A's matches the number of T's with a good accuracy, when counted over a single strand; obviously, similar proximal equity is observed for C's and G's. Finally, the second Chargaff's parity rule stipulates a proximal equity of frequencies of the strings comprising complementary palindrome: $f_\omega \approx f_{\bar\omega}$. Here ω and $\bar\omega$ are two strings counted over the same strand, so that they are read equally in opposite directions, with respect to the substitution rule, e.g., CTGA $\Leftrightarrow$ TCAG; see [13–18] for details.

Genomes differ in the figures of discrepancy of the second Chargaff's parity rule [19]; same is true for various parts of a genome. Thus, one can compare the transcriptomes in terms of this discrepancy. To do it, let's introduce that former:

$$\mu\left(W_3^{[1]}, W_3^{[2]}\right) = \frac{1}{64}\sqrt{\sum_{\omega=AAA}^{TTT} \left(f_\omega^{[1]} - f_{\bar\omega}^{[2]}\right)^2}, \qquad (2)$$

where ω and $\bar\omega$ are two triplets comprising complementary palindrome. Here we must take into account both couples: $f_\omega^{[1]} - f_{\bar\omega}^{[2]}$ and $f_\omega^{[2]} - f_{\bar\omega}^{[1]}$, since they exhibit different figures, in general.

Formula (2) measures a deviation between two frequency dictionaries; thus, one may expect that two dictionaries $W_3^{[1]}$ and $W_3^{[2]}$ may comprise the triplets from the opposite strands, if $\mu \to 0$. An inner discrepancy measure determined within a dictionary is another important characteristics of a dictionary. To measure it, one should change the formula (2) for

$$\xi\left(W_3\right) = \frac{1}{32}\sqrt{\sum_{\omega \in \Omega^*}\left(f_\omega - f_{\overline{\omega}}\right)^2}, \tag{3}$$

where Ω^* is the set of 32 couples of triplets comprising complementary palindromes. Obviously, here $|f_\omega - f_{\overline{\omega}}| \equiv |f_{\overline{\omega}} - f_\omega|$. We shall use the figures determined by (2) and (3) for transcriptome analysis.

2.4   Mutual Entropy to Measure the Quality of Total Transcriptome

The key aim of this paper is to compare tissue specific transcriptomes vs. the total one. To do it, we implemented a measure based on the mutual entropy calculation of the reads distribution over contigs of the total transcriptome. Describe this point in more detail. We used four tissues to get the tissue specific transcriptomes: needles, cambium, shoot and seedling. Surely, the abundance of the reads sets is different, for various tissues. So great difference in the abundances of the reads ensembles gathered for different tissues must be taken into account, and we have done it in the following way.

Table 1. Some figures characterizing transcriptomes. $L_{\min}$ is the minimal contig length, $L_{\max}$ the maximal contig length, $\langle L \rangle$ is average contig length, L_d is the selection length, and M_d is the abundance of contig set taken into consideration and N_R is the reads set abundance.

Transcriptome	$L_{\max}$	$\langle L \rangle$	L_d	M	M_d	N_R
Needles	9880	354	1000	59317	1851	2 504 853
Shoot	17893	532	5000	590240	1754	23 986 314
Seedlings	11008	455	2500	174805	1943	8 698 074
Cambium	20596	497	5000	628197	1455	9 563 901

Let N_{needles}, N_{shoot}, N_{seeding} and N_{cambium} be the numbers of the reads, in each read ensemble, respectively. Let then change the numbers for frequencies of the tissue specificity, as it occurs in the joint set of the reads:

$$f_{\text{needles}} = \frac{N_{\text{needles}}}{N_{\text{total}}}, f_{\text{shoot}} = \frac{N_{\text{shoot}}}{N_{\text{total}}}, f_{\text{seedings}} = \frac{N_{\text{seedings}}}{N_{\text{total}}}, f_{\text{cambium}} = \frac{N_{\text{cambium}}}{N_{\text{total}}},$$

where N_{total} is the sum of all N's shown above. The figures of f_{needles}, f_{shoot}, f_{seedings} and f_{cambium} provide the background to study the difference between total transcriptome and the specific ones.

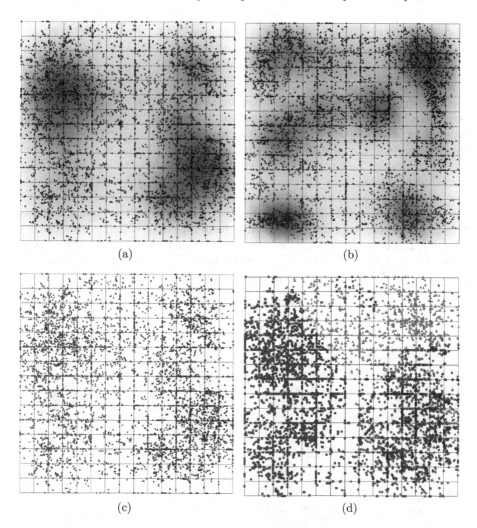

Fig. 1. Distribution of transcripts with greater mutual entropy (4) from total transcriptome; (a) is the case of $W_{(3,1)}$, (b) case of $W_{(3,3)}$. K-means is shown in (c) ($K = 2$) and in (d) $K = 3$; both cases are of $W_{(3,1)}$ type.

At the next stage, the numbers M_tissue (frequencies φ_tissue, respectively) of each tissue specific reads set observed over each transcript from the total transcriptome were obtained; to do it, we used back reads mapping over the total transcriptome transcripts. Thus, each transcript from total transcriptome was converted into a point in four-dimensional Euclidean space with the frequencies of tissue specific reads being the coordinates.

Finally, the mutual entropy

$$\overline{S}_k = \sum_{j=1}^{4} \varphi_j \cdot \ln\left(\frac{\varphi_j}{f_j}\right) \tag{4}$$

was determined for each transcript taken into consideration from the total transcriptome; here the index j enlists the tissues. The transcripts list was descending ordered, and the top part of the list has been analyzed. Index k in (4) enumerates the transcripts in the total transcriptome. Obviously, φ_j figures were determined for each transcript from the total transcriptome individually, while f figures were the same. Mutual entropy (4) measures a deviation of the distribution of the tissue specificity of reads observed within a transcript: if $\overline{S}_k = 0$, then the k-th transcript does not differ from the ensemble of the reads of the total transcriptome, and, in such capacity, is stipulated to be the most chimeric one. On the contrary, if a transcript yields the maximal deviation of (4) from zero, then it means the highest level of tissue specificity. It should be born in mind that the maximum of (4) depends on the specific tissue: in particular,

$$\max\left\{\overline{S}_k\right\} = -\ln f_k. \tag{5}$$

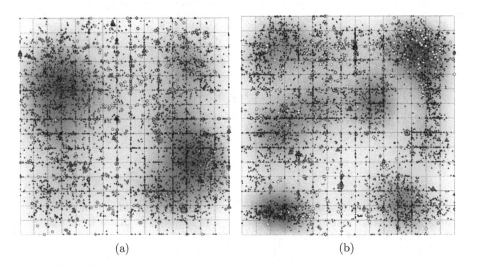

(a)   (b)

Fig. 2. Distribution of contigs with higher preference of the tissue specific reads occurrence; the case of $W_{(3,1)}$ is left and the case of $W_{(3,3)}$ is right. (Color figure online)

3   Results and Discussion

The visualization of the total transcriptome (via transformation of sufficiently long contigs into triplet frequency dictionaries $W_{(3,1)}$ and $W_{(3,3)}$) reveals a structuredness in that latter. Figure 1 shows the distribution of the contigs. Apparently, there are two clusters in the Fig. 1(a) and six clusters in Fig. 1(b). The clusters shown in Fig. 1 are provided by elastic map technique. Clustering with K-means for $K = 2$ and $K = 3$ is shown in Fig. 1(c) (two classes pattern) and Fig. 1(d) (three classes pattern). It should be said that these two patterns are very stable: more than 85 % of the runs of K-means converted to the same points distribution. The distributions provided by K-means with $K \geq 4$ were quite unstable.

Figure 2 answers the key question of the paper, whether the total transcriptome supports better assembling of tissue-specific ones, or not. Here we traced the distribution of the contigs with increased content of tissue-specific reads. To do that, we firstly identified the contigs with high level of mutual entropy (4), then checked what tissue reads prevail in a contig, and labeled it according to the tissue prevalence. Figure 2 shows the obtained distribution; here rosy circles represent cambium, green triangles represent needles and brown pentagons represent seedlings. Evidently, there is no preference in the tissue-specific enriched contigs over the clusters.

Also, Chargaff's discrepancies behaviour looks quite remarkable: for K-means classification with $K = 2$ the intraclasses discrepancies are $\xi_1 = 5.45 \times 10^{-4}$ and $\xi_1 = 5.90 \times 10^{-4}$, respectively, with the interclass discrepancy $\mu_{(1,2)} = 8.20 \times 10^{-4}$. Here the discrepancy between two classes seems to exceed those figures observed within a class. The situation is different, for $K = 3$. Here the intraclass discrepancies differ rather apparently, for three classes: $\xi_1 = 6.29 \times 10^{-4}$, $\xi_2 = 2.64 \times 10^{-4}$ and $\xi_3 = 5.91 \times 10^{-4}$, respectively. Obviously, the second class falls out of the general pattern of Chargaff's discrepancies. This fact may tell that the second class comprises the contigs from the opposite strands, unlike the first one and the third one. Same idea is supported by the figures of the interclass discrepancies; these are $\mu_{(1,2)} = 3.32 \times 10^{-4}$, $\mu_{(2,3)} = 4.27 \times 10^{-4}$, but $\mu_{(1,3)} = 3.71 \times 10^{-5}$.

That is a common place that a researcher is not guaranteed against the necessity to study total transcriptome, instead of a (tissue) specific one. Such situations may take place when a new (or rare) specimen is under analysis. Hence, one has to have a tool to evaluate the limits of knowledge that could be retrieved from the total transcriptome. Indeed, one may prefer to add sugar to a salty solution; others may want to add salt to a sweety sirup; nobody is able to distinguish the results. Meanwhile, significant number of chimeric transcripts may make a problem in analysis of a total transcriptome, say, in differential expression evaluation. If the tissue specificity of various reads is known á priori then one may eliminate the chimeric contigs from the ensemble due to specific entropy evaluation. The results presented above show some patterns revealed through clustering; this structuredness may be used for elimination of chimeric contigs. Nonetheless, the reliable approach to do it still awaits for further implementations.

References

1. Rahman, M.A., Muniyandi, R.C.: Review of GPU implementation to process of RNA sequence on cancer. Inf. Med. Unlocked **10**, 17–26 (2018)
2. Johnson, M.T.J., et al.: Evaluating methods for isolating total RNA and predicting the success of sequencing phylogenetically diverse plant transcriptomes. PLoS ONE **7**(11), 1–12 (2012)
3. Gorban, A., Popova, T., Zinovyev, A.: Codon usage trajectories and 7-cluster structure of 143 complete bacterial genomic sequences. Phys. A Stat. Mech. Appl. **353**, 365–387 (2005)
4. Gorban, A.N., Popova, T.G., Zinovyev, A.Y.: Seven clusters in genomic triplet distributions. Silico Biol. **3**(4), 471–482 (2003)
5. Sadovsky, M., Senashova, M., Malyshev, A.: Chloroplast genomes exhibit eight-cluster structuredness and mirror symmetry. In: Rojas, I., Ortuño, F. (eds.) Bioinformatics and Biomedical Engineering. LNCS, pp. 186–196. Springer International Publishing, Cham (2018). https://doi.org/10.1007/978-3-319-78723-7_16
6. Sadovsky, M.G., Senashova, M.Y., Putintseva, Y.A.: Chapter 2. In: Chloroplasts and Cytoplasm: Structure and Functions, pp. 25–95. Nova Science Publishers, Inc. (2018)
7. Fukunaga, K.: Introduction to Statistical Pattern Recognition. Academic Press, London (1990)
8. Gorban, A.N., Zinovyev, A.: Principal manifolds and graphs in practice: from molecular biology to dynamical systems. Int. J. Neural Syst. **20**(03), 219–232 (2010). PMID: 20556849
9. Mirkin, B.: The iterative extraction approach to clustering. In: Gorban, A.N., Kégl, B., Wunsch, D.C., Zinovyev, A.Y. (eds.) Principal Manifolds for Data Visualization and Dimension Reduction. LNCS, vol. 58. Springer, Heidelberg (2008). https://doi.org/10.1007/978-3-540-73750-6_6
10. Gorban, A.N., Zinovyev, A.Y.: Fast and user-friendly non-linear principal manifold learning by method of elastic maps. In: 2015 IEEE International Conference on Data Science and Advanced Analytics, DSAA 2015, Campus des Cordeliers, Paris, France, 19–21 October 2015, pp. 1–9 (2015)
11. Akinduko, A.A., Gorban, A.: Multiscale principal component analysis. J. Phys. Conf. Ser. **490**(1), 012081 (2014)
12. Mirkes, E.M., Zinovyev, A., Gorban, A.N.: Geometrical complexity of data approximators. In: Rojas, I., Joya, G., Gabestany, J. (eds.) Advances in Computational Intelligence. LNCS, pp. 500–509. Springer, Berlin (2013). https://doi.org/10.1007/978-3-642-38679-4_50
13. Mascher, M., Schubert, I., Scholz, U., Friedel, S.: Patterns of nucleotide asymmetries in plant and animal genomes. Biosystems **111**(3), 181–189 (2013)
14. Morton, B.R.: Strand asymmetry and codon usage bias in the chloroplast genome of Euglena gracilis. Proc. Nat. Acad. Sci. **96**(9), 5123–5128 (1999)
15. Forsdyke, D.R.: Symmetry observations in long nucleotide sequences: a commentary on the discovery note of Qi and Cuticchia. Bioinformatics **18**(1), 215–217 (2002)
16. Mitchell, D., Bridge, R.: A test of Chargaff's second rule. Biochem. Biophys. Res. Commun. **340**(1), 90–94 (2006)
17. Sobottka, M., Hart, A.G.: A model capturing novel strand symmetries in bacterial DNA. Biochem. Biophys. Res. Commun. **410**(4), 823–828 (2011)
18. Nikolaou, C., Almirantis, Y.: Deviations from Chargaff's second parity rule in organellar DNA: insights into the evolution of organellar genomes. Gene **381**, 34–41 (2006)
19. Albrecht-Buehler, G.: Fractal genome sequences. Gene **498**(1), 20–27 (2012)

A Hierarchical and Scalable Strategy for Protein Structural Classification

Vinício F. Mendes[✉], Cleiton R. Monteiro, Giovanni V. Comarela, and Sabrina A. Silveira

Department of Computer Science, Universidade Federal de Viçosa, Viçosa, Minas Gerais, Brazil
{vinicio.mendes,cleitom.monteiro,gcom,sabrina}@ufv.br

Abstract. Protein function prediction is a relevant but challenging task as protein structural data is a large and complex information. With the increase of biological data available there is a demand for computational methods to annotate and help us make sense of this data deluge. Here we propose a model and a data mining based strategy to perform protein structural classification. We are particularly interested in hierarchical classification schemes. To evaluate the proposed strategy, we conduct three experiments using as input protein structural data from biological databases (CATH, SCOPe and BRENDA). Each dataset is associated with a well known hierarchical classification scheme (CATH, SCOP, EC number). We show that our model accuracy ranges from 86% to 95% when predicting CATH, SCOP and EC Number levels respectively. To the best of our knowledge, ours is the first work to reach such high accuracy when dealing with very large data sets.

Keywords: Protein hierarchical classification · CATH · EC number · SCOP

1 Introduction

With the unprecedented increase in biological data generated over the last two decades, it is not possible to manually record the data that is being made available. We need robust and reliable computational techniques that can note the large volume of data available.

Proteins are the basic functional unit of the cells. Therefore, the academic community and the industry have spent a considerable amount of time studying techniques to understand the proteins composition, structure, and functionality. Applications range in several domains, for instance, developing a pest-resistant crop or a new treatment for a specific disease.

In this context, it may be valuable to a researcher, when studying a new (or less known) protein, to find other similar proteins, which have well-known characteristics. The reason is that based on such similarity, a lot can be learned about the protein under analysis. Therefore, developing new tools to predict the function of protein chains is a task worth pursuing.

© Springer Nature Switzerland AG 2019
I. Rojas et al. (Eds.): IWBBIO 2019, LNBI 11465, pp. 379–390, 2019.
https://doi.org/10.1007/978-3-030-17938-0_34

Although important, predicting the function of protein chains is a challenging task by itself, once it may depend on their sequence of amino acids and 3D structures [12,23], which are, in general, large and complex sets of information, possibly with the presence of noise [5]. Furthermore, there are several approaches to classify proteins according to their function, and for each one, there may be several thousand distinct classes. Hence, there is a computational challenge as well.

In this work, we are interested in the problem of performing structural classification of proteins. More specifically we are interested in protein hierarchical classification, which makes databases that implement hierarchical classification schemes particularly interesting to evaluate our strategy. Thus, we use CATH [13], SCOPe [2] and BRENDA [18] datasets and the classification schemes CATH, SCOP and EC number. We propose a hierarchical data mining strategy and, in order to overcome the issues related to working with structural data (e.g., complexity, a large amount of information and noise), we transform the raw data using Cuttof Scanning Matrix (CSM) [14], and reduce its size and noise via SVD (Singular Value Decomposition).

Our model has the advantage of being able to work with a large corpus of data (with more than 300,000 structures) and many distinct classes (approximately 2,000), which, in general, is a hard setup in data mining. Moreover, the model can be applied in other scenarios, as long as the classes are organized in a hierarchical fashion.

We conducted experiments in order to evaluate and validate our model. We show that the accuracy of the model ranges from 86% to 95% when predicting the first and all CATH, SCOP and EC number levels respectively. To the best of our knowledge, ours is the first work to reach such high accuracy when dealing with very large data sets.

The remainder of this article is organized as follows. In Sect. 2, we present our sources of data and data preprocessing steps we conducted and we describe our methodology, which is composed of data modeling and data reduction via SVD, and a hierarchical approach combined with machine learning techniques. In Sect. 3, we evaluate the model and show its usefulness when predicting the classes of the databases. We position our work in the literature in Sect. 1.1, and finally in the Sect. 4, we present final observations, limitations, and perspectives.

1.1   Related Work

In [3], the authors propose a strategy for class classification based on SCOP classification scheme. The experiments were performed on protein sequences using combined PSI-BLAST at the frequency of collocation of AA pairs. The work used classifiers implemented in Weka for class prediction, where they provided the final results with accuracy varying from 61 to 96%.

In [11], the authors developed a classification-based prediction model where data from protein backbone interpolations were used. In the training phase of the model, the descriptor was extracted based on proteins rays processed by the PDB files, and in the classification stage, the Fuzzy decision tree was used. The results obtained by classification vary from 96 to 99% accuracy, from the 6145 protein chains used.

In the work of [14], the technique called Cuttof Scanning Matrix (CSM) is proposed to predict families and superfamilies of proteins in large structural bases. The approach uses patterns of distances between residues as a way of capturing protein similarity. The results achieved by the research reached (on superfamilies) an average accuracy of 98.2%. We also use the CSM technique described in [14], but we propose here a hierarchical and scalar approach, capable of working at all levels of classification, with very large datasets.

In [15] developed a method to find proteins with similar structures and classify them from topological invariants [16]. In this method, the invariants of each chain are applied to a classification strategy, responsible for predicting its structural levels of CATH. In their results, a total of 96% of success was obtained when inferring CATH levels.

Another example is the work presented in [21], where the authors use SVM (Support Vector Machines) to predict CATH classes. The author processed the sequences of the protein chains in order to obtain protein sequences in a more regularly defined feature space. The results obtained show a range of 70 to 80%. However, in this work, the characteristics are related to sequences, not to protein structures.

In recent years, the number of protein structures has increased significantly. In [7], this increase is presented, with the various computational methods that were created in order to predict the structural class of proteins. These methods were often tested on small data sets, characterized by different sequence homologies, and were not reliably compared with other methods.

In the study [4], the automatic prediction of enzymatic functions is an important item of study due to the costs and lengthy nature of laboratory identification procedures. According to the authors, the hierarchical structure of the EC nomenclature is adequate for automatic function prediction, where several methods and tools have been proposed to classify enzymes. According to them, most classification studies are limited to specific classes or to specific levels of the hierarchy, providing a limitation on the data to be used.

"Evolution is a random generator of possible improvements in the face of the environmental challenges an organism experiences, where survival of the fittest ensures the retention of successful solutions into future generations" [22]. The authors state that minor changes in structure or context can have dramatic effects on functionality.

With the study carried out in [9] and [10], the search for protein function can be performed through dissimilarity between proteins and/or residues. The paper presents a way of calculating dissimilarity through contact maps from conserved amino acid residues.

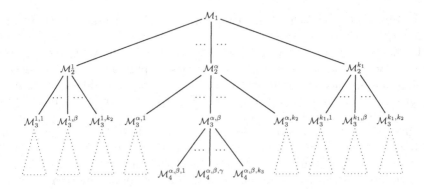

Fig. 1. Representation of the model $\mathcal{M}$ as a tree structure. The root (level 1) of the tree is responsible for classifying first level. The level i ($i > 1$) is responsible for classifying i-th CATH subclass. k_i ($1 \le i \le 3$) represents the number of distinct values of the i-th CATH subclass, which are not necessarily the same over all nodes of the same level. Dotted areas represent subtrees removed from the figure due to space constraints.

2   Methods

In this section, we describe our strategy to predict hierarchical levels for protein structures.

2.1   Data

Hierarchical Schemes. We used 3 different hierarchical classification schemes to validate the quality, generality and real-world applicability of our method. We performed hierarchical protein structural classification for CATH, SCOP, and EC number[1]. Protein structural data were obtained from Protein Data Bank (PDB) [17] and organized according to each scheme.

CATH [19] is organized in the levels Class, Architecture, Topology, and Homology. The first CATH level (Class) is the most general, which is responsible for describing the content of α-helices and β-sheets. The second level (Architecture) is responsible for describing the organization of secondary structures. The third level (Topology) differentiates proteins of same Class and Architecture based on their folds and their function. The fourth level (Homology) is related to structural and functional similarity.

SCOP (Structural Classification of Proteins) organizes protein domains into a hierarchy according to their structural and evolutionary relationships. SCOP levels are Family, Superfamily, Common Fold and Class [2].

Enzyme Commission number (EC number) is a classification scheme specific for enzymes and has 4 levels. From left to right, each level adds information about protein catalytic function. The database used for this methodology is BRENDA

---

[1] http://www.cathdb.info, http://scop.berkeley.edu/, https://www.brenda-enzymes. org.

(BRaunschweig ENzyme DAtabase), as a database containing comprehensive enzymatic and metabolic data [18].

Modeling. For simplicity, throughout this work, we will refer to different levels of the hierarchical classification schemes as *levels*. For instance, when we refer to level 1, it means the highest level of a hierarchy, level 2 means the second highest level of a hierarchy, and so on.

Levels with less than 10 examples were not considered, as we perform a 10 fold cross-validation. The resulting dataset has 97.11% of the records originally obtained, distributed in 3283 combinations of distinct CATH examples. For the SCOPe database, the results follow with 95.62% of the data and 4154 distinct examples; and in BRENDA it resulted in 97.96% of records and 1771 distinct examples.

Our data reduction approach is twofold. First, we transform the set of protein structures in the XYZ coordinate format into a much more compact representation, the Cutoff Scanning Matrix (CSM) [14]. We then further reduced the size of the dataset by applying SVD (Singular Value Decomposition) to the CSM.

In [14], the authors proposed the CSM, a matrix generated by the cumulative counts distributions based on the Euclidean distances of cumulative contact between the alpha carbons, making possible to represent a set of proteins. Each row in this matrix represents a chain of proteins through the distribution of contact pairs within a given distance.

In order to build the matrix, we vary the distance from 0 to 30 Å with a 0.2 Å step size, totaling 151 columns. It is important to mention that when computing the distance between residues, we considered the α-carbon atom as the residual's representative. We denote the resulting matrix by $\mathbf{X}$, with n rows and 151 columns.

Dimensionality Reduction. Dimensionality and noise of our data encoded as a distance matrix, $\mathbf{X}$, were reduced using Singular Value Decomposition (SVD) in a similar manner to [20]. We generated 100 approximations of our data representation with singular values ranging from 1 to 100, with step 5, and the matrix that resulted in the best classification model was selected. The reduced matrix is called or $\mathbf{Y}$.

2.2   A Hierarchical Model

An immediate way to classify proteins according to their classes would be the training model on the rows of $\mathbf{X}$ (or $\mathbf{Y}$) and their respective classes. Unfortunately, such an approach is not feasible because of nearly five thousand distinct classes in our largest dataset, which represents a major challenge for standard data mining.

To overcome this problem, we use hierarchically supervised learning methods, motivated by the fact that classes are actually hierarchical, four-level structures.

The main idea is that the i-th level of the model will be responsible for classifying the i-th level of the CATH, SCOP, or EC number.

For this purpose, we first created a $\mathcal{M}_1$ template to classify the first level. Similarly, we created a model $\mathcal{M}_2^a$ to classify the second level, since the first level is a. More generically, we created a model $\mathcal{M}_i^{a_1,\dots,a_{i-1}}$, $1 < i \le 4$, to classify the i-th level, given that the previous levels are $a_1, \dots, a_{i-1}$.

Overall, our model, denoted by $\mathcal{M}$, can be seen as a tree, which is depicted in Fig. 1.

Training the Model. A natural question that may arise at this point is: given $\mathbf{X}$ (or $\mathbf{Y}$) and the classes of each of its rows, how can the model $\mathcal{M}$ be trained?

We start by introducing some necessary notation. Let $\mathbf{Z}$ be a matrix ($\mathbf{Z}$ can be the CSM or the SVD-reduced CSM). Denote the rows of $\mathbf{Z}$ by a list $Z = (\mathbf{z}_1, \dots, \mathbf{z}_n)$, where $\mathbf{z}_i$ is associated to a protein with class c_i. Let $C = (c_1, \dots, c_n)$, $Z^{a_1,\dots,a_j}$ being the list of all elements of Z which have the first j subclasses equal to $a_1, \dots, a_j$, and $C^{a_1,\dots,a_j}$ their respective classes. Finally, for any $S \subseteq C$, we define $S(i)$ as a new list, which maps each element of S to its j-th subclass.

Then the procedure to train $\mathcal{M}$ is threefold: It starts by choosing a standard supervised learning technique, $\mathcal{A}$, to be used in each node of the tree; after the choice, train the model $\mathcal{M}_1$, using $\mathcal{A}$, on Z and $C(1)$; and at the end train the model $\mathcal{M}_i^{a_1,\dots,a_{i-1}}$, using $\mathcal{A}$, on $Z^{a_1,\dots,a_{i-1}}$ and $C^{a_1,\dots,a_{i-1}}(i)$.

The problem of training a model in a large corpus with many different classes is reduced to the problem of training several models, in smaller datasets, with considerably less distinct classes. In addition, another advantage of our approach is that the tree nodes are independent, that is, since the data is properly partitioned on all nodes, the models can be trained in parallel.

Classifying a New Instance. Once the model $\mathcal{M}$ is trained, given an unlabeled row of the data matrix, $\mathbf{z}$, the CATH, SCOP or EC number levels can be predicted by traversing the tree associated with the model, predicting one level of $\mathbf{z}$ per level of $\mathcal{M}$. More specifically, the prediction can be recursively performed as follows:

1. Predict the first level CATH, SCOP or EC number of $\mathbf{z}$, denoted by a_1, using $\mathcal{M}_1$;
2. Given that the first level $i-1$ CATH, SCOP or EC number of $\mathbf{z}$ were predicted as $a_1, \dots, a_{i-1}$, we predict the i-th level of $\mathbf{z}$, denoted by a_i, using $\mathcal{M}_i^{a_1,\dots,a_{i-1}}$.

3   Results

This section has two main objectives, both related to the model described in Sect. 2: first, to show that the model performs well when classifying the CATH, SCOP and EC number levels of protein structures; second, to show that the methodology is able to scale to large datasets. More specifically, we want to answer the following questions:

1. Which is an appropriate learning algorithm, $\mathcal{A}$, for each node of our hierarchical model, $\mathcal{M}$?
2. Is there a significant performance difference when training $\mathcal{M}$ with $\mathbf{X}$ (non reduced) or $\mathbf{Y}$ (SVD-reduced)?
3. How does $\mathcal{M}$ perform in different levels of the tree?
4. How does $\mathcal{M}$ perform in different nodes of the tree?

To answer those questions, Sect. 3.1 describes the setup of our experiments, while Sect. 3.2 presents the model evaluation.

3.1   Experimental Setup

We selected three candidates for the learning algorithm $\mathcal{A}$: NB (Gaussian Naive Bayes) [8], NN (Nearest Neighbors) [24] with $k = 3$, and RF (Random Forest) [1] with 500 decision trees. Those classifiers were chosen because they have better training efficiency on large datasets when compared to optimization-based approaches (e.g., Support Vector Machine [6]). The hyper-parameters of each model were selected by conducting initial experiments on a small random sample of the data using an exhaustive grid search and cross-validation.[2]

In Fig. 2, bars represent the metric average over the 10 testing folds, and error bars are 95% confidence intervals. All models were trained and tested with the SVD-reduced data matrix, $\mathbf{Y}$. The model based on Random Forests outperforms the others with respect to all metrics.

We evaluated our methodology using a stratified 10-fold cross-validation strategy. In other words, we randomly partitioned the rows of the data matrix (or SVD-reduced version) in 10 folds of equal size, keeping, for each fold, the corresponding CATH, SCOP or EC number classes of each line. Then, we conducted a 10-round experiment. In each round, we retained one fold for testing and the other nine for training the model. When testing the model, we compared the predicted and true classes using four different metrics: precision, recall, F1-Score, and accuracy. Finally, we computed the average and 95% confidence interval for each metric, over the results obtained in each one of the 10 folds. Since our work deals with a multi-class classification task, the metrics precision, recall, and F1-Score were computed for each class and then we took the weighted average of the results. Note that taking the weighted average over all classes may result in F1-Scores that are not between precision and recall (which does not happen on standard binary cases).

All the experiments were performed in a small-sized and shared computational infrastructure. Our jobs were allowed to use at most 45 GB of RAM and 5 CPU cores of an AMD Opteron 6376 processor. Although the models, which compose tree, can be trained in parallel, due to the hardware limitations we had to train each one individually. The whole training process, for the Random Forest classifier[3], took approximately 8 h. We would like to emphasize that by exploring

---

[2] We intend to make the code and details related to the models publicly available upon publication of this manuscript.

[3] Using the Python's Sklearn implementation (http://scikit-learn.org).

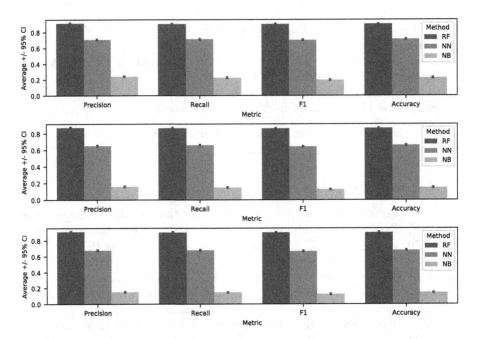

Fig. 2. Model performance for the three different classifiers in the databases used. On the *first line*, performance for CATH; on the *second line*, performance for SCOP; and on the *last line*, performance for BRENDA.

parallelism and more efficient implementation of the learning algorithms, such time can be significantly reduced.

3.2   Model Evaluation

Now we move to answer the four questions stated at the beginning of this section. First, we study the impact of different machine learning algorithms for training the model $\mathcal{M}$. Second, we analyze the impact of working with the original or SVD-reduced data matrix. Third, we compare the performance of the model in relation to the partial subclasses of CATH, SCOP, and EC number. Finally, we look at the results given by each node of the tree.

The Learning Algorithm. We begin by comparing the hierarchical models constructed with different algorithms of machine learning. The results obtained when training and testing the model in the SVD-reduced matrix are presented in Fig. 2, where it is possible to observe that the model based on Random Forests significantly outperforms models based on Gaussian Naive Bayes and Nearest Neighbors. In more detail, the results obtained by the CATH database reached in RF an F1 score of approximately 90%, NN and NB reached 70% and 19%, respectively. For this reason, from now on, we will report only the results related to the model based on the RF classifier. In the SCOPe and BRENDA databases,

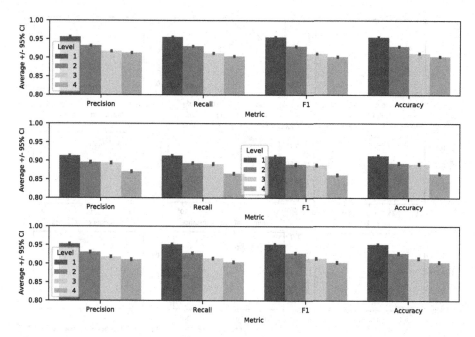

Fig. 3. Comparing the performance of the model at different levels at the databases utilized. On the *first line*, performance for CATH; on the *second line*, performance for SCOP; and on the *last line*, performance for BRENDA. Experiments performed with SVD-reduced data matrix and Random Forest Classifier. Overall, model performance degrades gracefully as more subclasses of CATH, SCOP, and EC number are predicted.

the results of F1 in RF reach approximately 86% and 94%, respectively. Although the results for RF are significantly better, it is important to notice that RF is a more complex model than the other two. For instance, while training RF took 8 h, the process of training NB and NN took no more than one hour each.

The Level of the Tree. Next, we evaluate the problem of understanding how the performance of the models, of each level of the tree, behaves in predicting partial levels. To this end, for each instance of the test set, we compare the expected and actual levels to the $i(1 \leq i \leq 4)$ level. For example, for the first level, we confront the evaluation metrics by comparing the predicted and actual level and for the second level, we proceed in the same way but considering the first two levels together.

The results for the above experiment are shown in Fig. 3. It can be observed that the performance of models at different levels degrades as levels move away from the root of the tree. Taking the precision score, for example, in the CATH database we have $\mathcal{M}_1$ scores 95% (significantly higher than previous results), while models in the second, third and fourth levels score 93%, 91.7%, and 91.3% respectively. For the SCOPe database, the precision reaches 91%, 89.5%, 89.3%, and 87% for the levels and in the BRENDA database with a 95%, 93%, 91.8% and 91% levels 1, 2, 3 and 4, respectively. This degrading performance can be

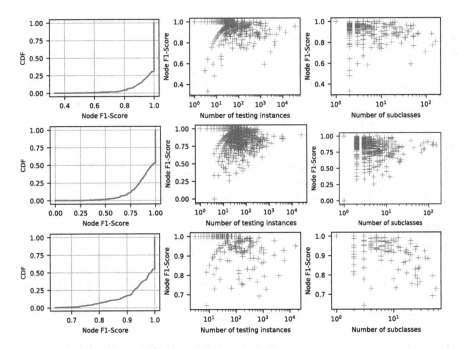

Fig. 4. Comparison of performance for CATH (first row), SCOP (second row) and EC numbers (last row), via F1 score, of different tree nodes. *Left Column:* distribution of the F1 scores of all nodes. *Central column:* F1-score *versus* the number of test instances sorted by each node. *Right Column:* F1-score *versus* the number of subclasses ordered by each node, i.e., their number of children.

explained due to the recursive nature of our model. Once an error is made at the i level, the test instance will be passed to the wrong model at the next level, increasing the chance of another error.

The Nodes of the Tree. In the last section, we showed that the level of the tree is an important factor when analyzing the accuracy of our model. In this section, we look at a different, but related issue, the performance of individual nodes of the tree. Our goal is to try to understand if the mistakes of our model happen in any particular type of situation. To achieve this goal, we analyze the performance of each node by predicting its corresponding subclass CATH, SCOP, and EC number. For simplicity, the results reported here are related to the weighted average of the 10 folds of our experiments, that is, they are the means along the 10 folds.

Figure 4 presents the results. First, it is possible to notice that 80% of the tree nodes have an F1-score of 100%, showing that most nodes actually perform considerably well. Approximately 20% of the nodes have and F1-score between 70% and 100%, and only a small fraction of nodes is related to lower F1-score levels.

Second, we try to shed light on the reason why some nodes are performing poorly. To that end, we try to correlate the F1-score of each node to two other variables: the number of testing instance in such node, and its number of sub-classes. Since the test set is a stratified random sample of the population, we conjecture that the larger the values of these two variables, the harder it is for the classifier. From Fig. 4, it is not possible to validate our conjecture. In other words, the poor performance of each node does not seem to be correlated with the number of testing instances or subclasses.

Investigating reasons for the poor performance of some nodes, and improving our model's predictions are directions for future research.

4   Conclusions

In this work, we propose a hierarchical strategy to perform protein structural classification. To validate our strategy, we used datasets of protein structures annotated with hierarchical schemes CATH, SCOP, and EC number to show that we are is able to predict their hierarchical levels.

Overall, the methodology presents a precision of 91% when predicting the four levels of CATH, 87% for SCOP and 91% for EC number, despite the challenges of working with more than 300,000 structures distributed in approximately two thousand different classes. In addition, the results are significantly better when looking at partial levels. For example, our model had a precision of 95% when predicting the first level of CATH (Class), 93% for the first two levels (Class + Architecture) and 91% for the first three (Class + Architecture + Topology).

Regarding the performance of the model, we observed that the Random Forest classifier presented better results when compared to Naive Bayes and Nearest Neighbor. In addition, we showed that the data matrix reduced by SVD improved the accuracy of the classification, thus that SVD helps to reduce the dimensionality and the noise in the data.

As future work, we are interested in improving the accuracy of the model exploring other machine learning techniques such as SVMs and Neural Networks. We intend to compensate for the slower training times, exploring parallelism, which can be easily achieved with our hierarchical model. In addition, we would like to investigate other representations for protein structural data.

Funding. This study was financed in part by the Coordenação de Aperfeiçoamento de Pessoal de Nível Superior - Brasil (CAPES) - Finance Code 001, Conselho Nacional de Desenvolvimento Científico e Tecnológico (CNPq) and Fundação de Amparo à Pesquisa do Estado de Minas Gerais (FAPEMIG).

References

1. Breiman, L.: Random forests. Mach. Learn. **45**(1), 5–32 (2001)
2. Chandonia, J.M., et al.: SCOPe: classification of large macromolecular structures in the structural classification of proteins-extended database. Nucleic Acids Res. **47**, D475–D481 (2018)

3. Chen, K.E., et al.: Prediction of protein structural class using novel evolutionary collocation based sequence representation. J. Comput. Chem. **29**(10), 1596–1604 (2008)

4. Dalkiran, A., et al.: ECPred: a tool for the prediction of the enzymatic functions of protein sequences based on the EC nomenclature. BMC Bioinform. **19**(1), 334 (2018)

5. Gu, J., et al.: Structural Bioinformatics, vol. 44. Wiley, London (2009)

6. Hearst, M.A., et al.: Support vector machines. IEEE Intell. Syst. Appl. **13**(4), 18–28 (1998)

7. Kedarisetti, K.D., et al.: Classifier ensembles for protein structural class prediction with varying homology. Biochem. Biophys. Res. Commun. **348**(3), 981–988 (2006)

8. McCallum, A., et al.: A comparison of event models for Naive Bayes text classification. In: AAAI/ICML-98 Workshop on Learning for Text Categorization, pp. 41–48 (1998)

9. Melo, R.C., et al.: A contact map matching approach to protein structure similarity analysis. Genet. Mol. Res. **5**(2), 284–308 (2006)

10. Melo, R.C., et al.: Finding protein-protein interaction patterns by contact map matching. Genet. Mol. Res. **6**(4), 946–963 (2007)

11. Mirceva, G., et al.: A novel approach for classifying protein structures based on fuzzy decision tree. In: 2018 2nd International Symposium on Multidisciplinary Studies and Innovative Technologies (ISMSIT), pp. 1–5. IEEE (2018)

12. Nelson, D.L., et al.: Lehninger Principles of Biochemistry, 6th edn. Macmillan Learning, New York (2013)

13. Pearl, F.M., et al.: The CATH database: an extended protein family resource for structural and functional genomics. Nucleic Acids Res. **31**(1), 452–455 (2003)

14. Pires, D.E., et al.: Cutoff Scanning Matrix (CSM): structural classification and function prediction by protein inter-residue distance patterns. BMC Genomics **12**(4), S12 (2011)

15. Rogen, P., et al.: Automatic classification of protein structure by using Gauss integrals. Proc. Nat. Acad. Sci. U.S.A. **100**(1), 119–124 (2003)

16. Rogen, P., et al.: A new family of global protein shape descriptors. Math. Biosci. **182**(2), 167–181 (2003)

17. Rose, P.W., et al.: The RCSB protein data bank: integrative view of protein, gene and 3D structural information. Nucleic Acids Res. **45**, D271–D281 (2016)

18. Schomburg, I., et al.: BRENDA, enzyme data and metabolic information. Nucleic Acids Res. **30**(1), 47–49 (2002)

19. Sillitoe, I., et al.: CATH: comprehensive structural and functional annotations for genome sequences. Nucleic Acids Res. **43**(D1), D376–D381 (2015)

20. Silveira, S.A., et al.: ENZYMAP: exploiting protein annotation for modeling and predicting EC number changes in UniProt/Swiss-Prot. PloS One **9**(2), e89162 (2014)

21. Sun, X.D., et al.: Prediction of protein structural classes using support vector machines. Amino Acids **30**(4), 469–475 (2006)

22. Tyzack, J.D., et al.: Understanding enzyme function evolution from a computational perspective. Curr. Opin. Struct. Biol. **47**, 131–139 (2017)

23. Wei, D., Xu, Q., Zhao, T., Dai, H. (eds.): Advance in Structural Bioinformatics. AEMB, vol. 827. Springer, Dordrecht (2015). https://doi.org/10.1007/978-94-017-9245-5

24. Weinberger, K., et al.: Distance metric learning for large margin nearest neighbor classification. Adv. Neural Inf. Process. Syst. **18**, 1473 (2006)

Protein Structural Signatures Revisited: Geometric Linearity of Main Chains are More Relevant to Classification Performance than Packing of Residues

João Arthur F. Gadelha Campelo[1,2(✉)], Cleiton Rodrigues Monteiro[3],
Carlos Henrique da Silveira[4], Sabrina de Azevedo Silveira[3],
and Raquel Cardoso de Melo-Minardi[1]

[1] Department of Computer Science, Universidade Federal de Minas Gerais,
Belo Horizonte, Brazil
gadelha@ufmg.br
raquelcm@dcc.ufmg.br
[2] Department of Biochemistry and Immunology,
Universidade Federal de Minas Gerais, Belo Horizonte, Brazil
[3] Department of Computer Science, Universidade Federal de Viçosa, Viçosa, Brazil
{cleiton.monteiro,sabrina}@ufv.br
[4] Advanced Campus at Itabira, Universidade Federal de Itajubá, Itajubá, Brazil
carlos.silveira@unifei.edu.br

Abstract. Structural signature is a set of characteristics that unequivocally identifies protein folding and the nature of interactions with other proteins or binding compounds. We investigate the use of the geometric linearity of the main chain as a key feature for structural classification. Using polypeptide main chain atoms as structural signature, we showed that this signature is better to precisely classify than using Cα only. Our results are equivalent in precision to a structural signature built including artificial points between Cαs and hence we believe this improvement in classification precision occurs due to the strengthening of geometric linearity.

Keywords: Structural signature · Protein main chain ·
Geometric linearity · Structure classification

1 Introduction

A high proportion of the PDB structures (PDB: Protein Data Bank - [3]) remains as hypothetical proteins or proteins of unknown functions because sequence

Electronic supplementary material The online version of this chapter (https://doi.org/10.1007/978-3-030-17938-0_35) contains supplementary material, which is available to authorized users.

I. Rojas et al. (Eds.): IWBBIO 2019, LNBI 11465, pp. 391–402, 2019.
https://doi.org/10.1007/978-3-030-17938-0_35

alignment-based methods have failed to match them to functionally characterized proteins [13]. Nevertheless, proteins whose 3D structures are known open up several possibilities of annotating their function based on structural comparisons.

Due to the breakthroughs in genome sequencing and structure resolution, the amount of sequenced proteins that have their structures determined grows rapidly [18]. Furthermore, the experimental elucidation of function of such structures is labour intensive and is a practical barrier to the use of this information [5]. In this context, the use of automated methods for function annotation becomes mandatory. It is now known that protein structures are more conserved than their sequences [9], and consequently, structure-based techniques have gained momentum [18].

Even though the function cannot be directly inferred from protein folding, structural data can be used to detect proteins with similar functions whose sequences have diverged throughout their evolution [14]. Nature uses only a few thousand types of foldings to create all known protein structures [6]. Chotia claims that all proteins of all kinds can be represented by about 1,000 different foldings [7].

Structural Classification of Proteins - SCOP - [1], which is used in this study, is the largest manually curated database on structural classification, based on the similarity of those structures and their amino acid sequences. SCOP classifies PDB proteins based on the following system: protein structures are divided into discrete domains that are hierarchically classified into levels of (1) class, (2) folds, (3) super-families and (4) families. In this classification scheme, proteins of the same family are the most structurally similar amongst themselves.

In a given set of structurally similar proteins, a possible approach to predict function is to define structural signatures, which are a set of characteristics that unequivocally identifies a protein folding and possibly interactions that can occur with other proteins or binding compounds. Such characteristics are concise representations of protein structures. We believe that their use in predicting protein function is a step further in comparison to the sequence-based methods. For example, Pires [15] investigates inter-residue distance patterns in a cutoff distance matrix for structural classification and function prediction, a technique called *CSM*.

CSM is the state-of-the-art in large-scale structure classification and is independent of structural alignment algorithms. It creates characteristics vectors that represent distance patterns between protein residues and uses those vectors as evidence for classification. *CSM* matrices were built from the geometric position of $C\alpha$ only. The authors report that they conducted experiments with other centroids rather than the $C\alpha$, such as the $C\beta$ or the last heavy atom (LHA) of the side chain. The alpha carbon ($C\alpha$) in organic molecules refers to the first carbon atom that attaches to a functional group, such as a carbonyl. The second carbon atom is called the beta carbon ($C\beta$). $C\alpha$ presented the best results in all experiments, a fact that they claimed "demanded deeper investigation". In this study, we investigate the use of the backbone atoms as a more discriminating structural signature. Our hypothesis was that the addition of intermediate points, by using other backbone atoms, strengthened the geometric linearity of the main chain and improved family differentiation using the *CSM* technique. In this context, we are calling geometric linearity the arrangement of points representing atomic positions along a polygonal in three-dimensional geometric space.

The contribution of this study is an investigation of the backbone linear geometric disposition (or of the Cα themselves, should fictional intermediate points be added between them) as a better structural signature. The discriminating capacity for classification and function prediction was compared in relation to the original CSM signature obtained using exclusively the Cα, and improvements were observed. Using main chain atoms as a structural signature, we were able to increase the accuracy of the *Full-SCOP* base family classification by up to 10.3% in relation to the accuracy of the original technique, which uses only Cα position.

2   Methods

To show that linearity of the polypeptide chain is preserved and more discriminating amongst families, we built classifiers following the CSM strategy. We chose this technique because it is the state-of-the-art in structural classification and independent on structural alignment algorithms (see Sect. 2.1). CSM results were used as a control group to compare with our results.

The organization of this section is as follows: Sect. 2.1 explains conceptual and operational details necessary to understand our method; Sect. 2.2 presents data sets used and Sect. 2.3 introduce the experiments performed.

2.1   Concepts

Cutoff Scanning Matrix - *CSM*. In this study, we adapted the CSM technique to find atomic distance patterns, and not just residue distance patterns. First, we calculated the Euclidian distance between all pairs of atoms of the set and defined a cutoff (distance threshold) to be considered and a distance step. We calculate the frequency of atomic pairs that are close to each other considering a distance threshold. Singular Value Decomposition - *SVD* - [4] is used as a pre-processing step to reduce dimensionality and noise.

Similarly to CSM paper, we varied the distance threshold from 0.0 Å to 30.0 Å in 0.2 Å increments, which generates a vector with 151 entries for each atomic set. Together, those vectors form the CSM. To summarize it, each line of the matrix represents a protein (or an atomic set derived from a protein), and each column represents the frequency of atomic pairs within a certain distance.

The intuition behind this method is that proteins with different foldings and functions present significantly different distances distribution for their residues-atoms or different packing patterns. On the other hand, it is expected that proteins with similar structures would have similar distance distributions for their residues-atoms, information which is captured by CSMs.

The cut variation (range) adds important information related to protein packing and captures, implicitly, the shape of proteins. This means that CSM manipulates two levels of crucial structural information: relevant non-local and local contacts. We can also observe that protein shapes directly interfere in the subjacent contact network, which reflects on protein folding, as reported by Soundararajan [17].

Linearity of the Main Chain and Intermediate Points. In this study, we investigate of the use of geometric linearity of the main chain (or of the $C\alpha$ themselves) and the addition of intermediate points between $C\alpha$s as a better discriminating structural signature. Our hypothesis was that the addition of this artificial intermediate points would strengthened linear characteristic of the main chain and improve classification precision.

Therefore, the relatively larger inter-point distance in the model that uses only $C\alpha$s (in general, the separation distance between $C\alpha$s in adjacent amino acids in a protein is about 3.8 Ås [12]), in comparison to the model that also uses C and N atoms ($C\alpha - C = 1.52$ Å, $C - N = 1.32$ Å e $N - C\alpha = 1.46$ Å [12]) brings uncertainty to the correct sequence of atoms in the set and, consequently, difficulty to the classification task. This very geometric linearity can be achieved (or simulated/strengthened) by adding artificial intermediate points between $C\alpha$s with no harm to the classification accuracy. Figure 1 illustrates this idea.

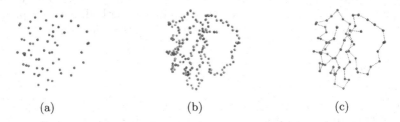

(a)   (b)   (c)

Fig. 1. Main chain atoms (PDB id *1TEC:I*). (a) $C\alpha$ atoms only. Geometrically, it is similar to a disordered set of points. (b) $C\alpha$, C and N atoms. The geometric linearity of the main chain is turned clearer. (c) A similar geometric linearity can be "achieved" by adding intermediate artificial points between the $C\alpha$ atoms.

In order to show the relevance of main chain geometric linearity information as a structural signature, we conducted experiments where artificial points were inserted between the $C\alpha$ points used in the model. We respect the order in which the atoms appear in the main chain (be it the $C\alpha$ or the complete $C\alpha - C - N$ sequence) because we believe that, in some way, this sequence information is indirectly captured by CSM. We conducted several tests varying the distance between intermediate points by 0.2 Å, 0.4 Å, 0.6 Å and 0.8 Å. We chose this interval (0.2 Å–0.8 Å) because the increment of CSM is 0.2 Å and, at the limit distance of 0.8 Å, it is still possible to include an intermediate point between backbone atoms relatively close to the middle point of those distances.

2.2   Data

Our datasets were built using the same protocol as the original CSM study. They used Gold-standard dataset, *SCOP* version 1.75 and *6SSE, 5SSE, 4SSE* and *3SSE* [10].

Initially, we included all the chains in those databases. We used only the first model for PDB files with several models. In files with more than one chain, the chains were separated according to *SCOP* classification and treated individually.

Some of those structures were discarded because they became deprecated or the chain was too short (less than 10 residues or atoms). We present in Supplementary Material a complete list of included chains, as well as the files that became exceptions (removed from the study) and the reasons therefor. Table 1 summarizes some statistics of the datasets.

Table 1. Statistics of the database chains used.

Dataset	Total	Used	% Discarded
Gold-Standard	899	895	0.44%
6SSE	2,315	2,303	0.52%
5SSE	2,930	2,853	2.63%
4SSE	1,756	1,720	2.05%
3SSE	880	866	1.59%
Full-SCOP 1.75	207,890	201,771	2.94%

2.3   Experimental Design

To show that our model that characterizes the linearity conformation of the main chain of protein structures is precise to discriminate families, we built classifiers and assessed their precision.

The methodology of this work was divided into two sets of experiments: the first aims to show that a model that uses atoms of the main chain is more discriminating amongst families than the model that uses $C\alpha$ atoms only. The second shows that the main factor that improves differentiation amongst families is not necessarily the main chain atoms, but rather the geometric linearity characteristic of the main chain. Another purpose of the second set of experiments is to show that the position of the atoms of the main chain and the angles that they form with each other (angles ϕ and ψ) only interfere with the discriminating capacity of the classifiers when intermediate points are added and their linear disposition is strengthened.

A different CSM was built for each experiment conducted. In CSM matrices, each line represents characteristics vector of the distance patterns between the atoms of the set of one of the protein chains. Those vectors (lines of the matrix) have 151 positions where the position i of the vector contains the frequency of atoms at a distance of 0.2 Å $\times i$ from each other. After CSM matrix was built, its dimensionality and noise were reduced by singular value decomposition (see Sect. 2.1). As in the original CSM work, we used 9 singular values. That vector set could then be used as input for a classification algorithm.

Evaluation. As in the original CSM study, all classifiers were tested using *Weka library* (http://www.cs.waikato.ac.nz/ml/weka/). We used the *k-nearest neighbors algorithm (k-NN)* [19] classifier - with best value of $k = 5$ (data not shown) - because it was reported as the highest performance in [15]. We used *10-fold cross-validation* and, therefore, only monomer groups with ten or more representatives were used. Classification performance was assessed with the following metrics: *precision* ($precision = TP/(TP+FP)$), *recall* ($recall = TP/(TP+FN)$), score *F1* (harmonic mean of precision and recall: $2(\frac{Precision \times Recall}{Precision + Recall})$) and area under a ROC curve (AUC).

In order to compare the mean values of the metrics more precisely, we ran each classifier 30 times (Supplementary Material). For each run with the same index (for example: tenth run), we used the same random seed for all classifiers. For runs with different indexes, we used different random seeds.

Since we made sure that the same seed was used for the same indexes and different seeds were used for different indexes, we were able to assertively compare the values of the results of each classifier with the same run index.

Statistically, we can interpret the results of same index classification for each classifier as the performance (metrics) of the same individual (same set of chains to be classified and same population division run by the *10-fold cross-validation*, since the random seed is the same), measured after receiving different treatments (each classification).

Consequently we were able to use a hypothesis test on paired samples and, by definition, the null hypothesis was that the mean value of the metrics of the 30 runs of each classifier would be identical. We used a *t-Student test on the paired samples*, with $30°$ of freedom, to find the respective *p-values*. We accepted the null hypothesis of identical means when $p > 0.05$. We rejected it otherwise.

The *p-value* tables are in the Supplementary Material (Supplementary Tables 7–30).

3   Results

The first set of experiments show that a model that uses atoms of the main chain is more preserved and discriminatory amongst families than the model that uses $C\alpha$ atoms only. In this set, four experiments were conducted and compared:

1. Using only alpha carbons of the monomer chains, an experiment we called $C\alpha$ in this study, to be used as a control group and enable comparison to the original CSM.
2. Using main chain atoms ($C\alpha$, C, N), an experiment we called *Backbone* in this study, to evaluate the conservation degree and inter-family discriminating ability. We suppressed the oxygen (O) of the carbonyl group because its geometric position is only one step removed from the geometric position of N.

3. Using only atoms of the side chains, an experiment we called *Side*, to enable comparison to the classification results using main chain atoms.
4. Using all atoms of the structure, an experiment we called *All*, to contrast the *Side* classification results to the *Backbone* classification results and illustrate the influence of the main chain on the discriminatory capacity of those two other classifiers.

Considering all datasets used, we had an accuracy improvement using atoms of the main chain compared to using Cα only (Tables 2, 3 and 4). Classification with Cα only was used as a control group because it was reported as the best result in [15].

It is noteworthy that:

1. Classification accuracy using backbone atoms is strictly dominant in relation to results using Cα. We believe this is due to their linear geometric disposition (Tables 2, 3, 4 and Sect. 2.1).
2. Classification accuracy using all atoms of the structure (*All* classifier) performed better than using only atoms of the side chain (*Side* classifier) (Tables 2, 3 and 4). Probably this is due to *All* set of points include atoms of the main chain. The worst classification performance of *Side* classifier compared to even the classification using Cα only corroborates our hypothesis since side chain atoms are not related to linear geometric disposition protein chain.

Table 2. Prediction of function for the gold-standard dataset. Control group Cα.

Superfamily	Cα				Backbone				All				Side			
	Prec	Recall	F1	ROC	Prec	Recall	F1	AUC	Prec	Recall	F1	AUC	Prec	Recall	F1	AUC
Amidohydrolase	0.991	0.986	0.980	0.976	+0.9%	+1.2%	+1.8%	+2.1%	+0.1%	+0.6%	+0.8%	+1.1%	-1.0%	-2.1%	-3.2%	-3.8%
Crotonase	0.982	0.984	0.982	0.969	+1.8%	+1.7%	+1.8%	+3.2%	+0.1%	-1.0%	-1.1%	-1.9%	+1.3%	+1.4%	+1.5%	+2.5%
Enolase	0.989	0.990	0.985	0.984	+0.7%	+0.8%	+1.2%	+1.4%	-0.2%	+0.1%	+0.2%	+0.5%	-3.7%	-2.8%	-4.3%	-3.7%
Haloacid dehalogenase	0.955	0.967	0.964	0.939	+4.8%	+3.4%	+3.7%	+6.3%	+4.8%	+2.5%	+2.7%	+4.5%	+4.5%	+2.4%	+2.6%	+4.4%
Isoprenoid synthase typeI	1.000	1.000	1.000	1.000	+0.0%	+0.0%	+0.0%	+0.0%	-3.2%	-4.7%	-4.9%	-9.0%	-8.9%	-4.9%	-5.0%	-9.0%
Vicinal oxygen chelate	0.996	0.998	0.998	0.997	+0.4%	+0.2%	+0.2%	+0.3%	-1.8%	-0.9%	-1.1%	-1.6%	-5.8%	-5.1%	-5.7%	-9.2%
All	0.988	0.988	0.988	0.991	+1.1%	+1.1%	+1.1%	+0.9%	+0.1%	+0.1%	+0.1%	+0.2%	-2.0%	-2.1%	-2.1%	-1.5%

Table 3. Structural classification for the Full-SCOP dataset. Control group Cα.

SCOP Level	Cα				Backbone				All				Side			
	Prec	Recall	F1	ROC	Prec	Recall	F1	AUC	Prec	Recall	F1	AUC	Prec	Recall	F1	AUC
Class	0.940	0.940	0.940	0.961	+3.8%	+3.8%	+3.8%	+2.4%	-1.4%	-1.4%	-1.4%	-0.9%	-5.4%	-5.4%	-5.4%	-3.5%
Fold	0.885	0.886	0.884	0.942	+7.7%	+7.6%	+7.7%	+3.6%	-0.7%	-0.7%	-0.7%	-0.3%	-8.8%	-8.8%	-8.8%	-4.2%
Superfamily	0.876	0.877	0.875	0.938	+8.4%	+8.2%	+8.4%	+3.9%	-0.4%	-0.4%	-0.4%	-0.2%	-8.8%	-8.7%	-8.8%	-4.1%
Family	0.829	0.831	0.828	0.915	+10.2%	+10.0%	+10.2%	+4.5%	+0.6%	+0.6%	+0.6%	+0.3%	-8.5%	-8.2%	-8.4%	-3.7%

Table 4. Structural classification for the SSEs datasets. Control group Cα.

DataSet	SCOP Level	Cα				Backbone				All				Side			
		Prec	Recall	F1	ROC	Prec	Recall	F1	AUC	Prec	Recall	F1	AUC	Prec	Recall	F1	AUC
3SSE	Class	0.970	0.970	0.969	0.980	+1.2%	+1.1%	+1.1%	+0.8%	-0.3%	-0.4%	-0.4%	-0.4%	-1.6%	-1.7%	-1.7%	-1.3%
	Fold	0.909	0.910	0.907	0.950	+3.9%	+3.6%	+3.9%	+2.0%	+2.4%	+2.1%	+2.4%	+1.1%	+1.8%	+1.5%	+1.8%	+0.8%
	Superfamily	0.913	0.913	0.910	0.952	+3.6%	+3.4%	+3.6%	+1.8%	+2.0%	+1.7%	+2.0%	+0.9%	+1.5%	+1.2%	+1.5%	+0.5%
	Family	0.887	0.888	0.884	0.940	+4.9%	+4.6%	+4.8%	+2.3%	+2.2%	+1.6%	+2.0%	+0.8%	+0.7%	+0.1%	+0.5%	-0.0%
4SSE	Class	0.974	0.973	0.973	0.983	+1.7%	+1.7%	+1.7%	+1.1%	-1.0%	-1.0%	-1.0%	-0.9%	-1.9%	-1.8%	-1.8%	-1.7%
	Fold	0.922	0.918	0.917	0.956	+5.2%	+5.5%	+5.5%	+2.8%	+1.3%	+1.4%	+1.4%	+0.8%	-0.1%	+0.1%	-0.0%	+0.1%
	Superfamily	0.919	0.915	0.913	0.955	+5.3%	+5.6%	+5.8%	+2.8%	+1.6%	+1.8%	+1.9%	+0.9%	+0.0%	+0.2%	+0.2%	+0.2%
	Family	0.902	0.898	0.896	0.947	+6.9%	+7.3%	+7.4%	+3.6%	+3.1%	+3.4%	+3.5%	+1.7%	+1.7%	+1.9%	+2.0%	+1.0%
5SSE	Class	0.957	0.957	0.957	0.970	+2.6%	+2.6%	+2.6%	+1.8%	+0.2%	+0.2%	+0.2%	+0.0%	-3.2%	-3.3%	-3.2%	-2.4%
	Fold	0.919	0.917	0.916	0.957	+5.2%	+5.4%	+5.4%	+2.7%	+1.8%	+1.9%	+1.9%	+1.0%	-6.0%	-6.0%	-6.1%	-3.0%
	Superfamily	0.912	0.910	0.908	0.954	+5.1%	+5.3%	+5.5%	+2.6%	+2.2%	+2.3%	+2.4%	+1.1%	-6.1%	-6.1%	-6.2%	-3.0%
	Family	0.911	0.908	0.905	0.953	+5.1%	+5.5%	+5.6%	+2.7%	+2.2%	+2.3%	+2.5%	+1.2%	-6.2%	-6.1%	-6.2%	-2.9%
6SSE	Class	0.975	0.975	0.975	0.984	+1.3%	+1.3%	+1.3%	+0.8%	+0.2%	+0.3%	+0.2%	-0.1%	-0.9%	-0.9%	-0.9%	-1.1%
	Fold	0.943	0.942	0.941	0.969	+3.0%	+3.1%	+3.1%	+1.5%	+1.4%	+1.4%	+1.5%	+0.7%	-0.1%	-0.0%	-0.1%	-0.0%
	Superfamily	0.940	0.938	0.937	0.968	+3.2%	+3.4%	+3.5%	+1.7%	+1.4%	+1.5%	+1.5%	+0.8%	+0.0%	+0.1%	+0.2%	+0.1%
	Family	0.930	0.928	0.927	0.963	+4.1%	+4.1%	+4.3%	+2.1%	+2.1%	+2.1%	+2.1%	+1.1%	+0.7%	+0.7%	+0.8%	+0.3%

With the results of the first set of experiments, in which the classification using main chain atoms (*Backbone*) performed better in almost all the metrics, we ran a second round of classification experiments to evaluate whether:

1. The main effect on the discriminating capacity of the classifiers is the geometric linearity of the main chain (due to the simple addition of intermediate points between the Cα) or;
2. The geometric position of C and N atoms are essential to explain the improvement in the discriminating capacity of the classifiers, since their positions could even capture, indirectly, angles ϕ and ψ.

The purpose of second set of experiments was to verify if the position of the atoms of the main chain is a crucial factor to improve the discriminating ability of the signature or if the linear geometric disposition characteristic of the chain would suffice. The following experiments were performed:

1. Adding fictional intermediate points between the Cα, at distances of 0.2 Å, 0.4 Å, 0.6 Å and 0.8 Å, which were herein called $C\alpha^{0.2}$, $C\alpha^{0.4}$, $C\alpha^{0.6}$ and $C\alpha^{0.8}$, respectively. The purpose was to get an "artificial" geometric linearity, similar to the geometric linearity obtained by adding C and N atoms in the *Backbone* experiment (see Sect. 2.1). That way, we get the geometric linearity effect of the main chain without the influence of the positions of C and N atoms and, indirectly, without the influence of angles ϕ and ψ. Improvement in classification accuracy was evaluated in comparison to the Cα control group.
2. Adding fictional intermediate points between the atoms of the main chain, at distances of 0.2 Å, 0.4 Å, 0.6 Å and 0.8 Å, which were herein called $Backbone^{0.2}$, $Backbone^{0.4}$, $Backbone^{0.6}$ and $Backbone^{0.8}$, respectively. The purpose was to evaluate the improvement in classification accuracy in relation to the Backbone control group.

Should the classifiers of the $C\alpha$ family ($C\alpha^{0.2}$, $C\alpha^{0.4}$, $C\alpha^{0.6}$ and $C\alpha^{0.8}$) have similar performance to the classifiers of the *Backbone* family (*Backbone*$^{0.2}$, *Backbone*$^{0.4}$, *Backbone*$^{0.6}$ and *Backbone*$^{0.8}$), that would suggest that is not the positioning of the atoms C and N the key factor to improve the accuracy of the classifiers.

With this second set of experiments, we showed that:

1. In general, the inter-point distance with better performance for adding inter-mediate points was 0.8 Å (Supplementary Material Tables 1–6). In most cases, classification results with the addition of intermediate points at distances of 0.8 Å ($C\alpha^{0.8}$ and *Backbone*$^{0.8}$) were shown to be dominant in relation to the classification results without the addition of intermediate points (Ca and *Backbone*, respectively). This pattern can be observed both for $C\alpha^{0.8}$ and for *Backbone*$^{0.8}$, being more evident for $C\alpha^{0.8}$ because the geometric linear-ity of the $C\alpha$ atomic group is lower than that of the *Backbone* group. With these experiments, we can see that adding intermediate points at inter-point distances of 0.8 Å was, in general, the best structural signature used in the present study.
2. There were no significant differences between the performances of $C\alpha^{0.8}$ and *Backbone* classifiers, nor between the $C\alpha^{0.8}$ and *Backbone*$^{0.8}$ classifiers (The best classifiers. See Tables 5, 6 and 7). Hence we could show that the positions of C and N atoms (which indirectly capture angles ϕ and ψ) are less deter-mining factors to improve the classifying quality than the linear geometric character of the main chain (see Sect. 2.1).

The results of the second set of experiments show that the positions of C and N atoms are determining factors to improve the discriminating degree in the model that uses *Backbone* atoms, but main chain linear geometry (see Sect. 2.1) is a predominant factor.

Analysis of the classification results suggests that linear geometry signature of the main chain of proteins of a particular family is a better classification differential than the signature that uses $C\alpha$ only. That means that the sequential character of the polypeptide chain and its geometric linearity are more relevant than just the residue packing as used in the original CSM.

Therefore, we conclude that the improvement in the classification accuracy of the CSM, when using a model with atoms of the main chain ($C\alpha$, C, N), in com-parison to the exclusive use of $C\alpha$, is due primarily to the exposure of the linear geometric character of the chain, which is obtained by adding C and N atoms. Therefore, the relatively larger inter-point distance in the model that uses only $C\alpha$, in comparison to the model that also uses C and N atoms, brings uncer-tainty to the correct sequence of atoms in the set and, consequently, difficulty to the classification task (Sect. 2.1 and Fig. 1).

This very linearity can be achieved (or simulated) by adding artificial inter-mediate points between $C\alpha$s, with no harm and with improvement in the clas-sification accuracy (in relation to the *Backbone*). We showed that it is not the position of C and N atoms (which indirectly captures angles ϕ and ψ) that is a key factor to improve the classification quality, but rather the strengthening of

Table 5. Prediction of function for the gold-standard dataset. Best results. Control group *Backbone*.

Superfamily	Backbone				Backbone$^{0.8}$				C_α				$C_\alpha^{0.8}$			
	Prec	Recall	F1	ROC	Prec	Recall	F1	AUC	Prec	Recall	F1	AUC	Prec	Recall	F1	AUC
Amidohydrolase	1.000	0.998	0.997	0.997	+0.0%	-0.2%	-0.4%	-0.5%	-0.9%	-1.2%	-1.8%	-2.1%	+0.0%	+0.1%	+0.1%	+0.2%
Crotonase	1.000	1.000	1.000	1.000	+0.0%	+0.0%	+0.0%	+0.0%	-1.8%	-1.6%	-1.8%	-3.1%	+0.0%	+0.0%	+0.0%	+0.0%
Enolase	0.997	0.998	0.998	0.998	-0.5%	-0.2%	-0.4%	-0.3%	-0.7%	-0.8%	-1.2%	-1.4%	-0.1%	-0.2%	-0.3%	-0.3%
Haloacid dehalogenase	1.000	1.000	1.000	0.999	-0.1%	+0.0%	+0.0%	-0.1%	-4.5%	-3.2%	-3.5%	-6.0%	+0.0%	+0.0%	+0.0%	+0.0%
Isoprenoid synthase typeI	1.000	1.000	1.000	1.000	+0.0%	+0.0%	+0.0%	+0.0%	+0.0%	+0.0%	+0.0%	+0.0%	-2.9%	-2.7%	-2.8%	-5.1%
Vicinal oxygen chelate	1.000	1.000	1.000	1.000	+0.0%	+0.0%	+0.0%	+0.0%	-0.4%	-0.2%	-0.2%	-0.3%	+0.0%	+0.0%	+0.0%	+0.0%
all	0.999	0.999	0.999	0.999	-0.2%	-0.2%	-0.2%	-0.2%	-1.1%	-1.1%	-1.1%	-0.8%	-0.1%	-0.1%	-0.1%	-0.1%

Table 6. Structural classification for the Full-SCOP dataset. Best results. Control group *Backbone*.

Level	Backbone				Backbone$^{0.8}$				C_α				$C_\alpha^{0.8}$			
	Prec	Recall	F1	ROC	Prec	Recall	F1	AUC	Prec	Recall	F1	AUC	Prec	Recall	F1	AUC
Class	0.976	0.976	0.976	0.985	+0.1%	+0.1%	+0.1%	+0.0%	-3.7%	-3.7%	-3.7%	-2.4%	+0.0%	+0.1%	+0.0%	+0.0%
Fold	0.953	0.953	0.953	0.976	+0.2%	+0.2%	+0.2%	+0.1%	-7.2%	-7.1%	-7.2%	-3.5%	+0.2%	+0.2%	+0.2%	+0.1%
Superfamily	0.949	0.949	0.949	0.975	+0.2%	+0.3%	+0.2%	+0.1%	-7.7%	-7.6%	-7.8%	-3.8%	+0.2%	+0.2%	+0.2%	+0.1%
Family	0.914	0.914	0.913	0.957	+0.2%	+0.2%	+0.2%	+0.1%	-9.3%	-9.1%	-9.3%	-4.3%	+0.1%	+0.1%	+0.1%	+0.1%

Table 7. Structural classification for the SSEs datasets. Best results. Control group *Backbone*.

SSE	Level	Backbone				Backbone$^{0.8}$				C_α				$C_\alpha^{0.8}$			
		Prec	Recall	F1	ROC	Prec	Recall	F1	AUC	Prec	Recall	F1	AUC	Prec	Recall	F1	AUC
3SSE	Class	0.981	0.980	0.980	0.988	+0.1%	+0.1%	+0.1%	+0.1%	-1.1%	-1.1%	-1.1%	-0.8%	-0.1%	-0.1%	-0.1%	-0.1%
	Fold	0.945	0.943	0.942	0.969	+0.3%	+0.5%	+0.5%	+0.3%	-3.7%	-3.5%	-3.7%	-1.9%	-0.6%	-0.4%	-0.5%	-0.3%
	Superfamily	0.946	0.943	0.943	0.970	+0.3%	+0.5%	+0.5%	+0.3%	-3.5%	-3.3%	-3.5%	-1.8%	-0.6%	-0.4%	-0.5%	-0.3%
	Family	0.930	0.928	0.927	0.961	-0.2%	-0.1%	-0.1%	+0.0%	-4.6%	-4.4%	-4.6%	-2.2%	+0.0%	+0.0%	+0.0%	+0.0%
4SSE	Class	0.990	0.990	0.990	0.994	+0.1%	+0.1%	+0.1%	+0.1%	-1.6%	-1.7%	-1.7%	-1.1%	-0.1%	-0.1%	-0.1%	-0.1%
	Fold	0.969	0.968	0.968	0.983	+0.9%	+0.9%	+0.9%	+0.5%	-4.9%	-5.2%	-5.2%	-2.7%	-0.1%	-0.1%	-0.1%	+0.0%
	Superfamily	0.967	0.966	0.966	0.982	+0.9%	+0.9%	+0.8%	+0.4%	-5.0%	-5.3%	-5.4%	-2.8%	+0.0%	+0.1%	+0.1%	+0.0%
	Family	0.964	0.963	0.962	0.980	+0.7%	+0.6%	+0.7%	+0.3%	-6.4%	-6.8%	-6.9%	-3.4%	+0.3%	+0.3%	+0.3%	+0.2%
5SSE	Class	0.982	0.982	0.982	0.987	+0.4%	+0.4%	+0.4%	+0.3%	-2.6%	-2.6%	-2.6%	-1.8%	+0.1%	+0.1%	+0.1%	+0.1%
	Fold	0.966	0.966	0.965	0.982	-0.2%	-0.2%	-0.2%	-0.1%	-4.9%	-5.1%	-5.1%	-2.6%	-0.2%	-0.3%	-0.2%	-0.1%
	Superfamily	0.958	0.958	0.957	0.979	+0.1%	+0.0%	+0.0%	+0.0%	-4.9%	-5.1%	-5.2%	-2.5%	+0.1%	+0.0%	+0.0%	+0.0%
	Family	0.958	0.958	0.956	0.978	-0.1%	-0.1%	-0.1%	+0.0%	-4.9%	-5.2%	-5.3%	-2.6%	+0.0%	+0.0%	-0.1%	+0.0%
6SSE	Class	0.988	0.988	0.988	0.992	+0.4%	+0.4%	+0.4%	+0.3%	-1.3%	-1.3%	-1.3%	-0.8%	+0.5%	+0.5%	+0.5%	+0.4%
	Fold	0.971	0.971	0.971	0.984	+0.5%	+0.5%	+0.5%	+0.3%	-2.9%	-3.0%	-3.0%	-1.5%	+0.7%	+0.7%	+0.7%	+0.4%
	Superfamily	0.970	0.969	0.969	0.984	+0.5%	+0.5%	+0.5%	+0.3%	-3.1%	-3.3%	-3.3%	-1.7%	+0.6%	+0.6%	+0.6%	+0.3%
	Family	0.968	0.967	0.967	0.983	+0.4%	+0.4%	+0.4%	+0.2%	-4.0%	-4.0%	-4.1%	-2.0%	+0.7%	+0.7%	+0.6%	+0.3%

the linear spatial character of the main chain. It was also shown that the degree of influence of angles ϕ and ψ on the discriminating capacity of the classifiers is not only less significant than the linear geometric disposition of the main chain, but also depends on the addition of artificial intermediate points to be precisely significant (see results of the *Backbone*$^{0.8}$ classifiers).

We have confidence to claim that we were able to improve the accuracy of the original CSM with the proposed new model because we performed statistical tests and obtained quite low *p-values* proving that our classification metrics are significantly better (see Sect. 2.3). As shown in *p-value* tables (Supplementary Tables 7–30), the null hypothesis was only accepted when the mean values of the metrics compared were very close. For all null hypotheses accepted, the percentage difference among mean values were less than or equal to 0.3%. In general, the *p-values* found were extremely low, and the null hypotheses that the means would be the same could be rejected.

Pires [15] reports that experiments were conducted with other centroids rather than the Cα, such as the Cβ or the last heavy atom (LHA) of the side chain. The Cα had the best performance in all experiments, a fact that he claimed "demanded deeper investigation". With this study, we believe that those centroids had the lowest performance exactly because they are even more distant from the linear geometric disposition of the main chain.

4 Conclusions

In this study, we revisited the problem of protein structure and function classification based on structural signatures. We compared the traditional state-of-the-art CSM method based on Cα to a novel model which is more focused on strengthening geometric linearity character of proteins main chain.

The purpose of our works was to identify conserved patterns in the main chains of protein families and use them to more precisely classify proteins. By analyzing several results of classification experiments, we concluded that main chain linear geometry patterns from the same protein family is a better classification differential than the original CSM signature based solely on Cα.

We concluded that the improvement in the classification performance when using a model with main chains atoms, in comparison to the exclusive use of Cαs is due, primarily, to the overemphasizing linear geometric character of the chain. This very linearity can also be achieved (or simulated) by adding artificial intermediate points between the Cα, with no harm and with improvement to classification accuracy. Therefore, this study suggests that it is not the position of C and N atoms (which indirectly captures angles ϕ and ψ) that is the key factor to improve classification performance, but rather the strengthening of the linear geometric character of the main chain.

As perspectives, we believe it would be interesting to verify weather this inclusion of intermediate points to strengthen the geometric characteristics could be used to improve the performance of CSM in other contexts it has been used so far as ligand prediction, mutation impact prediction, among others.

Funding. This work was supported by the Brazilian agencies Coordenação de Aperfeiçoamento de Pessoal de Nível Superior (CAPES) - Finance Code 51/2013 - 23038.004007/2014-82; Conselho Nacional de Desenvolvimento Científico e Tecnológico (CNPq) and Fundação de Amparo à Pesquisa do Estado de Minas Gerais (FAPEMIG).

References

1. Andreeva, A., Howorth, D., Brenner, S.E., Hubbard, T.J., Chothia, C., Murzin, A.G.: SCOP database in 2004: refinements integrate structure and sequence family data. Nucleic Acids Res. **32**(Suppl 1), 226–229 (2004)
2. Baker, D., Agard, D.A.: Kinetics versus thermodynamics in protein folding. Biochemistry **33**(24), 7505–7509 (1994)
3. Berman, H.M., et al.: The protein data bank. Nucleic Acids Res. **28**(1), 235–242 (2000)
4. Berry, M.W., Dumais, S.T., O'Brien, G.W.: Using linear algebra for intelligent information retrieval. SIAM Rev. **37**(4), 573–595 (1995)
5. Brown, S.D., Gerlt, J.A., Seffernick, J.L., Babbitt, P.C.: A gold standard set of mechanistically diverse enzyme superfamilies. Genome Biol. **7**(1), R8 (2006). https://doi.org/10.1186/gb-2006-7-1-r8
6. Choi, I.-G., Kim, S.-H.: Evolution of protein structural classes and protein sequence families. Proc. Natl. Acad. Sci. **103**(38), 14056–14061 (2006)
7. Chothia, C.: One thousand families for the molecular biologist. Nature. **357**(6379), 543–544 (1992)
8. Dill, K.A.: Polymer principles and protein folding. Protein Sci. **8**(06), 1166–1180 (1999)
9. Illergård, K., Ardell, D.H., Elofsson, A.: Structure is three to ten times more conserved than sequence - a study of structural response in protein cores. Proteins: Struct. Funct. Bioinformat. **77**(3), 499–508 (2009)
10. Jain, P., Hirst, J.D.: Automatic structure classification of small proteins using random forest. BMC Bioinform. **11**(364), 1–14 (2010)
11. Kauzmann, W.: Some factors in the interpretation of protein denaturation. Adv. Protein Chem. **14**, 1–63 (1959)
12. Laskowski, R.A., Moss, D.S., Thornton, J.M.: Main-chain bond lengths and bond angles in protein structures. J. Mol. Biol. **231**(4), 1049–1067 (1993)
13. Laskowski, R.A., Watson, J.D., Thornton, J.M.: ProFunc: a server for predicting protein function from 3D structure. Nucleic Acids Res. **33**(Suppl 2), W89–W93 (2005)
14. Lee, D., Redfern, O., Orengo, C.: Predicting protein function from sequence and structure. Nat. Rev. Mol. Cell Biol. **8**(12), 995–1005 (2007)
15. Pires, D.E., de Melo-Minardi, R.C., dos Santos, M.A., da Silveira, C.H., Santoro, M.M., Meira, W.: Cutoff Scanning Matrix (CSM): structural classification and function prediction by protein inter-residue distance patterns. BMC Genomics **12**(4), S12 (2011)
16. Privalov, P.L., Gill, S.J.: Stability of protein structure and hydrophobic interaction. Adv. Protein Chem. **39**, 191–234 (1988)
17. Soundararajan, V., Raman, R., Raguram, S., Sasisekharan, V., Sasisekharan, R.: Atomic interaction networks in the core of protein domains and their native folds. PLoS One **5**(2), e9391 (2010). https://doi.org/10.1371/journal.pone.0009391
18. Volkamer, A., Kuhn, D., Rippmann, F., Rarey, M.: Predicting enzymatic function from global binding site descriptors. Proteins: Struct. Funct. Bioinform. **81**(3), 479–489 (2013)
19. Altman, N.S.: An introduction to kernel and nearest-neighbor nonparametric regression. Am. Stat. **46**(3), 175–185 (1992)

Telemedicine for Smart Homes and Remote Monitoring

Positioning Method for Arterial Blood Pressure Monitoring Wearable Sensor

Viacheslav Antsiperov$^{(\boxtimes)}$ and Gennady Mansurov

Kotelnikov Institute of Radioengineering and Electronics of RAS,
Mokhovaya Street 11-7, 125009 Moscow, Russia
antciperov@cplire.ru, gmansurov@mail.ru
https://www.researchgate.net/profile/V_Antsiperov

Abstract. Measuring blood pressure in real time using wearable sensors mounted directly on the patient's body is promising tool for assessing the state of the cardiovascular system and signalling symptoms of cardiovascular diseases. To solve this problem, we developed a new type of wearable arterial blood pressure monitoring sensor. Constructively, this sensor can be embedded in a flexible bracelet for measuring the pressure in the underlying radial artery. Due to the very small measuring pads (less than $1\,\mathrm{mm}^2$) and, consequently, the ability to accurately position the contact pad directly over the artery, it is possible to ensure high quality of blood pressure measurement. However, since the artery itself is generally not visible, the correct positioning of the sensor is a non-trivial problem. In the paper we propose the solution of the problem – the positioning based on monitoring the pulse wave signals using three channels from closely spaced pads of a three-chamber pneumatic sensor.

Keywords: Non-invasive arterial blood pressure (ABP) monitoring ·
Pneumatic sensor · Sensor positioning problem ·
Multichannel measurements and control ·
Wearable medical sensors and devices ·
Embedded data processing (EDP)

1 Introduction

Non-invasive measurement of arterial blood pressure (ABP) using cuff instruments until recently has been most common. Since blood pressure indicators provide adequate data for many applications both in medicine and in medical research, their use has been stretched for almost a hundred years – during the XX century, despite the obvious inconvenience of cuff methods.

The inconvenience of cuff methods is associated with several circumstances. First, continuous measurement of blood pressure is impossible, since a pause of at least 1–$2\,\mathrm{min}$ between two blood pressure measurements is necessary to

The work is supported by the Russian Foundation for Basic Research (RFBR), grant N 17-07-00294 A.

I. Rojas et al. (Eds.): IWBBIO 2019, LNBI 11465, pp. 405–414, 2019.
https://doi.org/10.1007/978-3-030-17938-0_36

avoid the errors [1]. Therefore, short-term changes in blood pressure cannot be detected. Second, inflating the cuff is often a poorly controlled discontinuous process due to the need to create high pressure in a relatively large cuff volume over a relatively short period of time. So, pumping the cuff may disturb the patient and also cause the changes in blood pressure.

To solve the problem of discontinuity Peñáz in 1973 proposed, exploiting the idea of dynamic "unloading of vessel walls", a method of volume compensation for continuous ABP measurements [2]. The measurement unit of the Peñáz device is represented by a small finger cuff that contains a light source on one side and an IR receiver on the opposite side for controlling by a light absorption blood volume in the finger. The IR signal obtained in a such plethysmograph is then used in a feedback loop to regulate the pressure in the finger cuff. Proposed by Pressman and Newgard, the tonometric method for continuous ABP measurement is also based on the idea of compensation, but not of the blood volume, but of the blood pressure on artery wall [3]. Like the Peñáz method the arterial tonometry is based on pulse oscillation estimates, but the principle of arterial unloading is different. In this case the cuff is placed on the wrist, so the sensor is over the radial artery. The sensor presses the artery to the radial bone until it is flattened enough but not occluded. At this intermediate position arterial wall tension becomes parallel to the tonometer pad surface and does not affect the pressure measured by the sensor. Unfortunately, though Peñáz and Pressman–Newgard methods can, in principle, provide long-term continuous ABP monitoring, they are unsatisfactory candidates for wearable ABP devices because in order to avoid blood stasis in the limbs the cuff must be periodically relaxed.

However, such devices are of great importance for monitoring vital human data in a wireless LAN or GSM environment. The growing scope of application of smart phone technologies, their reduction in cost and increased ease of use in combination with parallel progress in sensing technology leads to a real transition from traditional medical care to mobile personal health monitors (PHM), among which the wearable ABP devices should play the central role [4].

To overcome the above-mentioned limitations some new approaches to a continuous, non-invasive measurement of ABP were recently proposed. Among them the approach proposed by Kaisti et al. should be mentioned. It is based on the use of MEMS pressure sensors and ABP waveform analysis for the non-invasive and continuous-time measurements [5]. The authors report that the use of modern MEMS sensor technologies has high performance in a compact package with low power consumption. They note that because the measurement procedure is sensitive to the exact placement of the sensor on an artery, it is preferable to have several sensors arranged into a grid and use an automated selection of the best MEMS sensor from the multiple of sensing elements. In the same article [5] one can find several references to other modern approaches to the development of wearable ABP monitoring sensors.

In order to avoid the problems associated with the cuff methods and with an eye to creating a mobile, wearable ABP sensors, several years ago we also attempted to develop a new approach to the non-invasive cuff-less continuous arterial blood pressure monitoring. But in contrast to well-known approaches [5] our method is based on another principle – the principle of local pressure compensation [6] (Fig. 1(A)).

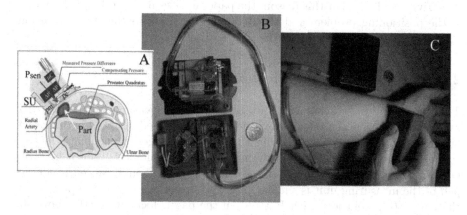

Fig. 1. The measurement principle of blood pressure due to local pressure compensation in the artery P_{art} by pressure in the measuring unit P_{sen} of the pneumatic sensor (A), pneumatic sensor design (B) and the appearance of the measuring unit applied to the patient's wrist (C).

The principle of local compensation implies that pressure compensation under the measuring contact pad is carried out not over the entire transverse perimeter of the artery (integrally), but on a small, local area of its wall. When the position of the pad is successfully chosen, right above the central axis of the artery (Fig. 1(A)) and its pressing leads to local flattening of the underlying artery wall, the pressure in the sensor chamber P_{sen} will be exactly equal to the pressure in artery P_{art} since the local elastic tension of the flat wall does not produce additional pressure in above/below rooms (the applanation principle [6]). Known methods of pressure compensation, including the method based on MEMS sensors, due to the comparable sizes of the measuring elements and arteries, should compensate besides ABP the additional (unknown) mean elastic pressure along the curved artery wall. For this reason, in contrast to the case of integral compensation, the case of local compensation gives us the possibility to carry out a direct, non-calibrated, without any corrections ABP measuring, which ensures its main advantage.

The practical implementation of local compensation method became possible thanks to previously developed unique technique of pressure compensating measurements on a very small working area ($1\,mm^2$ or less) [7]. Recalling, that an average diameter of a radial artery (in a flattened state) is $\sim 4\,mm$, we see that this is quite enough to realize the applanation conditions of measurement. Let us note, by the way, that each of the MEMS sensors has a size of $\sim 5\,mm$ (see [5]), so its data are integral over artery surface.

Due to the very small sensing contacts and, consequently, the ability to accurately position the measuring pad directly over the center of the artery, it become possible to ensure high quality blood pressure measurement. However, since the artery itself is generally not visible, the correct positioning of the sensor is a non-trivial problem. For this reason, the paper focuses in general on a discussion of the positioning problem and the development of approaches to its solution. This discussion is the subject of the next section.

2   Sensor Positioning Method

The high-performance characteristics of continuous measurement of the ABP pulse wave by proposed pneumatic sensor were discussed in detail in our previous work [6]. However, it turned out that the advantages of pneumatic locally-compensating ABP measurement are not gratis. One should pay for it by problems arising with the positioning of the sensor measuring unit. Since the contact pad of the measuring unit is smaller than the size of the artery, the pressure in the sensor P_{sen} coincides with P_{art} only if the pad is located exactly above the artery axis (see the discussion of local compensation principle above). Obviously, in the case of a patient's arm movement, which in the case of a wearable sensor will be a constant measurement artefact, the measuring unit can change its position, which will lead to the distortions. Changes in the shape and amplitude of the pulse wave of arterial pressure associated with a small displacements of the measuring pad are shown in Fig. 2(B).

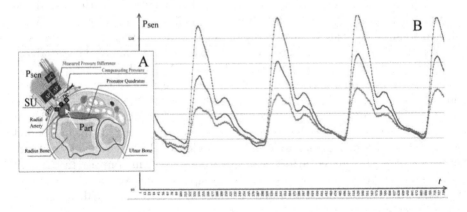

Fig. 2. Changes in the shape and amplitude of pulse wave signal (B) depending on changes of measuring pad position: • – pad is directly over the artery, ■, ♦ – pad is shifted to the left and to the right from the radial artery axis (A).

A detailed experimental study of the positioning problem revealed the following. In the position of the measuring pad just above the artery the P_{sen} signal has the greatest magnitude between the major maxima and minima and wherein both extrema are themselves more acute (see Fig. 2(B), plot •). The movement of the measuring pad to the left and right in the transverse direction of the artery axis is accompanied by a distortion of the shape and a decrease in the spread of the signal, and these changes are almost symmetrical with respect to left/right directions of displacement (in Fig. 2(B), plots ■, ♦ correspond to slightly asymmetric left/right positions). For the exactly symmetrical with respect to the artery axis left/right positions the pulse wave plots will substantially coincide.

This observation led us to the new design of pneumatic sensor for monitoring the arterial pressure. It should contain a measuring unit with three chambers (channels with its own independent pressure gauges) for three independent measuring pads. These pads should be arranged in a row so, that during the ABP measurement they are disposed in the direction transverse to the artery axis. The sizes of measuring pads ($\leq 1\,\text{mm}$) and the distances between them ($\sim 1.5\,\text{mm}$) are so small, that in certain position of the measuring unit the pads are simultaneously over the artery. Figure 3(A) gives a schematic view of the three-chambered pneumatic sensor and (B) gives results of simultaneously measured three-channel pulse wave signals at the proper upon the artery position. The details of the technical implementation of the sensor are reflected in the patent [7].

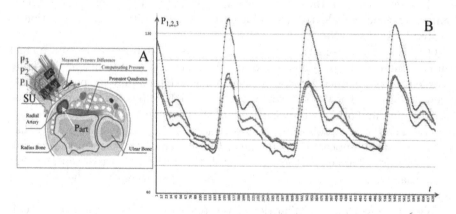

Fig. 3. Sensor with three-chamber measuring unit located exactly above the artery (A), coincidence of the side channel signals indicates the correct unit position (B) (in contrast to Fig. 2 all three channels are measured simultaneously).

In the claimed design of a three-chamber pneumatic sensor for measuring blood pressure, the main task of the additional (side) channels is to control the positioning of the measuring unit. That is, the correct location of the central pad corresponds to the maximum coincidence of signals from the side channels (see Fig. 3(B)). It is not significant that in these side channels complete "unloading" of the artery walls does not take place and, therefore, the signal of the arterial pressure in them is significantly distorted. It is important that when both side signals coincide, the central chamber is located exactly above the artery and in this position its signal will be an undistorted replica of the pressure in the artery.

3   Computer Procedure for Semi-automatic Rough Positioning

The methodology of ABP monitoring by a three-chamber pneumatic sensor, described above, assumes that all measuring unit pads "see" the artery, i.e. the signals in the corresponding channels represent a pulse wave (possibly of a distorted form) such as, for example, in Fig. 3. Let's refer this methodology as a precise positioning process. Obviously, this process must be preceded by a rough positioning process of searching such a position of the measuring unit in which a pulse wave would be presented in at least one channel (further in two, and then in all three channels). Until recently, the rough positioning process was carried out manually – by moving the measuring unit on the wrist near the position of the artery detected by palpation. It is clear that such manual positioning is not suitable for wearable sensors. For this reason, we made efforts to investigate approaches and to find possible solutions in the field of automatic (computer aided) positioning and keeping the three-chamber pneumatic sensor measuring unit in a proper (upon the artery) position—in the well position. Now we have found a satisfactory algorithmic solution for rough positioning, this section is devoted to its discussion.

Obviously, the automatic detection of the pulse wave presence in any channel can be formalized as the detection of some waveform repeatability in the signal recorded by the channel. Moreover, the confidence interval for the period of such a repeatability is known – it ranges from 0.5 to 1.5 s (excluding tachycardias, bradycardias, and other pathologies). Thus, the problem of rough positioning can be reformulated as a problem of such displacement of the measuring unit, that repeatability will appear firstly in one of the side channels, then in the central, and then in all channels of the sensor (with the initial absence of repeatability in the channels). The same applies to the loss of repeatability in the registered signals due to arm movement and other artefacts.

Without going into peculiarities of the technical mechanisms of the rough positioning implementation, or in specificity of the mechanical control of a such positioning, let us consider in more detail the actual identification of the presence or absence of repeatability in a particular channel at a certain time moment.

In order to solve some problems in the field of the nonstationary signal processing we, starting from the work [8], attempted to combine the main ideas of the wavelet-like (multiscale) and quadratic time-frequency analysis. The implementation of such a combination was achieved by the methodology of the non-stationary analysis for signals containing the quasi-periodic fragments. We named this methodology the multiscale correlative analysis (MCA) [8].

For biomedical signals demonstrating a generally nonstationary behavior, the presence of isolated fragments with clearly defined repeatability of some waveforms is the very usual situation. In speech signals, these are vocalized fragments (vowels), in EEG, for example, seizures. In ABP signals these are the fragments corresponding to a well positioned measuring unit, the duration of which we need to maintain as long as possible, see Fig. 4.

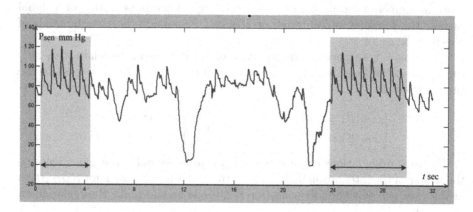

Fig. 4. Central channel registered signal during rough positioning process of measuring unit and marked fragments of its well positioned measurement states.

It is important to emphasize that when quasi-periodicity is mentioned, we are not talking about some kind of approximate periodicity, but about the properties of repeatability, more precisely, about approximate repeatability of any waveforms. For this reason, in the case of biomedical signals, that are essentially wideband signals, such fragments correspond more to models of regular consecutive pulses rather than to those consisting of a set of harmonic components. Because the analysis of such signals is more efficient within the time domain (some variants of the correlative processing or matched filtering), we proposed the synthesis of appropriate tools in the time-time scale domain, rather than in the time-frequency domain, as it is usual in traditional approaches, based on spectral analysis.

In the MCA the starting point is the definite signal time-scale representation as in the case of spectral analysis is some representation of the time-frequency type. In fact, the MCA time-scale representation is some special autocorrelation function estimate. Among numerous existing autocorrelation function estimates, the MCA representation $R_{MCA}(t, \tau)$ is chosen in the following form:

$$R_{MCA}(t, \tau) = \frac{1}{|\tau|} \int_{t-|\tau|/2}^{t+|\tau|/2} z(t' + \tau/2) z(t' - \tau/2) dt', \qquad (1)$$

where $z(t')$ is an ABP signal, τ – the time scale, also representing (variable) duration of adjacent to t correlated intervals of a signal, and t is a certain time moment at which the decision about the presence or absence of repeatability in $z(t')$ should be made. It is not difficult to demonstrate that averaging of (1) over the ensemble of implementations in the stationary case gives the exact theoretical autocorrelation function $\langle R_{MCA}(t,\tau) \rangle = R_z(\tau) = \langle z(t+\tau/2)z(t-\tau/2) \rangle$, i.e. (1) is unbiased estimate. It is symmetrical with respect to the parameter τ: $R_{MCA}(t,-\tau) = R_{MCA}(t,\tau)$. The distinctive feature of (1), which is of importance in segmentation problems, is the property of the preservation of a signal carrier, the property of vanishing of $R_{MCA}(t,\tau)$ outside the time intervals of non-zero values of $z(t)$.

For a more transparent interpretation of (1) it is useful to relate with $z(t')$ the signals of its local past $z_{Pt}(t')$ and local future $z_{Ft}(t')$ (with respect to the time moment t):

$$\begin{cases} z_{Ft}(t') = \theta(t')z(t+t') \\ z_{Pt}(t') = \theta(t')z(t-t') \end{cases}, \quad \theta(t') = \begin{cases} 1, & if \quad t' \geq 0 \\ 0, & if \quad t' < 0 \end{cases}. \quad (2)$$

By using designations (2), it is possible to eliminate in (1) the dependence on the scale τ of the integration limits. The corresponding expression for $R_{MCA}(t,\tau)$ in the case $\tau > 0$ is the following (for $\tau < 0$ one should make the replacement $\tau \to -\tau$):

$$R_{MCA}(t,\tau) = \frac{1}{\tau}\int_{-\infty}^{\infty} z_{Ft}(t')z_{Pt}(\tau-t')dt' = \frac{1}{\tau}\int_{-\infty}^{\infty} z_{Ft}(\tau-t')z_{Pt}(t')dt', \quad (3)$$

whence it follows that to within a multiplier $\frac{1}{\tau}$ representation $R_{MCA}(t,\tau)$ is the classical convolution of the signals $z_{Pt}(t')$ and $z_{Ft}(t')$ with effectively infinite limits of integration. To detect with the help of $R_{MCA}(t,\tau)$ (3) the quasi-periodic signal fragments, the following property of the theoretical autocorrelation $R_z(\tau)$ is usually used. It is well-known that the global maximum of any autocorrelation is reached at the beginning of scale coordinates, i.e. at $\tau = 0$. If the signal $z(t')$ has the quasi-period T, then $R_z(\tau)$ will, by definition, have the same quasi-period: $R_z(\tau+kT) = R_z(\tau), k = \pm 1, \pm 2, \dots$. Thus, in the points that are multiples of T, the quasi-periodic signal autocorrelation will have expressed (side) maxima. If the estimate $R_{MCA}(t,\tau)$ (1–3) at least approximately repeats the behaviour of the theoretical autocorrelation $R_z(\tau)$, then its side maxima will also indicate the quasi-period T and its multiples. This property serves as the basis for most estimates of the time scales of pulse repetitions in signals. Exactly as in the case of time-frequency analysis, where it is possible to estimate the frequencies of harmonic components using the maxima of the spectrum power, in the case of the time-scale signal representations $R_{MCA}(t,\tau)$ (3) it is possible using its maxima to estimate the quasi-period of signal pulses, i.e., the presence of local signal repeatability.

Figure 5 gives an example a two-dimensional representation (3) of a real ABP recording that contains a long well positioned measurement fragment. The trajectories of the first and second side maxima (bright strips along the t axis) correspond to the quasi-period $T \sim 0.76$ s and to the doubled quasi-period $2T \sim 1.52$ s. Note, that T fall in the mentioned above confidence interval for the quasi-period of pulse wave repeatability $(0.5, 1.5)$ s.

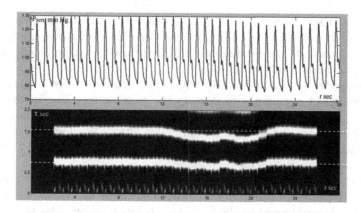

Fig. 5. Representation $R_{MCA}(t, \tau)$ (3) (bottom, in pseudo-color: white – max, black – min) of a real ABP signal that contains a fragment of a long well positioned measuring unit state. The trajectories of the first two side maxima (at $\sim$0.76 and $\sim$1.52 s) are well pronounced.

It follows from above discussion and it is illustrated by the Fig. 5, that the main difference between $R_{MCA}(t, \tau)$ for the fragments of well positioned measurements and the fragments of lost repeatability is the existence of the sequence of approximately equidistant, but may be slightly different in form maxima (peaks) of noticeable amplitudes. It is possible to exploit this property in many different ways to perform the actual identification of the presence or absence of repeatability. We selected the simplest criterion $C_{MCA}(t)$, consisting in checking the equidistance (multiplicity) of the locations of the first two side peaks:

$$C_{MCA}(t) = |T_2(t) - 2T_1(t)T_2(t)| < \Delta, \tag{4}$$

where $T_1(t)$ and $T_2(t)$ are the locations of two biggest in interval $(0, \tau_{max})$ side maxima of current (for time moment t) time-scale signal representation $R_{MCA}(t, \tau)$ (3), Δ – some small enough threshold $\sim$0.1.

4   Conclusions

The article proposes a new method of positioning a three-chamber pneumatic sensor for continuous measurement of blood pressure – a method based on local pressure compensation. It is shown that the basic idea of the method is quite simple and partly resembles the principle of pointing the radar antenna at a target by aligning the side lobes of the radiation power diagram. The method is intended to achieve stability in measuring the arterial blood pressure. In the absence of positioning unstable character of measurements could be caused by very small sizes (tens of microns) of the measuring unit contact pads. The technological solutions we found allow us to manufacture several working chambers in one measuring unit with a linear pitch from 1.5 mm to 2 mm in one unit.

According to the obtained results, the variant with three working chambers is optimal. We believe that this solution makes possible the implementation of the sensor with automatic measuring unit positioning on the artery, along with the already achieved manual positioning.

In the special section of the paper are discussed the theoretical and experimental results of our efforts to develop the possible solutions in the field of automatic (computer aided) positioning and keeping the measuring unit of three-chamber pneumatic sensor in found optimal position. We described a satisfactory algorithmic solution for rough positioning based on Multiscale Correlation Analysis (MCA) [8] time-scale representation.

In short, the results of testing the positioning of a three-chamber pneumatic sensor and the developed calibration methodology allow us to draw the following conclusions. The proposed method of rough positioning provides the ability to continuously measure blood pressure for a long time in a mobile environment. It also allows to perform synchronous transfer of measured data in a wireless LAN environment, for example, in Smart Homes, rehabilitation centers, in sports Halls, etc.

The most important task for the future is to replace the manual precise positioning of the measuring unit by automatic control and develop on this basis a mobile device for continuous monitoring of blood pressure parameters.

References

1. Campbell, N.R.C., Chockalingam, A., Fodor, J.G., McKay, D.W.: Accurate, reproducible measurement of blood pressure. Can. Med. Ass. J. **143**(1), 19–24 (1990)
2. Peñáz, J.: Photoelectric measurement of blood pressure, volume and flow in the finger. In: Digest 10th International Conference on Medical and Biological Engineering, p. 104. Dresden, Germany (1973)
3. Pressman, G.L., Newgard, P.M.: A transducer for the continuous external measurement of arterial blood pressure. IEEE Trans. Bio-med. Electron. BME **10**, 73–81 (1963)
4. Mena, L.J., Felix, V.G., Ostos, R., et al.: Mobile personal health system for ambulatory blood pressure monitoring. Comp. Math. Methods Med. **2013**, 1–13 (2013). Art. ID 598196
5. Kaisti, M., Leppanen, J., Lahdenoja, O., Kostiainen, P., et al.: Wearable pressure sensor array for health monitoring. In: Computing in Cardiology (CinC), Rennes, France, pp. 1–4 (2017)
6. Antsiperov, V., Mansurov, G.: Wearable pneumatic sensor for non-invasive continuous arterial blood pressure monitoring. In: Rojas, I., Ortuño, F. (eds.) IWBBIO 2018. LNCS, vol. 10814, pp. 383–394. Springer, Cham (2018). https://doi.org/10.1007/978-3-319-78759-6_35
7. Mansurov, G.K., et al.: Monolithic three-chambered pneumatic sensor with integrated channels for continuous non-invasive blood pressure measurement. Invention patent 2675066, Priority 26 February 2018. Bulletin No. 35 (2018). Patent - RU 2 675 066 C1, CPC - A61B 5/022 (2018.08); A61B 5/024 (2018.08)
8. Antsiperov, V.E.: Multiscale correlation analysis of nonstationary signals containing quasi-periodic fragments. J. Com. Technol. Electron. **53**(1), 65–77 (2008)

Study of the Detection of Falls Using the SVM Algorithm, Different Datasets of Movements and ANOVA

José Antonio Santoyo-Ramón$^{(\boxtimes)}$ ⓘ, Eduardo Casilari-Pérez ⓘ, and José Manuel Cano-García

Departamento de Tecnología Electrónica, Universidad de Málaga, ETSI Telecomunicación, 29071 Málaga, Spain
{jasantoyo, ecasilari, jcgarcia}@uma.es

Abstract. Falls are becoming a major public health problem, which is intensified by the aging of the population. Falls are one of the main causes of death among the elderly and in population groups that develop risk activities. In this sense, technologies can provide solutions to improve this situation. In this work we have analyzed different repositories of movements and falls designed to test decision algorithms in automatic fall detection systems. The objectives of the study are: firstly, to clarify what are the characteristics of the most significant accelerometry signals to identify a fall and secondly, to analyze the possibility of extrapolating the learning achieved with a certain database when tested with another one. As a novelty with respect to other works in the literature, the statistical significance of the results has been systematically evaluated by the analysis of variance (ANOVA).

Keywords: Fall detection system · Inertial sensors · Smartphone · Accelerometer · Machine learning · Supervised learning · ANOVA · Datasets of movements

1 Introduction

It is estimated that between 2015 and 2050 the world population aged over 60 will grow from 900 to 2000 million [1]. This remarkable demographic transformation will unquestionably cause a series of challenges in the health systems that must be faced in order to increase the quality of life of the population. The current study focuses on falls, one of the most relevant public health problems confronted by the world society. Falls are a major source of loss of autonomy, deaths and injuries among the elderly and have a noteworthy impact on the costs of national health systems. According to World Health Organization [2], 28–35% of the population older than 64 experience at least one fall every year, while fall-related injuries and hospitalization rates are expected to increase on average by 2% per year until 2030 [3]. Furthermore, there are other risk groups that are exposed to suffer severe falls during their work or leisure time (mountaineers, firemen, construction workers, antenna installers, window cleaners, cyclists, etc.).

I. Rojas et al. (Eds.): IWBBIO 2019, LNBI 11465, pp. 415–428, 2019.
https://doi.org/10.1007/978-3-030-17938-0_37

The World Health Organization has stated that falls are the second worldwide reason of mortality provoked by accidental or unintentional injuries: 37.3 million falls requiring medical attention occurs annually, while it has been estimated that around 646,000 people die every year due to falls [4]. Moreover, morbidity and mortality provoked by falls are closely associated to the speed of the medical response and first aid treatment after the incident [5]. Consequently, the analysis of cost-effective and automatic Fall Detection Systems (FDSs) has become a relevant research topic during the last decade.

2   Related Works on Wearable Fall Detection Systems

A Fall Detection System can be considered as a binary classification system designed to discriminate fall events from any other movement of the user (the so-called Activities of Daily Living or ADLs). The aim of a FDS is to maximize the possibility of detecting real falls, while simultaneously keeping to a minimum the number of false alarms. Thus, FDSs monitor the mobility of the users in order to automatically make their relatives or the medical staff aware whenever time a fall incidence is recognized.

FDSs can be classified into two generic groups depending on the nature of the employed sensors: context-aware systems and wearable systems.

Firstly, context-aware systems are based on sensors located in the environment around the user to be supervised like cameras, microphones, vibration sensors, etc. The detection decision is based on the signals captured by these sensors, which are placed in a definite area where the user is monitored. This kind of FDSs has some disadvantages related to the physical restrictions of the locations where the user can be monitored, as well as to their expensive installation and maintenance costs. Another serious inconvenience is their vulnerability to external events and interferences, such as noises, presence of another individual, falling objects, changes of the illumination level, etc. Besides, the user can feel their privacy is being compromised because of the non-stop use of audiovisual equipment.

On the other hand, wearable systems employ sensors that are integrated into the patient's clothing or are transported by the user within personal accessories. These sensors only monitor magnitudes associated to the patient mobility, such as the acceleration or angular velocity of the body. This type of FDSs offers some advantages over the context-aware solutions since they are more affordable, less intrusive and also less prone to the effects of external factors. Moreover, due to fact that the sensors always accompany the user, the limitations about the monitoring area are removed.

Currently, all the inertial sensors that are normally required by a wearable FDS (generally, an accelerometer and in some cases a gyroscope and a magnetometer) are integrated in most smartphones. FDS architectures based on smartphones present some benefits when compared to other specific commercial fall detection sensors, as these personal devices natively support multi-interface wireless communications (Wi-Fi, 3G/4G, Bluetooth, GPS). Therefore, users can be monitored almost ubiquitously at a low cost. In addition, at present, smartphones are omnipresent in the daily life of citizens, so they are not considered as an intrusive technology.

The penetration of smartphones among older people is obviously lower than in other age groups. Nevertheless, at least in western countries, this situation is swiftly changing. In [6], Deloitte predicts that the technological gap between generations will narrow over the following years to become almost negligible by 2023. Smartphone ownership among 55 to 75 years old is expected to reach 85% in western countries, a 10% increase over 2018. Moreover, owners will interact with their phones on average 65 times per day in 2023, a 20% increase over 2018.

On account of the easiness of developing a FDS on a smartphone, there are many works in the recent literature that have focused on the study of 'standalone' smartphone-based FDS, i.e., architectures that make use of the smartphone as the only element in the system, so that it acts as a sensing unit, data processor node and communication gateway [7–10]. Nevertheless, the position of the sensor that monitors the movements of the user is critical for the effectiveness of the FDS as it has been demonstrated by [11–13]. In this respect, the chest or the waist are the recommended locations to place the sensor. However, these are not comfortable and useful positions to put a smartphone, which is usually transported in a pocket or a hand-bag, where the phone may have a certain freedom of movements that can affect the representativeness of the mobility measurements. Another drawback related to the use of a smartphone is that sensors integrated in the smartphone were not originally conceived to quantify the intensity of the movements that a fall can produce, so their ranges are not always sufficient. Consequently, the solution is to employ specific sensors with a more ergonomic design that can be easily incorporated into the subjects' clothing or garments. Nowadays there is a great variety of low-cost sensing motes that can be used for this purpose. Many of these wearable motes embed low power wireless communications standards, such as Bluetooth, which is also supported by commercial smartphones, so that they can collaborate to produce the classification decision required by the FDS.

The use of Multisensor Body Area Networks (BANs) to characterize the human mobility has been extensively studied by the literature. In this regard, several works [14–17] have evaluated the proficiency of a set of 3 to 6 accelerometers distributed through strategic points of the human body to deploy automatic recognition systems of daily living activities. Nonetheless, in these works the network was never employed to identify falls. The system presented in [18] detects falls as abnormal human activities. The work by Özdemir and Barshan in [19] presented an architecture with multiple sensors for fall detection. The authors examined the performance of a FDS built on a BAN consisting of six wireless motes placed at six positions (head, chest, waist, wrist, thigh and ankle). From the developed testbed, a dataset of falls and movements of daily life executed by a group of experimental subjects was created. In [20] a repository of movements, known as SisFall, obtained from the activities produced by a group of volunteers was also published and analyzed. In this case, the measurements are executed by two sensors at the waist.

The present work is an extension of [11, 12, 21, 22]. A Bluetooth network is used in these investigations. The network is formed by four sensor nodes, each of them placed in a particular position on the body of the experimental subjects (chest, waist, wrist and ankle), as well as by a smartphone in the trouser pocket. Hence, a repository (called UMAFall) of falls and movements of daily life was generated. A series of falls detection algorithms based on thresholds and supervised learning strategies were evaluated. The combinations of sensor placements that produce the best detection performance were also identified and the selection of the acceleration statistics that characterize the body mobility was optimized. The final resulting dataset (UMAFall) containing all the log files has been made publicly available in Internet [23] as a benchmarking tool for the research on fall detection systems.

It should be noted that, in almost all the bibliography, the studies and analyses are simply based on the examination of the different means of the obtained magnitudes, without verifying their statistical significance, that is, without verifying that the variations in the obtained means are really significant when they are compared. In contrast, in [12, 22] we systematically evaluated the hybrid multisensor system developed for the UMAFall repository by using supervised learning algorithms and taking into account the statistical significance of the data by analysis of variance (ANOVA).

In this work, the process that was utilized in [12, 22] will be employed again to study UMAFall [11, 12, 21, 22], but now using a single sensor placed in the position for which the best performance was achieved: the waist, and making use of the SVM (Support Vector Machine) supervised learning. In addition, the evaluation will be extended to other two long public repository of movements: the database by Özdemir and Barshan [19] and the SisFall repository [20].

The goal of this study is twofold: firstly, to clarify what are the most significant characteristics of accelerometry signals for identifying a fall and then, to analyze the possibility of extrapolating the learning achieved with a certain database when the FDS is tested with another dataset.

3   Description of the Experimental Testbed

The methodology carried out in this study is the same considered by most authors devoted to the study of FDSs: monitoring a group of experimental subjects who carry an accelerometer and systematically perform a series of predefined movements, both Activities of Daily Living (ADL), as well as simulated falls on a protective surface. The basic information of the people participating in the analyzed databases and the movements is described in [11, 12, 19–21]. Subsequently, these sets of captured accelerometry samples are used offline to feed and test the SVM algorithm in a computer to assess the system performance.

It is important to emphasize that it is beyond the scope of this study to discuss the validity of using, as a framework for evaluation, the monitoring of activities and emulated falls carried out by volunteers (mostly young and healthy people), to test and track systems that are actually targeted to the real movements and real falls of elderly subjects (see [15, 24] for a further study on this controversial issue).

SVM algorithm is selected as the decision core of the FDS a it is one of the most used supervised learning methods in the field of fall detection systems and previously studies have demonstrated that the performance of this algorithm is accurate [12] when compared to other machine learning or threshold-based techniques.

In the next, the input features that are considered to feed the SVM will be detailed.

4 Machine Learning Algorithms and Selection of the Input Features

In the present work, the procedure to discern between fall and ADL is based on the use of a machine learning method, specifically a supervised learning algorithm. This type of architecture is able to perform predictions and classify input data based on a model created by the algorithm from a set of training data. In this way, the datasets of falls and movements of daily life are processed and successively employed to study the behavior of a supervised learning classification algorithm.

This kind of algorithms need to be trained before being applied to the test data. For this purpose, the algorithm will be provided with a series of training samples together with the respective real output decision (fall or ADL) that the system should generate. From the training phase, the algorithm builds a mathematical model that is then employed with the test data to categorize the movements, that is to say, to discriminate falls from ADLs [25]. This process is accomplished by feeding the algorithm with a series of input features (statistics that characterize the movements and which are computed from the mobility traces), which are extracted from the datasets.

The specific supervised learning algorithm that is used in this investigation is the SVM (Support Vector Machine) algorithm. This algorithm creates a hyperplane, based on the training data, which acts as a decision boundary to categorize and discriminate the samples. The hyperplane is built to be at the maximum possible distance from the points (of both types: falls and ADLs) closest to it. After this training phase, in order to study data for which the output decision is unknown, the model created by the algorithm is employed. So the decisions are directly based on the region (defined by the hyperplane) where the test sample is included [25]. In Fig. 1 an example of the operation of this algorithm for a two-dimensional space is illustrated graphically.

4.1 Feature Extraction: Selection of the Input Statistics of the Machine Learning Algorithms

A proper selection of the input characteristics in the development of any machine learning algorithm is a fundamental aspect. By processing the different samples of the studied datasets, a series of statistics are obtained. These statistics characterize each movement, both for the training samples and the study (or test) samples.

The statistical characterization of the movements is based on the measurements $(A_{X_i}, A_{Y_i}, A_{Z_i})$ of the acceleration monitored by the tri-axial accelerometer, as in most works in the literature. Falls begin with an initial free-fall period, which provokes a brusque decay of the acceleration components, which tend to 0 g, followed by one or several acceleration peaks caused by the impact against the floor [26]. Therefore, as in our previous studies [12, 22], we focus our analysis on the interval of the signal where the difference between the acceleration components suffer the highest variation.

Particularly, to analyze the acceleration measurements in every axis (x, y or z), we utilize a sliding window of duration $t_W = 0.5$ s or N_W samples.

$$N_W = \lfloor t_W \cdot f_s \rfloor \tag{1}$$

being f_s the sampling rate of the sensors.

This value of 0.5 s is consistent with other studies on FDS which also employ a sliding window to detect the accidents: 0.6 s [27], 0.75 s [28]. In [29] authors state that an analysis interval of 0.5 s is the best trade-off between efficacy, complexity and low power consumption to analyze the acceleration measurements in a FDS.

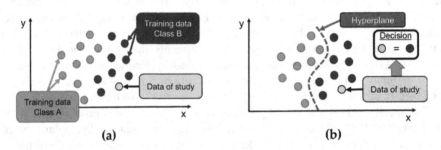

Fig. 1. Example of the performance of the SVM algorithm for a two-dimensional space (two input features: x and y): (a) distribution of the training data on the two-dimensional space (b) creation of the hyperplane and classification decision for a certain test sample (in yellow). (Color figure online)

In order to localize the interval where the acceleration components suffer the highest variation, we calculate for each window the module of the maximum variation of the acceleration components in the three axes. For the j-th window, this parameter ($A_{W_{diff}}(j)$) is calculated as:

$$A_{W_{diff}}(j) = \sqrt{\left(A_{X_{max_j}} - A_{X_{min_j}}\right)^2 + \left(A_{Y_{max_j}} - A_{Y_{min_j}}\right)^2 + \left(A_{Z_{max_j}} - A_{Z_{min_j}}\right)^2} \tag{2}$$

where $A_{X_{max_j}}, A_{Y_{max_j}}, A_{Z_{max_j}}$ indicate the maximum values of the acceleration components in the x, y and z-axes during the j-th sliding window. Thus, the analysis interval will be defined as the subset of consecutive samples $[k_0, k_0 + N_W - 1]$ where the maximum value of $A_{W_{diff}}(j)$ is found to be:

$$A_{W_{diff}}(k_0) = \max\left(A_{W_{(diff)}}(j)\right) \forall j \in [1, N - N_w - 1] \tag{3}$$

where k_0 is the first sample of the analysis interval while N indicates the total number of samples of the trace (for each axis).

The rest of the input features for the detection algorithms are computed just taking into account the values of the acceleration components during this analysis interval. We consider as possible features to the machine learning algorithms the next statistics (together with $A_{W_{diff}}$).

The Mean Signal Magnitude Vector (μ_{SMV}) describes the mean motion or agitation level of the body, this variable is calculable as the mean module of the acceleration vector during the analysis interval:

$$\mu_{SVM} = \frac{1}{N_W} \cdot \sum_{i=k_0}^{k_0+N_W-1} (SMV_i) \tag{4}$$

where SMV_i defines the Signal Magnitude Vector or acceleration module for the i-th measurement of the accelerometer:

$$SMV_i = \sqrt{A_{X_i}^2 + A_{y_i}^2 + A_{z_i}^2} \tag{5}$$

The standard deviation of the Signal Magnitude Vector (σ_{SVM}) informs about the variability of the acceleration, this statistical defines the variability of movements, that is, the existence of 'valleys' and 'peaks' in the evolution of the acceleration [30]:

$$\sigma_{SMV} = \sqrt{\frac{1}{N_W} \cdot \sum_{i=k_0}^{k_0+N_W-1} (SMV_i - \mu_{SVM})^2} \tag{6}$$

The mean absolute difference ($\mu_{SMV_{diff}}$) between consecutive samples of the acceleration module describes the abrupt fluctuation of the motion during a fall [30]:

$$\mu_{SMV_{diff}} = \frac{1}{N_W} \cdot \sum_{i=k_0}^{k_0+N_W-1} |SMV_{i+1} - SMV_i| \tag{7}$$

As a fall occurrence almost always implies a change in the orientation of the body, we also consider the mean rotation angle (μ_θ), computable as [40]:

$$\mu_\theta = \frac{1}{N_W} \sum_{i=k_0}^{k_0+N_W-1} \left(\cos^{-1} \left[\frac{A_{X_i} \cdot A_{X_{i+1}} + A_{Y_i} \cdot A_{Y_{i+1}} + A_{Z_i} \cdot A_{Z_{i+1}}}{SMV_i \cdot SMV_{i+1}} \right] \right) \tag{8}$$

The inclination of the body caused by the falls normally provokes a remarkable modification of the acceleration components that define the plane parallel to the floor when the subject is standing. When the subject remains in an upright position, the effect of the gravity strongly determines the value of the acceleration component in the direction that is perpendicular to the floor plane. Therefore, to characterize this phenomenon, we utilize as a new feature the mean module (μ_{Ap}) of these acceleration components, as it can be appreciated from the resting upright position illustrated in Fig. 2.

$$\mu_{Ap} = \frac{1}{N_w} \sum_{i=k_0}^{k_0+N_W-1} \left(\sqrt{A_{Y_i}^2 + A_{Z_i}^2} \right) \tag{9}$$

5   Results and Discussion

In this section, we firstly evaluate which of the previous input characteristics allow the SVM algorithm to achieve the best performance when discerning falls from ADLs. Secondly, we assess the detection efficacy of a system trained with a particular database when it is tested with the samples of another repository.

As performance metrics, we utilize the following parameters: sensitivity (Se) and the specificity (Sp), which are commonly computed to evaluate the effectiveness of binary classification systems. These metrics are not affected by the unbalance between the number of existing samples of each type (in this case, falls and ADLs) which are employed to test the detection algorithms.

The sensitivity describes the capacity of the classificatory system to correctly identify an event of the 'positive' class (here falls) when this event occurs. Sensibility can be calculated as the ratio between the number of true positives or TP (falls that were correctly identified) and the number of falls that actually occurred, that is to say, the sum of true positives and false negatives (FN) (or falls wrongly classified as ADLs).

$$Se = \frac{TP}{TP + FN} \tag{10}$$

The Specificity (Sp) can in turn be computed as the ratio between the measured amounts of TN or true negatives (properly identified ADLs) and the total number of executed ADLs, TN plus FP (False Positives or ADLs misidentified as falls).

$$Sp = \frac{TN}{TN + FP} \tag{11}$$

We consider the geometric mean of these two parameters ($\sqrt{S_p \cdot S_e}$) as the global metric to characterize the quality of the detection process. As in any system proposed for binary classification, a trade-off between these two parameters (Sp and Se) must be achieved as long as they are normally negatively correlated.

Fig. 2. Representation of the spatial reference system of the employed sensing devices, which are firmly attached to the subjects' body to guarantee that the reference system does not change during the experiments.

In addition, we utilize a 2k factorial design, where k designs the number of possible factors (selected input statistic) that can influence the detection process.

ANOVA test is used to investigate the statistical representativeness of the different results. This test allows deciding if the means of two or more different populations do differ. The ANOVA test determines if different treatments in an experiment cause significant differences in the final results, and consequently, it permits evaluating the importance of the different factors that may alter the operation of the fall detector [31].

To proceed with the supervised learning of the algorithms, the testbeds were randomly divided in a training dataset (around 200 samples) and a test dataset (with the remainder samples). To apply the ANOVA test, we also divided at random the test dataset into six different 'blocks' or 'subsets'. We decided to divide the samples into six subsets. This number allows representing the Gaussian nature of the residuals with six points while keeping a high population in each subset for an adequate estimation of the performance metric. The number of samples of the same movement type is almost homogeneous for the six sub-sets.

We firstly investigate which characteristics should be selected as input features of the machine learning algorithms to maximize the accuracy of the detection decision.

For that purpose, we compare the algorithm SVM by analyzing the performance metrics that are obtained when different sets of the six input features (described in the Sect. 4.1) are considered (number of possible combinations $= 2^6 - 1 = 63$).

We can validate the ANOVA analysis of the results based on the assumptions of normality and homogeneity of variance of the residuals. Moreover, the lack of homoscedasticity is not critical for the ANOVA test, as the experiments are balanced (all series have the same size) and that the ratio of the maximum to the minimum variances of the series does not exceed a proportion of 4 to 1 [31].

According to the ANOVA analysis, Figs. 4, 5 y 6 show the relative variation (expressed as a percentage) that each single input feature produces in the results, with respect to the global mean of the metric for every combination of training and test dataset. The figures also include the same values for some combinations that also produce a high variation.

In the Fig. 3 the confidence intervals of the geometric mean of sensitivity and specificity ($\sqrt{S_p \cdot S_e}$) with respect to the training and test datasets have been calculated. The utilized input features are $A_{w_{diff}}$ y μ_{Ap}, the input features with which the highest performance is achieved. This is utilized to analyze the possibility of extrapolating the learning achieved with a certain database when tested with another.

The final results enable quantifying the contribution of the different factors involved in the detection of a fall and discerning if they cause significant changes in the final performance of the system.

By analyzing the relative influence of the features, we observe that the most frequent statistics in the combinations that produce the highest effect on the algorithms when an accelerometer is located at the waist are $A_{w_{diff}}$ (values between 14.29% and 71.18% in 9 out of 9 tests), which is associated to the maximum variation of the acceleration components in the three axes, and μ_{Ap} (values over 11% in 7 out of 9 tests), which is linked to the changes in the perpendicularity of the body with respect to the floor plane. In addition, in some cases the combination of these features provokes a positive effect in the results too (Fig. 4).

On the other hand, as it can be seen in Fig. 3, the used training dataset is a factor that can significantly alter the performance of the fall detection algorithm. Thus, as expected, the best performance is reached when the training and test datasets come from the same repository. Although we can observe that Özdemir's training dataset yields a metric of 0.9 when UMAFall samples are used as the test samples, this value is not significantly different from the case where UMAFall samples are used for both training and testing the system. Nevertheless, it is noteworthy to remark how critical is to use as training datasets those repositories whose movements do not characterize the environment where the system will be used. The problem of this cross-validation can be observed in Fig. 3 where the SisFall dataset is employed as the training dataset to study the capability of the system to detect falls and ADLs with the samples of the others repositories.

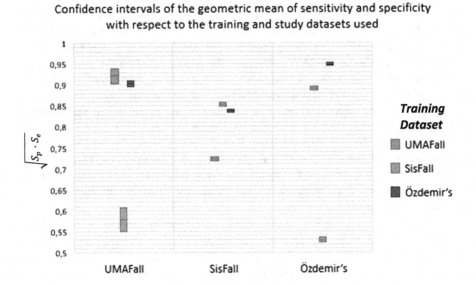

Fig. 3. Confidence intervals of the geometric mean of sensitivity and specificity with respect to the used training and test datasets used. The utilized input features are μ_{SVM}, SMV_{diff} y μ_{Ap}.

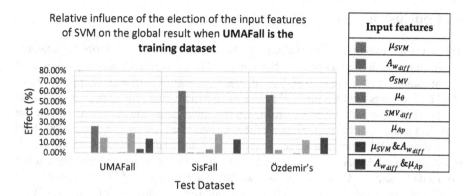

Fig. 4. Relative influence of the election of the input features on the global result when the SVM algorithm is applied and UMAFall is the training dataset.

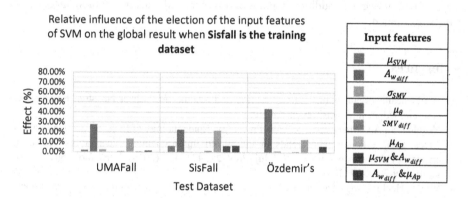

Fig. 5. Relative influence of the election of the input features on the global result when the SVM algorithm is applied and SisFall is the training dataset

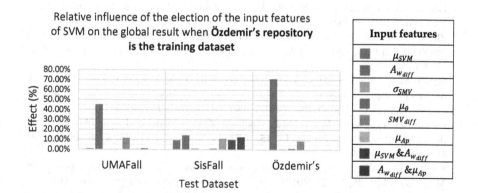

Fig. 6. Relative influence of the election of the input features on the global result when the SVM algorithm is applied and Özdemir's repository is the training dataset.

6 Conclusions

This study has two main contributions. On the one hand, we have described a wearable architecture for a Fall Detection System capable of detecting and reporting falls automatically. The system, based on a machine learning strategy (Support Vector Machine), obtains high geometrical means of specificity and sensitivity, using a single acceleration sensor placed at the waist. On the other hand, the ANOVA tool for the statistical analysis of the performance of the fall detector systems is utilized in this field of study. ANOVA test introduces a more rigorous analysis tool to decide which statistics allow a better characterization of the movements when discriminating a fall from a daily activity. Obtained results also show the limitations of FDSs when they are tested with movement samples captured in a different scenario from the datasets with which the system was trained.

Future research should contemplate other scenarios with different types of ADL movements and falls (for example, the cases in which subjects collapse very slowly and they do not fall over). In addition, the use of unsupervised machine learning algorithms will be examined.

References

1. World health organization: WHO—Ageing and health - Fact sheet No. 404. http://www.who.int/mediacentre/factsheets/fs404/en/
2. Yoshida, S.: A Global Report on Falls Prevention Epidemiology of Falls, Switzerland (2007)
3. Orces, C.H., Alamgir, H.: Trends in fall-related injuries among older adults treated in emergency departments in the USA. Inj. Prev. **20**, 421–423 (2014). https://doi.org/10.1136/injuryprev-2014-041268
4. World Health Organisation: Falls—Fact Sheet. https://www.who.int/en/news-room/fact-sheets/detail/falls
5. Noury, N., et al.: Fall detection–principles and methods. In: Annual International Conference of the IEEE Engineering in Medicine and Biology Society, pp. 1663–1666 (2007). https://doi.org/10.1109/IEMBS.2007.4352627
6. Lee, P., Duncan, S., Calugar-Pop, C.: Technology, Media and Telecommunications Predictions 2018. Deloitte (2018)
7. Zhang, T., Wang, J., Liu, P., Hou, J.: Fall detection by embedding an accelerometer in cellphone and using KFD algorithm. IJCSNS Int. J. Comput. Sci. Netw. Secur. **6**, 277–284 (2006)
8. Habib, M.A., Mohktar, M.S., Kamaruzzaman, S.B., Lim, K.S., Pin, T.M., Ibrahim, F.: Smartphone-based solutions for fall detection and prevention: challenges and open issues. Sensors (Basel) **14**, 7181–7208 (2014). https://doi.org/10.3390/s140407181
9. Casilari, E., Luque, R., Morón, M.J.: Analysis of Android device-based solutions for fall detection. Sensors (Switzerland) **15**, 17827–17894 (2015). https://doi.org/10.3390/s150817827
10. Khan, S.S., Hoey, J.: Review of fall detection techniques: a data availability perspective. Med. Eng. Phys. **39**, 12–22 (2017). https://doi.org/10.1016/j.medengphy.2016.10.014
11. Casilari, E., Santoyo-Ramón, J.A., Cano-García, J.M.: Analysis of a smartphone-based architecture with multiple mobility sensors for fall detection. PLoS One **11**, 1–17 (2016). https://doi.org/10.1371/journal.pone.0168069

12. Santoyo-Ramón, J.A., Casilari, E., Cano-García, J.M.: Analysis of a smartphone-based architecture with multiple mobility sensors for fall detection with supervised learning. Sensors **18**, 1155 (2018)
13. Fang, S.H., Liang, Y.C., Chiu, K.M.: Developing a mobile phone-based fall detection system on Android platform. In: 2012 Computing, Communications and Applications Conference, ComComAp 2012 (2012)
14. Bao, L., Intille, S.S.: Activity recognition from user-annotated acceleration data. In: Ferscha, A., Mattern, F. (eds.) Pervasive 2004. LNCS, vol. 3001, pp. 1–17. Springer, Heidelberg (2004). https://doi.org/10.1007/978-3-540-24646-6_1
15. Zhang, L., Liu, T., Zhu, S., Zhu, Z.: Human activity recognition based on triaxial accelerometer. In: 2012 7th International Conference on Computing and Convergence Technology, ICCCT, pp. 261–266 (2012)
16. Albinali, F., Goodwin, M.S., Intille, S.: Detecting stereotypical motor movements in the classroom using accelerometry and pattern recognition algorithms. Pervasive Mob. Comput. **8**, 103–114 (2012). https://doi.org/10.1016/j.pmcj.2011.04.006
17. Dong, B., Biswas, S.: Wearable networked sensing for human mobility and activity analytics: a systems study. In: 2012 4th International Conference on Communication Systems and Networks, COMSNETS 2012 (2012)
18. Yin, J., Yang, Q., Pan, J.J.: Sensor-based abnormal human-activity detection. IEEE Trans. Knowl. Data Eng. **20**, 1082–1090 (2008)
19. Özdemir, A.T.: An analysis on sensor locations of the human body for wearable fall detection devices: principles and practice. Sensors (Switzerland) **16**, 1161 (2016). https://doi.org/10.3390/s16081161
20. Sucerquia, A., López, J.D., Vargas-Bonilla, J.F.: SisFall: a fall and movement dataset. Sensors (Switzerland) **17**, 198 (2017). https://doi.org/10.3390/s17010198
21. Casilari, E., Santoyo-Ramón, J.A., Cano-García, J.M.: UMAFall: a multisensor dataset for the research on automatic fall detection. Proc. Comput. Sci. **110**, 32–39 (2017)
22. Santoyo-Ramón, J., Casilari, E., Cano-García, J.: Estudio de la detección de caídas utilizando el algoritmo SVM. In: Actas del XXXIII Simposium Nacional de la Unión Científica Internacional de Radio, URSI 2018, Granada, pp. 5–7 (2018)
23. Casilari, E., A. Santoyo-Ramón, J.: UMAFall: fall detection dataset. Universidad de Malaga. https://figshare.com/articles/UMA_ADL_FALL_Dataset_zip/4214283
24. Wang, L., Gu, T., Tao, X., Lu, J.: A hierarchical approach to real-time activity recognition in body sensor networks. Pervasive Mob. Comput. **8**, 115–130 (2012). https://doi.org/10.1016/j.pmcj.2010.12.001
25. Kotsiantis, S.B., Zaharakis, I.D., Pintelas, P.E.: Machine learning: a review of classification and combining techniques. Artif. Intell. Rev. **26**, 159–190 (2006). https://doi.org/10.1007/s10462-007-9052-3
26. Liu, S.H., Cheng, W.C.: Fall detection with the support vector machine during scripted and continuous unscripted activities. Sensors (Switzerland) **12**, 12301–12316 (2012). https://doi.org/10.3390/s120912301
27. Kangas, M., Konttila, A., Winblad, I., Jämsä, T.: Determination of simple thresholds for accelerometry-based parameters for fall detection. In: Proceedings of the Annual International Conference of the IEEE Engineering in Medicine and Biology, pp. 1367–1370 (2007)
28. Ngu, A., Wu, Y., Zare, H., Polican, A., Yarbrough, B., Yao, L.: Fall detection using smartwatch sensor data with accessor architecture. In: Chen, H., Zeng, D.D., Karahanna, E., Bardhan, I. (eds.) ICSH 2017. LNCS, vol. 10347, pp. 81–93. Springer, Cham (2017). https://doi.org/10.1007/978-3-319-67964-8_8

29. Lombardi, A., Ferri, M., Rescio, G., Grassi, M., Malcovati, P.: Wearable wireless accelerometer with embedded fall-detection logic for multi-sensor ambient assisted living applications. In: Proceedings of IEEE Sensors (2009)
30. Chen, K.-H., Yang, J.-J., Jaw, F.-S.: Accelerometer-based fall detection using feature extraction and support vector machine algorithms. Instrum. Sci. Technol. **44**, 333–342 (2016). https://doi.org/10.1080/10739149.2015.1123161
31. Montgomery, D.C.: Design and Analysis of Experiments, 5th edn. Wiley, Hoboken (2012)

Influence of Illuminance on Sleep Onset Latency in IoT Based Lighting System Environment

Mislav Jurić[1]([✉]), Maksym Gaiduk[1,2]([✉]), and Ralf Seepold[1,3]

[1] HTWG Konstanz, Alfred-Wachtel-Strasse 8, 78462 Konstanz, Germany
mislav.juric@gmail.com,
{maksym.gaiduk, ralf.seepold}@htwg-kosntanz.de
[2] Universidad de Sevilla, Avda. Reina Mercedes s/n, 41012 Seville, Spain
[3] Sechenov University, 2-4 Bolshaya Pirogovskaya st., 119991 Moscow, Russia

Abstract. The exposure to the light has a great influence on human beings in their everyday life. Various lighting sources produce light that reaches the human eye and influences a rhythmic release of melatonin hormone, that is a sleep promoting factor.

Since the development of new technologies provides more control over illuminance, this work uses an IoT based lighting system to set up dim and bright scenarios. A small study has been performed on the influence of illuminance on sleep latency. The system consists of different light bulbs, sensors and a central bridge which are interconnected like a mesh network. Also, a mobile app has been developed, that allows to adjust the lighting in various rooms. With the help of a ferro-electret sensor, like applied in sleep monitoring systems, a subject's sleep was monitored. The sensor is placed below the mattress and it collects data, which is stored and processed in a cloud or in other alternative locations.

The research was conducted on healthy young subjects after being previously exposed to the preconfigured illuminance for at least three hours before bedtime. The results indicate correlation between sleep onset latency and exposure to different illuminance before bedtime. In a dimmed environment, the subject fell asleep in average 28% faster compared to the brighter environment.

Keywords: Sleep latency · Ambient assisted living · Illuminance · Sleep quality

1 Introduction

Nowadays life of human beings in modern society is hardly imaginable without being daily exposed to various artificial lighting sources. It would be hard to perform routine activities if there were no additional lighting sources indoor or on the street. The research goal of this work is to determine the influence of different light scenarios before bedtime and the effect on sleep onset latency.

The development of IoT based technologies has enabled wide specter of their application. The Internet of Things term was firstly mentioned in 1985 and later coined

© Springer Nature Switzerland AG 2019
I. Rojas et al. (Eds.): IWBBIO 2019, LNBI 11465, pp. 429–438, 2019.
https://doi.org/10.1007/978-3-030-17938-0_38

by Kevin Ashton in 1999 [1]. Although the term and idea of interconnected minia-
turized devices existed in the past, and to certain extent the technology for development, the IoT has had the greatest impact relatively recently. Several key IoT enablers,
that occurred in the last decade, were crucial for the impact, such as smartphone
revolution and IoT components cost reduction [2]. The selected IoT lighting system has
the ability to set up different lighting environments. In one scenario, the illuminance is
set up to normal indoor lighting, up to 150 lx, whereas in the dim scenario it is reduced
to 10 lx. In both cases, a subject is performing similar routine activities while being
exposed to one of described lighting environments for at least three hours before
bedtime.

From the medical point of view, the secretion of melatonin hormone, a sleep
promoting factor, is stimulated because of the darkness and inhibited by the light that
reaches human eye. Photoreceptor cells located in the retina stimulate suprachiasmatic
nucleus (SCN), a part of hypothalamus, and location of the circadian biological clock.
As a response to a light or lack of it, the SCN starts several mechanisms to send
impulses to the pineal gland, which regulates the sleep-wake cycle [3]. The pineal
gland, often referred to as the 'third eye', is located near the center of the human brain
[4] and produces melatonin [5, 6]. When the eye is exposed to dim illumination, the
SCN sends more impulses to the pineal gland, producing a greater amount of
melatonin.

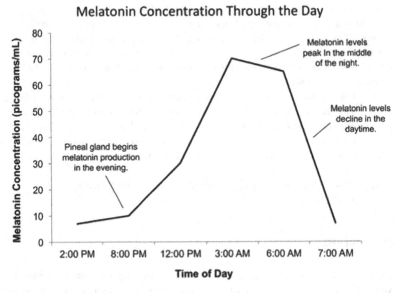

Fig. 1. Rhythmic release of melatonin [7]

Usually, it happens as the night falls and the light level drops, as shown in Fig. 1.
The produced melatonin modulates sleep and circadian phase by attaching to the SCN
melatonin receptors. Normally, during the daytime, a small amount of melatonin is

produced. With the fade of daylight, the secretion increases and reaches the peak in the middle of the night. As the night ends and by the time a person naturally wakes up, the secretion of the melatonin returns to its daytime level.

2 State of the Art

The technology used for the indoor lighting in this research [8] represents the state-of-the-art of currently available systems. The lighting system belongs to Philip Hue line [9]. Continuous improvement and development led to several innovations and upgrades, such as introduction of different sensors, brighter light and richer color of lightbulbs [10], becoming a global leader in lightning [11]. The mobile app integrates a Hue SDK and uses it for the communication with the lighting system. The lightbulbs allow fine adjustment of brightness while the light level sensors enable a user to monitor the brightness. With this help, a proper set up of the environment is possible.

In order to capture sleep data relevant for our investigation, the Emfit QS+Active system was used; it is monitoring a subject's sleep and determining sleep onset. It uses ferro-electret super sensitive sensor technology and relies on ballistocardiography, a technique that senses sudden ejections of blood into the great vessels with every heartbeat [12]. Therefore, it is able to detect the heartbeat even under a thick mattress. The sensor is contact-free, and the measuring process is done autonomously.

Several studies were performed to analyze the effect of light on the sleep and human health. The following paragraphs present just a few but relevant approaches for this project.

The conclusion of similar research [13] has reported longer sleep onset latency when a person is exposed to bright light before bedtime compared to dim light.

Another study [14] has demonstrated that negative influence of dim light at night is similar for male and female subjects: decreasing of total sleep time and sleep efficiency with simultaneous increasing of wake time after sleep onset.

Another investigation [15] has focused on the impact of light exposure during sleep on cardiometabolic function. Finally, the conclusion was done, that a light exposure during sleep increases insulin resistance due to higher insulin level in the groups sleeping with the light.

The effect of sleep with light on brain oscillations was a topic of research of [16]. The results have indicated the negative impact of light during the sleep on brain waves and on human health in general.

The results of presented articles demonstrate the importance of the further research on light influence on human's body.

3 System Architecture

IoT based systems are often explained trough reference models. Cisco has proposed a comprehensive seven-layer model to simplify complex systems. Dividing it into layers that are easier to understand and develop, the system can be optimized and standardized better for combining other products from a vendor [17].

The first layer, *Physical Devices and Controllers*, consists of a wide range of applicable edge devices, which represent 'things' in the IoT system. The second layer, *Connectivity*, is responsible for the information and commands transmission between edge nodes and the layer above. It relies on existing network technologies and ensures a reliable and timely communication. The third layer, *Edge Computing*, prepares the data for the accumulation in the upper, *Data Accumulation* layer, converting it from data in motion to data at rest. The fifth layer, *Data Abstraction*, simplifies further processing of data, while the sixth, *Application* layer, interprets information from lower layers. It also provides functionalities to operate the system. The last layer, *Collaboration and Processes*, includes human enrolment and collaboration, when the IoT system is supporting business processes [18].

3.1 System Architecture

In this work, the IoT based lighting system is applied in a home environment. This belongs to a specific type of smart home automation systems, or often called Domotics [19]. Physical devices and controllers of the lighting system consist of lightbulbs, various sensors, switches and a bridge. Mentioned devices have integrated microcontrollers and wireless communication controllers. The bridge is the main part of the system, abbreviating several layers from discussed reference model. It is responsible for the interaction between itself and other endpoint devices in a wireless sensor network (WSN), which is built on the Zigbee network [20]. The nodes communicate with the bridge in a mesh network, in which the bridge has the role of central gateway. The bridge operates the lights by sending certain parameter to a local URL, specific for the light whose parameter is being updated. Furthermore, the bridge handles data accumulation and abstraction of the devices' latest states, e.g., switches and lightbulbs. In that process, the data in motion is converted to data at rest.

The developed mobile app is written in Java programing language and it implements the Hue SDK. The bridge has a RESTful interface and must be connected to a router to be accessible as a web service. In order to run the system, the mobile app must be authorized in the bridge discovery process. The application was developed so that the lighting system could be managed and operated from a smartphone or tablet, establishing Human-Machine interaction (HMI) and enabling setup of Machine-to-Machine (M2M) communication. The great advantage of this system is the usage of already existing infrastructure.

The app provides functionalities to configure the home layout and positions of the lights within the rooms, as shown in Fig. 2. Each parameter of the lights can be automatically adjusted or manually configured, and sensors' latest readings are observed for further automatization. The lighting system was installed in a shared apartment with private room. The lightbulbs were placed in every room in which the subject spent the time until going to bed. Before conducting the research, the illuminance was checked using light level sensor. The sensor was pointed in various directions in the rooms, at the approximate height of the eyes in standing and sitting positions. The measurement discovered, that in certain areas, the bright illuminance benefits the most from ceiling light, due to its position. For that reason, in bright environment scenario, the ceiling light was used along with the lighting system. In

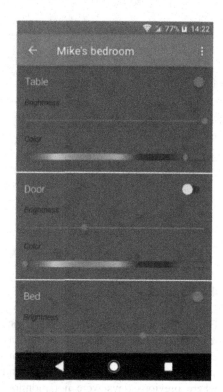

Fig. 2. Mobile app [8]

average, the illuminance was 70 lx. On the other hand, in dim illuminance, the system had the possibility to reduce the light level in average to 5 lx. It was sufficient to perform usual routines before bedtime. Measured illuminance in both scenarios corresponded to the desired values of the research.

3.2 Data Collection and Storage

In the morning, the subject observed sleep data was monitored by Emfit QS system, after previously noting the exact time of entering the bed in the evening. The Emfit QS system determined the time when subject fell asleep. The sleep onset latency was calculated as a difference between the time of going to bed and the exact time of falling asleep and stored on cloud using SleepStats web app. The web application, shown in Fig. 3, was developed in PHP programming language for the purpose of storing users' sleep data and statistical analysis. It was deployed within a cloud along with databases in order to support complete scalability.

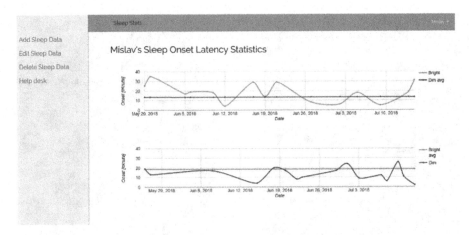

Fig. 3. SleepStats web app [8]

4 Results

During the measuring period, bright and dim environments were randomly chosen, resulting in various selection of dim and bright illuminance. In certain cases, dim environment preceded bright, in other it was opposite. Some days were monitored in same lighting setup for several days, and on the other hand there were isolated days, where previous days were not monitored due to subject's private obligations. The research had the goal to determine the overall difference between subject's sleep onset latencies, after being exposed to bright and dim illuminance for at least three hours before bedtime, regardless any other patterns. In total, the conducted research monitored 31 days, having 15 bright and 16 dim environments.

In the measuring period, including non-monitored days, the subject did not consume medical drugs or alcohol. The subject was a healthy young person, a nonsmoker, did not travel across time zones in several preceding months and during the measuring period and reported consuming moderate amount of coffee in the morning on non-measured days.

The average bedtime was 3:14 h am with standard deviation (SD) of 1:17 h. Although it was relatively late, the subject was consistent in both environment scenarios, displayed in Table 1. In bright illuminance, the average time of going to bed was 3:13 am, whereas in dim illuminance it was 3:16 am. The sleep onset latency results measured in minutes are shown in Table 2. Clearly, exposure to the dim environment before bedtime resulted in shorter sleep onset latency, as it was expected. In addition, the conclusion of shorter onset latency in dim illuminance compared to bright illuminance has already been reported [13].

Graphical comparison between sleep onset latencies in bright and dim scenarios, processed in the web app, is shown in Fig. 4. In the first line chart, the latencies in bright illuminance are compared to the average latency in dim illuminance and displayed throughout the measuring period. Reversely, the second chart displays the latencies in dim illuminance compared to the average latency in bright environment. In the first chart, 27% latencies in bright environment are shorter than the average dim latency. Averaged, these latencies are 57% shorter than the average sleep onset latency in dim illuminance. On the other hand, in the second chart, 25% latencies in dim environment are longer than the average latency in bright illuminance. In summary, these 25% latencies are 20% longer than the average sleep onset latency in a bright environment.

Table 1. Bedtime averages [8]

	Mean	SD
Dim	3:16 am	1:19 h
Bright	3:13 am	1:18 h
Overall	3:14 am	1:17 h

Table 2. Sleep onset latencies [8]

	Mean	SD
Dim	13.4 min	7.0 min
Bright	18.5 min	9.9 min

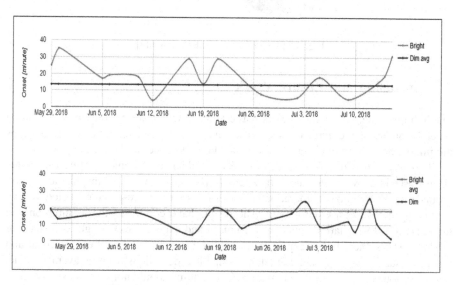

Fig. 4. Sleep onset latencies graphical results [8]

Therefore, it is noticeable from the graphs and table values that latency values are less dispersed in dim environment and the subject fell asleep with less deviations. It is also important to take into account that various other factors influence the sleep onset latency. Therefore, the exposure to the light before bedtime should not be taken as the only one. In Fig. 5, the overall sleep statistics are displayed in charts.

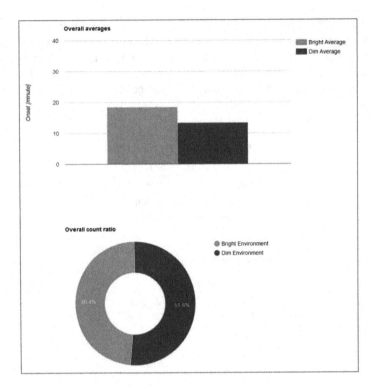

Fig. 5. Sleep onset overall statistics [8]

5   Conclusion and Future Work

The lighting system used in this work was combined with a sleep monitoring system, in order to conduct the research on sleep onset latency. The research in this paper analyzes the importance of the exposure to different illuminance before bedtime.

This research had a goal to determine the influence of home lighting on human sleep onset latency in dim and bright environments. Relying on the medical background and taking into consideration the lack of baseline, the research expected to have shorter latency in dim environment. The results have confirmed the assumption and implicate the close relation between sleep onset and exposure to the light before going to sleep. The subject fell asleep in average 28% faster in a dimmed environment and had more stable latencies, i.e., the latencies had less deviations from its average value. In conclusion, the exposure to the light before bedtime should not be exclusively taken as the only factor that influences the sleep onset, but it has firm indications and grounds of strong contribution.

In a next step, the measurement should be executed in an, for the subject, unaware measurement, and thus contributing to further results and providing a simpler conduction of the research. The developed web application entirely supports scalability and allocates needed resources on demand. Therefore, the research can be conducted on numerous subjects without additional changes to the current configuration.

Acknowledgements. This research was partially funded by the EU Interreg V-Program "Alpenrhein-Bodensee-Hochrhein": Project "IBH Living Lab Active and Assisted Living", grants ABH40 and ABH66.

References

1. Tomar, P., Kaur, G.: Examining Cloud Computing Technologies Through the Internet of Things. IGI Global, Pennsylvania (2017)
2. Presser, M.: The Rise of IoT – why today? IEEE IoT Initiative, 12 January 2016
3. Doghramji, K.: Melatonin and its receptors: a new class of sleep-promoting agents. J. Clin. Sleep Med.: JCSM: Official publ. Am. Acad. Sleep Med. **3**(5 Suppl), S17–S23 (2007)
4. Brun, J., Chazot, G.: The basic physiology and pathophysiology of melatonin. Sleep Med. Rev. **9**(1), 11–24 (2005). https://doi.org/10.1016/j.smrv.2004.08.001
5. Watson, R.R.: Melatonin in the Promotion of Health, 2nd edn. CRC Press, Boca Raton (2011)
6. Pandi-Perumal, S.R., Cardinali, D.P.: Melatonin: From Molecules to Therapy. Nova Publishers, New York (2007)
7. Liu, S., Madu, C.O., Lu, Y.: The role of melatonin in cancer development. Oncomedicine **3** (1), 37–47 (2018). https://doi.org/10.7150/oncm.25566
8. Jurić, M.: IoT based intelligent lighting system. University of Zagreb & University of Applied Sciences Konstanz (2018)
9. Hue Home Lighting: Philips Hue Release Timeline & History, 17 January 2018. https:// huehomelighting.com/philips-hue-release-timeline/. Accessed 22 June 2018
10. Hue Home Lighting: The Key Differences Between 1st, 2nd & 3rd Gen Philips Hue Bulbs? 16 December 2017. https://huehomelighting.com/key-differences-1st-2nd-3rd-gen-philips-hue-bulbs/. Accessed 22 June 2018
11. Philips Lighting: Philips Lighting First Quarter Results 2018 report, 26 April 2018. https:// www.signify.com/static/quarterlyresults/2018/q1_2018/philips-lighting-first-quarter-results-2018-report.pdf. Accessed 22 June 2018
12. Kortelainen, J.M., Mendez, M.O., Bianchi, A.M., Matteucci, M., Cerutti, S.: Sleep staging based on signals acquired through bed sensor. IEEE Trans. Inf Technol. Biomed. **14**(3), 776–785 (2010). https://doi.org/10.1109/TITB.2010
13. Burgess, H.J., Molina, T.A.: Home lighting before usual bedtime impacts circadian timing: a field study. Photochem. Photobiol. **90**(3), 723–726 (2014)
14. Cho, C.H., Yoon, H.K., Kang, S.G., Kim, L., Lee, E.I., Lee, H.J.: Impact of exposure to dim light at night on sleep in female and comparison with male subjects. Psychiatry invest. **15**(5), 520–530 (2018)
15. Mason, I., Grimaldi, D., Malkani, R.G., Reid, K.J., Zee, P.C.: Impact of light exposure during sleep on cardiometabolic function. Sleep **41**(suppl 1), A46 (2018). https://doi.org/10. 1093/sleep/zsy061.116
16. Cho, J.R., Joo, E.Y., Koo, D.L., Hong, S.B.: Let there be no light: the effect of bedside light on sleep quality and background electroencephalographic rhythms. Sleep Med. **14**(12), 1422–1425 (2013). https://doi.org/10.1016/j.sleep.2013.09.007. ISSN 1389-9457
17. Cisco Systems Inc.: The Internet of Things reference model, 4 June 2014. http://cdn.iotwf. com/resources/71/IoT_Reference_Model_White_Paper_June_4_2014.pdf. Accessed 6 June 2018

18. Kumar, R.P., Smys, S.: A novel report on architecture, protocols and applications in Internet of Things (IoT). In: 2018 2nd International Conference on Inventive Systems and Control (ICISC), Coimbatore, India, pp. 1156–1161, 19–20 January 2018. https://doi.org/10.1109/icisc.2018.8398986
19. Govindraj, V., Sathiyanarayanan, M., Abubakar, B.: Customary homes to smart homes using Internet of Things (IoT) and mobile application. In: 2017 International Conference On Smart Technologies For Smart Nation (SmartTechCon), Bangalore, India, pp. 1059–1063, 17–19 August 2017. https://doi.org/10.1109/smarttechcon.2017.8358532
20. Kinney, P.: Zigbee: wireless control that simply works. In: Proceedings of Communications Design Conference (2003)

Clustering and Analysis of Biological
Sequences with Optimization Algorithms

Efficient Online Laplacian Eigenmap Computation for Dimensionality Reduction in Molecular Phylogeny via Optimisation on the Sphere

Stéphane Chrétien[1]([✉]) and Christophe Guyeux[2]

[1] National Physical Laboratory, Hampton Road, Teddington TW11 0LW, UK
stephane.chretien@npl.co.uk
[2] Femto-ST Institute, UMR 6174 CNRS, Université de Bourgogne Franche-Comté, Besançon, France
christophe.guyeux@femto-st.fr

Abstract. Reconstructing the phylogeny of large groups of large divergent genomes remains a difficult problem to solve, whatever the methods considered. Methods based on distance matrices are blocked due to the calculation of these matrices that is impossible in practice, when Bayesian inference or maximum likelihood methods presuppose multiple alignment of the genomes, which is itself difficult to achieve if precision is required. In this paper, we propose to calculate new distances for randomly selected couples of species over iterations, and then to map the biological sequences in a space of small dimension based on the partial knowledge of this genome similarity matrix. This mapping is then used to obtain a complete graph from which a minimum spanning tree representing the phylogenetic links between species is extracted. This new online Newton method for the computation of eigenvectors that solves the problem of constructing the Laplacian eigenmap for molecular phylogeny is finally applied on a set of more than two thousand complete chloroplasts.

Keywords: Nonlinear dimentionality reduction ·
Laplacian eigenmap · Online matrix completion ·
Biomolecular phylogeny

1 Introduction

Molecular phylogenetics is the science of analysing genetic molecular differences in DNA sequences, in order to gain information on an organism's evolution, with the goal to better understand the process of biodiversity. It has been a topic of extensive interest for the bio-informatics community for many decades. Using statistical and computational tools, the result of the molecular phylogenetic analysis is the computation of a phylogenetic tree, hence giving access to

© Springer Nature Switzerland AG 2019
I. Rojas et al. (Eds.): IWBBIO 2019, LNBI 11465, pp. 441–452, 2019.
https://doi.org/10.1007/978-3-030-17938-0_39

possible inference of the old DNA sequence of their last common ancestor. The analysis begins with a phase of multiple alignment of biological sequences consisting, *e.g.*, of nucleotides or amino acids. The alignment then shows the modifications undergone by the sequences over time: a column showing, for example, a polymorphism indicates a mutation, when a gap is a sign of an insertion or a deletion of a sub pattern. From an evolution model (mutation matrix), the goal is then to find the evolutionary tree that maximizes the likelihood of having the evolution indicated by the multiple alignment, under hypothesis of the chosen evolution model.

In order to produce the optimal multiple alignment of a set of sequences, one considers a relevant collection of authorized editing operations (for example, for biological sequences: changing a letter, creating a gap, and increasing a gap), each having a cost, and one looks for the smallest succession of editing operations allowing to pass from one sequence to another in the set. The underlying assumption is that nature is parsimonious, but the associated optimization problem is known to belong to the class of NP-hard optimisation problems. Multiple alignment being fundamental in any molecular phylogeny study, various methods have therefore been proposed in order to produce a "good" alignment, if not optimal, by increasing the alignment as and when, by e.g. adding a new sequence to be aligned at each iterate. Quality of the alignment is systematically "inverse proportional" to the computation time. Based on these alignments, the biological data can be transformed into numbers, and further analysis can be put to work. In particular, the work in [2] demonstrates that using PCA and clustering [9] can be instrumental in the investigation of phylogenetic data by providing a clear and rigorous picture of the underlying structure of the dataset. Other, more sophisticated tools such as the recent nonlinear dimensionality reduction techniques [19] can be employed, but have not yet gained sufficient appeal among data analytics practitioners in the community.

One of these methods, the Laplacian eigenmaps [3], has a great potential for improving the statistical analysis of phylogenetic data by accounting for their non-linear (potentially) low dimensional structure. One main drawback of such methods is that all pairwise distances between genomes are implicitly assumed available, which, due to the computational burden of estimating the alignments, is a very complicated issue that hinders the wider application of such refined methods. On the other hand, Laplacian eigenmap computation being as simple to perform as the PCA, online approaches [11] that only need a small proportion of the pairwise distances have a great potential for overcoming these computational issues. Such online algorithms progressively estimate the principal eigenvectors without having to wait for the full matrix to be known. This problem is very much related to the online matrix completion problems.

Our goal in the present paper is to provide an efficient online optimisation technique for the computation of the Laplacian eigenmap [3] for the embedding and analysis of phylogenetic data, and to demonstrate the applicability of the approach to the analysis of real data. Application of Laplacian eigenmaps to gene sequence analysis and clustering was first proposed in [5]. The main novelty of our work is to propose a principled approach to reducing the number of pairwise affinities that need to be computed in the context of gene sequences. Moreover, we devise a new stochastic gradient algorithm for computing the most significant eigenvectors based on optimisation on manifolds [1,17].

2   Background on the Laplacian Eigenmap

The Laplacian eigenmap [3] is based on the construction of a similarity matrix W. This matrix is intended to measure the similarity between each pair of sequences by providing a number ranging between 0 and 1. The main assumption on W is that the greater the similarity is, the closer are the sequences to each other.

In order to create this similarity matrix, a multiple global alignment of the DNA sequences is performed using the MUSCLE (Multiple Sequence Comparison by Log-Expectation [8]) software. Then, an ad hoc Needleman Wunsch distance [14] is computed for each pair of aligned sequence, and with the EDNA-FULL scoring matrix. This distance takes into account that DNA sequences usually face mutations and insertion/deletion. Note that, by using MUSCLE as first stage of this matrix computation, we operate only one (multiple) sequence alignment, instead of $\frac{n(n-1)}{2}$ (pairwise) alignments in the classical Needleman Wunsch algorithm (that usually contains two stages: finding the best pairwise alignment, and then compute the edit distance).

Let us denote by M the distance matrix obtained by this way. M is then divided by the largest distance value, so that all its coefficients are between 0 and 1. W can finally be obtained as follows:

$$\forall i, j \in [\![1, n]\!], \qquad W_{i,j} = 1 - M_{i,j},$$

in such a way that $W_{i,j}$ represents the similarity score between sequences i and j. Once the similarity matrix has been constructed, the next step is to create the normalized Laplacian matrix, as follows:

$$L = D^{-1/2}(D - W)D^{-1/2},$$

where W is the similarity matrix defined previously and D is the degree matrix of W. That is to say, D is the diagonal matrix defined by:

$$\forall i \in [\![1, n]\!], D_{i,i} = \sum_{j=1}^{n} W_{i,j}.$$

L being symmetric and real, it is diagonalisable in a basis of pairwise orthogonal eigenvectors $\{\phi_1, ..., \phi_n\}$ associated with eigenvalues $0 = \lambda_1 \leqslant \lambda_2 \leqslant ... \leqslant \lambda_n$. The Laplacian Eigenmap consists in considering the following embedding function:

$$c_{k_1}(i) = \begin{pmatrix} \phi_2(i) \\ \phi_3(i) \\ \vdots \\ \phi_{k_1+1}(i) \end{pmatrix} \in \mathbb{R}^{k_1},$$

where $c_{k_1}(i)$ is the coordinate vector of the point corresponding to the ith sequence. In other words, the coordinate vector of the point corresponding to the ith sequence is composed of the ith coordinate of each of the k_1 first eigenvectors, ordered according to the size of their eigenvalues. The next section addresses the problems of getting around the computation of all pairwise affinities.

3 The Online Newton Method on the Sphere for Computing the Laplacian Eigenmap

In this section, we introduce our online Newton algorithm for computing the eigenvectors of the Laplacian matrix. The computation of the main eigenvector is equivalent to a maximisation problem on the unit sphere:

$$\max_{\|x\|_2=1} \quad v^t L v. \tag{1}$$

Taking the spherical constraint into account is crucial in practice, although not usually discussed in the literature; see [17].

3.1 Background on Eigenvector Computation with Partially Observed Matrices

One particular problem which has recently attracted a lot of interest is the one of matrix completion, which asks whether one can recover the eigenvectors of a matrix based on a small fraction of the entries only. Our eigenvector computation for Laplacian eigenmap embedding is directly related to that problem.

It is well known in particular that matrix completion can be solved under low rank assumptions, even with very few queries of the matrix entrees [6]. This observation raised the question of understanding if practical progressive estimation of the principal eigenvectors of an unknown low rank matrix can be efficiently performed. This problem was recently studied in [7] for positive semi-definite matrices.

The approach of [7] uses a deflation approach and a lacunary gradient method. Their analysis is based on a non trivial extension of the arguments for the convergence analysis of the plain stochastic gradient algorithm of [16] for PCA, where it was shown that convergence of the method does not depend on the spectral gap.

3.2 Our Online Newton Algorithm

Our method is an improvement of [7]. In the full observation setting, descent methods on manifolds provide some of the fastest methods for eigendecomposition [1,4,17]. However, to the best of our knowledge, no stochastic variant has been proposed in the literature. Our approach is thus the first to fill this gap, and we will apply it to the relevant problem of molecular phylogenetics, where computing pairwise affinities is prohibitively expensive.

The standard Newton method on the sphere reads:

- Compute $y^{(l)} = (L - x^t L x \ I)^{-1} x^{(l)}$
- Set $\alpha^{(l)} = 1/x^{{(l)}^t} y^{(l)}$, $w^{(l)} = -x^{(l)} + \alpha^{(l)} \ y^{(l)}$, and $\theta^{(l)} = \|w^{(l)}\|_2$.
- Update $x^{(l+1)} = x^{(l)} \cos(\theta^{(l)} + \sin(\theta^{(l)})/\theta^{(l)} \ w^{(l)}$.

The online version of this method is based on replacing the matrix L with a matrix filled with zeros in the places where the pairwise affinity has not been computed at the current iterate. The main trick is to replace this sparse matrix with a low rank approximation obtained using a singular value decomposition. Algorithm 1 provides the details of this method, in which normal_matrix and zeros means matrices with parameter size, and respectively normaly distributed or equal to 0. qr() returns the QR decomposition of a provided matrix while svd() stands for the singular value decomposition. randint returns integers uniformly distributed between the two parameters, and M^t is for the transposition of a matrix M.

Data: Number of sequences N, targeted dimension r, Number of
    iterations L
Result: the largest eigenvalue and its eigenvector
Initialization;
/*Compute a random orthonormal matrix */
Q = normal_matrix(N,5);
$Q = QQ^t$;
X = normal_matrix(N,r);
$X,\_ = \mathrm{qr}(X)$;
SX=zeros(N, r)
Main loop;
for $l = 1, ..., L{+}1$ **do**
    Qstoch = zeros(N,N);
    for $n = 0, ..., 5000$ **do**
        i = randint(N);
        j = randint(N);
        $Qstoch_{i,j} = Q_{i,j}$;
    end
    $QQstoch = (I_N{-}X.X^t)*Qstoch*X$;
    U, S, V = svd($QQstoch$);
    U = first column of U;
    S = diagonal matrix whose first component is the first component of
    S;
    V =first column of V;
    $X = XV^t cos(\frac{10}{l}S)V + U sin(\frac{10}{l}S)V$;
    $X, RR = \mathrm{qr}(X)$;
    $SX = SX + l.X$;
    $XX = SX/l^2$;
    XX=normalization of XX;
end

      Algorithm 1. The online Newton method on the sphere

A typical convergence behavior of the associated eigenvalue with the present method is presented in Fig. 1 below (here, the largest one, the other ones being computed using a deflation approach).

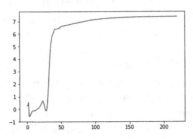

Fig. 1. Typical convergence behavior for our online Newton method for the computation of the eigenvalue associated with the eigenvector of interest (here, the largest one, the other ones being computed using a deflation approach.

4   Application in Molecular Phylogeny

4.1   General Presentation

Any molecular phylogeny study begins with a phase of multiple alignment of biological sequences consisting, *e.g.*, of nucleotides or amino acids [12]. The alignment then shows the modifications undergone by the sequences over time: a column having, for example, a polymorphism indicates a mutation, when a gap is a sign of an insertion or a deletion of a sub pattern. From an evolution model (mutation matrix), the goal is then to find the evolutionary tree that maximizes the likelihood of having the evolution indicated by the multiple alignment, under hypothesis of the chosen evolution model.

The running time to compute the alignment between two sequences of respective lengths m and n being equal to $O(mn/\log n)$ by using Needleman Wunsch algorithm, various approaches propose to use a quick approximation of the latter to more efficiently fill the distance matrix equivalent to multiple alignment (and which basically requires $\frac{N(N-1)}{2}$ distance calculations for a set of N sequences). In view of this observation, we propose to reconstruct the phylogenetic link from an incomplete estimate of the distance matrix. Following the online descent on the sphere presented previously, we can estimate one by one all the eigenvectors of the distance matrix.

The Laplacian eigenmaps applied by using the eigenvectors associated to the three largest absolute eigenvalues leads to an embedding of the N sequences in points belonging in a space of dimension 3. A complete undirected graph can be deduced, in which each sequence occupies a vertex of the graph, and for which the edge between nodes i and j is weighted by the Euclidean distance between points i and j associated with the sequences thus labelled. The extraction of a covering tree of minimal weight, for example with the Kruskal algorithm, allows to infer a phylogenic relationship between the original sequences without having to calculate multiple alignment, and knowing only a small part of the distance matrix. Such an approach is applied to a concrete dataset in the following section.

4.2   Data Collection and Analysis

Thousands of complete genomes of chloroplasts are available now, which can be found for instance on the NCBI website. A Python script was written that automatically downloads all complete sequences of chloroplasts currently available on this website, which amounts to 2,112 genomes of average size: 151,067 nucleotides (ranging from 51,673 to 289,394 nuc.). They represent the global diversity of plants as a whole.

Even though they all derive from a common ancestor (probably a cyanobacterium), this ancestor dates back to such a time that the genomes are very divergent from each other. Each gene in the core genome of chloroplasts therefore corresponds to potentially very different nucleotide sequences between two very distant plants. Also, if calculating the distance of a couple of representatives of a given gene is quite feasible, aligning the thousand DNA sequences of any core gene is very difficult, and leads to an extremely noisy alignment. Multiple alignment tools such as Muscle [8] take several hours to a day of calculations even for small core genes, while requiring a large amount of memory. And the alignment does not ultimately resemble much, so that the phylogenetic tree built from this alignment has many badly supported branches, and leads to obvious inconsistencies in view of taxonomy. The data set is much too large for T-Coffee [15], when ClustalW [13] allows, by its various modes, either to obtain in a reasonable time a very noisy alignment, or gets lost in endless calculations.

One way to obtain a phylogenetic tree well supported on a substantial part of the core genome of these chloroplasts would consist in calculating separately, for each order or family of plants, a multiple alignment followed by a phylogenetic inference. Then, to group this forest of trees in a supertree, by means of an ad hoc algorithm. Although feasible, such an approach has two important limitations. On the one hand, branch support information is lost when the super tree is built. On the other hand, the number of trees to calculate in the forest increases exponentially with the taxonomic level chosen to separate species, and if the calculation time for each tree is reduced, this reduction is compensated by the number of trees to calculate. Conversely, the approach detailed in this article allowed us to reconstruct a reliable phylogeny in a reasonable time, see below.

4.3   Experimental Results

The 2,112 complete sequences have been automatically annotated by Dogma [20] and GeSeq [18], two web services specifically designed for gene prediction in chloroplastic genomes. This latter has outperformed the former in terms of accuracy, when considering their ability to recover well the annotations of some reference genomes. Such a result is not surprising, taken into account the fact that

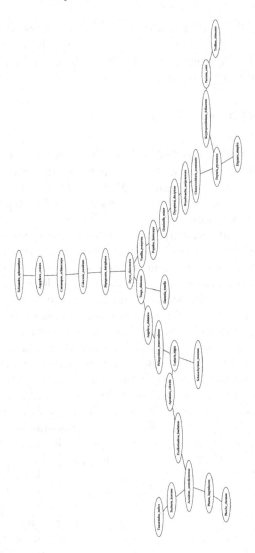

Fig. 2. Obtained phylogeny with RPL2 gene (extraction from the big graph)

Dogma has been released almost 2 decades ago while GeSeq is a brand new algorithm: to make its predictions, GeSeq relies on a basis of knowledge that is much more recent and complete than the one of Dogma, which was therefore abandoned in the remainder of the study.

According to GeSeq annotations, each genome as 81.86 genes in average, the smallest genome exhibiting 32 genes while the largest one has 92 genes. The pan genome has 92 genes, while the core genome is constituted by RPL2, RPS2, RRN16, and RRN23. Being everywhere, these 4 genes can be used to compute the phylogeny of the 2,112 genomes. Each core gene leads to a distance matrix of size 2,112 × 2,112 to estimate, thus to 2,112 eigenvectors on which to apply the Laplacian eigenmap technique, and then to infer the tree.

Our approach allows to infer a phylogenetic tree of the set of all available complete chloroplasts in a very reasonable time. The latter is a function of the hyperparameter L setting the stop criterion in the loop determining a new eigenvalue, which measures the variation in the estimate of a given eigenvalue: when it is below the threshold set by the user, the estimate is returned and the next eigenvalue is considered by investigating the subspace orthogonal to the previously obtained eigenvectors.

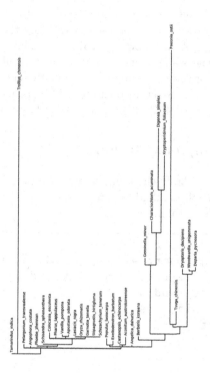

Fig. 3. Phylogenetic tree using Muscle and RAxML

Our proposal has been fully designed using Python language, and the networkx library [10] has been used to compute the covering tree of minimal weight: Euclidian distance between each resulting couple of 3D points has led to a complete graph, whose covering tree of minimal weight has been computed with Kruskal. To validate the obtained tree and for the sake of illustration, we focused on a small subset of 30 divergent sequences of RPL2 gene, investigating whether the phylogenetic relationships extracted from the big tree with 2,112 species are in agreement with the taxonomy obtained with a more classical approach, still applicable for this small collection of sequences.

Our obtained phylogenetic is represented in Fig. 2, while the tree inferred with RAxML (multi-alignment using Muscle, GTR+Gamma evolutionary model) is provided in Fig. 3. As can be seen on this small randomly extracted sub-set, the phylogenetic reconstruction is coherent and broadly sensible, despite the fact that the tree was reconstructed over a small part of the Needleman-Wunch distance matrix. The errors that can be detected in our tree can be reduced by using the hyperparameter values: a compromise must be found to obtain an efficient and accurate calculation.

5 Conclusion

In this paper, we proposed a new online Newton method for the computation of eigenvectors that solves the problem of constructing the Laplacian eigenmap for molecular phylogeny. As a follow up project, we plan to study the problem of active learning in the same framework in a future publication, in order to optimise the selection of the pairs on which the alignment is performed. Extensible backend for hyperparameter auto-tuning will be provided, and the scalable phylogenetic tool will be applied on genome sets of large scale.

A   A Python Implementation

The following code gives the Python implementation of the method for the more general case of the Stiefel manifold, a generalisation of the sphere. (The case of the sphere corresponds to taking $r = 1$.)

```python
from numpy.random import normal
from numpy.linalg import eig, qr, svd, norm
from numpy import matrix, zeros, eye, diag
from random import randint
from math import sin, cos
from pylab import plot, show

N, r = 10, 1

Q = matrix(normal(0,1,(N,5)))
Q = Q*Q.T

umax,lambmax = eig(Q)
lambmax = lambmax[:,umax.argmax()]
umax = max(umax)

X = matrix(normal(0,1,(N,r)))
X,R = qr(X)

SX=matrix(zeros(X.shape))
L = 1000
lamb, scal = [], []
```

```
for l in range(1,L+1):
    Qstoch = matrix(zeros(Q.shape))
    for ll in range(0,5000):
        i=randint(0,N-1)
        j=randint(0,N-1)
        Qstoch[i,j]=Q[i,j]
    QQstoch = (eye(N)-X*X.T)*Qstoch*X
    U,S,V = svd(QQstoch)
    U=U[:,0:r]
    S=diag(diag(S[0:r]))
    V=V[0:r,0:r]
    X = X*V.T*cos(10./(l**1)*S)*V+U*sin(10./(l**1)*S)*V
    X,RR = qr(X)
    SX=SX+l*X
    XX=SX/l**2
    XX=XX/norm(XX);
    lamb.append(max(diag(XX.T*Q*XX)))
    scal.append(abs(X.T*umax))


plot(range(len(lamb)),lamb)
show()
```

References

1. Absil, P.-A., Mahony, R., Sepulchre, R.: Optimization Algorithms on Matrix Manifolds. Princeton University Press, Princeton (2009)
2. Alexe, G., et al.: PCA and clustering reveal alternate mtDNA phylogeny of N and M clades. J. Mol. Evol. **67**(5), 465–487 (2008)
3. Belkin, M., Niyogi, P.: Laplacian eigenmaps for dimensionality reduction and data representation. Neural Comput. **15**(6), 1373–1396 (2003)
4. Boumal, N., Mishra, B., Absil, P.-A., Sepulchre, R.: Manopt, a matlab toolbox for optimization on manifolds. J. Mach. Learn. Res. **15**(1), 1455–1459 (2014)
5. Bruneau, M., et al.: A clustering package for nucleotide sequences using Laplacian Eigenmaps and Gaussian mixture model. Comput. Biol. Med. **93**, 66–74 (2018)
6. Candes, E.J., Plan, Y.: Matrix completion with noise. Proc. IEEE **98**(6), 925–936 (2010)
7. Chretien, S., Guyeux, C., Ho, Z-W.O.: Average performance analysis of the stochastic gradient method for online PCA. arXiv preprint arXiv:1804.01071 (2018)
8. Edgar, R.C.: MUSCLE: multiple sequence alignment with high accuracy and high throughput. Nucleic Acids Res. **32**(5), 1792–1797 (2004)
9. Friedman, J., Hastie, T., Tibshirani, R.: The Elements of Statistical Learning. Springer Series in Statistics, vol. 1. Springer, New York (2001). https://doi.org/10.1007/978-0-387-21606-5
10. Hagberg, A., Swart, P., Chult, D.S.: Exploring network structure, dynamics, and function using networkx. Technical report, Los Alamos National Lab (LANL), Los Alamos, NM, USA (2008)

11. Hazan, E., et al.: Introduction to online convex optimization. Found. Trends® Optim. **2**(3–4), 157–325 (2016)
12. Huson, D.H., Rupp, R., Scornavacca, C.: Phylogenetic Networks: Concepts, Algorithms and Applications. Cambridge University Press, Cambridge (2010)
13. Li, K.-B.: ClustalW-MPI: ClustalW analysis using distributed and parallel computing. Bioinformatics **19**(12), 1585–1586 (2003)
14. Needleman, S.B., Wunsch, C.D.: A general method applicable to the search for similarities in the amino acid sequence of two proteins. J. Mol. Biol. **48**(3), 443–453 (1970)
15. Notredame, C., Higgins, D.G., Heringa, J.: T-coffee: a novel method for fast and accurate multiple sequence alignment. J. Mol. Biol. **302**(1), 205–217 (2000)
16. Shamir, O.: Convergence of stochastic gradient descent for PCA. In: International Conference on Machine Learning, pp. 257–265 (2016)
17. Smith, S.T.: Optimization techniques on riemannian manifolds. Fields Inst. Commun. **3**(3), 113–135 (1994)
18. Tillich, M., et al.: GeSeq-versatile and accurate annotation of organelle genomes. Nucleic Acids Res. **45**(W1), W6–W11 (2017)
19. Van Der Maaten, L., Postma, E., Van den Herik, J.: Dimensionality reduction: a comparative. J. Mach. Learn. Res. **10**, 66–71 (2009)
20. Wyman, S.K., Jansen, R.K., Boore, J.L.: Automatic annotation of organellar genomes with DOGMA. Bioinformatics **20**(17), 3252–3255 (2004)

PROcket, an Efficient Algorithm to Predict Protein Ligand Binding Site

Rahul Semwal[1], Imlimaong Aier[1], Pritish Kumar Varadwaj[1(✉)], and Slava Antsiperov[2]

[1] Indian Institute of Information Technology-Allahabad, Allahabad, India
{pbi2014001, rss2016503, pritish}@iiita.ac.in
[2] Kotelnikov Institute of Radio Engineering and Electronics,
Russian Academy of Science, Moscow, Russia
antciperov@cplire.ru

Abstract. To carry out functional annotation of proteins, the most crucial step is to identify the ligand binding site (LBS) information. Although several algorithms have been reported to identify the LBS, most have limited accuracy and efficiency while considering the number and type of geometrical and physio-chemical features used for such predictions. In this proposed work, a fast and accurate algorithm "PROcket" has been implemented and discussed. The algorithm uses grid-based approach to cluster the local residue neighbors that are present on the solvent accessible surface of proteins. Further with inclusion of selected physio-chemical properties and phylogenetically conserved residues, the algorithm enables accurate detection of the LBS. A comparative study with well-known tools; LIGSITE, LIGSITECS, PASS and CASTptool was performed to analyze the performance of our tool. A set of 48 ligand-bound protein structures from different families were used to compare the performance of the tools. The PROcket algorithm outperformed the existing methods in terms of quality and processing speed with 91% accuracy while considering top 3 rank pockets and 98% accuracy considering top 5 rank pockets.

Keywords: Protein · Ligand binding site · Functional annotation · Phylogenetic · Grid

1 Introduction

High resolution protein structural data has been rapidly increasing due to progress in computational and experimental methodologies [1]. Experimental advancements have made it possible to obtain multiple protein structures with different resolution for the same protein, and it is also more often the case where multiple conformation of particular protein is generated, to know the dynamic behavior of protein, using computational approaches. The protein structure guides researchers to design specific molecules for the analysis of protein-molecular interaction. Moreover, the availability of large conformational data helps to explore additional chemical spaces [2]. The interaction between proteins and other molecules define how it performs biological function, such as protein-ligand interaction, protein-DNA interaction, and protein-protein interaction.

© Springer Nature Switzerland AG 2019
I. Rojas et al. (Eds.): IWBBIO 2019, LNBI 11465, pp. 453–461, 2019.
https://doi.org/10.1007/978-3-030-17938-0_40

One of the major factors for interaction to happen is by identification of shape complementary region or binding site region corresponding to the binding molecule [3–6]. The binding site region in protein corresponds to the pocket or cavity located on the protein surface. Identification of such regions on the protein surface is therefore a crucial step towards the protein-ligand docking and structure-based drug designing.

Several studies have been carried out for the characterization and identification of cavities or pockets in protein surface. In order to identify pockets, one has to consider several aspects associated with it:

1. Identification of the pocket itself [6–24]. In this, an approach is required which limits itself on protein surface for pocket identification and has the ability to bind small molecules.
2. Ranking of predicted pocket, for instance, based on their likeliness to bind to small molecules. Since several pockets are detected on the protein surface, there must be some scoring criteria to select the relevant pockets. Generally, it has been observed that the largest pocket has high frequency corresponding to the ligand binding site [21].
3. Last, but not the least, corresponds to the induced fit model which specifies the three-dimension shape of the pocket corresponding to the ligand [25–28].

The last aspect of the pocket has several issues corresponding to the scoring criteria and pocket shape. Since scoring criteria is strongly dependent on the quality of the identified pocket, and in absence of ligand, the method may or may not predict the relevant pocket. Here, we primarily focus on potential pocket identification from three-dimension protein structure.

Generally, two types of approaches, grid-based approach and grid free approach, are used to predict ligand binding pocket [6–19, 21, 23, 24]. In grid-based approach, a protein structure is initially projected onto three dimensional grids. Grid points that are on the protein surface are then identified based on certain conditions. For instance, in POCKET [7] the grid points are divided into two classes; solvent-accessible grid points and solvent-inaccessible grid points. This method searches for cavity along the x, y and z-axis to locate solvent-accessible grid points that are enclosed by solvent-inaccessible grid points. However, the result obtained via this method is not satisfactory as it is unable to detect pockets with 45° orientation with orthogonal axes. To overcome this problem, LIGSITE [13] was developed which extends its searching process to four cubic diagonals for better prediction of pocket and is independent of protein orientation onto the grid. The extension of LIGSITE is the LIGSITECS approach [14] which uses Connolly surface of the protein to detect binding site pocket. On the other hand, grid free methods either make use of probe-based approaches or Voronoi diagram-based approaches. In probe-based approach, the probe position is first identified with respect to the surface of the protein, and then based on some criteria, the probe cluster is identified with respect to the candidate cavity. For instance, PASS [18] program iteratively places probes on the protein to detect the surface and then searches the cavity probes based on "burial count", which counts the protein atom within 8 Å radius. A "probe weight" is assigned to each cavity probe based on the burial count and neighboring probes. Finally, to detect binding pocket, "active site points" is calculated for cavity probes. Another probe-based approach was proposed by Kawabata and Go [23], and Nayal et al. [21]. They made use of two different size probes to detect the ligand binding pocket. The smaller probes were used to detect the surface of the protein, whereas larger probes were used to detect collections of small probes on the surface depression or cavity of the protein. The other approach related to Voronoi diagram is CAST [17]. In this approach, a Voronoi diagram is constructed from the protein atoms. The Voronoi diagram includes Voronoi cells which contain one protein atom and controls other neighboring atoms in space. The Voronoi diagram computes the Delaunay triangulation of protein surface atom to predict ligand binding sites.

In this study, we present a new grid-based approach to predict ligand binding sites. The method starts with grid initialization and mapping of protein atoms onto the grid. To detect ligand binding pockets on the protein surface, a sophisticated scanning process is used to identify highly enclosed grid points belongs to pocket. These grid points help in determining the shape of the pocket. Finally, a clustering algorithm clusters all the enclosed grid points belonging to the pocket with appropriate shape. At last, the candidate pockets are ranked based on their enclosed score and cluster size to detect the ligand binding pocket.

2 Methodology

PROcket approach is a multistep process that can be described as follows:

2.1 Grid Initialization and Mapping

A cubic three-dimensional rectangular regular grid is constructed to map the three-dimension protein structure. For this, the Eigen vector of the given protein structure was used to transform the protein's molecule around principle axis [2, 29]. The transformed coordinate system was used for cubic grid construction. After grid construction, the transformed coordinate and corresponding van der Waals radii (Table 1) was used to map all protein atoms on to the grid.

Table 1. Atom's name and corresponding van der Waals radii in Angstroms

Atom name	Radius
C (Carbon)	1.6 Å
H (Hydrogen)	1.2 Å
O (Oxygen)	1.52 Å
N (Nitrogen)	1.55 Å
P (Phosphorus)	1.8 Å
S (Sulphur)	1.8 Å
F (Fluorine)	1.47 Å

The mapping process divides the grid region in two parts: occupied region and unoccupied region. To distinguish between the occupied and unoccupied region, a

Boolean Flag variable is used with each grid point. The occupied region contains the protein atoms and its Flag status is set to true. The unoccupied region represents solvent exposed area and its Flag status is set to be false.

2.2   Grid Cubes Credibility Determination for Cavity

In grid-based approach, cavity in a protein can be defined as a series of unoccupied grid points bound by protein atoms (occupied region). To discover such points, the unoccupied region was divided into two parts: surface sub-region and non-surface sub-region (Fig. 1). The surface sub-region contains those unoccupied grid points which are within 4 Å distance from the surface atoms of protein. The non-surface sub-region contains unoccupied points. The enclosed score of unoccupied grid points were calculated in their respective sub region, which demonstrates whether the points are considered to be a part of the cavity.

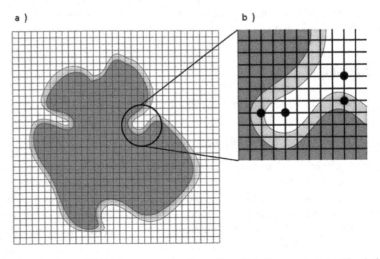

Fig. 1. Semantic view of PROcket process to detect ligand binding pocket. (a) The dark region in grid represents the occupied region of grid. The gray region represents the surface sub region and white area represents the non-surface region of unoccupied region.

To calculate enclosed score of unoccupied grid point, a fourteen-dimension vector was defined which scans unoccupied grid cubes in fourteen directions (six along the positive and negative x, y and z axis and eight along grid directions). A maximum scanning distance threshold was defined for all direction which is to be set by the user. In our algorithm, the default value of scanning distance is set to be 10. During scanning process of unoccupied grid points, if an occupied region is encountered, the enclosed score for that unoccupied grid point was incremented by 1. The value of enclosed score represents the degree of coverage of unoccupied point within cavity or in surface depression. The enclosed score is calculated for all unoccupied points in their respective sub regions. To filter out the grid points which are not enclosed enough, an enclosed threshold for surface and non-surface sub regions was dynamically defined as follows:

$$\mu_j = \frac{\sum_{i=1}^{N_j} Enclose\_score_{i_j}}{N_j} \tag{1}$$

$$SD_j = \sqrt{\frac{\sum_{i=1}^{N_j} \left| Enclose\_score_{i_j} - \mu_j \right|^2}{N_j}} \tag{2}$$

$$enclosedness_{threshold_j} = \mu_j + SD_j \tag{3}$$

where μ_j, SD_j represents the mean and standard deviation of j-th sub region, $Enclose\_score_{i_j}$ represents the i-th grid point enclose score in j-th sub region, N_j represents the size of j-th sub region and j either represents surface sub region or non-surface sub region.

The unoccupied points with enclosed score less than the enclosed threshold of respective sub-region were removed from the corresponding sub-regions. The remaining unoccupied grid cubes from two sub regions were combined together into enclosed cubes list for further investigation of the neighborhood. The neighborhood demonstrates how well an unoccupied-enclosed cube is surrounded by other well unoccupied-enclosed points, and how it controls the shape of the respective cavity. To investigate the neighborhood of unoccupied-enclosed grid points, a twenty-six-dimension vector was defined, which scan the unoccupied-enclosed grid points in twenty-six directions (using all possible combination of 1, 0, 1 along x, y and z axis and avoiding the current cube position) to include other unoccupied-enclose grid point neighbors. The neighborhood score was calculated for all unoccupied-enclosed points. To filter out unoccupied-enclosed points which did not have sufficient neighbor, a neighborhood threshold was dynamically defined as follows:

$$\mu_n = \frac{\sum_{i=1}^{M} Neighbourhood_i}{M} \tag{4}$$

$$SD_n = \sqrt{\frac{\sum_{i=1}^{M} \left| Neighbourhood_i - \mu_n \right|^2}{M}} \tag{5}$$

$$Neighbourhood_{threshold_n} = \mu_n + SD_n \tag{6}$$

where μ_n, SD_n represents the mean and standard deviation of unoccupied-enclosed point's neighborhood, $Neighbourhood_i$ represents the i-th unoccupied-enclosed point neighborhood score, M represents the number of unoccupied-enclosed points.

The unoccupied-enclosed grid cubes having neighboring score less than the neighborhood threshold were removed from the enclosed cubes list. The remaining unoccupied-enclosed grid points were considered to be part of cavity and promoted for clustering process.

2.3 Clustering and Interface Residue Calculation

The process of combining the unoccupied-enclosed grid points of respective cavity into a single group is known as clustering. The clustering process describes the size of cavity in term of number of highly enclosed unoccupied points having sufficient number of neighbors, which is analogous to the volume of the cavity. Figure 2 describes the pseudo code of clustering process.

```
for i=1 to enclosed_cubes_list.length:
        cluster[i].add(enclosed_cube_list[i])
        neighbor_list =enclosed_list_cubes[i].get_neighbours()
        enclosed_cubes_list.remove(enclosed_cube_list[i])
        while neighbor_list.length > 0:
                new_neighbour_list={}
                for j =1 to neighbor_list.length:
                        new_neighbor_list.add(neighbor_list[j].get_neighbour())
                cluster[i].add(neighbor_list[j])
                enclosed_cubes_list.remove(neighbor_list[j])
                new_neighbour_list= new_neighbour_list - cluster[i]
                neighbor_list = new_neighbour_list
```

Fig. 2. Algorithm to cluster enclosed grid points belongs to candidate cavity.

After performing clustering, the size of each cavity was determined. Higher value of cluster size represents larger cavities, while the smaller value represents smaller cavities in the protein. Generally, a highly enclosed (buried/depth) cavity with high volume is considered to be the ligand binding side [21]. The above stated criterion was used to rank each obtained cavity of protein. To this, a cavity rank variable (CR) was used, which is equal to the average of maximum enclosed grid point (ECC) and cluster size (CS) of the respective cavity, and its mathematical description was defined in equation:

$$CR_i = avg(max(ECC_i) + CS_i) \tag{7}$$

Where CR_i represents the rank of i-th cavity, ECC_i represents unoccupied enclosed grid point list of i-th cavity and CS_i represents the cluster size of i-th cavity of protein. After ranking the cavity, the residue forming the cavity has been identified and projected as output.

3   Result and Discussion

In order to evaluate the prediction quality of our proposed approach, PROcket, a test data set of forty-eight ligand bound complex was extracted from the public domain database RCSB protein database (PDB) [30]. The above test data set collection was used in earlier studies [14] to compare the prediction quality of various approaches such as LIGSITE, LIGSITE^{CS}, CAST and PASS. The dataset collection was used to validate the prediction quality of our approach. The ligand information denoted by HET (heteroatom) identifier was excluded from the forty-eight PDB-files prior to computation. To determine whether a cavity is the ligand binding site, the geometric center of each outputted cavity is computed. If any atom of the ligand is within 4 Å from this geometric center, the cavity was considered as the ligand binding site. To verify the predicted ligand binding site with actual ligand binding sites, we performed the structural comparison between the predicted and actual ligand binding sites using the open source software Chimera [31].

Table 2 describes the predicted ligand binding sites within top three results obtained by each of the five considered approaches, matched with known ligand binding site. At rank 1, the success rate of our approach is only 64% which is slightly lower than the success rate of CAST, LIGSITE, and LIGSITE^{CS}, 69%, 69%, and 67%, respectively, but slightly higher than that of PASS success rate at 63%. However, at rank 3, the success rate of our approach is 91%, which is much better than the success rate of other approaches.

Table 2. Comparison of success rates for 48 complexed protein structures in percentage.

S. No	Software	Rank 1	Rank 3
1	PROcket	66	91
2	LIGSITE^{CS}	69	87
3	LIGSITE	69	87
4	CAST	67	83
5	PASS	63	81

The results obtained by our approach, PROcket, are classified into six classes: first, second, third, fourth, fifth ligand binding site, and none of these. Table 3 shows the percentage of these six classes. The results indicate the efficiency of PROcket in detecting ligand binding pocket, and further filtering, such as scoring based on drugability could improve the result.

Table 3. Number of proteins in each class for 48 bound structures.

Class	No of proteins (as %)
Class 1: Binding site in largest Pocket	(32/48) = 66
Class 2: Binding site in second largest Pocket	(10/48) = 20
Class 3: Binding site in third largest Pocket	(2/48) = 4
Class 4: Binding site in fourth largest Pocket	(2/48) = 4
Class 5: Binding site in fifth largest Pocket	(1/48) = 2
Class 6: Binding site in none of the above	(1/48) = 2

4 Conclusions

Several methods have been developed to identify pockets or cavities on protein surfaces, and to describe the relationship between the predicted pocket and ligand binding sites. In this paper, we propose an automated method, PROcket, to detect ligand binding site on protein surface. We compared our approach with LIGSITE, LIGSITE^{CS}, PASS, and CAST on 48 ligand bound protein dataset. The result shows that our method is capable of predicting ligand binding site with high success rate (91%), compared to other methods, within top three results and 98% within top five results.

References

1. Dutta, S., et al.: Data deposition and annotation at the worldwide protein data bank. Mol. Biotechnol. **42**(1), 1–13 (2009)
2. Craig, I.R., Pfleger, C., Gohlke, H., Essex, J.W., Spiegel, K.: Pocket-space maps to identify novel binding-site conformations in proteins. J. Chem. Inf. Model. **51**(10), 2666–2679 (2011)
3. Katchalski-Katzir, E., Shariv, I., Eisenstein, M., Friesem, A.A., Aflalo, C., Vakser, I.A.: Molecular surface recognition: determination of geometric fit between proteins and their ligands by correlation techniques. Proc. Natl. Acad. Sci. **89**(6), 2195–2199 (1992)
4. Jones, S., Thornton, J.M.: Principles of protein-protein interactions. Proc. Natl. Acad. Sci. **93**(1), 13–20 (1996)
5. Heifetz, A., Katchalski-Katzir, E., Eisenstein, M.: Electrostatics in protein–protein docking. Protein Sci. **11**(3), 571–587 (2002)
6. Halperin, I., Ma, B., Wolfson, H., Nussinov, R.: Principles of docking: an overview of search algorithms and a guide to scoring functions. Proteins: Struct. Funct. Bioinf. **47**(4), 409–443 (2002)
7. Levitt, D.G., Banaszak, L.J.: POCKET: a computer graphies method for identifying and displaying protein cavities and their surrounding amino acids. J. Mol. Graph. **10**(4), 229–234 (1992)
8. Delaney, J.S.: Finding and filling protein cavities using cellular logic operations. J. Mol. Graph. **10**(3), 174–177 (1992)
9. Del Carpio, C.A., Takahashi, Y., Sasaki, S.I.: A new approach to the automatic identification of candidates for ligand receptor sites in proteins: (I) search for pocket regions. J. Mol. Graph. **11**(1), 23–29 (1993)
10. Kleywegt, G.J., Jones, T.A.: Detection, delineation, measurement and display of cavities in macromolecular structures. Acta Crystallogr. Sect. D: Biol. Crystallogr. **50**(2), 178–185 (1994)
11. Masuya, M., Doi, J.: Detection and geometric modeling of molecular surfaces and cavities using digital mathematical morphological operations. J. Mol. Graph. **13**(6), 331–336 (1995)
12. Peters, K.P., Fauck, J., Frömmel, C.: The automatic search for ligand binding sites in proteins of known three-dimensional structure using only geometric criteria. J. Mol. Biol. **256**(1), 201–213 (1996)
13. Hendlich, M., Rippmann, F., Barnickel, G.: LIGSITE: automatic and efficient detection of potential small molecule-binding sites in proteins. J. Mol. Graph. Model. **15**(6), 359–363 (1997)

14. Huang, B., Schroeder, M.: LIGSITE^{csc}: predicting ligand binding sites using the Connolly surface and degree of conservation. BMC Struct. Biol. **6**(1), 19 (2006)
15. Ruppert, J., Welch, W., Jain, A.N.: Automatic identification and representation of protein binding sites for molecular docking. Protein Sci. **6**(3), 524–533 (1997)
16. Liang, J., Woodward, C., Edelsbrunner, H.: Anatomy of protein pockets and cavities: measurement of binding site geometry and implications for ligand design. Protein Sci. **7**(9), 1884–1897 (1998)
17. Dundas, J., Ouyang, Z., Tseng, J., Binkowski, A., Turpaz, Y., Liang, J.: CASTp: computed atlas of surface topography of proteins with structural and topographical mapping of functionally annotated residues. Nucleic Acids Res. **34**(Suppl_2), W116–W118 (2006)
18. Brady, G.P., Stouten, P.F.: Fast prediction and visualization of protein binding pockets with PASS. J. Comput. Aided Mol. Des. **14**(4), 383–401 (2000)
19. Venkatachalam, C.M., Jiang, X., Oldfield, T., Waldman, M.: LigandFit: a novel method for the shape-directed rapid docking of ligands to protein active sites. J. Mol. Graph. Model. **21**(4), 289–307 (2003)
20. An, J., Totrov, M., Abagyan, R.: Pocketome via comprehensive identification and classification of ligand binding envelopes. Mol. Cell. Proteomics **4**(6), 752–761 (2005)
21. Nayal, M., Honig, B.: On the nature of cavities on protein surfaces: application to the identification of drug-binding sites. Proteins: Struct. Funct. Bioinf. **63**(4), 892–906 (2006)
22. Glaser, F., Morris, R.J., Najmanovich, R.J., Laskowski, R.A., Thornton, J.M.: A method for localizing ligand binding pockets in protein structures. PROTEINS: Struct. Funct. Bioinf. **62**(2), 479–488 (2006)
23. Kawabata, T., Go, N.: Detection of pockets on protein surfaces using small and large probe spheres to find putative ligand binding sites. Proteins: Struct. Funct. Bioinf. **68**(2), 516–529 (2007)
24. Kim, D., Cho, C.H., Cho, Y., Ryu, J., Bhak, J., Kim, D.S.: Pocket extraction on proteins via the Voronoi diagram of spheres. J. Mol. Graph. Model. **26**(7), 1104–1112 (2008)
25. McGovern, S.L., Shoichet, B.K.: Information decay in molecular docking screens against holo, apo, and modeled conformations of enzymes. J. Med. Chem. **46**(14), 2895–2907 (2003)
26. Bhinge, A., Chakrabarti, P., Uthanumallian, K., Bajaj, K., Chakraborty, K., Varadarajan, R.: Accurate detection of protein: ligand binding sites using molecular dynamics simulations. Structure **12**(11), 1989–1999 (2004)
27. Yang, A.Y.C., Källblad, P., Mancera, R.L.: Molecular modelling prediction of ligand binding site flexibility. J. Comput. Aided Mol. Des. **18**(4), 235–250 (2004)
28. Murga, L.F., Ondrechen, M.J., Ringe, D.: Prediction of interaction sites from apo 3D structures when the holo conformation is different. Proteins: Struct. Funct. Bioinf. **72**(3), 980–992 (2008)
29. Foote, J., Raman, A.: A relation between the principal axes of inertia and ligand binding. Proc. Natl. Acad. Sci. **97**(3), 978–983 (2000)
30. Berman, H.M., et al.: The protein data bank. Nucleic Acids Res. **28**(1), 235–242 (2000)
31. Pettersen, E.F., et al.: UCSF Chimera—a visualization system for exploratory research and analysis. J. Comput. Chem. **25**(13), 1605–1612 (2004)

Gene Expression High-Dimensional Clustering Towards a Novel, Robust, Clinically Relevant and Highly Compact Cancer Signature

Enzo Battistella[1,2,3,4(✉)], Maria Vakalopoulou[1,2,4], Théo Estienne[1,2,3,4], Marvin Lerousseau[1,2,3,4], Roger Sun[1,2,3,4], Charlotte Robert[1,2,3], Nikos Paragios[1], and Eric Deutsch[1,2,3]

[1] Gustave Roussy-CentraleSupélec-TheraPanacea Center of Artificial Intelligence in Radiation Therapy and Oncology, Gustave Roussy Cancer Campus, Villejuif, France
enzo.battistella@gustaveroussy.fr
[2] INSERM, U1030 Paris, France
[3] Université Paris Sud, UFR de Médecine, Paris, France
[4] CVN, CentraleSupélec, Université Paris-Saclay and INRIA Saclay, Gif-sur-Yvette, France

Abstract. Precision medicine, a highly disruptive paradigm shift in healthcare targeting the personalizing treatment, heavily relies on genomic data. However, the complexity of the biological interactions, the important number of genes as well as the lack of substantial patient's clinical data consist a tremendous bottleneck on the clinical implementation of precision medicine. In this work, we introduce a generic, low dimensional gene signature that represents adequately the tumor type. Our gene signature is produced using LP-stability algorithm, a high dimensional center-based unsupervised clustering algorithm working in the dual domain, and is very versatile as it can consider any arbitrary distance metric between genes. The gene signature produced by LP-stability reports at least 10 times better statistical significance and 35% better biological significance than the ones produced by two referential unsupervised clustering methods. Moreover, our experiments demonstrate that our low dimensional biomarker (27 genes) surpass significantly existing state of the art methods both in terms of qualitative and quantitative assessment while providing better associations to tumor types than methods widely used in the literature that rely on several omics data.

Keywords: Clustering · Predictive signature · Biomarkers · Genomics

1 Introduction

Advances in omics data interpretation such as genomics, transcriptomics, proteomics and metabolomics contributed to the development of personalized

© Springer Nature Switzerland AG 2019
I. Rojas et al. (Eds.): IWBBIO 2019, LNBI 11465, pp. 462–474, 2019.
https://doi.org/10.1007/978-3-030-17938-0_41

medicine at an extraordinarily detailed molecular level [7]. Major advances in sequencing techniques [15] as well as increasing availability of patients which gave access to a big amount of data are the backbones of precision medicine paradigm shift. Among them, the first omics discipline, genomics, focuses on the study of entire genomes as opposed to 'genetics' that interrogated individual variants or single genes [8]. Genomic studies investigate frameworks for studying specific variants of genes, producing robust biomarkers that contribute to both complex and mendelian diseases [5] as well as the response of patients to treatment [21]. However, these studies suffer from the curse of dimensionality and face several statistical limits reporting instead of causality, random correlations leading to false biomarker discoveries as stated in [4]. For these reasons the largest topics of research on genomics is the development of robust clustering techniques that are able to reduce the dimensionality of the genetic data, while maintaining the important information that they contain [18,19].

Clustering algorithms are commonly used with big data sets to identify groups of similar observations, discovering invisible to the human eye patterns and correlations between them [6]. Cluster analysis, primitive exploration with little or no prior knowledge, has been a prolific topic of research [23]. It aims to group the variables in the best way that minimizes the variation within the groups while maximizing the distance between the different groups. Among a variety of methods, some of the most commonly used are the K-Means [17], the agglomerative hierarchical clustering [20] and the spectral clustering [16].

Cluster analysis on RNA-seq transcriptomes is a wide spread technique [2] aiming to identify clusters or modules of genes that have similar expression profiles. The main goal of such techniques is to propose groups of genes which are biologically informative such as containing genes coding for proteins interacting together or participating to a same biological process [3]. Several studies have investigated the use of machine learning algorithms towards powerful, compact and predictive genes signatures [5] as biomarkers associated to e.g. tumor types. However, most of them rely on a priori knowledge to choose the genes of the signatures leading to redundancy and loss of information, where evidence based methods as well as the ability to determine unknown to the humans higher order correlations could have tremendous diagnostic, prognostic and treatment selection impact. In [18], the authors propose a clustering algorithm, CorEx algorithm [22], to design from scratch a predictive gene signature evaluated for ovarian tumors. Even if this study showed that powerful gene biomarkers can be generated, it has a lot of limitations such as the association with only one specific tumor type and a signature with several hundred genes.

A very important step towards the generation of informative clusters is their evaluation with independent and reliable measures for the comparison of the parameters and methods. This task is very challenging in the case of genomic clustering, as the clusters should also contain biological information. There are variety of metrics that can assess the quality of the clusters in a statistical matter as the Silhouette Value [10], the Dunn's Index [14] or more recently the Diversity Method [12]. As a complement, the Protein-Protein Interaction (PPI) and the

CorEx Algorithm. CorEx [22] was successfully applied on various fields and, also, on genes [18]. The algorithm finds a set S' of k latent factors that describe the data set S in the best way. Formally, let us consider the Total Correlation of discrete random variables $X^1, ..., X^p$ as

$$TC(X^1, ..., X^p) = \sum_{1 \leq i \leq p} H(X^i) - H(X^1, ..., X^p) \qquad (2)$$

and the Mutual Information of two discrete random variables X^i, X^j as

$$MI(X^i, X^j) = \sum_{X_p^i \in X^i} \sum_{X_q^j \in X^j} P(X_p^i, X_i^q) \log \frac{P(X_p^i, X_q^j)}{P(X_p^i)P(X_q^j)} \qquad (3)$$

where $P(X_p^i, X_q^j)$ is the joint probability function and $P(X_p^i), P(X_q^j)$ are marginal probability functions. The algorithm minimizes the Total Correlation $TC(S|S')$. Then, the clusters are defined by assigning each data point x^p to the latent factor f maximizing the mutual information $MI(X^p, f)$. The algorithm requires as an input the number k of latent factors corresponding to the number of clusters.

2.2 LP-stability Clustering Algorithm

We present here the evaluated LP-stability clustering [13] which is a linear programming algorithm that has been successfully used on variety of problems. It aims to optimize the following linear system

$$PRIMAL \equiv \min_C \sum_{p,q} d(x^p, x^q) C(p, q)$$

$$s.t. \sum_q C(p, q) = 1 \qquad (4)$$

$$C(p, q) \leq C(q, q)$$

$$C(p, q) \geq 0.$$

where $C(p, q)$ represents the fact that x^p belongs to the cluster of center x^q. To decide which points will be used as centers, the notion of stability is defined as

$$S(q) = \inf\{s, \; d(q, q) + s \; \text{PRIMAL has no optimal solution with } C(q, q) > 0\}.$$

Let us denote Q the set of stable clusters centers. The algorithm solves the clustering using the DUAL problem

$$DUAL \equiv \max_D D(h) = \sum_{p \in V} h^p$$

$$s.t. \; h^p = \min_{q \in V} h(p, q) \qquad (5)$$

$$\sum_{p \in V} h(p, q) = \sum_{p \in V} d(x^p, x^q)$$

$$h(p, q) \geq d(x^p, x^q).$$

$h(p, q)$ corresponds here to the minimal pseudo-distance between x^p and x^q, h^p corresponds to the one from x^p. In particular, the algorithm formulates the computation of clusters as

$$DUAL_Q = \max DUAL \text{ s.t. } h_{pq} = d_{pq}, \forall \{p, q\} \cap Q \neq \emptyset. \tag{6}$$

The proposed clustering approach is metric free (it can integrate any distance function), does not make any prior assumption on the number of clusters and their distribution, and solves the problem in a global manner seeking for an automatic selection of the cluster centers as well as the assignments of each observation to the most appropriate cluster. Only one parameter has to be defined, the penalty vector v, that turns $d(q, q)$ in $d'(q, q) = d(q, q) + v_q$ in PRIMAL, influencing the number of clusters.

To cope with the dimensionality of the observations as well as the low ratio between samples and dimensions of each sample, a robust statistical distance was adopted for our experiments. It comes from Kendall's rank correlation [11]:

$$Kendall(x^p, x^q) = 2\frac{N_C - N_D}{n(n-1)} \tag{7}$$

where N_C is the number of concordant pairs and N_D the number of discordant pairs. A pair of observations (x_u^p, x_v^q) and (x_u^p, x_v^q) is considered as concordant if their ranks agree i.e. $x_u^p > x_v^p \Leftrightarrow x_u^q > x_v^q$. They are considered as discordant if $x_u^p > x_v^p \Leftrightarrow x_u^q < x_v^q$.

The distance is then defined as: $d(x^p, x^q) = \sqrt{2(1 - Kendall(x^p, x^q))}$.

3   Experimental Results

3.1   Evaluation Criteria

In order to assess the performance of the proposed solution, we have adopted joint qualitative/quantitative assessment. Biological relevance of the proposed solution was used to assess the quality of the results, while well known statistical methods were adopted to determine the appropriateness of the proposed solution from mathematical view point. In particular, the criteria used are the following:

- **Enrichment Score:** To assess the biological information of the clusters, enrichment is one of the most popular metrics used in the literature [18]. Enrichment corresponds to the probability of obtaining a random cluster presenting the same amount of occurrences of a given event as in the assessed cluster. This event for our experiments was defined as the number of PPI. In particular, for each cluster the p-value of the enrichment is calculated and the cluster is defined as enriched if the p-value is below a given threshold. The enrichment score corresponds to the proportion of enriched clusters.
- **Dunn's Index:** The Dunn's Index [14] assesses if the clusters have a small inter-cluster variance compared to the intra-cluster variance. Formally,

$$Dunn(\mathcal{C}) = \frac{\min_{1 \le i,j \le k} \delta(C_i, C_j)}{\max_{1 \le i \le k} \Delta(C_i)}$$ where $\delta(C_1, C_2)$ is the distance between the
two closest points of the clusters C_i and C_j, $\Delta(C_i)$ is the diameter of the
cluster *i.e.* the distance between the two farthest points of the cluster C_i.
Even if Dunn's Index is one of the commonly used metrics for evaluating the
quality of the clustering it can varies dramatically even if only one cluster is
not well formed. However, we chose this metric over the various existing ones
to show the importance of having homogeneously well formed clusters.

To assess the relevance of the results obtained, we compared the clustering
with the methods presented in Sects. 2.1 and 2.2 but also with the performance
of random clusters. This comparison is very important to prove that the information captured by the clusters is associated with the gene interactions and it
cannot be achieved by a random selection of genes.

3.2 Data Set

For our experiments we used a data set from the TCGA data portal [1]
with tumor types that can be treated by radiotherapy and/or immunotherapy
(Table 1). It contains **4615** samples well distributed among all the ten different tumor types. In particular, we investigate the following types of tumors,
namely: Urothelial Bladder Carcinoma (BLCA), Breast Invasive Carcinoma
(BRCA), Cervical Squamous Cell Carcinoma and Endocervical Adenocarcinoma
(CESC), Glioblastoma multiforme (GBM), Head and Neck Squamous Cell Carcinoma (HNSC), Liver Hepatocellular Carcinoma (LIHC), Rectum Adenocarcinoma (READ), Lung adenocarcinoma (LUAD), Lung Squamous Cell Carcinoma
(LUSC) and Ovarian Cancer (OV). For each sample, we had the RNA-seq values
of **20 365** genes normalized by reads per kilobase per million (RPKM).

Table 1. Number of the different samples used per tumor type.

Tumor type	BLCA	BRCA	CESC	GBM	HNSC	LIHC	READ	LUAD	LUSC	OV
# of Samples	427	1212	309	171	566	423	72	576	552	307

3.3 Implementation Details

The optimization and selection of parameters per algorithm has been performed
by grid search, for a wide range of values. In particular, for the random clustering
and K-Means algorithm, we studied the following numbers of clusters: 5, 10,
15, 20, 25 and between 30 and 100 with an increasing step of 10 and with an
increment of 25 for CorEx algorithm because of its computational complexity.
For the LP-stability algorithm, as the number of clusters is not directly specified,
we gave the same penalty value for all the genes. We used penalty values such
that we have numbers of clusters comparable to the ones of the other algorithms.

For the enrichment score we performed evaluations with different thresholds values *i.e.* 0.005, 0.025, 0.05 and 0.1. Moreover, for the Dunn's Index, we used the same distance as the one used to compute each of the clustering to have the best related score to the clustering metric.

To evaluate the clusters that we have obtained from the proposed method, together with the other baseline algorithms, we performed sample clustering using an automatically determined reduced number of genes. In particular, for each method, we produced a gene signature from its best clustering by selecting as representatives of each cluster its center. For the LP-stability clustering, the centers were defined as the actual stable center genes computed by the algorithm. However, for the rest of the clustering methods, we selected the medoid gene *i.e.* the gene the closest to the centroid of the cluster. The sample clustering was performed using K-Medoids method, a variant of K-Means algorithm, coupled with Kendall's rank correlation to determine a distance between patients according to the genes of the signature. The evaluation of those sample clustering was performed by assessing the distribution of the tumor types across the clusters.

3.4   Results and Discussion

In Fig. 1 and Table 2, we summarize the performance of LP-stability and the baseline algorithms using both the enrichment and the Dunn's Index metrics. The Table 2 reports for each method its best clustering according respectively to the enrichment and the average enrichment with threshold 0.005, the Dunn's Index and the number of clusters. We chose this threshold value because it is the most restrictive one. In general, the evaluated algorithms reports their best scores with a relatively small amount of clusters (less than 30).

Starting with the enrichment score, one can observe that for a small number of clusters the enrichment is very high, reaching 100%, even in the case of the random clustering. This can be justified by the fact that a low number of clusters contains a large number of interactions between genes, leading to a near perfect enrichment without any statistical significance. However, when the number of clusters increases, in the case of the random clustering, the enrichment is dramatically decreased, while for the rest of the algorithms remains more stable. At this point, it should be noted that the LP-stability method outperforms the other algorithms in terms of enrichment, reporting very high and stable enrichment, which is more than 90% for all cases. On the other hand, the random clustering reports the lowest enrichment scores for more than 30 clusters, while K-Means reports the lowest enrichment compare to the other algorithms. This poor, worse than random performance for low number of clusters can be explained by the very unbalanced clusters produced by K-Means in this case, for instance for the clustering of 5 clusters, one of the cluster contain 20217 genes over 20365 and 3 clusters contain less than 10 genes. Moreover, CorEx reports high enrichment, however is not as stable as LP-stability as it is decreased for more than 20 clusters. The stability of LP-stability is also indicated from the average enrichment for a threshold 0.005 in Table 2, where one can observe that it reports 96% while CorEx reaches only 71%.

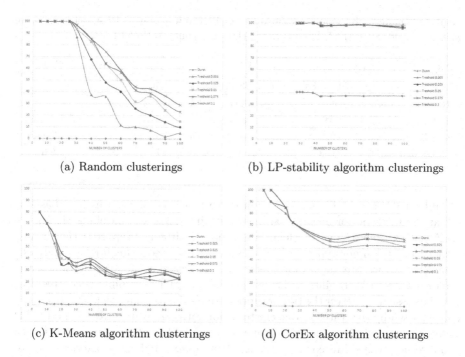

(a) Random clusterings

(b) LP-stability algorithm clusterings

(c) K-Means algorithm clusterings

(d) CorEx algorithm clusterings

Fig. 1. Graphs indicating the PPI enrichment with the different thresholds and the Dunn's Index according to the number of clusters for each clustering method.

Concerning the Dunn's Index, LP-stability outperforms the other algorithms reporting a score always above 30%, that corresponds to one order of magnitude improvement. For the other methods, Dunn's Index is very low, under 5%, indicating either that at least one cluster is poorly defined with high variance, or that at least a pair of clusters is very close to each other. Thus, LP-stability seems to define a solution without extreme ill-defined clusters. One can notice that the best Dunn's Index is in agreement with the best enrichment score indicating that the most biologically informative clusters are obtained for well-defined ones.

To assess even further the performance of each clustering method, we evaluate the expression power of each signature by associating it with tumor types (Table 1). The evaluation is performed by assessing the distribution of the tumors across the clusters. As our goal is to associate 10 tumor types, we used the best gene signature for each of the algorithms to cluster our cohort into 10 groups, in a fully unsupervised manner. In Fig. 2, we present the distribution of the tumor types per algorithm into the 10 clusters. The signatures from the baselines methods fail to define clusters associated to tumor types. This is certainly due to the very small number of clusters, only 5, that the signature depends on. On the other hand, LP-stability, with only 27 genes, reports very high associations with tumor types. That proves the superiority of LP-stability to define the right number of clusters allowing a low dimensional signature minimizing

Table 2. Quantitative evaluation in terms of PPI and average PPI enrichment score with threshold 0.005 (ES), Dunn's Index (DI) and computational time.

Method	Best ES			Best DI			Average ES (%)	Time
	ES (%)	DI (%)	Clusters	ES (%)	DI (%)	Clusters		
Random	100	1.1	10	100	1.1	10	54	-
K-Means	80	2.9	5	80	2.9	5	37	3h
CorEx	100	2.4	5	100	2.4	5	71	>5 days
LP-stability	100	**40.6**	27	100	**40.6**	27	**96**	**1.5 h**

the information loss. To better compare the proposed signatures to a baseline signature we so performed the sample clustering using the baselines signatures of 25 and 30 genes. The K-Means signature of 30 genes reported the highest associations to tumor types and for this reason we used it for further analysis.

In Table 3 we present a more detailed comparison of the distribution of the tumor types for LP-stability and K-Means. In general, LP-stability generates clusters that associate better the tumor types than K-Means. In particular, LIHC type was successfully separated in one cluster from both signatures. LUSC and LUAD were also successfully associated in one cluster related to lung tumors (clusters 3 and 4 respectively). Moreover, both signatures associated two clusters related to squamous tumors containing mainly BLCA CESC, LUSC and HNSC types (clusters 0 & 8 and 1 & 8 respectively). Concerning the BRCA type, K-Means signature clustered the most of the samples in one group, however the rest of the samples, were grouped in unrelated types such as the GBM type. Whereas, LP-stability signature clustered the BRCA samples in several small clusters that may relate to the various molecular types of BRCA, and grouped the remaining BRCA with the OV type which are related (cluster 3). Finally, both signatures have a cluster including only tumors that can be smoking related containing mainly CESC, HNSC, READ, LUSC and LUAD (clusters 8 & 7 respectively).

These two sample clusterings show promising results as we can relate them to the ones obtained in [9], reporting the same kind of clusters by performing sample clustering on a very large set of omics data. They indeed reported, as we do, pan-squamous clusters (LUSC, HNSC, CESC, BLCA), but also pan-gynecology clusters (BRCA, OV) and pan-lung clusters (LUAD, LUSC). They also noticed the separation of BRCA in several clusters that they linked to basal, luminal, Chr 8q amp or HER2-amp subtypes. However, they obtained only one third of mostly homogeneous clusters, and even reported clusters mixing up to 75% of the total number of tumors types they considered.

Computational Complexity and Running Times: The computation time is an important parameter playing a significant role for the selection of an algorithm. For each algorithm the approximate average time needed for the clustering is presented in Table 2. The different computation time have been computed using Intel(R) Xeon(R) CPU E5-4650 v2 @ 2.40 GHz cores. In general, the computational time augments with an increasing number of clusters. However, for

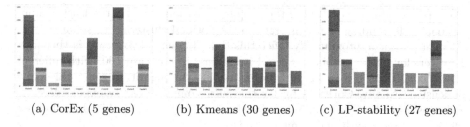

(a) CorEx (5 genes)   (b) Kmeans (30 genes)   (c) LP-stability (27 genes)

Fig. 2. Evaluation of the produced signature in association with the tumor types

Table 3. Proportion of each tumor type per cluster which is higher than 10% is reported from the LP-stability and Kmeans algorithms.

Tumor types	LP-stability (27 genes)	K-means (30 genes)	Best
BLCA	57% BLCA ⇒ 33% cluster 8 26% BLCA ⇒ 10% cluster 0 <10% BLCA ⇒ clusters 1, 3, 7	54% BLCA ⇒ 59% cluster 7 18% BLCA ⇒ 22% cluster 1 14% BLCA ⇒ 7% cluster 8 <10% BLCA ⇒ cluster 2, 4, 9	~
BRCA	26% BRCA ⇒ 75% cluster 1 20% BRCA ⇒ 100% cluster 2 19% BRCA ⇒ 100% cluster 6 18% BRCA ⇒ 100% cluster 9 10% BRCA ⇒ 20% cluster 3 **Clusters with related types**	55% BRCA ⇒ 98% cluster 0 27% BRCA ⇒ 20% cluster 4 <10% BRCA ⇒ clusters 1, 2, 7 **Clusters unrelated to GBM type**	LP
CESC	58% CESC ⇒ 15% cluster s0 38% CESC ⇒ 16% cluster 8 **Squamous related clusters**	54% CESC ⇒ 15% cluster 8 25% CESC ⇒ 16% cluster 1 16% CESC ⇒ 16% cluster 7 **Squamous mixed with non squamous**	LP
GBM	100% GBM ⇒ 79% cluster 7	98% GBM ⇒ 57% cluster 2 **Mixed with unrelated BRCA types**	LP
HNSC	89% HNSC ⇒ 43% cluster 0 10% HNSC ⇒ 7% cluster 8 **Squamous related clusters**	86% HNSC ⇒ 62% cluster 8 11% HNSC ⇒ 18% cluster 1 **Squamous related clusters**	~
LIHC	90% LIHC ⇒ 100% cluster 5	98% LIHC ⇒ 98% cluster 5	~
READ	82% READ ⇒ 9% cluster 8 **Smoking related**	55% READ ⇒ 10% cluster 7 32% READ ⇒ 5% cluster 4 **Smoking related**	~
LUAD	80% LUAD ⇒ 85% cluster 4 **Lung cluster**	93% LUAD ⇒ 83% cluster 3 **Lung cluster**	~
LUSC	54% LUSC ⇒ 25% cluster 0 23% LUSC ⇒ 18% cluster 8 15% LUSC ⇒ 15% cluster 4 **Squamous and lung clusters**	53% LUSC ⇒ 97% cluster 6 20% LUSC ⇒ 17% cluster 3 11% LUSC ⇒ 21% cluster 1 **Squamous and lung clusters**	K-Means
OV	92% OV ⇒ 60% cluster 3 <5% OV ⇒ clusters 1, 8 **Cluster with related BRCA**	71% OV ⇒ 86% cluster 9 15% OV ⇒ 10% cluster 4 10% OV ⇒ 7% cluster 7 <10% OV ⇒ clusters 0, 2 **Mixed clusters**	LP

the reported clusters of Table 2 the proposed method is by far the least computationally demanding as it converges to the optimal clustering in about 90 min.
K-Means needs approximately twice this time. In general, k-means is very fast,
however, for better stability, several iterations, in our case 100, with different
initial conditions has to be performed, making the algorithm computationally
expensive. Finally, CorEx is by far the most computationally expensive algorithm as it needs more than 5 days for the clustering, making this algorithm not
efficient for data with high dimensionality.

In order to assess the significance of the results and provide a fair comparison
with the state of the art and the baseline methods a spider chart summary is
presented in Fig. 3 where six criteria were considered: (i) the clinical relevance
of the outcome with the number of tumor types where the method signature
performed best, (ii) the statistical relevance of the outcome with the average
enrichment score, (iii) the mathematical relevance of the outcome with the best
Dunn's Index (iv) the biological relevance of the outcome with the best enrichment score, (v) the running time and (vi) the compactness of the signature.
Towards eliminating the bias introduce from the compactness of the signature,
we have also compared our approach with signatures of similar compactness generated by the baseline and the state of the art method. It is clearly shown that
our approach outperforms by at least a margin of magnitude in all aspects.

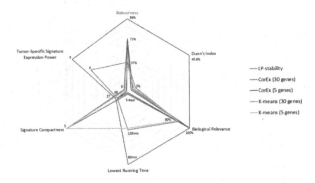

Fig. 3. Spider graph comparing the different methods

4   Conclusion

In this paper we presented and compared, LP-stability algorithm, a powerful
center-based clustering algorithm towards a low-dimensional, robust, genetic
signature/biomarker shown to be highly biologically relevant. The algorithm
outperforms the baseline methods both in terms of computational time, quantitative and qualitative metrics. Moreover, the obtained clusters formulate a gene
signature which has been evaluated for ten different tumor locations, proving

causality and strong associations with them similar to the ones reported in the literature by using a large set of omics data. In the future, we aim to extend the proposed method towards discovering stronger gene dependencies through higher-order correlations between gene expression data, as well as using this biomarker for therapeutic treatment selection in the context of cancer.

Acknowledgements. We would like to acknowledge the partial support of Amazon Web Services and Pr. Stefano Soatto for fruitful discussions. We also thank Y. Boursin, M. Azoulay and Gustave Roussy Cancer Campus DTNSI team for providing the infrastructure resources used in this work. This work was supported by the Fondation pour la Recherche Médicale (FRM; no. DIC20161236437).

References

1. Center BITGDA: Analysis-ready standardized TCGA data from broad GDAC firehose 2016_01_28 run (2016)
2. Cowen, L., Ideker, T., Raphael, B.J., Sharan, R.: Network propagation: a universal amplifier of genetic associations. Nat. Rev. Genet. **18**(9), 551–562 (2017)
3. van Dam, S., Võsa, U., van der Graaf, A., Franke, L., de Magalhães, J.P.: Gene co-expression analysis for functional classification and gene-disease predictions. Brief. Bioinf. **19**(4), 575–592 (2018). bbw139
4. Drucker, E., Krapfenbauer, K.: Pitfalls and limitations in translation from biomarker discovery to clinical utility in predictive and personalised medicine. EPMA J. **4**(1), 7 (2013)
5. Dunne, P.D., et al.: Cancer-cell intrinsic gene expression signatures overcome intra-tumoural heterogeneity bias in colorectal cancer patient classification. Nat. Commun. **8**, 15657 (2017)
6. Halkidi, M., Batistakis, Y., Vazirgiannis, M.: On clustering validation techniques. J. Intell. Inf. Syst. **17**(2), 107–145 (2001)
7. Hanahan, D., Weinberg, R.A.: Hallmarks of cancer: the next generation. Cell **144**(5), 646–674 (2011)
8. Hasin, Y., Seldin, M., Lusis, A.: Multi-omics approaches to disease. Genome Biol. **18**(1), 83 (2017)
9. Hoadley, K.A., et al.: Cell-of-origin patterns dominate the molecular classification of 10, 000 tumors from 33 types of cancer. Cell **173**, 291–304 (2018)
10. Kaufman, L., Rousseeuw, P.: Clustering by Means of Medoids. In: Dodge, Y. (ed.) Proceedings of the Statistical Data Analysis Based on the L1 Norm Conference, Neuchatel, 1987. North-Holland (1987)
11. Kendall, M.G.: A new measure of rank correlation. Biometrika **30**, 81–93 (1938)
12. Kingrani, S.K., Levene, M., Zhang, D.: Estimating the number of clusters using diversity. Artif. Intell. Res. **7**(1), 15 (2017)
13. Komodakis, N., Paragios, N., Tziritas, G.: Clustering via LP-based stabilities. In: Koller, D., Schuurmans, D., Bengio, Y., Bottou, L. (eds.) Advances in Neural Information Processing Systems, vol. 21, pp. 865–872. Curran Associates, Inc., New York (2009)
14. Kovács, F., Legány, C., Babos, A.: Cluster validity measurement techniques. In: 6th International Symposium of Hungarian Researchers on Computational Intelligence. Citeseer (2005)

15. Kurian, A.W., et al.: Clinical evaluation of a multiple-gene sequencing panel for hereditary cancer risk assessment. J. Clin. Oncol. **32**(19), 2001–2009 (2014)
16. Luxburg, U.V.: A tutorial on spectral clustering. Stat. Comput. **17**, 395–416 (2007)
17. MacQueen, J.: Some methods for classification and analysis of multivariate observations. In: Proceedings of the Fifth Berkeley Symposium on Mathematical Statistics and Probability, Volume 1: Statistics. University of California Press (1967)
18. Pepke, S., Steeg, G.V.: Comprehensive discovery of subsample gene expression components by information explanation: therapeutic implications in cancer. BMC Med. Genom. **10**(1), 12 (2017)
19. Ramaswamy, S., et al.: Multiclass cancer diagnosis using tumor gene expression signatures. Proc. Natl. Acad. Sci. **98**(26), 15149–15154 (2001)
20. Sibson, R.: SLINK: an optimally efficient algorithm for the single-link cluster method. Comput. J. **16**(1), 30–34 (1973)
21. Sun, R., et al.: A radiomics approach to assess tumour-infiltrating CD 8 cells and response to anti-PD-1 or anti-PD-l1 immunotherapy: an imaging biomarker, retrospective multicohort study. Lancet Oncol. **19**(9), 1180–1191 (2018)
22. Ver Steeg, G., Galstyan, A.: Discovering structure in high-dimensional data through correlation explanation. In: Advances in Neural Information Processing Systems, pp. 577–585 (2014)
23. Xu, R., Wunsch II, D.: Survey of clustering algorithms. Trans. Neur. Netw. **16**(3), 645–678 (2005)

Computational Approaches for Drug Repurposing and Personalized Medicine

When Mathematics Outsmarts Cancer

Somnath Tagore[1] and Milana Frenkel-Morgenstern[2(✉)]

[1] Department of Systems Biology, Columbia University Medical Center,
Herbert Irving Cancer Research Center, 1130 Street Nicholas Avenue,
New York, NY 10032, USA
st3179@cumc.columbia.edu
[2] The Azrieli Faculty of Medicine, Bar-Ilan University,
8 Henrietta Szold Street, 13195 Safed, Israel
milana.morgenstern@biu.ac.il

Abstract. Mathematics has become essential in cancer biology. Recent developments in high-throughput molecular profiling techniques enable assessing molecular states of tumors in great detail. Cancer genome data are collected at a large scale in numerous clinical studies and in international consortia, such as The Cancer Genome Atlas and the International Cancer Genome Consortium. Developing mathematical models that are consistent with and predictive of the true underlying biological mechanisms is a central goal of cancer biology. In this work, we used percolations and power-law models to study protein-protein interactions in cancer fusions. We used site-directed knockouts to understand the modular components of fusion protein-protein interaction networks, thereby providing models for target-based drug predictions.

Keywords: Fusion proteins · Protein-protein interaction networks ·
Site-directed percolations

1 Some Old School of Thoughts

Cancer results due to the accumulation of multiple alterations in a single transformed cell [1]. Even if the probability of transformation is extremely low for a single cell, cancer could arise by chance within a lifetime if many cells are at risk. Moreover, the number of cells at risk and transformed can be inferred from cancer epidemiology. Several common cancers have been shown to exhibit increased incidence with age and can be described by a simple equation [2].

$$p = bt^k \tag{1}$$

Here, p is the probability of cancer, b is a constant, t is an individual's age and k is the number of rate-limiting stages. For instance, Eq. 1 fits the epidemiology of colorectal cancer when k is 5 or 6, where k corresponds to the number of rate-limiting mutations. But, Eq. 1 does not include biological parameters and these are incorporated into b. Arguably, cancer incidence should increase with greater numbers of cells at risk, with greater numbers of cell divisions, and with higher mutation rates. Likewise,

I. Rojas et al. (Eds.): IWBBIO 2019, LNBI 11465, pp. 477–485, 2019.
https://doi.org/10.1007/978-3-030-17938-0_42

normal mutation rates have been observed to be low and around one mutation per billion bases per division [3], which extrapolates to a probability of mutating a single specific gene of $\approx 1,000$ base pairs in a single division as 10^-. Thus, the probability of cancer, p, after a single division is extremely low if six rate-limiting, k, mutations are required.

$$p = \left(u^k \right) \tag{2}$$

Thus, in Eq. 2, the probability of cancer is 10^{-36} when the mutation rate is $u = 10^{-6}$ mutations per gene per division and $k = 6$. Further, it is highly improbable that cancer arises in a single cell after a single division.

A better calculation is the probability of cancer after multiple divisions wherein just one of the many cells is at risk in the body. Thus, the probability of not accumulating a critical mutation, $1 - u$, in one cell lineage after a certain number of divisions, d, is represented in Eq. 3:

$$p = (1 - u)^d \tag{3}$$

Here, p defines 'no mutation in one critical gene'. With more divisions, the probability of no mutation decreases. It follows that the probability of mutation after d divisions is represented in Eq. 4:

$$p = 1 - (1 - u)^d \tag{4}$$

Here, p defines 'mutation in one critical gene'. For multiple k genes, Eq. 5:

$$p = \left(1 - (1 - u)^d \right)^k \tag{5}$$

Here, p defines 'mutation in all k critical genes'. Equation 5 calculates the probability of a single cell accumulating all k driver mutations after d divisions. It follows that the probability of not accumulating all k mutations in a single cell after d divisions is represented in Eq. 6:

$$p = 1 - \left(1 - (1 - u)^d \right)^k \tag{6}$$

Here, p defines 'not all k critical genes mutated'.

The probability that a single cell accumulates six driver mutations is low. However, cancer arises when the first cell out of many at risk within an individual transforms; this occurs considerably earlier than for the average cell. For an organ, the probability of cancer depends on the number of cells at risk, which is fewer than the total number of cells, because mutations can only accumulate in long-lived stem cell lineages. For instance, in the colon, the number of cells at risk is the number of stem cells per crypt (N) multiplied by the total number of clonal units or crypts (m), as in Eq. 7:

$$p = \left(1 - \left(1 - (1-u)^d\right)^k\right)^N m \qquad (7)$$

Here, p defines 'no stem cell with all k critical mutations in the colon'. It follows that the probability of cancer (p) for a single individual is represented in Eq. 8:

$$p = 1 - \left(1 - \left(1 - (1-u)^d\right)^k\right)^N m \qquad (8)$$

Equation 8 is an algebraic representation or a probabilistic model of colorectal cancer that starts from birth and ends when the first stem cell (out of many at risk) accumulates a critical number of k rate-limiting driver mutations. The model assumes all mutations (drivers and passengers) are initially selectively neutral and arise as replication errors. Thus, Eq. 8 illustrates that age-related increases in cancer frequencies may result from relatively normal division and mutation rates.

2   Origin of Cancers

A deterministic model for cancer origin can provide insights into how hierarchical tissue structures affect cancer risk and treatment effects [4]. Stochastic modelling, in contrast, has proved useful for determining whether a stem cell, a transit-amplifying cell (also known as a progenitor cell) or a terminally differentiated cell is more likely to serve as the cell of origin of a particular tumour type. Using such approaches for hematopoietic malignancies [5], a progenitor cell was found to be more likely to initiate tumorigenesis than a stem cell, as the large number of progenitor cells can compensate for the need to accumulate a larger number of mutations. The probability of cancer initiation was found to be highest when progenitor cells first acquire an oncogenic mutation and then gain self-renewal capabilities.

2.1   The Onco-Tree Model

The onco-tree model is based on a probabilistic phylogenetic tree approach. It relaxes the assumption of a strict sequential order of the linear genetic model and permits multiple paths to full transformation [6]. The temporal order of events is computed as a function of the distance of an event from the root node (that is, the time between initiation and the event) [7]. The relative position of each node on the onco-tree is then constructed using co-occurrence frequencies of mutations across tumors. The onco-tree methodology still has restrictions, as it imposes one single onco-tree structure per data set [8]. Therefore, mixtures of onco-tree models were later introduced to combine multiple independent tree structures, and were applied to various cancer types, for example, nasopharyngeal carcinoma and oral cancer [9]. To overcome the limitation of tree-structured models, namely, the absence of ancestors for multiple leaves, directed acyclic graphical models were developed [10]. These models determine the order of somatic alterations from cross-sectional data sets, at the cost of a larger computational burden owing to increased model complexity. A possible solution to this problem is to decrease modelling resolution, and focus on pathway-level events instead of investigating individual mutations [11].

2.2 Evolutionary Dynamics Models

The onco-tree model is based on a probabilistic phylogenetic tree approach. It relaxes the assumption of a strict sequential order of the linear genetic model and permits multiple paths to full transformation [6]. The temporal order of events is computed as a function of the distance of an event from the root node (that is, the time between initiation and the event) [7]. The relative position of each node on the onco-tree is then constructed using co-occurrence frequencies of mutations across tumors. The onco-tree methodology still has restrictions, as it imposes one single onco-tree structure per data set [8]. Therefore, mixtures of onco-tree models were later introduced to combine multiple independent tree structures, and were applied to various cancer types, for example, nasopharyngeal carcinoma and oral cancer [9]. To overcome the limitation of tree-structured models, namely, the absence of ancestors for multiple leaves, directed acyclic graphical models were developed [10]. These models determine the order of somatic alterations from cross-sectional data sets, at the cost of a larger computational burden owing to increased model complexity. A possible solution to this problem is to decrease modelling resolution, and focus on pathway-level events instead of investigating individual mutations [11].

3 Applying Power-Laws to Study Protein-Protein Interactions of Cancer Fusions

3.1 Power-Laws

Gene fusions have been recognized as important diagnostic and prognostic biomarkers in malignant hematological disorders and childhood sarcomas [12]. Recently, their biological and clinical impact in solid tumors has also been appreciated [12]. Fusions in cancer are usually produced by chromosomal translocation; and incorporate parts of two different parental proteins [14]. Arguably, the best-known example is the BCR-ABL fusion, which is an oncogenic fusion protein, considered to be the primary driver of chronic myelogenous leukemia [15], BCAS3-BCAS4 in breast cancer [16] and EWSR1-ETV4 in Ewing sarcoma [17]. Further, for understanding the complex activities and dynamics of fusions in various cancer phenotypes, the knowledge of their ongoing protein-protein interactions (PPIs) is essential. Likewise, these interconnectivities imply that the impact of a specific genetic abnormality is not locally restricted to a specific protein, but applicable along the links of the network, in that the abnormality alters the activity of gene products that are otherwise found in healthy individuals [18].

   To study the behavior of PPI networks of all cancer subtypes, specifically, breakdown against targeted knockouts of proteins, we used the Power-Law model.

A power-law is a relationship between two entities, e.g., frequency of proteins vs. their total occurrence in each PPI network. Power law distributions are long-tailed distributions, in that if we plot the number of nodes with degree d against d, we find that the area under the tail of the distribution is polynomial. In contrast, the area under the tail of the Gaussian (normal) distribution is exponentially small. We introduce here the concept of chimeric protein-protein interactions (ChiPPI [19]), which use domain-domain co-occurrence scores to identify preserved interactors of chimeric proteins. Essentially, we plot the power-law for each fusion PPI, with the degree distribution of proteins vs. the ChiPPI-score for each protein, using a log-log scale on both axes. We considered the Barab´asi–Albert model for studying the PPI networks of individual fusion proteins and determined whether they behave as scale-free networks using a preferential attachment mechanism. We observed that most fusion PPI networks can be categorized as either scale-free networks, having power-law degree distributions, or hierarchical networks; while some PPI networks exhibit random graph models like the Erdős–Rényi model and the Watts–Strogatz model, which do not exhibit power laws [18].

For the training part of this study, we randomly selected 150 fusions and their parental proteins from the ChiTaRS-3.1 database [13, 20, 21]. We predicted the corresponding interactors for each fusion using the previously developed ChiPPI method [19]. For testing, we considered 672 unique aliquot IDs and their corresponding 3091 unique fusions from The Cancer Genome Atlas, for 25 cancer subtypes. For the training phase, we generated the Power-Law models for 300 parental and 150 fusion proteins in leukemia, lymphoma (LL), sarcoma (SC) and solid tumors (ST). We found that more PPI-networks for parental and fusion proteins belong to the random category in SC and ST, due to less connectivity among vertices in SC and ST compared to LL, or perhaps due to the lesser presence of communities and hubs, which are the source of major interactions. We observed that if the PPI-networks of parental proteins belong to a network category, the fusions also belong to the same, with some exceptions. Thus, in LL, the PPI-networks of BCR, ABL1 and BCR-ABL1 are scale-free; as well as KMT2A, KMT2A and MLLT10 are scale-free, whereas the MLLT10 network follows the hierarchical category. These findings indicate that the majority of parental and fusion proteins belong to the scale-free category (118 cases), followed by the hierarchical (24 cases) and random categories (8 cases).

Similarly, in SC, the number of PPI-networks for parental and fusion proteins belonging to the random category was high compared to LL (30 networks), i.e. almost like the hierarchical category (31 cases), but much less than the scale-free category (89). Lastly, in ST, the observations were dramatically different, for the scale-free (68), hierarchical (34) and random categories (48). For the test study, we considered 672 aliquot IDs belonging to 25 cancer subtypes, having 3091 fusions and 6182 parental proteins. Of these, 18 cancer subtypes belong to CA (previously ST), two to GL, one to ME and LK (previously LL) and two to LY (previously LL), whereas one belongs to SC. Finally, we found that the network categorization for LK, LY, SC, CA, ME and GL fusions display a Poisson distribution. These findings indeed coincide with the observation that fusions are mostly drivers in LK, LY, ME and GL, but not in CA. The reason for this might be that in CA, the number of proteins with lesser interactions (lower connectivity) is greater than the number with higher interactions. Consequently, the function of fusion PPIs depends mostly on the ubiquitous hubs, rather than on the central hubs. For example, in prostate cancer, most fusion proteins act as passenger mutations.

We considered the higher-degree hubs of all PPI networks in *LK, LY, SC, CA, ME* and *GL* for analyzing whether they are drivers of lesser interacting protein functions. The number of higher-degree hubs in communities were more in *SC* than in *LK, LY, ME* and *GL*, where there was an overall even distribution. Further, due to the lower average degree in *CA*, PPI-networks in the random category increased. In *LK, LY, ME* and *GL*, higher-degree hubs ranged from 315 (FUS in FUS-ERG fusion) to 23 (CBFB in CBFB-MYH11); in *SC*, from 315 (FUS in FUS-CREB3L1) to 4 (SYT4 in SSX1-SYT4); and in *CA*, from 134 (PRMT1 in BCL2L12-PRMT1) to 1 (EPHA6 in EPHA6-CNTN6). These results indicate that more networks are affected in *LK, LY, ME* and *GL* than in *SC* and *CA*. Similarly, the number of ubiquitous hubs increased in *CA*, compared to *LK, LY, ME, GL* and *SA*. The lowest number of ubiquitous hubs was observed in *CLLE-ES (LK)*, and the highest in *RECA-EU (CA)*. This coincides with our hypothesis that the distribution of ubiquitous hubs is more in *CA* than in other cancers (Fig. 1).

3.2   Percolations

The robustness of a PPI network can be measured by its percolation threshold, which may be interpreted as the critical number of link or node removals that must occur for the network to shift from a regime where it is only slightly perturbed to a regime in which it is fragmented into small disconnected clusters. The study of robustness encounters two variants: first, the robustness of the topologies (maintenance of topological connectivity) of networks, called "structural robustness", against failures of nodes or links; and second, the robustness of the dynamical processes (maintenance of dynamical processes) running on networks, referred to as "dynamical robustness". In our study, we consider that each interaction (or protein) in the PPI network is occupied with probability p. A cluster of interactions is defined as the set of neighboring occupied interactions. If the size of the PPI network approaches infinity, the transition from an unconnected to a connected network occurs sharply when p crosses a critical threshold called the percolation threshold. Whatever property an interaction represents, this property percolates through the network and the emergence of the percolating cluster represents a phase transition. In this study, we performed both site-directed and bootstrap percolation for targeted knockouts of proteins from the PPI network. Site-directed percolation specifies targeted removal of proteins from the PPI network, whereas bootstrap percolation indicates pruning out less significant links and identifying communities [17, 18].

For site-directed percolation, we initially selected 1% of the proteins, removed them and calculated the size of the largest connected component; if 50% of the proteins was reached, we stopped. Proteins are selected according to their proportion in the network, and are thus targeted. The functioning of complex networks such as the internet and social networks has been shown to be crucially dependent on the interconnections between network nodes. These interconnections are such that when some nodes in the network fail, others that are connected through them to the network also become disabled and the entire network may collapse. Thus, to understand robustness of these complex PPI networks, we need to know whether such networks can continue to function after a fraction of their proteins is removed by site-directed knockouts. Accordingly, the robustness of PPI networks under attack is dependent upon the structure of the underlying network and the nature of the attack. Previous research on complex networks has focused on two types of initial attack: random attack and hub-targeted attack. In a random attack, each node in the network is attacked with the same probability [18]. In a hub-targeted attack, the probability that high-degree nodes will be attacked is higher than for low-degree nodes. Moreover, random and hub-targeted attacks apparently do not adequately describe many real-world scenarios in which complex networks are impaired by damage that is localized, i.e., a node is affected, then its neighbor's node, and then their neighbors' nodes. However, in our study on fusion PPI networks, we observed that hub-targeted attack necessarily destroys the structure of the network, resulting in power-law breakdown. Thus, we found that high robustness and resilience of the PPI networks of LK, LY, ME and GL are due to strong clusters, compared to SC and CA. The upshot is that proteins belonging to larger clusters that result in the strong robustness and resilience of these PPI networks should be critically analyzed.

4 Discussion

The amount and breadth of tumor molecular profiling has increased tremendously in recent years, mainly due to the advent of cost-effective high throughput sequencing technologies. Genomic data on cancer stems from a variety of sources, including (i) cell lines cultivated in laboratories, (ii) xenografts derived from patient tumors and engrafted into model organisms like mice, and (iii) clinical patient samples from biopsies. Recent advancements in high-throughput molecular profiling techniques enable assessing molecular states of tumors in great detail. Cancer genome data are collected at a large scale in numerous clinical studies and in international consortia, such as The Cancer Genome Atlas and the International Cancer Genome Consortium. However, cancer is not only a disease of the genome, but of abnormal cellular interactions in the tumor tissue. For example, the fitness of a clone depends on its genotype and the tissue environment of the cells. The tissue micro-environment is a complex dynamical system with multiple cellular components that can influence cancer progression and evolution. Once the technological hurdles of single-cell genomic profiling, such as inefficient and unbiased genome amplification, are overcome and individual cancer genomes can be identified reliably at a larger scale, tumor evolution can be studied more precisely and in greater detail. Novel and more powerful probabilistic models for these data will be required and are already being developed. These will need to account for the spatial dynamics of tumors to enable a systems view on cancer progression. Additionally, they will need to account for cancer-specific properties, such as generally nonhomogeneous rates of evolution. Developing mathematical models that are consistent with and predictive of the true underlying biological mechanisms is a central goal of cancer biology. Experimental design and perturbations have been shown to have major influence on parameter estimation, and subsequently on the output and accuracy of the computational model. Graphical model network inference is subject to a large proportion of false positive edges. Environmental and experimental design factors that are not accounted for can further misguide models. Assessing and improving the utility of mathematical models in the context of cancer biology will continue to be an active area of cancer research.

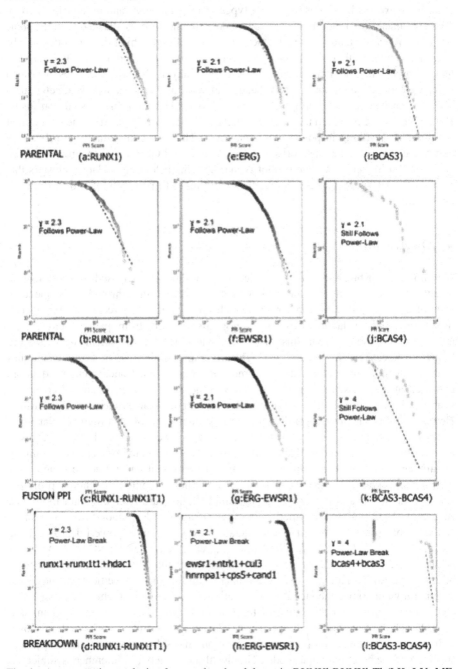

Fig. 1. A comparative analysis of power-law breakdown in RUNXl-RUNXl Tl (LK, LY, ME and GL), EWSRl-ERG (SC) and BCAS3-BCAS4 (CA)

References

1. Nowell, P.C.: The clonal evolution of tumor cell populations. Science **194**, 23–28 (1976)
2. Armitage, P., et al.: The age distribution of cancer and multistage theory of carcino-genesis. Br. J. Cancer **1**, 1–2 (1954)
3. Jones, S., et al.: Comparative lesion sequencing provides insights into tumor evolution. Proc. Natl. Acad. Sci. U.S.A. **105**, 4283–4288 (2008)
4. Werner, B., et al.: A deterministic model for the occurrence and dynamics of multiple mutations in hierarchically organized tissues. J. R. Soc. Interface **10**, 20130349 (2013)
5. Haeno, H., et al.: A progenitor cell origin of myeloid malignancies. Proc. Natl. Acad. Sci. USA **106**, 16616–16621 (2009)
6. Desper, R., et al.: Inferring tree models for oncogenesis from comparative genome hybridization data. J. Comput. Biol. **6**, 37–51 (1999)
7. Desper, R., et al.: Distance-based reconstruction of tree models for oncogenesis. J. Comput. Biol. **7**, 789–803 (2000)
8. Hoglund, M., et al.: Statistical behavior of complex cancer karyotypes. Genes Chromosomes Cancer **42**, 327–341 (2005)
9. Huang, Z., et al.: Construction of tree models for pathogenesis of nasopharyngeal carcinoma. Genes Chromosomes Cancer **40**, 307–315 (2004)
10. Pathare, S., et al.: Construction of oncogenetic tree models reveals multiple pathways of oral cancer progression. Int. J. Cancer **124**, 2864–2871 (2009)
11. Hjelm, M., et al.: New probabilistic network models and algorithms for oncogenesis. J. Comput. Biol. **13**, 853–865 (2006)
12. Mitelman, F., et al.: The impact of translocations and gene fusions on cancer causation. Nat. Rev. Cancer **7**, 233–245 (2007)
13. Gorohovski, A., et al.: ChiTaRS-3.1-the enhanced chimeric transcripts and RNA- seq database matched with protein-protein interactions. Nucleic Acids Res. **45**(D1), D790–D795 (2017)
14. Frenkel-Morgenstern, M., et al.: Chimeras taking shape: potential functions of proteins encoded by chimeric RNA transcripts. Genome Res. **22**(7), 1231–1242 (2012)
15. Bentz, M., et al.: Detection of chimeric BCR-ABL genes on bone marrow samples and blood smears in chronic myeloid and acute lymphoblastic leukemia by in situ hybridization. Blood **83**, 1922–1928 (1994)
16. Bärlund, M., et al.: Cloning of BCAS3 (17q23) and BCAS4 (20q13) genes that undergo amplification, overexpression, and fusion in breast cancer. Genes Chromosomes Cancer **35**(4), 311–317 (2002)
17. Ishida, S., et al.: The genomic breakpoint and chimeric transcripts in the EWSR1-ETV4/E1AF gene fusion in Ewing sarcoma. Cytogenet Cell Genet. **82**(3–4), 278–283 (1998)
18. Barabási, A.L.: Scale-free networks: a decade and beyond. Science **325**(5939), 412–413 (2009)
19. Frenkel-Morgenstern, M., et al.: ChiPPI: a novel method for mapping chimeric protein-protein interactions uncovers selection principles of protein fusion events in cancer. Nucleic Acids Res. **45**(12), 7094–7105 (2017)
20. Frenkel-Morgenstern, M., et al.: ChiTaRS: a database of human, mouse and fruit fly chimeric transcripts and RNA-sequencing data. Nucleic Acids Res. **41**(Database issue), D142–D151 (2013)
21. Frenkel-Morgenstern, M., et al.: ChiTaRS 2.1–an improved database of the chimeric transcripts and RNA-seq data with novel sense-antisense chimeric RNA transcripts. Nucleic Acids Res. **43**(Database issue), D68–D75 (2015)

Influence of the Stochasticity in the Model on the Certain Drugs Pharmacodynamics

Krzysztof Puszynski$^{(\boxtimes)}$ (iD)

Institute of Automatic Control, Silesian University of Technology,
Akademicka 2A, 44-100 Gliwice, Poland
Krzysztof.Puszynski@polsl.pl

Abstract. In this paper I analyze the impact of the stochasticity on the three different levels (genes, mRNA and protein) on the of drug pharmacodynamics of a large class of drugs. I focus on the basic mechanisms underlying the dose-response curves considering two elementary molecular circuits. Both consist in the gene activation/deactivation, then gene transcription and following translation into the corresponding protein. In the first circuit gene activation and deactivation are spontaneous whereas gene deactivation rate in the second circuit depends on the protein level introducing negative feedback. In both cases drug is assumed to enhance the protein degradation level and the success of the therapy is considered as lowering the protein level below given threshold for given time. My numerical simulation shows that the level on which the stochasticity is introduced to the model (none, genes, mRNA, protein) influences not only the shape of dose-response curves but also the value of the critical dose i.e. the dose which causes of the positive response to the therapy in at least half of the cells.

Keywords: Biological model · Minimal dose therapy ·
Stochastic models

1 Introduction

In the recent years more and more effort is put for better understanding of the molecular basis of the various diseases including two most popular: cancer and HIV. With this knowledge new, promising drugs are developed and tested. Two main drug-related mechanisms which must be investigated during the drug development are the drug pharmacokinetics and pharmacodynamics. The drug pharmacokinetics basically responds to the question "what the body do with the drug". It describes how the drug is distributed in the various body compartments such as stomach, blood, liver, brain etc after the oral or injection administration and how it is removed from body. The variable considered in pharmacokinetic is easy to define and it is the drug concentration in the particular body compartment. The drug pharmacodynamics responds to the question "what the drug do

I. Rojas et al. (Eds.): IWBBIO 2019, LNBI 11465, pp. 486–497, 2019.
https://doi.org/10.1007/978-3-030-17938-0_43

with the body". It is much harder not only to investigate but even to define. For many drugs we know the final effect but we are still unable to respond how they work on the molecular level. Also the variable behind drug pharmacodynamics is not easy to define because it requires the answer of question: "How is the effect of a drug measured?" [1]. However for some important drugs we can clearly define the drug effect and its measurement method, therefore the variable behind. For example the function of some antibiotics and anti-tumor drugs is to kill target cells and we can measure the percentage of cells killed in targeted population, other example are the cytostatics anti-tumor drugs which role is to block proliferation of the targeted cells and we can measure how many of the treated cells will proliferate after the therapy. The cell death or proliferation block may be received by lowering the level of some proteins or its functionality as for example in the case of drug called Nutlin which block the Mdm2-p53 complexes creation by attaching to the Mdm2 in the specific domain in which p53 suppose to join [2]. This could be also considered as the lowering of the functional Mdm2 level in the cell. In the present work I consider this type of drug influence on the targeted cells.

For better understanding of the molecular background of the many diseases such as cancer as well as molecular mechanisms behind the drug influence on the living cells systems engineering, especially mathematical methods such as modeling and simulation are incorporated [3]. Various approaches to the modeling and model simulation analysis are considered: cellular automata [4], linear with switchings [5], stochastic [6] or [7] but still the most common approach is to use the Ordinary Differential Equations (ODE). ODE based modeling and simulation has their undisputed advantages, it is easy to develop and analyze and ODE numerical simulation are very fast. But as we showed in [2,8] the deterministic approach through ODE may be not sufficient when the drug pharmacodynamics is considered. As mentioned in [8] "if the dynamics of the intracellular biomolecular network would be deterministic, the experimental in vitro dose-response curves would necessarily be of the type all or nothing". In reality the experimentally observed curves are sigmoidal and in the mentioned work we postulate that the mechanism behind it may be the stochastic gene switching thus stochastic approach is required.

When the mathematical model of the intracellular processes is developed the scientist has to decide between simple, minimal model which catch the main dynamics of the system and neglect many other aspects [8] or complex model which catch as many interactions influencing the studied network dynamics as possible [9]. The first approach gives models which are easier to simulate and analyze and much easier to understand by the other. The disadvantages of such models is that they catch only some general dynamics which may never be observed as long as single cell is considered. Thus the conclusions made based on their analysis e.g. the optimal therapy protocol may be far from the reality. The complex models are usually better to catch the real dynamics of the considered pathways but are harder to develop, which results for example from the much larger set of parameters whose value has to be determined, harder to simulate, because the simulation takes much more time and the outputs data are much bigger, and analysis. But what seems to be one of the main disadvantages of such models they are much harder to fully understand by the other scientist and because of that, omitted in their research.

The third possible approach is to develop the in-the-middle models which do not take into account all the possible interactions but omit or simplify some of them. For example when full protein production is considered one has to build the proper equations for gene activation/deactivation process then mRNA production and degradation and finally protein production and degradation including all other possible influences on these three levels. The simplification of such model may be done for example by omitting the genes or genes and mRNA levels. Usually it is done by replacing the variable describing the gene state or number of mRNAs in the mRNA or protein production terms respectively by their expected or mean values. To receive the dose - response curves one has to introduce the stochasticity to the model. It can be done by assuming that all reactions in the system are stochastic and performing the simulation according to the Gillespie algorithm or one of its modifications. In the full gene-mRNA-protein models we have three levels of stochasticity in the simplified only two (mRNA-protein) or one (protein). The question arises how the different levels of stochasticity influence the received drug-response curves during the drug pharmacodynamics investigation.

This question is one of the most important when so-called minimal-dose therapy is considered. In the recent years more and more attention is put to the drug side effects problem and the drug dosage. The common approach "to use the maximal allowed dose" becomes replaced by the "use minimal necessary dose". This not only lowers the sometimes serious side effect of the drugs but also is more comfortable for patients and cheaper to apply. This approach resulted for example in the metronomics therapies development [10]. Of course the determination of the minimal dose necessary to obtain therapeutically desired results is more complicated than maximum dose which not cause lethal response from patients. It is the field where mathematical modeling and simulation may help which is one of the reason that this field of science continuously grows. In the current paper I will show one of the possible problem related to such research.

1.1   Considered Models

To investigate the influence of the stochasticity in the model on the different levels on the drug phamacodynamics I considered two simple models. The first one is straight-forward model in which active gene produces mRNA which in turn produces protein. The second one is negative-feedback loop model in which the protein plays a role of its own transcription inhibitor. In both cases I consider a simple therapy in which the drug is introduced to the system as a constant input such as in the many in vitro experiments. The drug function is to enhance the drug degradation rate and the successful therapy is considered as the lowering of the protein level below the given threshold continuously for a given time. Numerical simulations of 1000 cells in each case were performed and the number of responding, this is fulfilling the assumption of successful therapy, cells was measured. The considered models and the results are described below.

Model 1. First model consist of three ODE (Eqs. 1–3). First one describes spontaneous genes activation and deactivation, assuming that we have $N_A = 2$ genes in total. The second describes mRNA production through transcription of the active genes and its degradation. The last one stays for protein production through mRNA translation, spontaneous degradation and finally drug caused degradation.

$$\frac{dG}{dt} = q_a * (N_A - G(t)) - q_d * G(t). \tag{1}$$

$$\frac{dmRNA}{dt} = t_1 * G(t) - d1 * mRNA(t). \tag{2}$$

$$\frac{dA}{dt} = t_2 * mRNA(t) - d_2 * A(t) - d_3 * DRUG * A. \tag{3}$$

When the stochasticity is introduced into the system the corresponding equation is replaced by the reaction propensities and the whole model is simulate by the mixed stochastic-deterministic approach as in our previous works [2] or [11]. The introduced propensities are as follow:

For the gene activation and deactivation:

$$\mu_1 = q_a * (N_A - G(t)). \tag{4}$$

$$\mu_2 = q_d * G(t). \tag{5}$$

For the mRNA production and degradation:

$$\mu_3 = t_1 * G(t). \tag{6}$$

$$\mu_4 = d1 * mRNA(t). \tag{7}$$

For the protein production, spontaneous and drug caused degradation:

$$\mu_5 = t_2 * mRNA(t). \tag{8}$$

$$\mu_6 = d_2 * A(t). \tag{9}$$

$$\mu_7 = d_3 * DRUG * A. \tag{10}$$

Model parameters were so choosen that they are inside experimentally observed range and in the case of no drug gives one active gene, around 333 mRNA molecules and around 160000 protein molecules. Parameters values of model 1 are presented in Table 1.

Model 2. The second model differs from first one in the gene deactivation rate as now it depends on the protein concentration. The second and third equations are the same as in model 1. The modified equation presents genes spontaneous activation and drug driven deactivation:

$$\frac{dG}{dt} = q_a * (N_A - G(t)) - q_d * G(t) * A(t). \tag{11}$$

Following, the gene deactivation propensity is different in model 2:

$$\mu_2 = q_d * G(t) * A(t). \tag{12}$$

The rest of the propensities is the same as in case of model 1. Also in the case of parameters most of them is the same as in model 1. The only different parameter value is for the gene deactivation rate which in case of model 2 is equal $1.7375 * 10^{-9}$.

Table 1. The values of the models parameters

Parameter	Description	Value	Unit
q_a	Gene activation rate	$2.78 * 10^{-4}$	1/sec
q_d	Gene deactivation rate	$2.78 * 10^{-4}$	1/sec
t_1	mRNA transcription rate	0.05	Molecules/sec
t_2	Protein translation rate	0.1	1/sec
d_1	mRNA degradation rate	$1.5 * 10^{-4}$	1/sec
d_2	Protein degradation rate	$2.0822 * 10^{-4}$	1/sec
d_3	Drug caused protein degradation rate	$2.084 * 10^{-4}$	1/sec
N_A	Number of alleles	2	Molecules

In both cases I want the protein level to stay for 12 h below the threshold set as the half of the initial protein amount that is the threshold value was set to be $Th = 80000$ molecules.

2   Results

2.1   Simulation Protocol

The 1000 single cell simulations of the both model, for all considered cases were performed. In each, cells starts from the same initial conditions which are close to the steady state (the closest integer value was chosen). Then 24 h simulations of each cell before the drug introduction was performed to ensure the initial population heterogeneity. Next the drug was introduced to the system with the given dose and the whole system was simulated for additional 72 h. Finally the percentage of the cells in which the protein level was continuously below the given threshold for the 12 h. This cells were considered as responding cells.

2.2   Model 1

In this model the drug range from 0 to 2 a.u. was considered. The drug caused protein degradation rate d_3 was so chosen that the drug dose 1 a.u. is the dose in which the pure deterministic model shows positive response to the therapy. It means that if all the intracellular processes were deterministic then all the cells in the population will respond to the therapy when dose is higher or equal 1 a.u. and not a single cell will respond for the doses lower than 1 a.u. (Fig. 1).

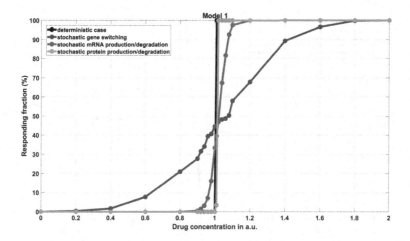

Fig. 1. Dose-response curves of the model 1. Please notice the switch-like shape of purely deterministic case (black line), almost switch-like for stochasticity-in-protein case (blue) and sigmoidal for remaining cases. The dots represents the actual simulation points while lines are drown to connect the dots. (Color figure online)

As one may expect the results differ for the stochastic cases. The percentage of the responding cells is no longer 0% or 100% with the single threshold at 1 a.u. but becomes sigmoidal with the numbers of responding cells growing with the growing dose. This effect is especially visible when the stochasticity is put at the gene switching level. In this case responding cells appear at the dose of 0.2 a.u. (0.4%) reaching 99.8% for the dose 1.8 a.u and 100% for 2 a.u. When the model stochasticity is introduced to the model at the level of mRNA production and degradation the sigmoidal shape of the dose-response curve is less visible (Fig. 1). It is because the responding cells appear only at the dose of 0.9 a.u. and reach the 100% at 1.2 a.u. (Fig. 2). When the model stochasticity is put at the protein production/degradation level, the range of the doses in which the percentage of responding cells goes from 0% to 100% is very short, between 1 and 1.02 a.u. which makes the sigmoidal shape of the dose-response curve even less visible (Figs. 1 and 2).

One can notice that not only the shape of the curve changes but also the dose at which the percentage of the responding cells reaches 50%. This dose, called critical dose is very crucial for the proper therapy development indicating the dose at which it becomes more probable that the cells respond to the therapy than not respond. As mentioned before the critical dose is equal 1 a.u. for purely deterministic case. In the case of stochastic genes switching it is placed close to the 1.08 a.u. whereas for the case with stochastic mRNA production/degradation close to 1.05 a.u. The last case, the closest to the deterministic is stochastic protein production/degradation in which the critical dose takes value close to the 1.02 a.u.

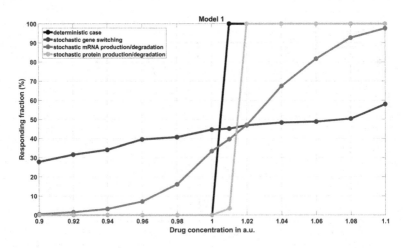

Fig. 2. Dose-response curves of the model 1. Amplification of the curves from Fig. 1 at the range 0.9–1.1 a.u. The dots represents the actual simulation points while lines are drown to connect the dots. (Color figure online)

2.3   Model 2

Model 2 is the one with the negative feedback loop so the lowering of the protein level by drug will cause the lowering of the probability of gene deactivation/gene deactivation rate depending of the stochasticity status at gene level. This will enhance the expected value/mean value of the gene state and in turn mRNA level and finally protein level. So one can expect that in this case higher doses will be required to keep the protein level below the given threshold. Because of that I investigated drug dose range from 0 to 3 a.u (Figs. 3 and 4).

Critical dose in the pure deterministic model is equal 1.68 a.u. and the dose-response curve as in the model 1 has the switch-like shape. Similarly to the Model 1 case with the introduction of the stochasticity to the model, the dose-response curve changes its shape to the sigmoidal one and also the value of critical doses changes. In the case when gene state switching is stochastic the responding cells starts to appear (0.4%) with the dose 0.8 a.u. which, as expected is much higher

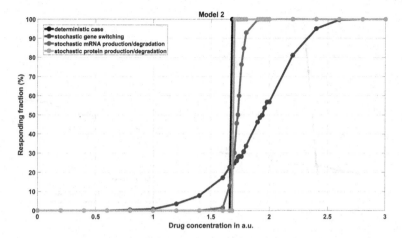

Fig. 3. Dose-response curves of the model 2. Please notice the switch-like shape of purely deterministic case (black line), almost switch-like for stochasticity-in-protein case (blue) and sigmoidal for remaining cases. The dots represents the actual simulation points while lines are drown to connect the dots. (Color figure online)

than in the model without negative feedback (0.2 a.u.). Also the doses at which the percentage of responding cells reaches 100% is higher: 2.8 a.u. compared to 1.8 a.u. The critical dose in this case is close to 1.94 a.u. which is much higher than in deterministic case and also higher then in model 1 with stochasticity in genes switching. With the stochasticity at mRNA level responding cells starts to appear with the dose 1.6 a.u. and reach 100% with the 1.92 a.u. Critical dose in this case is close to 1.73 a.u. The last case in which stochasticity is related to protein production/degradation processes gives the almost switch-like shape with critical dose 1.7 a.u. One can notice that also in this two last cases the doses at which response of the cells appears, reach the 100% and the value of critical dose is higher than in deterministic case and in the respective cases of model 1.

2.4 Time Courses

When we look at the time courses of the median, 1st and 3rd quartile of the protein level (Fig. 5), we can notice that the value spread between median and quartiles is the highest in the case of stochastic gene switching (Fig. 5 first row), lower in the case of stochasticity at mRNA level (Fig. 5 middle row) and the lowest in the case of stochasticity at protein level (Fig. 5 third row). One can also notice the difference between the response of model 1 (Fig. 5 first column) and model 2 (Fig. 5 second and third column). In the model 1, after the drug introduction to the system in $t = 24$ h protein level drops asymptotically to the new value. In case of model 2 we can notice that first the level rapidly drops below the final value and then slowly raise to the final level. This over-reaction

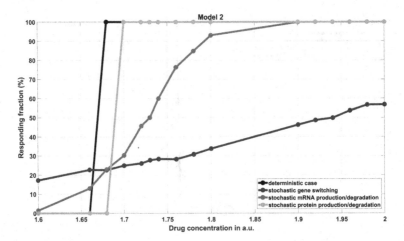

Fig. 4. Dose-response curves of the model 2. Amplification of the curves from Fig. 3 at the range 1.6–2 a.u. The dots represents the actual simulation points while lines are drown to connect the dots. (Color figure online)

on the plot is caused by the negative feedback loop. Drug introduction causes protein level to drop rapidly but then, lower protein level enhances the gene state level which in turn enhances mRNA level and finally after some time the protein level slightly increases. It is important to notice, that the time interval between the protein level drop and following time, in which the protein level reaches its final value, is much smaller than the time interval in which the protein level has to be below the assumed threshold thus it the observed over-reaction will not cause the cells to be considered as responding to the therapy.

In the case of model 1 median protein level is slightly below the threshold (green line) in all considered stochasticity cases (Fig. 5 first column). One can notice that the 1st quartile is above the given threshold in the case when stochasticity is in the gene switching or mRNA production/degradation but below in case of protein production/degradation. This indicates that in the first two cases it may often happens that protein level will jump above the threshold and thus makes the drug ineffective. In the last case protein level also fluctuates but the fluctuations do not cross the threshold and drug is effective. This is reflected by the responding cells percentage which for dose 1.02 a.u. is equal: 46.9%, 47.1% and 100% respectively (Fig. 2).

When the model 2 is considered similar dependencies may be observed (Fig. 5 second and third column). With the drug dose equal 1.73 a.u. the median and quartile of the protein level are lower compared to the dose 1.68 a.u. for all stochasticity cases. Depending on the spread between the median and quartiles this will made less or more significant impact on the fraction of the responding cells. With the high spread the impact on cells response is small. For the stochasticity-in-genes case the fraction of responding cells increase from 22.6% to 27.8% (Fig. 4). The stochasticity-in-mRNA case has smaller spread and thus

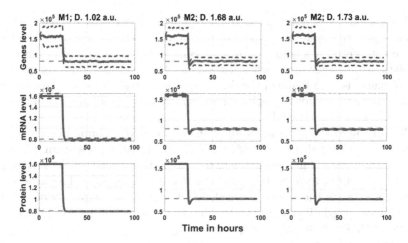

Fig. 5. Time courses of the median level of protein and its 1st and 3rd quartiles. The first row presents the results in the case when the stochasticity is in the gene state change. The second row is for the case when stochasticity is in the mRNA production and degradation processes. The last row is for the case when stochasticity is put to the protein production and degradation. First column shows the results of model 1 and dose 1.02 a.u. while remaining of model 2 and doses 1.68 a.u. and 1.73 a.u. respectively. The solid line stays for median while dashed for 1st and 3rd quartile. (Color figure online)

more visible change in the cells response, from 22.9% to 50.1%. The smallest spread is in the stochasticity-in-protein case which in turn shows most significant change in the responding cells fraction, from 0% to 100%.

3 Discussion

One can notice the general rule in the results. As long as we consider gene − > mRNA − > Protein production chain then, the higher is the stochasticity located (gene is the highest) the higher impact it has on the results received, compared to the pure deterministic case. It includes the shape of the dose-response curves which changes from the switch-like to sigmoidal and location of the critical dose. It results from the high signal amplification which occurs in the protein production chain. The single active gene which is transcribed results in the production of hundreds of mRNA molecules. The single mRNA molecule may be used to produce hundreds of proteins through translation. So the single stochastic event at the gene level may change the protein level by tens of thousands of protein molecules when single stochastic event at the mRNA level only by few hundreds, finally at the protein level only by single protein.

The presented results are especially important when the minimal-dose therapies are considered, especially those which depends on the lowering or enhancing the level of targeted proteins. My results suggest that if they are developed based only on the experimental results received at the population level, which corresponds to the deterministic simulation and/or were tested by using numerical simulations but the simulations were deterministic or even the stochasticity was introduced to the system but on low level (e.g. protein) then these therapies may be not effective as high as expected. Simply as I show the deterministic or stochastic at low level approaches will predict to low critical dose compared to the real one. Also the shape of the dose-response curves will be far from the real one. This aspects cannot be neglected and should be carefully investigated when such therapies are developed.

The presented work considers simple models and very simple therapy. Many aspects were neglected such as full ADME implementation, post-transcriptional and post-translational modifications of the products, influence of the gene on/off switching time or mRNA/protein production/degradation time. This aspects are worth to explore in the future. Nevertheless I expect that they may change the size of the observed differences but the differences itself will be still present and noticeable, thus not should be neglected.

Acknowledgments. This work was partially supported by the grant number 2016/23/B/ST6/03455 founded by National Science Centre, Poland.

References

1. Mortensen, S.B., Jonsdottir, A.H., Klim, S., Madsen, H.: Introduction to PK/PD modelling with focus on PK and stochastic differential equations. IMM-Technical Report-2008-16
2. Puszynski, K., Gandolfi, A., d'Onofrio, A.: The pharmacodynamics of the p53-Mdm2 targeting drug nutlin: the role of gene-switching noise. PLOS Comput. Biol. **10**(12), e1003991 (2014)
3. Świerniak, A., Kimmel, M., Smieja, J., Puszynski, K., Psiuk-Maksymowicz, K.: System Engineering Approach to Planning Anticancer Therapies. Springer, Cham (2016). https://doi.org/10.1007/978-3-319-28095-0
4. Hogeweg, P.: Multilevel cellular automata as a tool for studying bioinformatic processes. In: Kroc, J., Sloot, P.M.A., Hoekstra, A.G. (eds.) Simulating Complex Systems by Cellular Automata. UCS, pp. 19–28. Springer, Heidelberg (2010). https://doi.org/10.1007/978-3-642-12203-3_2
5. Ochab, M., Puszynski, K., Swierniak, A.: Influence of parameter perturbations on the reachability of therapeutic target in systems with switchings. Biomed. Eng. Online **16**(1), 77 (2017)
6. Kristensen, N.R., Madsen, H., Ingwersen, S.H.: Using stochastic differential equations for PK-PD model development. J. Pharmacokinet Pharmacodyn **32**, 109–141 (2005)
7. Rodriguez-Brenes, I.A., Komarova, N.L., Wodarz, D.: The role of telomere shortening in carcinogenesis: a hybrid stochastic-deterministic approach. J. Theor. Biol. **460**(7), 144–152 (2019)
8. Puszynski, K., Gandolfi, A., d'Onofrio, A.: The role of stochastic gene switching in determining the pharmacodynamics of certain drugs: basic mechanisms. J. Pharmacokinetics Pharmacodynamics **43**(4), 395–410 (2016)
9. Jonak, K., Kurpas, M., Szoltysek, K., Janus, P., Abramowicz, A., Puszynski, K.: A novel mathematical model of ATM/p53/NF-kB pathways points to the importance of the DDR switch-off mechanisms. BMC Syst. Biol. **10**(1), 1–12 (2016). https://doi.org/10.1186/s12918-016-0293-0

10. Scharovsky, O.G., Mainetti, L.E., Rozados, V.R.: Metronomic chemotherapy: changing the paradigm that more is better. Curr. Oncol. **16**(2), 7–15 (2009)
11. Kozłowska, E., Puszynski, K.: Application of bifurcation theory and siRNA-based control signal to restore the proper response of cancer cells to DNA damage. J. Theor. Biol. **408**, 213–221 (2016)

Graph Model for the Identification of Multi-target Drug Information for Culinary Herbs

Suganya Chandrababu$^{(\boxtimes)}$ and Dhundy Bastola$^{(\boxtimes)}$

School of Interdisciplinary Informatics,
University of Nebraska at Omaha, Omaha, USA
{schandrababu,dkbastola}@unomaha.edu

Abstract. Drug discovery strategies based on natural products are re-emerging as a promising approach. Due to its multi-target therapeutic properties, natural compounds in herbs produce greater levels of efficacy with fewer adverse effects and toxicity than monotherapies using synthetic compounds. However, the study of these medicinal herbs featuring multi-components and multi-targets requires an understanding of complex relationships, which is one of the fundamental goals in the discovery of drugs using natural products. Relational database systems such as the MySQL and Oracle store data in multiple tables, which are less efficient when data such as the one from natural compounds contain many relationships requiring several joins of large tables. Recently, there has been a noticeable shift in paradigm to NoSQL databases, especially graph databases, which was developed to natively represent complex high throughput dynamic relations. In this paper, we demonstrate the feasibility of using a graph-based database to capture the dynamic biological relationships of natural plant products by comparing the performance of MySQL and one of the most widely used NoSQL graph databases called Neo4j. Using this approach we have developed a graph database HerbMicrobeDB (HbMDB), and integrated herbal drug information, herb-targets, metabolic pathways, gut-microbial interactions and bacterial-genome information, from several existing resources. This NoSQL database contains 1,975,863 nodes, 3,548,314 properties and 2,511,747 edges. While probing the database and testing complex query execution performance of MySQL versus Neo4j, the latter outperformed MySQL and exhibited a very fast response for complex queries, whereas MySQL displayed latent or unfinished responses for complex queries with multiple-join statements. We discuss information convergence of pharmacochemistry, bioactivities, drug targets, and interaction networks for 24 culinary herbs and human gut microbiome. It is seen that all the herbs studied contain compounds capable of targeting a minimum of 55 enzymes and a maximum of 250 enzymes involved in biochemical pathways important in disease pathology.

Keywords: Drug discovery · Graph databases · Neo4j ·
Relational databases · Herbal medicines · Multi-targets · Drug-targets

© Springer Nature Switzerland AG 2019
I. Rojas et al. (Eds.): IWBBIO 2019, LNBI 11465, pp. 498–512, 2019.
https://doi.org/10.1007/978-3-030-17938-0_44

1   Introduction

Multi-target drugs and combinatorial therapies have seen increased focus in the past decade due to their advantages in therapeutic efficacy, particularly in complex diseases including HIV, cancer and diabetes [1]. Naturally occurring compounds found in herb have been shown to have potential interaction effects, including mutual enhancement, mutual assistance, mutual restraint and mutual antagonism [2]. Although information on single herbal-target has important meaning, biological characteristics and knowledge regarding molecular mechanisms of drug action are limiting and mainly include complex interactions among various herbal components and targets [3]. A model for developing new synergistic combinations against multiple protein targets based on a rational and systematic drug design strategy is greatly in need.

One of the fundamental aims in multi-target combinatorial therapies is to understand complex relationships among heterogeneous biological data, which contribute to our understanding of the functions of a living cell in health and diseased condition. However, understanding of such complex interactions among heterogeneous biological data is very difficult due to composite relationships that exist between them. Over the past few decades, a wealth of information regarding biochemical properties of herbal compounds, its location and their biological activities have been accumulated. They have been made available through a number of different public and private repositories. However, such scattered data from thousands of experiments, publications and other diverse sources make the storage and retrieval process not only cumbersome but unusable and inaccessible for data mining. Additionally, to discover the synergy associated with interacting targets, pathways, and diseases using data dispersed across many resources would be highly impractical. To overcome such limitations, various techniques have been developed with biological networks to understand the fundamental mechanisms that control dynamic cell organization and molecular mechanisms [4].

Although widely used among many biological and metabolic pathway knowledgebases for data storage, retrieval and management, the network of hyperlinks connecting all the temporal data on a biological network was deemed complex and very difficult to model competently in a relational database [5]. Alternatively, using network pharmacology strategy and known prior knowledge relating to natural compound's enzyme target and its association with biosynthetic or metabolic pathways, through graph database model that captures nodes and edge relationships, has been successful in revealing the underlying synergistic details more efficiently [6]. Therefore, in the current study we assess the usefulness of the graph database, HbMDB [7,8] that we have developed. This graph database contains a working list of natural compounds from 24 culinary herbs; a pilot to help us explore the molecular mechanism of multi-target drug compounds of therapeutic value. We employed Neo4j, one of the most commonly used graph databases for building HbMDB and compared its performance with MySQL in diverse situations by probing the database and using many use cases. Additionally, the effectiveness of HbMDB in exploring the interactions between the small molecules in herbal medicines and gut microbes has been presented. This computational framework helps to capture multiple pieces of biological knowledge relating to drugs including a detailed representation of cellular processes as an ordered network of molecular reactions, interconnecting various entities like compounds, target enzymes, associated disease pathways and interactions with the gut microbiome.

In the rest of the article, we emphasize the motivation behind our adoption of a graph database and show how HbMDB benefits from this change in data integration, traversal and retrieval, and how it overcomes the limitations imposed by relational databases. The HbMDB graph database use-cases elucidate the power of NoSQL database engines like Neo4j in the investigation of complex biological data types. The results suggest Neo4j is superior to MySQL in querying complex relationships among heterogeneous data and is a useful platform in the study of complex biological entities such as the multi-target drugs from a natural source like plants.

2   Why Graph Databases for Biological Networks?

The information about cellular interactions, domain knowledge from researchers, biologists and physicians, etc., are critical for the understanding of complex mechanisms like the drug-target interactions, cellular functions of genes, and gene-disease interactions. This will require one to analyze a wide variety of clinical and experimental data. Such requirements demand a flexible and powerful biological data management system that allows data model transformation and integration, semantic mapping, data conversion and integration, and conflict resolution. The graph structure supports network data storage and application of graph analysis algorithms to facilitate a biologist-friendly visual graph query system. Since many graph database systems use the property graph model, nodes and edges can hold properties associated with them. They can be used to hold names, unique identifiers and other information about the entity. We chose to implement in Neo4j graph platform for its superior performance relating to (a) speed and performance, (b) data visualization, (c) flexibility and stability and (d) easy programmability [5].

3   Paradigm Shift from Relational to Graph Model

Below, we have described the motivation behind our adoption of a graph database, and we have shown how HbMDB benefits from this change in the underlying hardware/software and storage technology for overcoming the previously mentioned limitations imposed by relational databases.

3.1   Hardware and Software Setup

MySQL Database: To begin with, the MySQL storage engine has to be set up by trained programmers and parameters have to be tuned to improve the performance. MyISAM operates much faster than InnoDB, however, it does not guarantee data integrity. Also, MyISAM is prone to table locking issues which frequently occurs when more than 5 million data are processed in the indexed state, thereby deteriorating the retrieval performance.

Neo4j Database: *Neo4j Desktop* is the new mission control centre launched especially for developers. It is free with registration, and it includes a development license for Enterprise Edition as well as an installer for getting access to the APOC library. It can also be easily connected to the production servers, and eventually, it also makes installing other components like the graph algorithms or Java upgrades easy.

3.2   Memory Configuration

MySQL Database: MySQL allocates buffers and cache for improving the performance of database operations. However, the memory engines used in MySQL does not provide transaction support, and so we will have to manage transactional integrity and referential integrity by writing additional lines of code wherever it is needed being a lot less efficient than letting the DB do this for us. Usage of many stored procedures will substantially increase the memory usage of every connection that is using those stored procedures. Also, stored procedures pose a lot of debugging difficulties. Very few database management systems allow debugging of stored procedures and unfortunately, MySQL does not fall under that category.

Neo4j Database: Neo4j offers the lightning-fast read and write performance needed in big biological networks, while still protecting data integrity. It is the only enterprise-strength graph database that combines the advantages of native graph storage, scalable architecture optimized for speed, and ACID compliance, thereby ensuring predictability of relationship-based queries. Neo4j allows staggering loading speed of huge data sizes, with a very low memory footprint. The memory configuration includes the following four steps: (1) OS memory sizing, (2) page cache memory sizing, (3) heap memory sizing, and (4) transaction state (Fig. 1).

3.3   Optimization of Disk I/O

MySQL Database: Disk searching is a huge performance bottleneck in MySQL databases. This problem becomes more apparent when the amount of data keeps evolving and becomes too large in order to effectively cache it. Usage of disks with low seek time is required for overcoming this issue. It also requires optimization of many size parameters like **innodb_buffer_pool_size** parameter, **innodb_log_file_size** parameter, **tmp_table_size** and **max_heap_table_size** parameter, from their default values.

Table 1. List of public databases used for building HbMDB

Name	Description	Reference
PubMed, MEDLINE	Information on most commonly used culinary and medicinal herbs	[9]
PhytoChemNAL	In-depth plant, chemical compounds and bioactivity information	[10]
ChEMBL	Information of bioactive drug-like small molecules	[11]
PubChem	Information of the chemical molecules, their activities against biological assays	[12]
DrugBank	Information of drugs and drug-targets	[13]
TOXNET	Information of biochemical, pharmacological, physiological, and toxicological effects of drugs and other chemicals	[14]
BRENDA	Comprehensive enzyme and enzymatic reactions and functions information	[15]
KEGG	Information of genomes, biological pathways, drugs and chemical substances	[16]
MetaCyc	Information of metabolic pathways	[17]
UniProtKB	Information of proteins like protein name or entry number, genes and enzymes	[18]
GenBank	Genome, gene and transcript sequence data of gut microbes	[19]

Neo4j Database: In Neo4j by default, most Linux distributions schedule I/O requests through the Completely Fair Queuing (CFQ) algorithm, to arrive at a good balance between throughput and latency. However, in most cases, the particular I/O workload of a given database is better served and managed by the "Deadline scheduler". The Deadline scheduler provides a higher preference to read requests and processes them as soon as they are received, thereby decreasing the latency of reads, and increasing the latency of writes.

4   Design

4.1   Collection of Diverse Information for HbMDB

The HbMDB data model naturally forms a large interconnected network that can be seen as a directed graph, comprising of a set of nodes and directed edges connecting ordered pairs of nodes. Storing HbMDB data in its natural form has multiple benefits. Most considerably, it does not require any transformation of data into a flat or deformalized table format. The resulting database is easier to maintain as new data can be added by simply writing the respective Cypher queries, without the need for changing the schema and also avoiding writing complex algorithms. The information relating to culinary and medicinal herbs, and gut microbes was obtained from multiple sources as shown in Table 1.

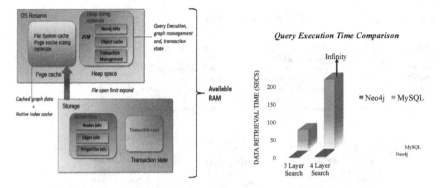

Fig. 1. Memory configuration in Neo4j

Fig. 2. Comparison of the performance of query execution between relational and graph databases.

5 Results and Discussion

5.1 Comparison of Neo4j with Relational Model

1. Schema and Query Performance: HbMDB was probed with many use cases (Fig. 3), where the use of a graph model together with a query language like Cypher greatly improved the response times and simplified the code necessary for querying the database. Some of the use cases that were probed by traversing the HbMDB graph database included (a) retrieval of all target enzymes for a specific or all compounds in a given herb (b) the metabolic or biosynthesis pathways in which these enzymes participate and (c) identification of key players in a given disease pathway by deconstruction of a complex or a set into its participating molecules.

A comparison of relational and graph database schema with a simplified example for herb-disease pathway metabolic data is shown (Fig. 3). In the relational use case, four junction tables, Compound, Activity, Compound_TARGET_GENE and Compound_TARGET_ENZYME are required to model these many-to-many relationships (Fig. 3a). Each junction table contains primary keys and associated foreign keys to other tables. The SQL query to input and output entities of a given herb-disease pathway combination requires five join operations per junction table (Fig. 3b). In the first stage of its execution, each join operation forms a Cartesian product between the tables and, during the filtering process, all rows of the result set which are not of interest are discarded. The same structure of herb-disease pathway metabolic data with inputs and outputs is modelled in a simpler way with Neo4j as exemplified by the schema presented in Fig. 3c. The *Herb* node, exhibits named "Contains" relationships to the corresponding *Compound* node and which in turn contains two outgoing relationships named "Exhibits" and "Targets" to the nodes *Activity* and *Target Gene* node respectively. The *Target Gene* node exhibits named "Is_ A" relationships to corresponding *Target Enzyme* node and *Target Enzyme* exhibits named

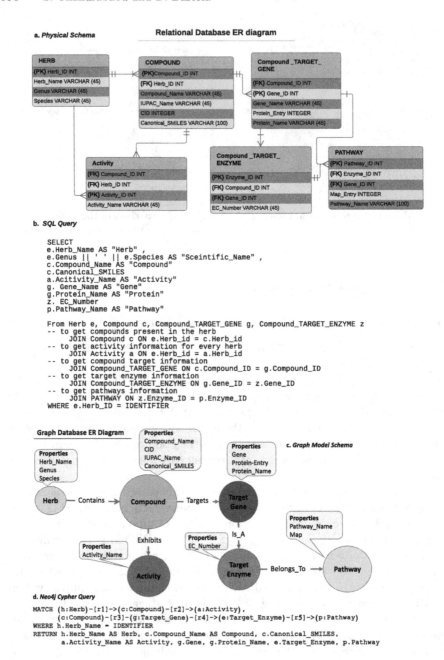

Fig. 3. A simplified example where herbs are mapped to the respective metabolic pathways of a target enzyme is shown. (a) In the relational use case, four junction tables are required to model these many-to-many relationships. (b) SQL query for retrieving pathway entries for a given herb using multiple "joins" is shown. (c) The same schema modeled as graph. (d) The same query written using Cypher, in a shorter but more intuitive manner.

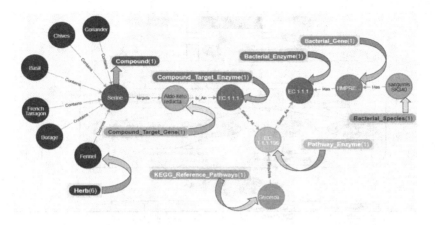

```
MATCH (h:Herb)-[r1]->(c:Compound)-[r2]->(t:Compound_Target_Gene), (t:Compound_Target_Gene)-[r3]-
    (e:Compound_Target_Enzyme)-[r4]->(p:Pathway_Enzyme)-[r5]->(k:KEGG_Reference_Pathways),
    (p:Pathway_Enzyme)-[r6]-(b:Bacterial_Enzyme),(g:Bacterial_Gene)-[r7]->(b:Bacterial_Enzyme),
    (s:Bacterial_Species)-[r8]->(g:Bacterial_Gene),(j:Bacterial_Genus)-[r9]->(s:Bacterial_Species)

WHERE k.Pathway = "Glycerolipid metabolism"
RETURN h, c, t, e, p, k, b, g, s LIMIT 6
```

Fig. 4. Representation of the query in Neo4j. (a) Sample output for retrieving 6 herbs which have enzymes taking part in the "Glycerolipid metabolism" pathway (b) Cypher query for retrieving the result shown above.

"Belongs_To" relationships to *Pathway* node. All the nodes have their associated properties (Fig. 3c), which can be viewed by hovering over the respective node. Taking advantage of Cypher, the same query executed using SQL, is written in a shorter but more intuitive manner using Neo4j (Fig. 3d). Finally, all nodes matching the specified pattern are returned. A sample query for retrieving information regarding a particular pathway, Glycerolipid metabolism and the result obtained using Neo4j is shown in Fig. 4. This exemplifies how query performance can be improved by employing graph databases, such as Neo4j, that offer a more appropriate alternative for cases with highly interconnected data.

2. Performance Analysis Based on Speed: The relational models, as discussed previously, make use of multiple "joins" to infer relationships among different tables, requiring significant execution times. With "joins", every table included multiplies to the row size of the final table operating at a Big-O of exponential growth [5]. Consequently, the size of the input must remain as small as possible, else the time complexity becomes so large that it never ends. To exemplify this time comparison between the two models with same kinds and levels of data, we performed a query search and retrieval operation, in MySQL and Neo4j databases, for traversing through three layers of data. We also used Neo4j based HbMDB for retrieving all the compound-target gene information for a given herb, which is a 3-layer search traversal of data. Neo4j retrieved

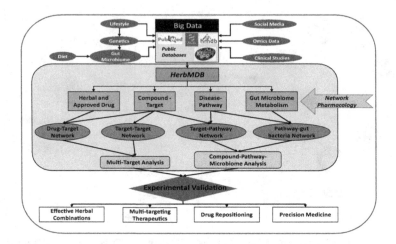

Fig. 5. Central illustration: HbMDB in multi-target analysis and drug design

the results in 0.58 s, while MySQL took 58.325 s for this 3-layer search. When a 4-layer search was conducted for retrieving more than 5000 rows of data using both the databases, MySQL was unable to return a result while Neo4j outperformed MySQL in this case too by retrieving results in 5.028 s (Fig. 2).

5.2   Use of HbMDB in Drug Discovery and Precision Medicine

1. Multi-target Analysis: HbMDB is a resource instrumental in the identification of bioactive compounds in herbs with multi-target capabilities. It is an integrated systems pharmacology platform useful in the development of new drugs used in the treatment of chronic diseases (Fig. 5). To evaluate various features of this resource, we carefully analyzed 24 culinary herbs, 371 bioactive compounds and 847 compound target genes in the HbMDB. With a subset of data, we constructed an herb-target network (HT), which is a bipartite graph consisting of two disjoint sets of nodes, with one set for herbs and the other for target enzymes (Fig. 6). Investigation of all the 24 culinary herbs shows that it consists of compounds that target more than one enzyme (Fig. 7). In addition to targeting many proteins (genes), the compounds from culinary herbs also target proteins with enzymatic functions in bacteria (Fig. 8). The smallest number of enzymes targeted by the herb *Lovage* is 55, while *Cloves* consisted of 157 molecular targets and 33 bioactive compounds (Fig. 8). Many of these compounds are known to be useful in the treatment of cancer [20]. For instance, eugenol which is an active compound present in *Cloves* was selected as a potential molecule capable of interfering with several cell-signaling pathways, specifically the nuclear factor kappa B (NFKB) responsible for cancer [20]. This factor is activated by free radicals which results in the expression of genes that can suppress apoptosis and induce cellular transformation, proliferation, invasion, metastasis among cancer patients [20]. The result shows that the herbs can be broadly grouped into

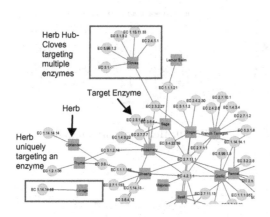

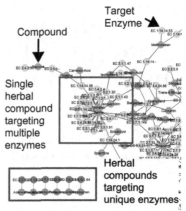

Fig. 6. Herb-Target (HT) sub-network. The green nodes represent the herbs and the orange nodes represent the target enzymes. The network shows herbs capable of targeting multiple enzymes and unique enzymes. (Color figure online)

Fig. 7. Compound-Target (CT) sub-network. The pink nodes represent the herbal compounds and the purple nodes represent the target enzymes. The network shows compounds capable of targeting multiple enzymes and unique enzymes. (Color figure online)

three types. The first type includes herbs that have compounds with few active targets; the second type were those with few compounds but many targets; and finally, the third type, which was of our primary interest, included herbs which consisted of many compounds and all have many protein targets (Fig. 9).

Serine and Threonine are the two compounds with 250 molecular targets. Anomalous serine/threonine phosphatase activity has been associated with several pathological states including diabetes, cardiovascular disorders, cancer, and Alzheimer's disease [21]. Therefore, the pharmacological manipulation of serine and threonine along with their phosphatase enzyme activity is an attractive strategy for the treatment of many pathological conditions. Both these compounds are worthy of clinical testing for these disease conditions. Whether these bioactive compounds identified here have a positive or negative role on the net functional outcome is a crucial next step that needs further study, which we plan to pursue.

2. Graph Network for Multi-agent Drug Discovery: Multiple agents in one combination of drugs were usually endued with different roles, and so we first tried to identify the major herbal compounds which play a dominant role in treating diseases. The hub proteins in HT (Fig. 10) and CT (Fig. 11), show how multiple enzymes can be targeted by compounds shared by different herbs and may play central roles in multi-target therapeutics. These are potential high-value targets in drug development and should be given more attention.

Fig. 8. Herb-target analysis. List of target genes and enzymes targeted by herbs

Fig. 9. Herb category analysis.

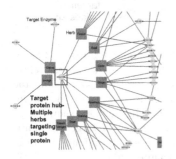

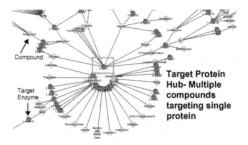

Fig. 10. Radial view of Herb-Target (HT) sub-network. The green nodes represent the herbs and the orange nodes represent the target enzymes. The network shows multiple herbs targeting a single enzyme (Color figure online)

Fig. 11. Radial view of Compound-Target (CT) sub-network. The pink nodes represent the herbal compounds and the purple nodes represent the target enzymes. The network shows multiple compounds targeting a single enzyme. (Color figure online)

For instance, the high degree enzyme node ("EC 1.1.1.21 - *aldehyde reductase*"), shows node degree of 24 in HT and 37 in CT network, suggesting a central role in multiple pathways and multi-target therapeutics. *Aldehyde reductase* is known to be involved in many metabolic pathways including Pentose and glucuronate interconversions, Fructose and mannose metabolism, Galactose metabolism, and the biosynthesis of Folate. Indeed, *aldehyde reductases* play well-ascribed roles in the lipid metabolism pathway, Glycerolipid metabolism, and has been recognized as a therapeutic target for a variety of clinical conditions including Acquired Immunodeficiency Syndrome, Adenocarcinoma, Breast Neoplasms, Carcinogenesis and heart failure, etc. [22]. Henceforth, our results from HbMDB suggest that herbs which usually target a group of proteins and pathways for the specific function are likely to work synergistically on a given disease condition.

Table 2. List of herbs and herbal compounds exhibiting anti-inflammatory effect

Herbs	Compounds	Target enzymes	Pathway
Cinnamon	[(+)-Catechin, Caffeic-Acid, (-)-EPICATECHIN, Cinnamaldehyde, Eugenol, Cinnamic-Acid]	[EC 1.1.1.-, EC 1.1.1.21, EC 1.2.1.-, EC 2.3.1.-, EC 3.1.-.-]	Glycerolipid metabolism
Thyme	[Apigenin, Ursolic-Acid, Chlorogenic-Acid, Luteolin, Oleanolic-Acid, Caffeic-Acid, Linoleic-Acid, Gallic-Acid, Rosmarinic-Acid, Kaempferol, Eugenol, Cinnamic-Acid, Oleic-Acid]	[EC 1.1.1.-, EC 1.-.-.-, EC 1.1.1.21, EC 1.2.1.-, EC 2.3.1.-, EC 2.4.1.-, EC 3.1.-.-, EC 3.1.1.3]	Glycerolipid metabolism
Lovage	[Caffeic-Acid, Bergapten, Eugenol]	[EC 1.1.1.-, EC 1.1.1.21, EC 1.2.1.-]	Glycerolipid metabolism

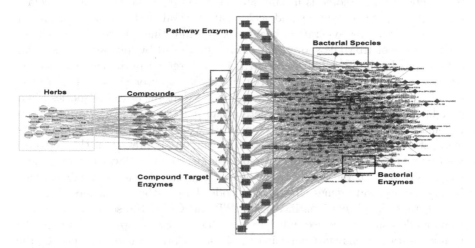

Fig. 12. Heterogeneous sub-network showing herbs which are capable of targeting bacterial enzymes which belong to the pathway "Glycerolipid metabolism"

3. Herbs for Alternative Therapy: The growing concern associated with the excessive use of inappropriate antibiotics has greatly encouraged the use of probiotics and plant natural products as an alternative therapy to maintain colon health. Compared to a single chemical with a single target for most antibiotics, the benefit of using herbs is that the complex chemical composition of a living plant is uniquely capable of interacting with the human body in an equally complex way. Herbs and spices have traditionally been used in the diet, not as a source of nutrition but to enhance flavor and increase organoleptic properties. Additionally, they are regarded as a rich source of anti-oxidant, anti-inflammatory or anti-microbial compounds. The examination of HbMDB shows that culinary herbs are a good source of phytochemicals; many of which possess anti-inflammatory activity and target enzymes in the Glycerolipid metabolism pathway (Table 2 showing top 3 results), key in metabolic signal generation.

These compounds and metabolites identified from an untargeted analysis, and mapped to enzymes or genes provide links to biochemical pathways of interest, such as the glycerol metabolism pathway noted earlier. With the help of these links, routes were traced to the gut microbiota. Our result suggests that 1693 bacterial enzymes are potential targets for the compounds present in 24 culinary herbs. Herbal compounds are well known for their interactions with gut microbes and their capabilities to alter the microbial metabolites including short-chain fatty acids (SCFAs), bile acids (BAs) and lipopolysaccharides (LPS). These interactions are correlated with metabolic diseases such as dysbiosis, type 2 diabetes (T2D), inflammatory bowel disease (IBD), obesity and non-alcoholic fatty liver disease (NAFLD). Dysbiosis, which is the imbalance in the composition of biome is usually an indication of colon health and a cofactor in causing inflammation and leaky gut. Several herbal drugs are known to aid in promoting synergistic healing of the intestine affected by dysbiosis. Investigation of HbMDB shows that enzymes encoded by four major microbial enterotypes known to inhabit the gut, can be targeted by herbal compounds. These enzymes are the potential drug targets to monitor the dysbiosis of the gut microbiome. Through HbMDB it was found that gut bacterial species could encode 43 enzymes from the lipid metabolism pathway. On tracing routes from HT and CT to the bacterial-enzymes network we arrive at a heterogeneous network, starting from the herb nodes to the compounds present in the herbs, to the target enzymes targeted by those compounds, to the pathway to which those enzymes belong to, and finally to the bacterial species which are capable of harboring the necessary genes which can encode those pathway enzymes (Fig. 12). The therapeutic effects identified using HT and CT of the selected herbs and compounds can now be extended to a pathway-focused approach, where the compound-target enzymes participating in a pathway can be mapped to the metabolic models of the gut-microbiome. This way, we can determine the organism(s) harboring necessary genes that are capable of producing the metabolites of interest through pathways and thereby avoid the occurrence of dysbiosis.

6   Conclusion

In conclusion, through the adoption of the Neo4j graph database, and by harnessing the power of the Cypher query language, HbMDB proved to be an effective graph model over the relational model based on a number of performance criteria including search speed. As a result of this shift in the underlying data storage platform from relational to a graph, the average query time was reduced by up to 93%. Using multiple use cases, we demonstrated the use of HbMDB in identifying natural compounds present in culinary herbs capable of targeting multiple proteins/enzymes, which are of value in drug design as well as its use for therapeutic intervention in precision medicine. Interestingly, herbs with common protein targets that are responsible for different cellular function (anti-inflammatory and antioxidant) may work synergistically. Network analysis also shows that bioactive compounds in herbs have enzyme targets in bacteria, which inhabit a gut, which helps to develop alternative therapies that use dysbiosis of gut microbiome as a potential therapeutic strategy. Most importantly, the developed resource now provides a means to study the mechanistic role of bioactive compounds in the treatment of complex diseases. In the near future, we plan to upgrade the features and services of HbMDB by providing web access to the database and also making it publicly available to leverage its full potential.

References

1. Galsky, M., Vogelzang, N.: Docetaxel-based combination therapy for castration-resistant prostate cancer. Ann. Oncol. **21**(11), 2135–2144 (2010)
2. Patwardhan, B., Gautam, M.: Botanical immunodrugs: scope and opportunities. Drug Discov. Today **10**(7), 495–502 (2005)
3. Barabasi, A.-L., Oltvai, Z.N.: Network biology: understanding the cell's functional organization. Nat. Rev. Genet. **5**(2), 101 (2004)
4. Li, J., Zhao, P.X.: Mining functional modules in heterogeneous biological networks using multiplex pagerank approach. Front. Plant Sci. **7**, 903 (2016)
5. Chandrababu, S., Bastola, D.R.: Comparative analysis of graph and relational databases using HerbMicrobeDB. In: 2018 IEEE International Conference on Healthcare Informatics Workshop (ICHI-W), pp. 19–28. IEEE (2018)
6. Barabási, A.-L., Gulbahce, N., Loscalzo, J.: Network medicine: a network-based approach to human disease. Nat. Rev. Genet. **12**(1), 56 (2011)
7. Chandrababu, S., Bastola, D.R.: CuHerbDB-for pharmacogenomics and study of phytochemicals in culinary and medicinal herbs. In: 2017 IEEE International Conference on Bioinformatics and Biomedicine (BIBM), pp. 1787–1794. IEEE (2017)
8. Chandrababu, S., Bastola, D.: An integrated approach to recognize potential protective effects of culinary herbs against chronic diseases. J. Healthcare Inform. Res. 1–16 (2018)
9. Kilicoglu, H., Fiszman, M., Rodriguez, A., Shin, D., Ripple, A., Rindflesch, T.C.: Semantic MEDLINE: a web application for managing the results of PubMed searches. In: Proceedings of the Third International Symposium for Semantic Mining in Biomedicine, vol. 2008, pp. 69–76. Citeseer (2008)
10. Duke, J., Bogenschutz, M.J.: Dr. Duke's phytochemical and ethnobotanical databases. USDA, Agricultural Research Service (1994)
11. Bento, A.P., et al.: The chembl bioactivity database: an update. Nucleic Acids Res. **42**(D1), D1083–D1090 (2014)
12. Kim, S., et al.: Pubchem substance and compound databases. Nucleic Acids Res. **44**(D1), D1202–D1213 (2015)
13. Wishart, D.S., et al.: DrugBank 5.0: a major update to the drugbank database for 2018. Nucleic Acids Res. **46**(D1), D1074–D1082 (2017)
14. Fowler, S., Schnall, J.G.: TOXNET: information on toxicology and environmental health. AJN Am. J. Nurs. **114**(2), 61–63 (2014)
15. Scheer, M., et al.: BRENDA, the enzyme information system in 2011. Nucleic Acids Res. **39**(Suppl. 1), D670–D676 (2010)
16. Kanehisa, M., Goto, S., Sato, Y., Furumichi, M., Tanabe, M.: KEGG for integration and interpretation of large-scale molecular data sets. Nucleic Acids Res. **40**(D1), D109–D114 (2011)
17. Caspi, R., et al.: The MetaCyc database of metabolic pathways and enzymes and the BioCyc collection of pathway/genome databases. Nucleic Acids Res. **42**(D1), D459–D471 (2013)

18. UniProt Consortium: Activities at the universal protein resource (UniProt). Nucleic Acids Res. **42**(D1), D191–D198 (2013)

19. Wheeler, D.L., et al.: Database resources of the national center for biotechnology information. Nucleic Acids Res. **35**(suppl_1), D5–D12 (2005)

20. Aggarwal, B.B., Shishodia, S.: Molecular targets of dietary agents for prevention and therapy of cancer. Biochem. Pharmacol. **71**(10), 1397–1421 (2006)

21. McConnell, J.L., Wadzinski, B.E.: Targeting protein serine/threonine phosphatases for drug development. Mol. Pharmacol. **75**, 1249–1261 (2009)

22. Moon, J., Liu, Z.L.: Direct enzyme assay evidence confirms aldehyde reductase function of Ydr541cp and Ygl039wp from saccharomyces cerevisiae. Yeast **32**(4), 399–407 (2015)

On Identifying Candidates for Drug Repurposing for the Treatment of Ulcerative Colitis using Gene Expression Data

Suyeon Kim, Ishwor Thapa, Ling Zhang, and Hesham Ali[✉]

College of Information Science and Technology,
University of Nebraska at Omaha, Omaha, NE 68182, USA
{suyeonkim,ithapa,lzhang,hali}@unomaha.edu

Abstract. The notion of repurposing of existing drugs to treat both common and rare diseases has gained traction from both academia and pharmaceutical companies. Given the high attrition rates, massive time, money, and effort of brand-new drug development, the advantages of drug repurposing in terms of lower costs and shorter development time have become more appealing. Computational drug repurposing is promising approach and has shown great potential in tailoring genomic findings to the development of treatments for diseases. However, there are still challenges involved in building a standard computational drug repurposing solution for high-throughput analysis and the implementation to clinical practice. In this study, we applied the computational drug repurposing approaches for Ulcerative Colitis (UC) patients to provide better treatment for this disabling disease. Repositioning drug candidates were identified, and these findings provide a potentially effective therapeutics for the treatment of UC patients. This preliminary computational drug repurposing pipeline will be extended in the near future to help realize the full potential of drug repurposing.

Keywords: Drug repurposing · Computational tools ·
Gene expression data · Systems biology · Next generation healthcare

1 Introduction

The application of approved drugs to identify novel disease indications, known as drug repurposing/repositioning, provides several opportunities over traditional drug discovery. Traditional drug development is a time-consuming and costly process that takes an average of around 13–15 years and costs more than 2 billion. In addition, safety concerns or lack of efficacy maximize the risk of failure. Drug repurposing can provide solutions to these issues faced by pharmaceutical companies. This repurposing holds the promise of clinical trials that are cost-effective and faster and have lower failure rates than in the traditional drug development

© Springer Nature Switzerland AG 2019
I. Rojas et al. (Eds.): IWBBIO 2019, LNBI 11465, pp. 513–521, 2019.
https://doi.org/10.1007/978-3-030-17938-0_45

pipeline. The advent of genomic technologies has enabled researchers to compare the large-scale patient samples at different molecular levels with molecular changes based upon drug treatment. In order to find the drugs for treatment of different disease by integration of various molecular features, systems biology approaches play a role in the discovery new therapeutics. Gene expression signatures have been the most widely used in the systematic approach, among other molecular features [1,12,17]. For example, one of the most used approaches starts with a disease gene expression signature. This can be utilized to identify candidate drugs that have a reversal relationship with the disease by comparing disease and control groups. Although a number of previous studies have demonstrated its potential in drug discovery, they suffer from limitations. This preliminary computational drug repurposing pipeline is currently being extended to address other diseases and help realize the full potential of drug repurposing.

Ulcerative Colitis (UC) is a chronic disorder disease that is concomitant with an increased risk of colorectal cancer due to chronic inflammation [2,13]. UC is a subgroup of Inflammatory Bowel Disease (IBD) which also includes Crohn's disease. There is no cure for UC and the currently available treatment can only alleviate the symptoms and also has dangerous side effects [2,7]. Thus, the need for new therapeutics options for UC, using a systematic computational approach for drug repurposing, has been brought to attention.

The drug repositioning work has contributed in a variety of computational methods for the identification of new therapeutics with the use of existing drugs. For an example, signature-based drug repurposing methods have gained attention along with the development of high-throughput sequencing technologies. Signature-based drug repurposing methods make use of gene expression signatures to identify drugs having similar gene expression profiles in cell lines treated with approved drugs or having opposite gene expression profiles to that of diseased samples. Connectivity Map (CMap), Gene Expression Omnibus (GEO), and the recent Library of Integrated Network-based Cellular Signatures (LINCS) datasets [11] are explored to develop and apply novel computational approaches that can enable informed drug repositioning.

In this study, we performed a computational pipeline for drug repositioning employed by Sham et al. [18], integrating public gene expression signatures of drugs and Ulcerative Colitis patients. The pipeline for drug repurposing has previously been shown to be successful when applied to psychiatric disorders [18]. Using this approach, we found a number of repositioning candidates, such as Sirolimus, Trichostatin A, Vorinostat, Wortmannin, and LY-294002 as top lists. Some of these candidate drugs were found to have a potency to be used as Ulcerative Colitis treatment.

The established groundwork created by our preliminary research allows us to expand our work in multiple directions. Our preliminary results have demonstrated which candidate drugs can have a reversal role in changing the gene expression of UC samples compared to control. It would be the natural next steps to identify the minimal group of genes associated with the candidate drugs that can reduce the side effects of the disease. From systems biology approach point of view, the attempt to leverage drug-drug interaction data would be beneficial to understand possible physiological effects or targets of drugs.

2 Methods

In this section, we describe the methods applied in order to obtain new candidate drugs for treating Ulcerative Colitis (*UC*) using the drug re-purposing pipeline employed by Sham et al. [18]. However, we applied slight modifications to the original pipeline in order to best make use of the publicly available data for *UC*. While Sham et al. used GWAS data to impute gene expression data for different disease conditions, we start with gene expression data that is already published. We believe that the use of expression data directly removes the imputation bias.

2.1 Datasets

We primarily used three different datasets (a) gene expression, (b) connectivity map and (c) KEGG disease database.

Gene Expression Dataset. We obtained GSE92415 from the Gene Expression Omnibus (GEO) and selected samples from three different groups. We obtained microarray gene expression data for UC samples with no treatment (n $=$ 87) and a healthy control (n $=$ 21). The probe ids from the expression data were converted to gene symbols using the GPL13158 platform file. Average gene expression values were computed for the genes mapped with multiple probe ids. The z-scores were calculated for each gene using the expression values from two groups (UC samples and healthy control).

Connectivity Map (CMap). From the Connectivity Map, or CMap resource, we utilize gene expression signatures that arise from treatment of small molecules (drugs) [12]. There are about 6000 drug induced expression signatures for more than 1300 drugs. We have used the rank matrix file provided by CMap database to compare it with the gene expression profile from GEO.

KEGG Disease Database. We use the KEGG disease database [9] to obtain a list of drugs which have been approved for treatment of UC in Japan, US and Europe. The *H01466* entry in KEGG disease refers to the UC disease. The list of drugs being used for UC are shown in Table 1.

2.2 Drug Re-purposing Pipeline

Here we briefly summarize the drug re-purposing pipeline described by Sham et al. [18] to compare gene expression profiles of drugs obtained from CMap (rank matrix) to the gene expression profiles obtained from (a) disease samples and (b) healthy control. In contrast to the pipeline used by Sham et al., we used

Table 1. Drugs shown in KEGG disease database for treatment of UC. The column 'Drug Name' lists the KEGG names of the drugs and 'Entry ID' represents the KEGG Drug Bank ID

Drug name	Entry ID
Triamcinolone acetonide	DR:D00983
Dexamethasone	DR:D00292
Dexamethasone sodium phosphate	DR:D00975
Hydrocortisone	DR:D00088
Hydrocortisone acetate	DR:D00165
Hydrocortisone sodium succinate	DR:D00978
Prednisolone	DR:D00472
Prednisolone sodium phosphate	DR:D00981
Prednisone	DR:D00473
Budesonide	DR:D00246
Sulfasalazine	DR:D00448
Mesalamine	DR:D00377
Olsalazine sodium	DR:D00727
Balsalazide disodium	DR:D02715
Methylprednisolone	DR:D00407
Methylprednisolone sodium succinate	DR:D00751
Methylprednisolone acetate	DR:D00979
Cortisone acetate	DR:D00973
Vedolizumab	DR:D08083
Infliximab	DR:D02598
Adalimumab	DR:D02597
Golimumab	DR:D04358

the gene expression data from GEO instead of imputing the expression profile from GWAS data. In the next step, the Spearman and Pearson correlations were measured to find reverse patterns of expression between the Cmap and UC sample gene expression datasets. The Kolomogorov-Smirnov (KS) test was applied to evaluate if a set of disease related genes are ranked higher or lower than expected in the list of genes obtained from drug induced expression levels. In addition to whole dataset, the correlation tests and KS test were performed for top 50, 100, 250 and 500 ranked genes based on average expression among samples from the same group. The results from all these tests were used to compute average rank for each drug perturbation. We also performed 100 permutation tests for simulated datasets generated by randomly assigning gene expression to a gene in disease expression data. Then we applied KS and correlation tests on

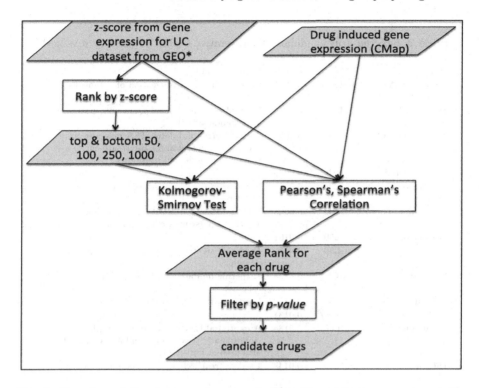

Fig. 1. Overview of the drug repurposing pipeline. The white boxes represent the processes and grey ones represent various input and output data formats.

the permuted data and calculated *p-value*. Figure 1 shows the overview of the steps performed in this study.

3   Results and Discussion

This study aimed to use drug repurposing pipeline to identify potential candidates for drug repurposing in leveraging gene expression data and Cmap database. We applied this pipeline to UC patients who haven't started medication for the treatment and identified a number of interesting candidates for repurposing. Several repeated drugs and small molecules were observed in top lists of our 20 drug candidates. These top 20 candidate drugs from 78 statistically significant candidate drugs (p value < 0.05) were included in the Table 2.

Multiple occurrences of the same drugs and small molecules such as Sirolimus, Trichostatin A, Vorinostat, Wortmannin, and LY-294002 are identified (noted by *). In particular, these repeated drugs and small molecules indicated that there are possibly true positive results. Table 2 shows the details of drug descriptions obtained from public databases such as DrugBank and National Center for Advancing Translational Sciences. Next, we summarize the drugs that appeared

Table 2. Selected drug candidates for Ulcerative Colitis (ordered by p values)

Drug	Cell line	p value	Brief description
Vorinostat**	MCF7	0.0031	Currently under investigation of cutaneous T cell lymphoma
Sirolimus**	MCF7	0.0048	Isolated as an antifungal agent with potent anticandida activity
Trichostatin A***	MCF7	0.0054	An antifungal antibiotic. TSA may also have therapeutic potential for the treatment of a variety of genetic and infectious
Trichostatin A***	MCF7	0.0065	
Pioglitazone	MCF7	0.0029	For used in the treatment of type 2 diabetes mellitus
Wortmannin***	HL60	0.0098	Inhibit cancer cell growth and has shown activity against mouse and human tumor
Wortmannin***	HL60	0.0101	
Selegiline	HL60	0.011	Used as a treatment for the major depressive disorder
LY-294002***	MCF7	0.0128	A specific inhibitor of phosphatidylinositol 3-kinase
Etynodiol	HL60	0.013	Used as a hormonal contraceptive drug
Vorinostat**	MCF7	0.0131	
Amantadine	HL60	0.0134	An antiviral that is used in the prophylactic or symptomatic treatment of influenza A. And also used as anti-parkinsonian agent
Diflorasone	HL60	0.0143	Used to treat itching and inflammation of the skin
Aciclovir	MCF7	0.0147	An antiviral agent only after it is phosphorylated in infected cells by a viral-induced thymidine kinase
Trichostatin A***	MCF7	0.0162	
Ellipticine	HL60	0.0189	
Acepromazine	HL60	0.0197	One of the phenothiazine derivative psychotropic drugs
Sirolimus**	MCF7	0.0205	
Piperlongumine	HL60	0.0215	This compound is easily available, inexpensive, and has therapeutic effects against cancer, heart disease, intestinal diseases, diabetes, obesity, joint pain and other conditions in Chinese Herbal and Indian Ayurvedic medicine
LY-294002***	MCF7	0.0221	
Pinacidil	HL60	0.0221	A clinically effective vasodilator used for the treatment of hypertension
LY-294002***	MCF7	0.0247	
Valproic acid	MCF7	0.0249	Used in the treatment of epilepsy
Heliotrine	HL60	0.025	No description
Wortmannin***	HL60	0.0251	
Sulfachlorpyridazine	HL60	0.0265	An antimicrobial used for urinary tract infections. Highly effective against diseases caused by *Escherichia coli*
Benzamil	HL60	0.0268	Useful sodium channel blocker for the long-term treatment of the biochemical defect in the lungs of patients with cystic fibrosis

Note that * is number of times same drug is shown in the top 20 list.

multiple times as shown in Table 2. Sirolimus (rapamycin) is an immunosuppressive agent, originally isolated as anifungal agent, is used in post-transplanation management. Several case reports have been published on the use of Sirolimus as a new therapy in mostly Crohn's disease and Ulcerative Colitis (UC) [14,15,19]. Trichostatin A (TSA) is an antifungal antibiotic, a potent and specific inhibitor of histone deacetylase (HDAC) activity. It manifests very effective anticancer activity and have been studied for potential anti-inflammatory properties [4]. In a similar manner, Vorinostat (or suberoylanilide hydroxamic acid (SAHA)) inhibits the HDAC1, HDAC2 and HDAC3 from the lysine residues of histone proteins. Currently, Vorinostat is marketed under the name Zolinza from Merck pharmaceutical company which is used for the treatment of cutaneous T-cell lymphoma. Wortmannin is a fungal metabolite with anti-inflammatory properties, which was identified as an inhibitor of phosphoinositide 3-kinase (PI3Ks), while LY294002 was known as the first synthetic PI3K inhibitor [16].

3.1 Sirolimus

Sirolimus is also known as rapamycin, obtained by a strain of *Streptomyces hygroscopicus*. Although it possesses both antifungal and antineoplastic properties, Sirolimus has been successfully used in a number of gastrointestinal inflammatory disorders. Massey et al., used Sirolimus to treat refractory Crohn's disease in adult patients [14]. Mutalib et al., provides evidence on the use of Sirolimus as effective therapy in a subgroup of children with severe refractory IBD [15]. Yin et al., studied the therapeutic effect of Sirolimus in mice after the induction of TNBS colitis [19]. The results of experiment have shown that Sirolimus-treated mice regained healthy conditions similar to control group of mice with 85% of survival rate. It provides a strong evidence of the therapeutic effect of Sirolimus on the experimental colitis. Therefore, these aforementioned studies suggest that Sirolimus may provide a promising drug alternative to the current approaches for managing IBD.

3.2 Trichostatin A and Vorinostat

Histone deacetylase (HDAC) are key enzymes regulating important cell processes such as cell-cycle progression and apoptosis. Although HDACs, especially Vorinostat, is already marketed for cancer treatment, several studies demonstrated a connection between HDACs and intestinal inflammation to analyze the effects of HDACi on animal models of colitis [5,6]. Glauben et al., indicated that suberoylanilide hydroxamid acid (SAHA) or Vorinostat resulted in a significant reduction in inflammation in both destran sulphate sodium (DSS)- and trini-trobenzene sulfonic acid (TNMS)-induced colitis. De et al. also provided an evidence that pan-HDACi such as Trichostatin-A and SAHA, which are clinically approved drugs, showed that they could ameliorate development of DSS colitis [3]. Both groups strongly suggest that HDAC inhibitors might be served as a potential therapeutic target for the modulation of macrophage responses in inflammatory bowel disease. Thus, the roles of HDACi in colitis models as described by these studies may suggest that they can be strong therapeutic candidates for the treatment of IBD.

3.3   Wortmannin and LY-294002

Wortmannin and LY-294002 are chemical compounds that has shown to act as potent inhibitors of phosphoinositide 3-kinase (PI3Ks). Both of them are initially used to inhibit cell proliferation in cancer treatment by inhibiting the PI3K signaling pathway. In Huang et al.'s study, they investigated the role of PI3k signaling pathway in pathogenesis of Ulcerative Colitis (UC) [8]. In their result, Wortmannin was shown to significantly alleviate the inflammation of colitis as assessed by disease activity index and histological score in DSS-induced mice. These results indicated that the PI3k signaling pathway plays a role in the occurrence of UC. As LY-294002 is also a type of PI3K inhibitor, one study demonstrated LY-294002 targets the PI3K pathway and hampers progressive colitis [10]. These findings show the role of PI3K inhibitor on the impact of colitis induced model. Hence, potent inhibitor of PI3Ks such as Wortmannin and LY-294002 can be the future therapeutics of UC.

To summarize, we have identified several viable candidate drugs for the potential treatment of UC using expression data. This proposed study highlights the importance of employing a computational drug repurposing pipeline to extract information from gene expression datasets and reliably identify new effective treatment option for various diseases using known drugs. This approach, coupled with proper clinical trials, could lead to more effective treatments at reduced cost.

4   Conclusion

The development of advanced computational bioinformatics tools continues to significantly impact biomedical research. Using the growing biological data to identify new ways to benefit from existing drugs has the potential to improve the quality and affordability of future medicine. In this study, we develop a drug repurposing pipeline to propose new usages for candidate drugs with the help of the available gene expression data. We identify several candidate drugs to be repurposed and used for the treatment of Ulcerative Colitis (UC) patients. The drugs we propose as drug candidates are widely used in experimental study of colitis induced mice model. Our results show the positive potency of using these drugs for future therapeutics of UC. This preliminary study can be considered as a first step that can be expanded multiple directions to advance computational drug repurposing. Advanced modeling approaches are needed to integrate different types of biomedical data to further improve the drug repurposing process. This is particularly critical when more than one drug is needed to treat certain conditions and data associated with known drug-drug interactions need to be incorporated in the repurposing pipeline. We anticipate that computational tools for drug repurposing will play a major role in the important domain of drug design in the near future.

References

1. Chen, B., et al.: Reversal of cancer gene expression correlates with drug efficacy and reveals therapeutic targets. Nat. Commun. **8**, ncomms16022 (2017)
2. Chumanevich, A.A., et al.: Repurposing the anti-malarial drug, quinacrine: new anti-colitis properties. Oncotarget **7**(33), 52928 (2016)
3. De Zoeten, E.F., Wang, L., Sai, H., Dillmann, W.H., Hancock, W.W.: Inhibition of HDAC9 increases T regulatory cell function and prevents colitis in mice. Gastroenterology **138**(2), 583–594 (2010)
4. Edwards, A.J., Pender, S.L.: Histone deacetylase inhibitors and their potential role in inflammatory bowel diseases (2011)
5. Glauben, R., et al.: Histone hyperacetylation is associated with amelioration of experimental colitis in mice. J. Immunol. **176**(8), 5015–5022 (2006)
6. Glauben, R., et al.: Histone deacetylases: novel targets for prevention of colitis-associated cancer in mice. Gut **57**(5), 613–622 (2008)
7. Grenier, L., Hu, P.: Computational drug repurposing for inflammatory bowel disease using genetic information. Comput. Struct. Biotechnol. J. **17**, 127–135 (2019)
8. Huang, X.L., et al.: PI3K/Akt signaling pathway is involved in the pathogenesis of ulcerative colitis. Inflamm. Res. **60**(8), 727–734 (2011)
9. Kanehisa, M., Goto, S., Furumichi, M., Tanabe, M., Hirakawa, M.: KEGG for representation and analysis of molecular networks involving diseases and drugs. Nucleic Acids Res. **38**(Suppl. 1), D355–D360 (2009)
10. Khan, M.W., et al.: PI3K/Akt signaling is essential for communication between tissue-infiltrating mast cells, macrophages, and epithelial cells in colitis-induced cancer. Clin. Cancer Res. **19**(9), 2342–2354 (2013)
11. Koleti, A., et al.: Data portal for the library of integrated network-based cellular signatures (LINCS) program: integrated access to diverse large-scale cellular perturbation response data. Nucleic Acids Res. **46**(D1), D558–D566 (2017)
12. Lamb, J., et al.: The connectivity map: using gene-expression signatures to connect small molecules, genes, and disease. Science **313**(5795), 1929–1935 (2006)
13. Lin, L., Sun, Y., Wang, D., Zheng, S., Zhang, J., Zheng, C.: Celastrol ameliorates ulcerative colitis-related colorectal cancer in mice via suppressing inflammatory responses and epithelial-mesenchymal transition. Front. Pharmacol. **6**, 320 (2016)
14. Massey, D., Bredin, F., Parkes, M.: Use of sirolimus (rapamycin) to treat refractory Crohn's disease. Gut **57**(9), 1294–1296 (2008)
15. Mutalib, M., et al.: The use of sirolimus (rapamycin) in the management of refractory inflammatory bowel disease in children. J. Crohn's Colitis **8**(12), 1730–1734 (2014)
16. Sheridan, C., Downward, J.: Inhibiting the RAS-PI3K pathway in cancer therapy. In: The Enzymes, vol. 34, pp. 107–136. Elsevier (2013)
17. Sirota, M., et al.: Discovery and preclinical validation of drug indications using compendia of public gene expression data. Sci. Transl. Med. **3**(96), 96ra77 (2011)
18. So, H.C., et al.: Analysis of genome-wide association data highlights candidates for drug repositioning in psychiatry. Nat. Neurosci. **20**(10), 1342 (2017)
19. Yin, H., et al.: Sirolimus ameliorates inflammatory responses by switching the regulatory T/T helper type 17 profile in murine colitis. Immunology **139**(4), 494–502 (2013)

Author Index

Printed in the United States
By Bookmasters